IDRC-118e

EARTHQUAKE HISTORY of ETHIOPIA and the HORN OF AFRICA

Pierre Gouin
Director of the Geophysical Observatory
University of Addis Ababa, Addis Ababa, Ethiopia
From 1957 to 1978

This monograph is a companion volume to "Seismic Zoning in Ethiopia" published by the Geophysical Observatory, University of Addis Ababa. It gives the primary and secondary sources as well as the interpretation on which the statistical analyses of regional seismic and volcanic hazards were based. Throughout the text, mention of political boundaries is intended only to orient the reader and has been resorted to only when other well-known reference points were unavailable. They in no way should be considered as reflecting exact international boundaries.

In addition to this publication, IDRC sponsored part of the research between 1974 and 1978 in cooperation with the Geophysical Observatory of the University of Addis Ababa. The views expressed are those of the author and do not necessarily represent the views of the International Development Research Centre.

The International Development Research Centre is a public corporation created by the Parliament of Canada in 1970 to support research designed to adapt science and technology to the needs of developing countries. The Centre's activity is concentrated in five sectors: agriculture, food and nutrition sciences; health sciences; information sciences; social sciences; and communications. IDRC is financed solely by the Government of Canada; its policies, however, are set by an international Board of Governors. The Centre's headquarters are in Ottawa, Canada. Regional offices are located in Africa, Asia, Latin America, and the Middle East.

Postal Address: Box 8500, Ottawa, Canada K1G 3H9
Head Office: 60 Queen Street, Ottawa

Gouin, P.

IDRC-118e

Earthquake history of Ethiopia and the Horn of Africa. Ottawa, Ont., IDRC, 1979, 259 p. : ill.

/IDRC publication/, /seisms/, /chronology/, /Ethiopia/, /East Africa/ — /seismology/, /volcanic eruptions/, /spatial analysis/, /damage/, /maps/.

UCD: 550.34(63) ISBN: 0-88936-194-0

Microfiche edition available

Contents

ለተማሪዎች።

ይህ፡ መጽሐፍ፡ ለጂኦፊዚካል፡ ኦብሰርቫቶሪ፡ ሠራተኞች፡ ለቀድሞዎቹ፣ ተማሪዎቼ፣ እንዲሁም፡ የኢትዮጵያን፡ ምድር፡ በሳይንስ፡ አማካይነት፡ በጥበብ፡ ለሚመራመሩት፡ ሳይንቲስቶቿ፡ ሁሉ፡ በትህትና፡ አበረክታለሁ።

ይህን፡ መጽሐፍ፡ በሰበሰብኳቸውና፡ በታረምኳቸው፡ ምክንያቶች፡ ካዩ፡ መሠረቴ፡ ይታተም፡ ዘንድ፡ ሳቅድ፡ ያልተሟላ፡ ሥራ፡ ሆኖ፡ እንደሚቀርብ፡ በፍጹም፡ አልዘነጋሁም። ለዚህም፡ ያለመሟላት፡ ምክንያቶች፤ አንደኛ፦ ለዚህ፡ መጽሐፍ፡ ጥናት፡ ፍሬ፡ ነገሮች፡ በቂ፡ ማሠረጃ፡ የሚሆኑ፡ ጽሑፎች፡ ባለማግኘቴ፤ ሁለተኛ፡ እኔው፡ ራሴ፡ የሰበሰብኳቸው፡ ማስረጃዎች፡ ከመብዛታቸው፡ የተነሣ፡ ሁሉንም፡ አቀነባብሮ፡ አንድ፡ መጽሐፍ፡ ለማቅረብ፡ አዳጋች፡ በመሆኑና፣ ሦስተኛ፡ ይህን፡ ጥናት፡ ለማካሄድ፡ የተገለገልኩባቸው፡ መሣሪያዎች፡ የተወሰኑ፡ ስለነበሩ፡ ነው።

ምንም፡ እንኳን፡ ይህ፡ መጽሐፍ፡ ያልተሟላ፡ ቢሆንም፡ ለኢትዮጵያ፡ ሳይንቲስቶች፡ ጠቃሚ፡ ሆኖ፡ ሊገኝ፡ ይችላል። ዋናው፡ ዓላማዬ፡ ስለ፡ መሬት፡ መንቀጥቀጥ፡ ለሚመራመሩት፡ ኢትዮጵያውያን፡ መጽሐፉን፡ እንዲያሻሽሉት፣ እንዲያርሙት፡ አበረክታለሁ።

Pierre Gouin

Acknowledgments

My gratitude goes first to the staff of the Geophysical Observatory, University of Addis Ababa. It was my pleasure to work with them from the opening of the observatory in 1957 until 1978. Without their continuous and friendly cooperation, this survey could not have been conducted. For this particular research, I owe sincere thanks to Ato Yirga Tesfaye, the late Emile Cambron, Laika Mariam Asfaw, Paul A. Mohr, Frances M. Dakin, Ato Theodros Tsigé, and Wozerit Bekelech Degefe.

In the process of securing original and often unpublished documents, precious aid was supplied by the following: Sergew G. Sellassie, Director of the Ethiopian Church Microfilm Library; Richard Pankhurst, former Director of the Institute of Ethiopian Studies at the University of Addis Ababa; J-P. Rothé, former Director of the Institut de Physique du Globe, Strasbourg; Prof G. Marinelli, University of Pisa, Italy; R. J. Brazee of the Seismology Division, NOAA; J. C. Ruegg and J. C. Lépine, Institut de Physique du Globe, Paris and Arta; W. Rigden, Durham University, England; A. A. Hughes, Deputy Director of the International Seismological Centre, Edinburgh, Scotland; M. Pearlman and P. Clark, Smithsonian Astrophysical Observatory, Cambridge, USA; the Director of the Station Météorologique in Djibouti; and the Editor of the *Réveil de Djibouti* for the authorization to consult their files.

Computation facilities were kindly provided by: the Highway Authority in Addis Ababa; R. Lilwall at the Institute of Geological Sciences in Edinburgh; Lynn Sykes at Lamont-Doherty Geological Observatory of Columbia University; W. G. Milne and G. A. McMechan at the Pacific Geoscience Centre in Victoria, Canada; R. W. Girdler at Newcastle upon Tyne University; J. D. Fairhead of Leeds University; and Honeywell Informatique, Montreal, Canada.

Special thanks go to Anne Stevens of the Division of Seismology and Geothermal Studies, Department of Energy, Mines, and Resources, Ottawa, for evaluating the first draft of this earthquake history and recommending its publication, and to Anis Samaan-Hanna for his assistance in the translation of foreign texts. Gérard Despatie, M. S. Dubos, and Jérome Gagnier kindly read parts of the manuscript. Authorization to use copyright material is mentioned in appropriate entries.

Finally, acknowledgments are also due for a 2-week stay at the Institute of Geological Sciences in Edinburgh authorized by P. L. Willmore; a 3-month stay at Victoria Observatory authorized by K. Whitham and W. G. Milne of the Earth Physics Branch, Department of Energy, Mines, and Resources, Canada; and a 1-year fellowship at Brebeuf College Observatory in Montreal graciously offered by its Director, Father Maurice Buist, S.J.

I apologize to those whose names I may have inadvertently omitted; their aid is also gratefully acknowledged.

This research was sponsored by the Faculty of Science, University of Addis Ababa, and by a grant from the International Development Research Centre, Canada.

Pierre Gouin

Introduction

Historical records of the last six centuries and more recent instrumental observations reveal that Ethiopia and the northern sector of the Horn of Africa have been almost continuously jolted by earth tremors and, less frequently, by outbursts of volcanic activity. A highly dissected surface geology, expressed by complicated networks of spectacular faults with vertical throws, often scaled in thousands of metres, and altimetric profiles ranging from below sea level to 3500 metres above it, bear witness that telluric activity is not a phenomenon of recent historical — as opposed to geological — origin (Fig. 1). If we restrict the telluric activity in this region of northeastern Africa to the rifting process only, it dates back to the fissuring of the Nubian swell, which created the three-branch rift system comprised of the Ethiopian, the Red Sea, and the Gulf of Aden rifts.

Rifting in northeastern Africa started some 30 million years ago when the earth's crust, uplifted by a lithothermal system of forces (Gass 1970a,b) generated by a stationary hot spot in the underlying mantle (Burek 1975; Burke and Whiteman 1973; Burke and Wilson 1976), reached the breaking point and fissured along the axes of maximum uplift. During these 30 million years, three major epirogenic episodes can be identified (Gass 1970a,b; Mohr 1971a,b,c). However, the background seismic and volcanic noise caused by crustal readjustments to the vertical forces of alternate doming and subsiding, and to horizontal tensions and shears progressively accumulating along the boundaries of newly formed tectonic plates, must have been practically continuous during these millions of years and will continue for many centuries to come.

Earthquakes and volcanic activity are therefore a fact of life in northeastern Africa, a fact that stems from the geographical location of that region: Ethiopia and the northern sector of the Horn of Africa border the main oceanic boundary between two major tectonic plates, the African and Arabian plates, and also straddle what has been recently termed "a failed rift" that marks the incipient

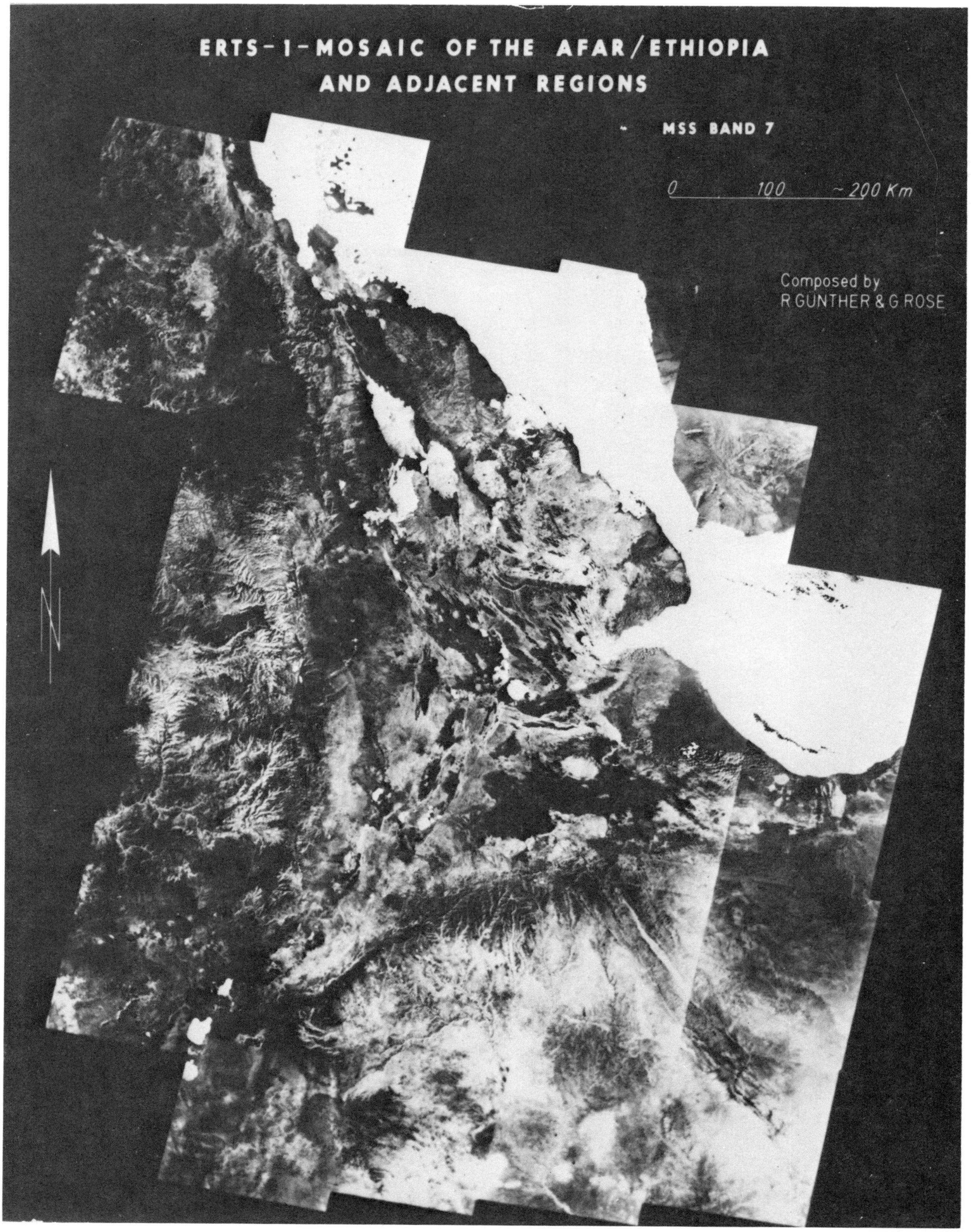

Fig. 1A. ERTS-1. Photomosaic of Ethiopia and the Horn of Africa, after Kronberg et al. 1975 (reproduced by permission of the authors and the Schweizerbart'sche Verlagsbuckhanlung, Stuttgart).

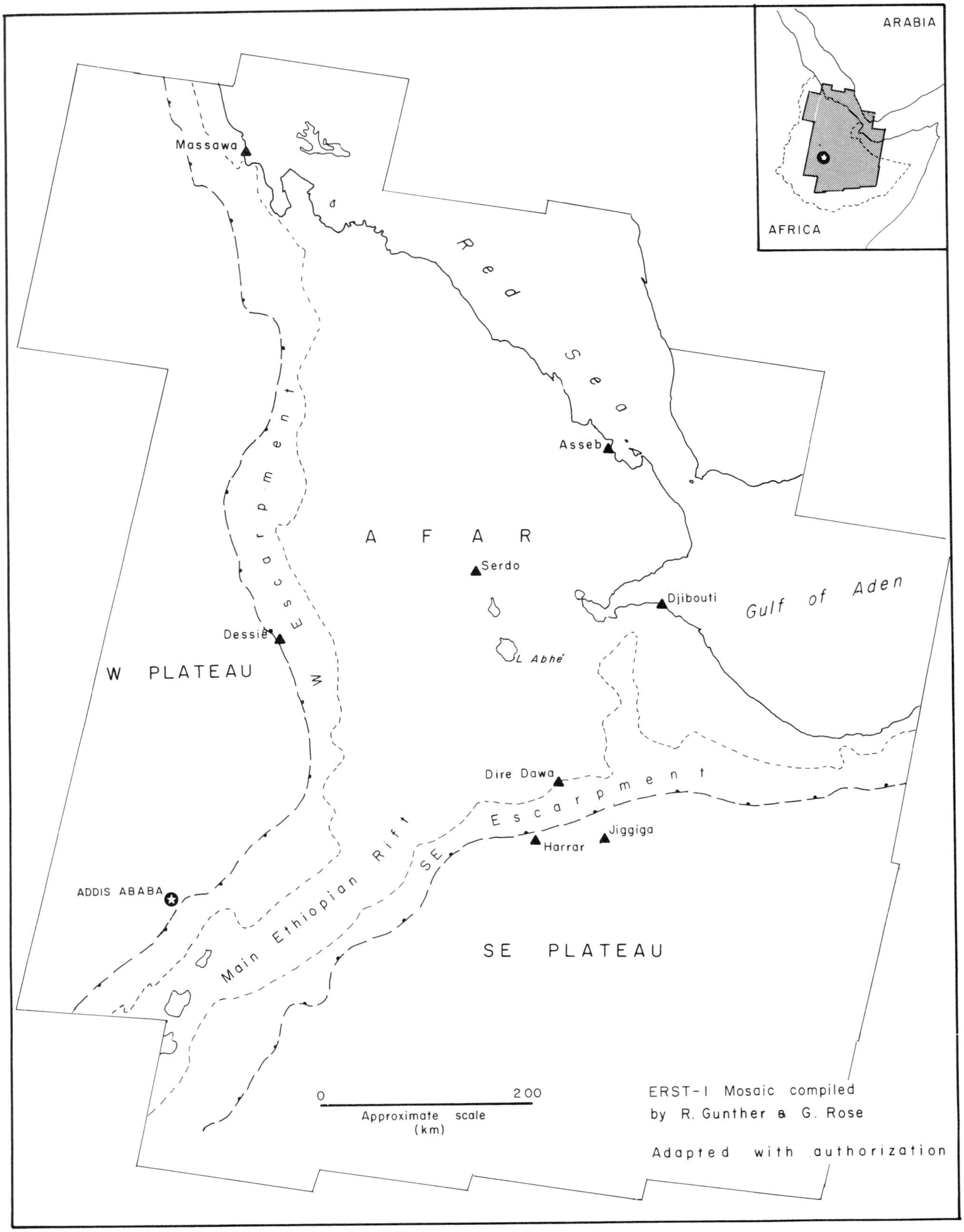

Fig. 1B. Identification map of the important structural elements and regions shown on the photomosaic (Fig. 1A).

breaking away of the East African minor continental plate from the rest of the African continent. The contiguous major oceanic boundary is identified by the Gulf of Aden and the Red Sea rifts; the "failed rift" is the Ethiopian rift valley southwardly prolonged by the East African rift system.

This fact of life has been recognized by Ethiopian authorities. To cope with it in a logical manner, a historical survey of the localization, recurrence, and amplitude of the hazards to life and property due to earthquakes and volcanic activity has been carried out, statistical projections have been made concerning the maximum amplitudes to be expected at 650 grid points (including 32 population centres and 60 existing or projected power dams and reservoirs), and seismicity maps have been produced on which Ethiopia was divided into four danger zones of relatively comparable risks (see Table 1).

The results of the survey were presented in July 1976 to the Ministry of Public Works and Water Resources and to the Planning Commission, Addis Ababa, in a communication entitled *Seismic Zoning in Ethiopia* (Gouin 1976).

Table 1. Definition of seismic zones in Ethiopia.

Zone[a]	Danger rating	Intensity range[b] (MM)[c]	Ground acceleration range (%g)[b]
0	No	$4.5 > I_{100}$	$1 > \alpha_{100}$
1	Minor	$6.4 > I_{100} \geqslant 4.5$	$4.5 > \alpha_{100} \geqslant 1$
2	Moderate	$7.4 > I_{100} \geqslant 6.5$	$10 > \alpha_{100} \geqslant 4.6$
4	Major	$I_{100} \geqslant 7.5$	$\alpha_{100} \geqslant 10$

[a]The zone number classifies the zone and corresponds to the seismic factor R in the formula used by engineers in the computation of lateral seismic forces (V) on structures. This formula is in the form of: V = 1/4 RKCIFW, where: V = lateral seismic forces at the base of the structure; K,C,F, and W = structural factors; and I = the *importance factor* assigned to different types of buildings according to their importance for the survival of the community.

[b]The subscript 100 indicates the predicted time period, in this case 100 years.

[c]MM = Mercalli-Modified Intensity Scale (1931).

Aim and Scope of Monograph

Because of its objective, namely to be a practical reference paper for engineers, the publication *Seismic Zoning in Ethiopia* had to be restricted in scope to sets of numerical values based on personal interpretations of data for which no unique, straightforward, and unambiguous solutions were often possible. Other seismologists, especially those who will be responsible for either regionalizing in greater detail the seismic zoning maps of Ethiopia or updating the chapter of the security building code on "Structural Design in Earthquake Areas," will need to consult primary sources often unpublished or not locally accessible. They will be given, in some cases, the reasons for my particular choice of epicentral parameters and date of occurrence for the events that took place in the regions they will have to study. This book tries to answer that need. To achieve that purpose, it is conceived as an annotated and edited version of part of my data file on the seismic and volcanic activity in northeastern Africa.

The file opens with the first known written document on telluric activity in Ethiopia, namely, the eruption accompanied by violent earthquakes of volcano Dubbi near the city of Asseb on the Red Sea coast in A.D. 1400 (Region C, entry 1400); it should close with the events of December 1974, but a few subsequent earthquakes have been added. Five hundred and seventy five years of written history are covered; needless to say, the coverage is temporally, spatially, and informatively uneven. Temporally, the distribution of the reported events range from scanty references by chronicle writers and occasional travelers to modern continuous instrumental monitoring. Spatially, the density of historic written documents is heavily biased. It is biased toward the north and northwest of the country where the administration was usually located and where monastic libraries succeeded in preserving the manuscripts through centuries often scarred by wars and devastation. Moreover, at the beginning of this century, the interest of a group of first-rate earth scientists, such as G. Dainelli, O. Marinelli, L. Palazzo, etc., focussed on the geology and telluric activity in northern and northeastern Ethiopia. Their field studies were supplemented by the opening of a seismic station in Asmara in 1913 (Palazzo 1913). World War I put an end to the operation of that station. Details on the Asmara seismic station are given later in this survey. Since 1957, with the opening of the seismic station at Addis Ababa, the density of instrumentally recorded shocks of lesser magnitude has increased toward the geometric centre of Ethiopia, and since 1972–73, the Djibouti seismic network operated by the Institut de Physique du Globe, Paris, has shifted the apparent gravity centre of the low magnitude earthquakes from Central Ethiopia to the Gulf of Tadjoura.

For the computation of seismic hazards out of these 575 years of documents, of disparate and unequal quality, a restricted sampling period had to be chosen. During this period the spatial and temporal distributions of the information were judged sufficiently homogeneous and adequate to warrant statistical treatment within reasonable limits of confidence. Statistical completeness tests were performed (see Discussions and Conclusions): the results showed that the last 75 years of information on Ethiopian earthquakes were sufficiently complete to be considered representative of the present and near-future seismic activity in Ethiopia and the Horn of Africa. *Seismic Zoning in Ethiopia* (Gouin 1976) was based on the analysis of the seismic activity during these 75 years. There is now a need to insert this 75-year sampling window into a larger historical perspective to determine: (1) if we have been dealing with an anomalous period of seismic activity; and (2) if the statistical predictions based on more recent observations are borne out by historical events of amplitudes equivalent to those predicted.

It is logical to assume that Ethiopian historians reported the largest tremors, those important enough to impress people to the point of using a particular earthquake as a landmark in history to date a battle, a coronation, an astronomical phenomenon such as a solar eclipse or a comet, or even their own age. For instance, among the Guraghe nation, which is on the western shoulders of the Ethiopian rift valley, elderly people still reckon their age as so many years *after the earthquakes that were thought to mark the end of the world*, meaning the earthquakes of August 1906 (see Region A, entry 1906/Shoa and Region C, entry 1906–07 Afar Depression). There is a high probability, therefore, that at least the largest earthquakes that occurred in the northern sector of Ethiopia during the last four or five centuries have been reported. Their estimated amplitude, plotted on the maximum amplitude-versus-time of occurrence curves, is therefore a check on the validity of the statistical projections based on shorter sampling periods. On the other hand, it should be noted that the paucity of documents from the southern half of Ethiopia should in no way be interpreted as a sign of absence or of lesser seismic activity. "No documents" never equates to "no earthquakes." This statement is supported by local Galla legends about former localities presumably destroyed by seismic tremors and now covered up by lakes. The legends are heard in the rift valley from Lake Zwai to Abbaya, and on its western escarpment in Gemu-Gofa.

Outline of Geological History

At the end of the pre-Cambrian era, some 570 million years ago, the crystalline rocks that formed the basement complex of the present Afro-Arabian swell had been above sea level for a long time; they remained in this condition for another 350 million years until the end of the Paleozoic. Such a long period of erosion and denudation left the earth's surface of East Africa, Ethiopia, and Arabia almost as completely peneplained as the rest of the African continent.

Two hundred and twenty five million years ago, crustal motion started. During the late Triassic and early Mesozoic, a regional epirogenic sinking of the crust commenced causing a progressive transgression of the ocean from the southeast northwestward, that is from the Indian Ocean coast of present Somalia in the general direction of Lake Tana in northwest Ethiopia. This downward crustal movement, concomitant with a sedimentation process, was active during the first half of the Mesozoic era. It came to a stop in the middle of the same era, reversed its direction into an upward motion during the Upper Jurassic, brought the crust's surface up to sea level by the end of the Mesozoic, 65 million years ago, and finally uplifted it through intermittent but progressive phases throughout Tertiary and Quaternary times to the present level of more than 3000 m above sea level.

Under the excessive tensions created within the earth's crust by this doming process, some 60 million years ago, the crystalline and sedimentary rock layers fissured mostly along or in the vicinity of the zones of maximum uplift, thus allowing enormous outpourings of molten lava to cover the older rock layers. These flows, identified as the volcanic trap series, covered most of Central Ethiopia to a thickness in some places exceeding 3000 m.

The major rifting episode — the one referred to earlier in this text — occurred during Middle Miocene and initiated the birth of what Gregory in 1896 designated as the "Great Rift Valley," that is a system of rifts that extends over 6000 km from the Dead Sea in the Middle East to Mozambique on the southeastern coast of Africa (Fig. 2). At about its midlength, this "Great Rift" cut through the Afro-Arabian swell and dissected it in a Y-pattern, thus creating the Red Sea, the Gulf of Aden, and the Ethiopian rifts, and consequently, three clear-cut dome segments separated by the expanse of the rifts floors (see Fig. 6). The northwest and southeast plateau segments of the dome and the separating rift floors have been retained for the present study. The approximate limits of the region they occupy are marked by a 270° arc-segment in Fig. 2.

Because volcanism is intimately linked with crustal and lithospheric zones of weakness, volcanoes have been abundant in the three branches of the Afro-Arabian sector of the "Great Rift" since Miocene, when Hawaiian-type shield units pierced through the flows of flood basalts deposited in Early Miocene. Figure 3 is a distribution map of the major volcanic units as they can be identified today: some are dead; some are dormant; some are still active. For the last 2.5 million years volcanic activity has been largely restricted to the rifts and Afar floors. The latest explosive and destructive volcanic manifestation dates from 1861, when the Dubbi volcano near Asseb on the Red Sea coast erupted and reportedly caused over 100 casualties. The presence for an unknown number of years

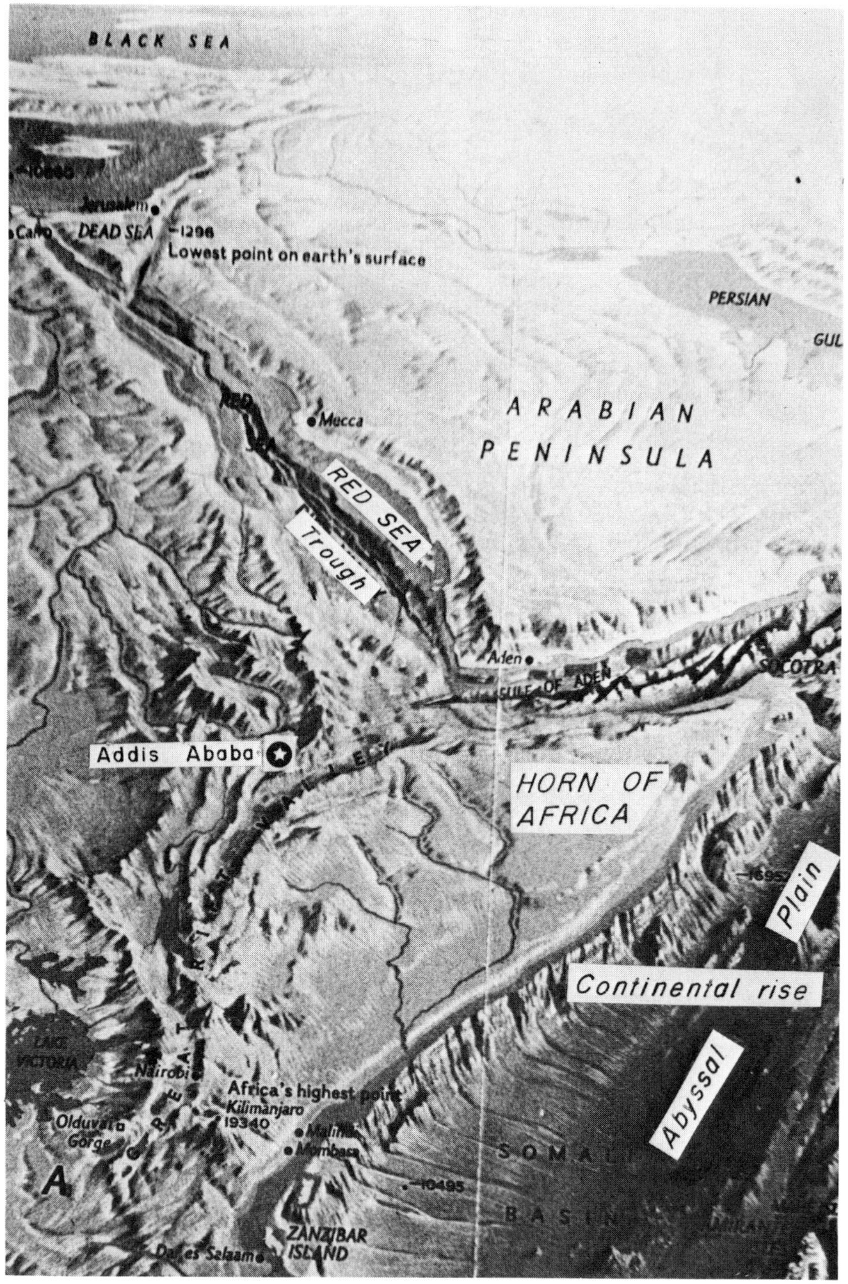

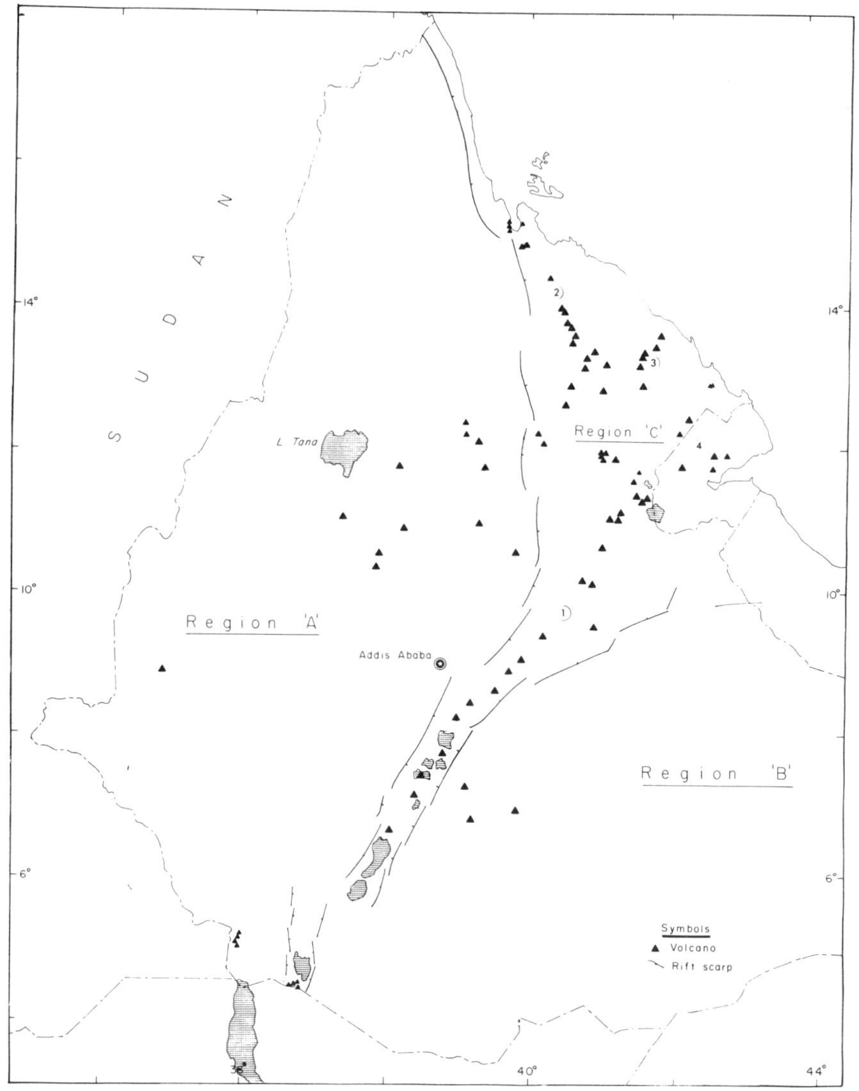

Fig. 3. Location map of important volcanic centres in Ethiopia and in the Republic of Djibouti.

← *photo mosaic of same area.*

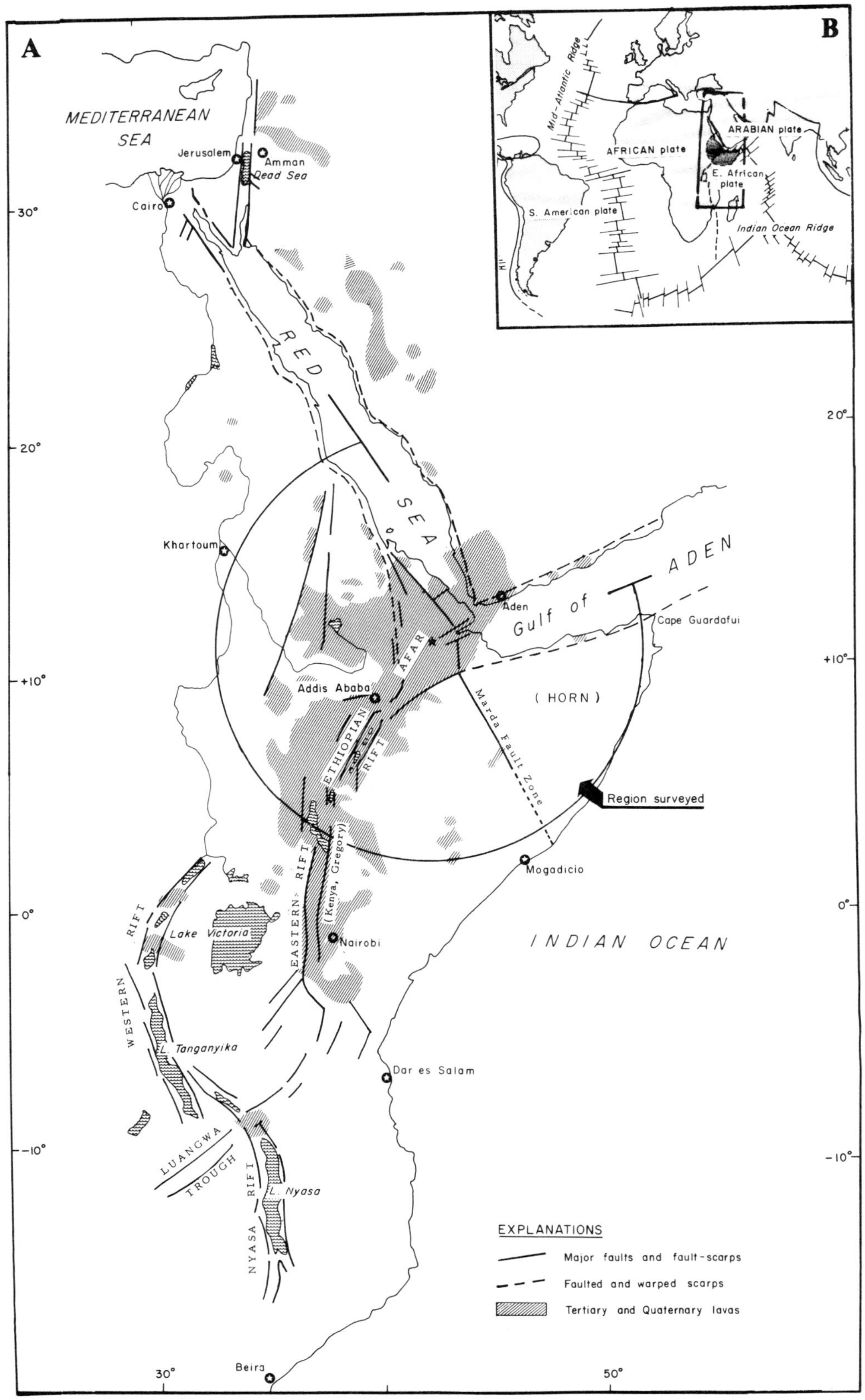

Fig. 2. Outline of the Middle East and East African rift systems. Fig. 2A is a topographic map that underlines with a vertical exaggeration of 100:1 the important structural elements of the systems from the north of the Mediterranean Sea to Mozambique and Madagascar (adapted by permission from an original map of the Indian Ocean produced by the National Geographic). Fig. 2B identifies by name the important sectors of the Great Rift and its proper location in the tectonic plate pattern around Africa.

of a “boiling lava” lake in the caldera of the Erta’Ale volcano in the Danakil depression is a reminder that volcanic activity is a fact of life in Ethiopia (Fig. 4).

The Area Surveyed

The region presently surveyed for its seismic and volcanic activity is geographically located in northeastern Africa, southwest of the junction between the southern Red Sea and the Gulf of Aden. It is diamond-shaped with the apexes oriented toward the four main points of the compass; its approximate centre is N 10°, E 40°. The natural boundaries of this area, clearly discernible on topographic and bathymetric maps, are: (1) to the northeast, the central trough of the southern Red Sea; (2) to the west, the edge of the Ethiopian Plateau almost coincident with the international boundary between Ethiopia and the Republic of the Sudan; (3) to the south, the rather ill-defined southern limits of the Ethiopian swell. It is taken in this survey as the south end of Lake Turkana (formerly, Lake Rudolf) and the north end of the Kenya swell; (4) to the southeast, the East African continental rise in the Indian Ocean off the coast of Somalia; and (5) to the north-northeast, the axial rift of the Gulf of Aden.

As illustrated in Fig. 2, this sector of the African continent is a well-defined structural unit, characterized by an imposing uplift of the earth’s crust whose plateaus reach 3300 m, that contrasts strongly with the relatively low and uniform peneplains of the continent. For the geographer, this region is identified as Ethiopia and the Horn of Africa, although geographers have never clearly defined the western and southern borders of the Horn; for the geologist,[1] it is the African segment of the Afro-Arabian or Nubian swell; and for the seismologist, this region falls almost in its entirety under the international classification code: Seismic Region 37: Africa, Geographic Region 558: Ethiopia.

In mapping the seismicity of a region, the seismically active surrounding areas within a minimum distance of 200 km must be included for two reasons: (1) the tremors, even if they originate outside the international boundaries of the region studied, could reach and affect the region of interest; and (2) in the regional mapping of seismic parameters, such as intensity, ground accelerations, etc., the isolines would fold over themselves near the borders if the surrounding levels of activity were not

[1]In his book *The Geology of Ethiopia,* Mohr (1962, p. 10), considered the whole region described above as the Horn of Africa. Personally, I prefer to see the term applied to either the Somali Peninsula, limited to the southwest by the Marda Fault Zone and its extension towards the Indian Ocean, or to the East African Plate, which is the region east of the Ethiopian and East African rifts from Cape Guardafui at the extreme northeast tip of the Peninsula to Dar es Salaam in Tanzania (Figs. 2 and 5).

Fig. 4. The Erta’Ale caldera: (1) the two northern pits as they appeared in February 1971, each containing a molten lava lake; (2) the northernmost pit crater as seen from the east. Most of its surface was apparently solidified; only a smaller lava lake was active at the north end and manifested intense fountaining activity. In January 1975, the whole lake surface was boiling (photos J. Varet).

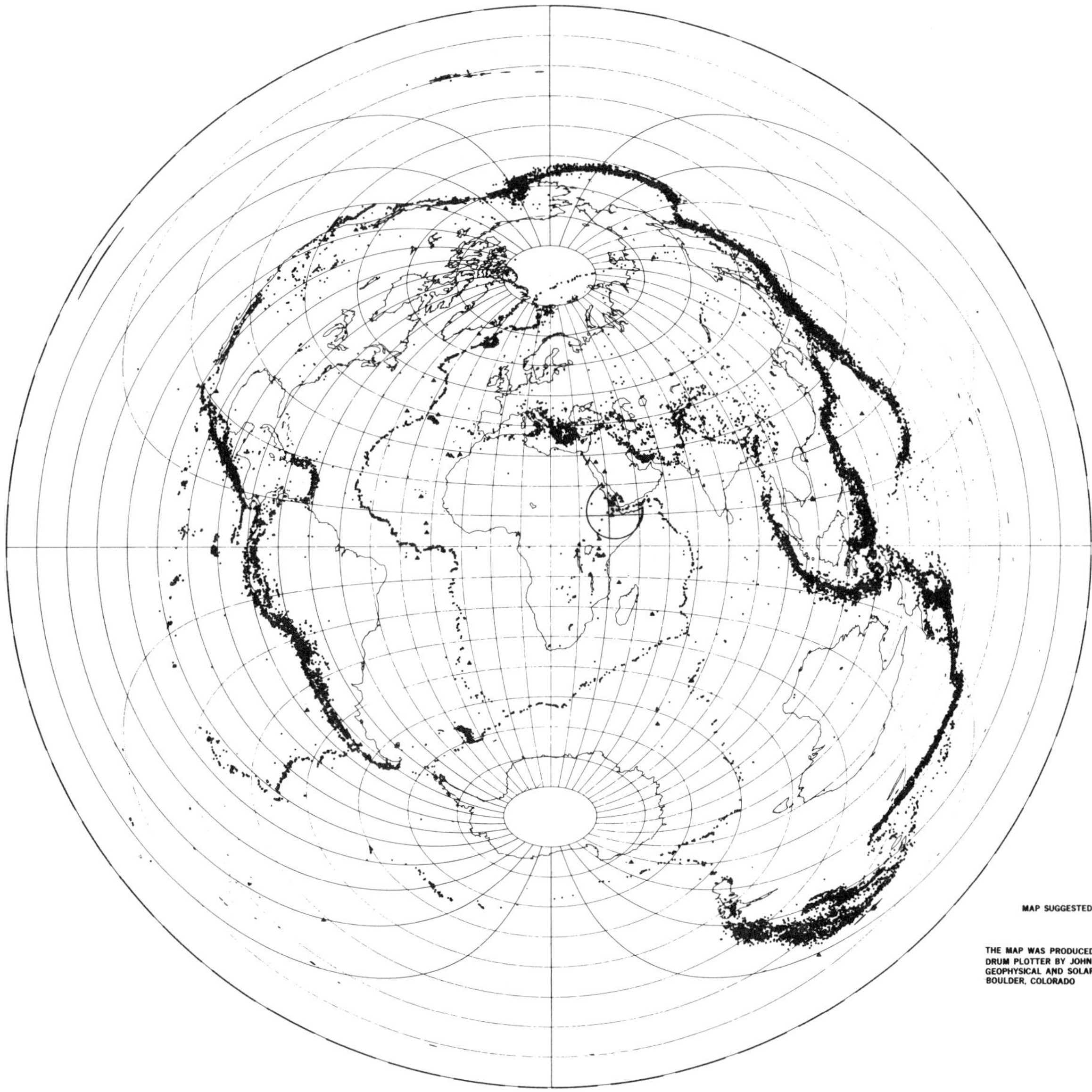

MAP SUGGESTED BY DR. ATHELSTAN SPHILHAUS.

THE MAP WAS PRODUCED ON A CALCOMP 925/936 DRUM PLOTTER BY JOHN O. WARD OF THE NATIONAL GEOPHYSICAL AND SOLAR-TERRESTRIAL DATA CENTER, BOULDER, COLORADO

Fig. 5. World epicentre map of earthquakes, with a magnitude larger than 4.5, that occurred during the period 1963–74 and were reported to the U.S. National Oceanographic and Atmospheric Administration by more than 10 seismic stations. The region presently surveyed is defined on the map by the segment of the circle. This map is reproduced by courtesy of the Environmental Hazard Branch of the U.S. Geological Survey (NOAA/EDS/NGSDC). The map projection is an azimuthal equidistant centred on the equator and 20° E; all distances from the centre are true.

included in the computing process. For this reason, a border zone around Ethiopia and overlapping the Republic of the Sudan, Kenya, Somalia, and the whole Republic of Djibouti was included. The whole area is roughly encompassed by the segment of a circle running counterclockwise over Africa from the central axis of the Red Sea to the central axis of the Gulf of Aden (see Figs. 2 and 5).

On all the world's epicentre distribution maps, even on the older ones when epicentral determinations were less accurate and showed a higher location scatter (see Bellamy 1939), when the number of recording stations was much lower than the present worldwide network and their location unevenly distributed, and when the sensitivity of the instruments was low and therefore only larger events were reported, the region presently studied clearly stood out at the junction of three active belts apparently coinciding with the Red Sea, the Gulf of Aden, and the north end of the East African Rift (Fig. 5). The present monograph is a supplement to these seismicity maps. It relates historic events that took place before the advent of the seismographs, presents details on regional earthquakes of smaller amplitudes than those reported by international agencies, and most of all, it tries to place the modern but often anonymous instrumental information within its human and social Ethiopian context.

The Data File

The data file is comprised of two main complementary sections: a descriptive section; and a numerical listing of the important seismic parameters.

The first part contains the historical, cultural, instrumental, and geological information deemed necessary for a better understanding of the seismic events known from written documents to have occurred in Ethiopia and in the Horn of Africa. In this section, the events are grouped according to the regions that naturally emerged from the breakup of the African segment of the Nubian swell: the two highland or plateau regions and the rift units that dissected them. Figure 6 diagrammatically identifies the selected regions as: Region A, the Ethiopian western plateau; Region B, the southeastern plateau; Region C, the main Ethiopian rift, Afar, and the southern Red Sea; Region D, the Gulf of Aden western sector; and Region E, the Gemu-Gofa and Turkana rifts.

It should be noted that, contrary to what one would expect, the plateau-rift escarpments, which are geologically structural parts of the rifts, have been included in Regions A and B rather than in Region C. The reasons are practical and sociological. Because more than 90% of the seismic and volcanic activity is connected with the rifts, Regions A and B would otherwise have almost no place in this catalogue. On the other hand, however, the aim of this study is the evaluation of the seismic hazards to life

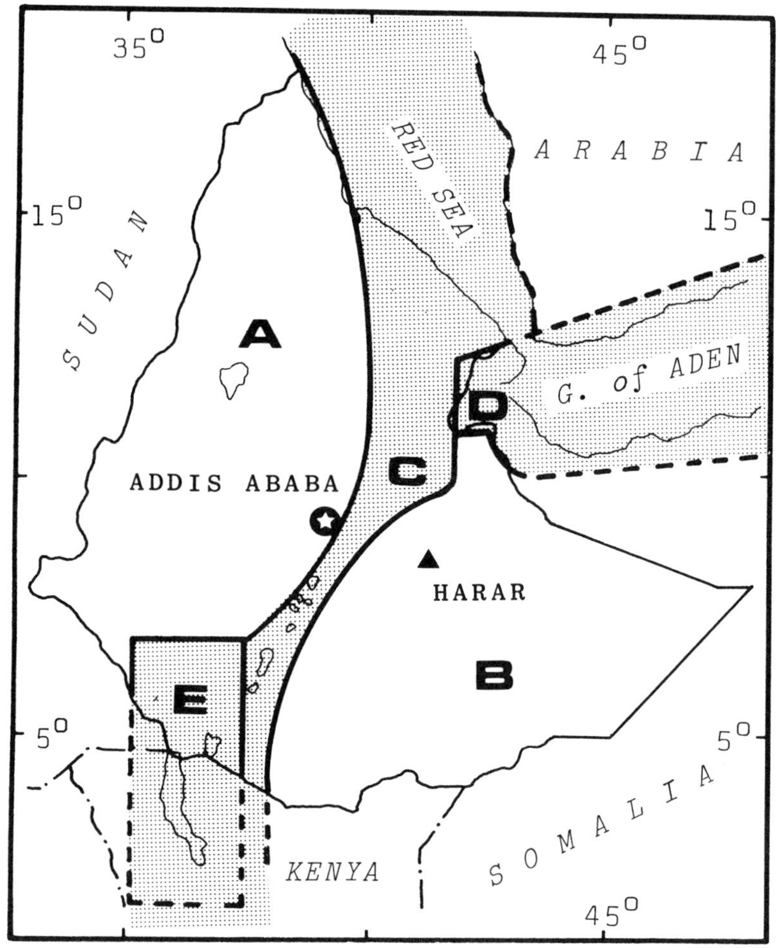

Fig. 6. Natural subdivision of Ethiopia and of the Horn of Africa into five distinct structural units or seismic regions.

and property, and although most shocks originate along rift structures, the greatest damaging effects are found on the plateaus where the majority of the population resides and where most documents have been written. Therefore, the division of Ethiopia into seismic regions has been influenced by the location of the seismic effects on society rather than strictly by geological norms.

The second section is a nonregionalized chronological computer listing of the information in the data file that could be transposed into digital form. Because of the restrictions imposed by a tabular presentation the information is incomplete, especially where confidence limits are concerned. For such details, reference will have to be made to the regions mentioned in the first section.

Structure of the Data File

The earthquake history of Ethiopia and the Horn of Africa is presented in the form of a series of individual entries containing information on single earthquakes or on a whole period of seismic activity. Each entry is indexed and classified by date of occurrence; the index code is:

Year/Month/Day e.g. 1875/XI/24

where the year(s) and day(s) are always given in Arabic and the month(s) in Roman numerals. A dash indicates that a sequence of events lasted over the period of time hyphenated; a comma identifies discrete dates. For instance:

1972/IX–X reads: sequence of events that occurred during the months of September and October 1972;

1973/VII/13, 23 reads: independent events that occurred on 13 and 23 July 1973, respectively.

In the first part, where the information has been subdivided according to five seismic regions, the classification is chronological within each region; in part two, the classification is strictly chronological.

Reference Map of Ethiopia

Figure 7 presents a political map of Ethiopia and the contiguous countries and seas. Its layout is based on standard maps published by Ethiopian government agencies. The provincial boundaries of the 14 administrative regions are those of early 1970. In the text, if a historical change in the name of a locality had occurred, which could lead to a mislocation or the erroneous identification of a site, the relevant information is indicated.

Mapping boundaries, especially international boundaries, is a rather delicate problem, especially when the boundaries have changed with the times, are not always under pleasant political circumstances, or may still be contested. A seismologist could do away with such problems by locating all the sites by their geographic or geocentric coordinates. One has to admit, however, that even if such descriptions of an earthquake damaged site could be extremely accurate, one important parameter would be missing, the human and social element, especially when the survey deals with hazards to human life and property.

In some cases, for the sake of clarity, international boundaries were plotted on the maps and reference was made to these boundaries when no

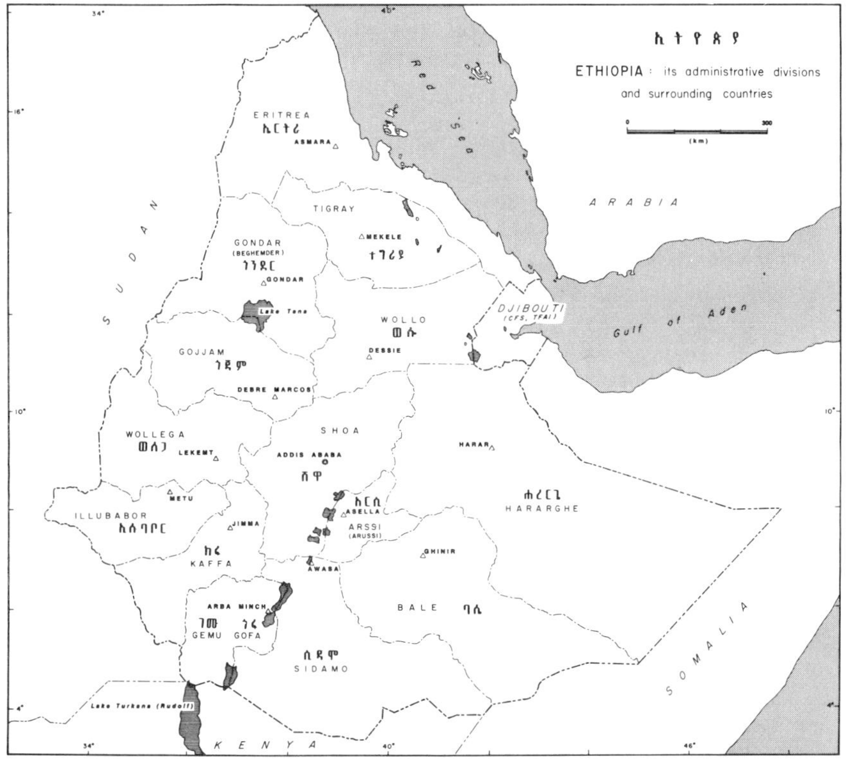

Fig. 7. Administrative divisions of Ethiopia. All boundaries are approximate.

other well-known reference features could be used. Such cases have been reduced to a minimum and the qualifying expression “approximate international boundaries” is used.

Sources of Information and Selection of Data

The Sources

This survey is not the first attempt to collate, classify, and evaluate earthquakes, volcanic activity, and other natural phenomena of the same category that have caused physical havoc in Ethiopia and the Horn of Africa. Before it, two surveys of such historical phenomena were published: one on Eritrean earthquakes published in Florence in 1912 by Marinelli and Dainelli and a second by Palazzo (1915) on earthquakes in present Ethiopia and the Gulf of Tadjoura region.

For historical events, Palazzo's work was the starting point of my research. All the primary sources quoted in his work were reexamined, translated into English, and reevaluated. Translations of the texts written in languages unfamiliar to me were controlled by experts. If I do not always accept Palazzo's interpretations, it is simply because I had more information on hand than he had 60 years earlier.

To Palazzo's sources were added those collected at world data centres and in various libraries of Ethiopia, Europe, and America; either personally or through the help of scholars in Ethiopian history. Relevant primary and secondary sources are given for each entry; whenever useful, the secondary sources are parenthesized.

Selection of the Data

Three main priority criteria were used to select relevant data for each entry from the bulk of the information available: (1) events reported in the Ethiopian literature or by the local media — whether objectively important or not, these events are known to the public and therefore need to be evaluated; (2) instrumental information reported by the international seismological agencies or published in scientific papers; and (3) from the volume of instrumental data compiled by the Geophysical Observatory at Addis Ababa: (a) shocks that were reported and could have caused some damage; and (b) other groups of smaller events of lesser amplitude, such as 'swarms' for instance, when they could be well located — although their amplitudes are below damage threshold, the information on their location in a specific region or with reference to a particular tectonic structure is of importance for the analysis of regional seismicity.

Dating the Local Historical Events

The date of occurrence used for indexing the entries is the date of the calendar commonly known as the Western or Gregorian calendar and abbreviated by the acronym of the Latin expression *Anno Domini,* A.D. In this chronology, the dates before 5 October 1582 refer to the Julian calendar; the later ones, to the Gregorian calendar.

There are reasons for being specific about chronologies used in dating historical events that took place in regions of culture different from those of the author or reader. A good example was given by Ambraseys (1962), when he brought to the attention of seismologists the fact that many modern earthquake catalogues for the Middle East region carried dating errors as large as 622 years or more. Such errors were introduced because one author (Willis 1928) overlooked the chronology used in the original Arabic primary sources and inadvertently took Islamic dates (Hegira calendar, A.H.) for Gregorian dates (A.D.) As the Islamic chronology starts with the flight of Mohammed from Mecca on the date corresponding to 16 July A.D. 622 and its calendar is lunar, the minimum error in Willis' catalogue is 622 years. More serious errors could easily be made when dealing with Ethiopian chronologies.

In this survey, only Ethiopian chronologies known from written documents and directly related to earth tremor reports are discussed. The other chronologies transmitted orally by the elders, such as the Borana calendar in southern Ethiopia, are not mentioned even if they are remarkably accurate and based on astronomical observations.

Structure of the Ethiopian Calendar

The Year — The Ethiopian year is a Julian year; its length is 364.25 solar days. In all Ethiopian chronologies, the first day of the year (Meskerem 1) originally corresponded to 29 August of the Julian calendar. Because Ethiopia did not adopt the Gregorian calendar reform of A.D. 1582, which is basically the 10-day correction from 5 to 15 October A.D. 1582, and the drop of the leap day in 1700, 1800, and 1900, Day One (Meskerem 1) advanced in time with respect to Gregorian dates from the original 29 August (or 30 after a leap year) to 11 or 12 September as it is at present. The table below indicates the days in the Gregorian calendar that correspond to the first day of the Ethiopian year (Meskerem 1) during different periods of history:

From A.D. 1583 to 1699	Meskerem 1 = September 8 (9)
From A.D. 1700 to 1799	Meskerem 1 = September 9 (10)
From A.D. 1800 to 1899	Meskerem 1 = September 10 (11)
From A.D. 1900 to 2099	Meskerem 1 = September 11 (12)

NOTE: The day indicated in parenthesis corresponds to Meskerem 1 following a leap year.

Ethiopian years are grouped in cycles. One of these cycles lasts 4 years, each bearing the name of an evangelist and given in the same order as the books in the New Testament, that is, Matthew, Mark, Luke, and John. The year of Luke is the leap year of the cycle and the extra day is added at the very end of the year. As a rule, leap years in the Ethiopian chronologies are the years whose number divided by four leaves three as the remainder.

The Month — The Ethiopian year in calendar science is called "embolismic" because it is subdivided into 13 months: 12 months of 30 days each and one shorter month of 5 days during regular years and 6 during leap years. The Ethiopian months are: 1 Meskerem; 2 Tekemt; 3 Hédar; 4 Tahsas; 5 Ter; 6 Yekatit; 7 Megabit; 8 Myazya; 9 Genbot; 10 Sené; 11 Hamlé; 12 Nahassié; 13 Pagoumié.

The Day — At present, the civil day — although I do not think it has been legally defined — could be considered to run from midnight to midnight. Historically, however, the Ethiopian day was, and still is by the Church, reckoned from sunset to sunset as was customary with the Greeks, the Romans, the Italians up to 19th century, and the Muslims. The hours are counted from 0 at sunset to 12 at sunrise and again from 0 at sunrise to 12 at sunset, so that, during the day, 3 o'clock Ethiopian time is 9 a.m. and 9 o'clock is 3 p.m. Before the advent of personal watches, the time-of-the-day was reckoned in feet of shadow cast by a man or a standard reference pole (for more information on the subject see Makarios 1967).

Ethiopian Eras and Corresponding Chronologies

As is customary in all civilizations, outstanding historical events were chosen, often spontaneously, as the beginning of new eras — with each of them a new chronology came to life. Among the wealth of Ethiopian chronologies, those used by the authors of the documents relating each tremor and discussed in this survey are as follows: The Era of the Creation of the World (*Amet'Alem*); The Christian Era (*Anno Domini*); The Era of the Incarnation (in Ghe'ez *Amete Sigawe*, in Amharic *Amete Meheret*); The Diocletian Era or the Era of the Martyrs (*Amete Semaetat*); and The Islamic Period (for the Arabs *Hegira*, for the Ethiopians *Amete Tembalat*).

To understand the structure and often the reason for the choice of Year 1 of many of these eras, it has to be recalled that extensive use was made of the 19, 28, 76, and 532 year lunar–solar cycles. Briefly:

(1) 19-year lunar cycle (*Ne'us Qamar*) or the Golden Number — cycle that restores the phases of the moon to the same days of the civil month;

(2) 28-year solar cycle (*Awde Tsehaye*) — cycle that restores the first day of the year to the same day of the week. In the Ethiopian chronology system, the original cycle started on a Tuesday during a year of St. Matthew; all the following cycles have to start on the same day, a Tuesday, hence also all the preceding cycles;

(3) 76-year lunar cycle (*Ma'ckalawi Qamar*) (19 × 4 Evangelists) — a cycle that restores the lunar age of each year of the cycle of the Evangelists to the same days of the civil month; and

(4) 532-year lunar–solar cycle (*'Aby Qamar*) also known as the Dionysian or Victorian period (19 × 28 years) — it is the period when the phases of the moon repeat not only at the same date, but also on the same day of the week. This cycle was conceived by the Egyptian monk Annianus in A.D. 412 to facilitate the computation of the date of Easter.

Keeping in mind that the above mentioned lunar and solar cycles are the basic elements in the Ethiopian chronologic structures, we now define the eras mentioned above.

The Era of Creation (Amet' Alem) — All civilizations, ancient and modern, have tried to date the event that is of the greatest importance to man: the creation or beginning of the world he lives in. The first permanent reference points man could find were astronomical observations and among these, the easiest ones to make were the observations of lunar and solar cycles. From these observations they built their chronologies. When these chronologies had to deal with the distant past, assuming no change in celestial motions and a beginning at zero for each series of cycles, the chronology architects had to conclude that the creation of the world coincided with the zero of one of these cycles. For them, the most important one was the 19-year lunar cycle. Logically therefore, the world had to have started on a multiple of 19 years. The problem was the choice of the "multiple" of 19.

For the Hebrews, the "multiple" is 216 (216 × 19 = 4104 years B.C.). In converting to the actual Jewish calendar date, one has to take into consideration that the Jewish year is lunar and therefore about 10 days shorter than the solar year.

For the Ethiopians, the "multiple" is 290 (290 × 19 = 5492 years). This is an approximation worked out in A.D. 412 by the Egyptian monk Annianus from the information he collected from the Old Testament. His calculations yielded 25 March for the beginning of the lunar cycle and 25 December 5501 of Creation for the birth of Christ.

A contemporary of his, Panodorus, from the same data determined 5493 years for the Creation and December 5502 for the first Christmas; a discrepancy of one calendar year. To explain that discrepancy, I strongly suspect that since the beginning of the year was fixed by law to 29 August, Annianus considered the intervening period from March to August as belonging to Year Zero, while Panodorus considered them as part of Year One. Whatever the reasons, this one-year discrepancy was carried on through the centuries by different schools of thought, namely, the schools of Debre Bizen and Debre Damo in northern Ethiopia. The only way to decipher which school of thought an author belonged to is to cross-reference the dates he indicated with those of known solar eclipses, if he happened to mention any.

The Christian Era (Anno Domini) — This is the era that gave the world its present international calendar. The period before Christ is abbreviated B.C. introduced after the year's number; the following period is abbreviated A.D. and placed before the year's number (e.g. 330 B.C.; A.D. 1979). For that second period (A.D.), the Julian calendar was used until 5(15) October 1582 and was then replaced by the Gregorian calendar. Note that all countries did not adopt the Gregorian reform at the same time.

The Christian Era is reckoned by Ethiopians as the year 5492(3) of Creation.

The Era of Incarnation (Amete Sigawe; Amete Meheret) — This is the current official chronology of Ethiopia and is referred to as *Amete Meheret*, the Amharic expression for "Year of Mercy." Meskerem 1 of Year 1 (A.M.) corresponds to 29 August 5501 of Creation (mathematically, 1 Meskerem of Year 0 corresponds to 29 August 5500 of Creation). Hence, the difference of 8 years with the Gregorian calendar from

September until the end of December and 7 years from January until next September.

It will be noticed, from the documents quoted further on in Part I of this survey, that the name *Amete Meheret* is given indiscriminately to different chronologies. In this survey, *Amete Meheret* is abbreviated A.M. and refers only to the current official Ethiopian calendar. In the descriptive catalogue explanations are given when necessary.

The Era of the Martyrs (Amete Semaetat) — The historical occasion that gave birth to this era was the persecution of Diocletian, which was so bloody in Egypt that the Christians decided to perpetuate its memory by inaugurating a new era, the Era of the Martyrs. It started on 1 Meskerem (29 August) of the 305th 19-year lunar cycle, therefore 1 Meskerem, 5777 from Creation. This date does not correspond to any particular event related to the persecution; it is the first day of the 19-year cycle nearest to the coronation of Diocletian. Here again, a one-year ambiguity exists for the beginning of the era and therefore for the beginning of each of its 532-year cycles.

The Era of the Martyrs is practically the only one that is divided into cycles of 532 years and where each cycle generates a new calendar. During the 14th century, the official chronology was the calendar of the third 532-year cycle of the Era; it commenced on 1 Meskerem 1340(1341) A.M. and was called *Amete Meheret*, for the Era of Mercy. For clarity sake in this text, whenever this third cycle is used, the date is followed by the acronym A.S. for *Amete Semaetat*.

Here again, all ambiguous dates were ascertained by cross-references with astronomical observations. Details of the computations are given in relevant entries.

The Islamic Period (Hegira; Amete Tembalat) — The historical event at the start of this period is the departure of Mohammed from Mecca on 16 July A.D. 622. In 1639, Calif 'Umar proclaimed that date as Day One of the Muslim Era, that is 1 Muharram, A.H. 1. The conversion of Hegira dates into Julian/Gregorian dates is not easy because: (1) the Muslim year is lunar and the Julian/Gregorian is solar; and (2) the Muslim year is centred on the fasting month of Ramadam, which has to coincide with the dead season in a pastoral culture. Because a lunar year is shorter than a solar year by some 10 days and the seasons follow a solar cycle, the month of Ramadam did not keep its relative position with respect to the seasons — it had to be brought back to a useful position occasionally. Unfortunately, this readjustment was not done in a regular manner. This is remedied in the modern Islamic calendar: each month now starts at New Moon and the readjustment is made over a period of 30 years by adding an extra day to the years 2, 5, 7, 10, 13, 16, 18, 21, 24, 26, and 29. In doing so, 34 Hegira years equals 33 solar years. In this survey, for the conversion of the Hegira dates into Julian/Gregorian dates, the *Tables de concordance des Eres Chrétiennes et Hégiriennes* (Cattenoz 1961) were used.

In Ethiopian manuscripts, the Islamic Period is referred to as *Amete Tembalat* and is assigned a beginning either in A.D. 667 or in A.D. 679 instead of the official date of A.D. 622. These differences are unexplained. Fortunately, in many of the Ethiopian manuscripts, events are often dated in more than one calendar.

Each of the chronologies mentioned above has been identified, and its correlation with the others and with the Gregorian calendar cross-checked by reference to the solar eclipses mentioned in the primary Ethiopian sources. The four eclipses used are those of 7 June 1434; 18 May 1528; 14 January 1553; and 4 November 1668. Figure 8 illustrates the path of each eclipse and the percentage of its totality observed in Ethiopia.

Identification of Localities

Historical documents are seldom precise about the locations where tremors were felt or where damage was caused. They often identify the sites only by reference to localities that might not at present bear the same name, or by reference to realms and provinces that have either disappeared or whose boundaries have fluctuated with time. Thorough investigations have been conducted among Ethiopian scholars and elders to identify doubtful sites. To avoid errors of interpretation, the estimated geographic coordinates of the sites are given along with a descriptive location within the province or administrative region to which they belong. In almost every entry, detailed location maps supplement the general political map of Ethiopia presented in Fig. 7.

Another problem is the transliteration of locality names from Ethiopic to English because many sounds are particular to the local languages and the scripts different from the Latin alphabet. One has only to compare the spelling of the name of a harbour like Massawa on various geographical maps (Massawa, Massoua, Mazua, Maçua, Mesewwa, Macaww'a, Massowah, Missiwa, Mitsewa, and Mits'ewa) to realize that a unique transliteration of each word is impossible unless an official decree determines which phonetic system is to be used. Throughout history, writers who were not linguists by profession reproduced Ethiopic sounds the best they could by matching them with the nearest corresponding sounds in their own language. A perfect match is impossible because there are about 80 languages in Ethiopia many of whose sounds, such as those which are explosive or glottalized, have no exact correspondance in European languages and as mentioned above, cannot be rendered with the Latin alphabet. It is no wonder, then, that Buxton mentions in the preface of his book, *The Abyssinians* (1970, p. 16), that he found 13 different spellings for the locality of Wokru, which is written with only two syllabic symbols in Amharic and Tigrigna (Agroo, Corov, Overo, Ougro, Uero, Ouagero,

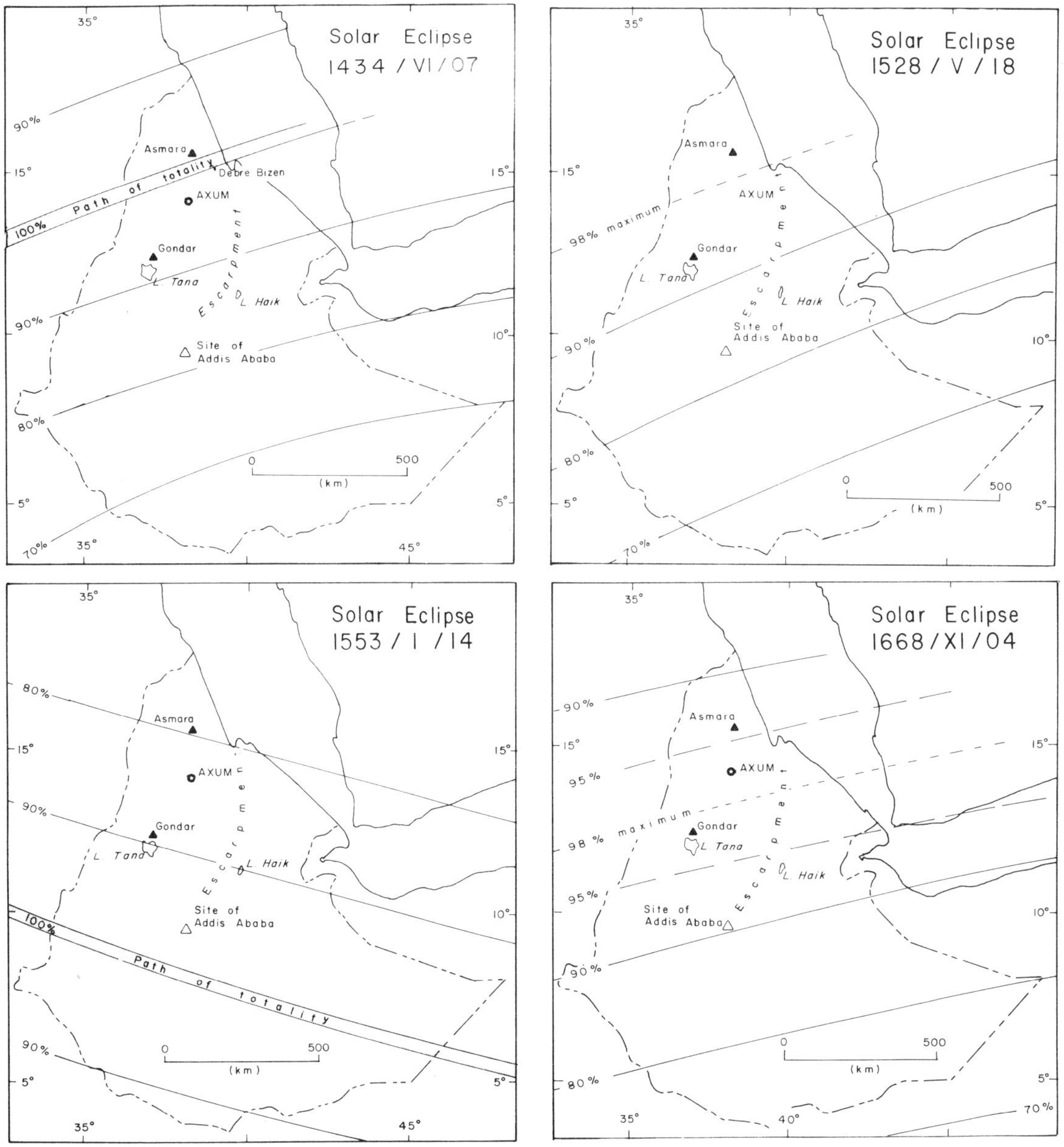

Fig. 8. Path of the four historical eclipses used to identify older Ethiopian calendars and to date earthquakes. The computation of the visibility distribution was graciously performed by Jean Meeus, Vereniging voor Sterrenkunde, Belgium.

Oukero, Ouogro,Uogoro, Woghuro, Wogro, Wagro, and Weqro). By the way, Buxton, not satisfied with a choice of 13 spellings, produced a 14th (Wuqro). The Shell map of Ethiopia (copyright 1973) beat him to it by finding a 15th spelling (Wikro).

In this survey, the transliteration of Ethiopic words is not based on any particular linguistic system. The present text is addressed to seismologists, not to linguists and the names are given to locate, as accurately as possible, sites where an earthquake occurred or volcanic activity was observed. In this case, the spelling of the site-name is somewhat immaterial provided the geographic coordinates are accurate. Consequently, the spelling adopted is the one most commonly used in World Atlases, but some variants are often given in parentheses.

The only geographic coordinates system used in the final location of sites is the Greenwich system. In the primary sources, however, these other reference-meridians are utilized: Paris in France; Ferro in the Canary Islands at 17°37′45″ west of Greenwich (not to be confused with the neighbouring Ile-de-Fer); Pulkovo in Russia; and Rome in Italy.

Earthquake Amplitude Parameters

Intensity

The parameter "intensity" expresses the visual appraisal of the amplitude or force of earth shaking at different sites within the perceptive area of an earthquake. It is a complex noninstrumental parameter combining psychological, geological, and engineering observations. Many intensity scales have been proposed since the first one by Gastaldi in 1564. Even today, not all countries utilize the same scale.

In the older Ethiopian documents, the amplitude of damage or strength of earth shaking was expressed either in Mercalli, Cancani, or Sieberg intensity grades. Even at present, the intensities reported by the Meteorological Office in Djibouti are quoted in Rossi-Forel grades. Description of these intensity scales along with 35 other older scales is given in two papers by Davidson (1921, 1953). UNESCO proposed an international scale, coded MSK 1964 in a document UNESCO/NS/SEISM/33 dated Paris, 26 October 1965. Since the Building Code of Ethiopia (revised draft Building Code 1974, Addis Ababa, Part E, p. 2) specifies that seismic hazards in this country should be evaluated on the Mercalli-Modified 1931 scale (Wood and Newmann 1931) all intensity evaluations have been converted to that scale of 12 grades, abbreviated M.M. in this text. Added to it was the classification of masonry structures suggested by Richter (1958, p. 136–137), which reads as follows and has no connection with the convential classes A, B, and C used by construction engineers.

Class A — Good workmanship, mortar, and design; reinforced, especially laterally, and bound together by using steel, concrete, etc. Designed to resist lateral forces. Few buildings erected before 1960 can be classified in this category.

Class B — Good workmanship and mortar; reinforced but not designed to resist lateral forces.

Class C — Ordinary workmanship and mortar; no extreme weakness like failing to tie at corners, but neither reinforced nor designed against horizontal forces.

Class D — Low standard of workmanship; weak materials such as mud bricks and packed earth (*tchika*); horizontally weak.

As readers might not have on hand the description of each grade of the intensity scales, abridged versions of the two most important ones, The Rossi-Forel and Mercalli-Modified, are reproduced in Tables 2 and 3 (also see Fig. 9).

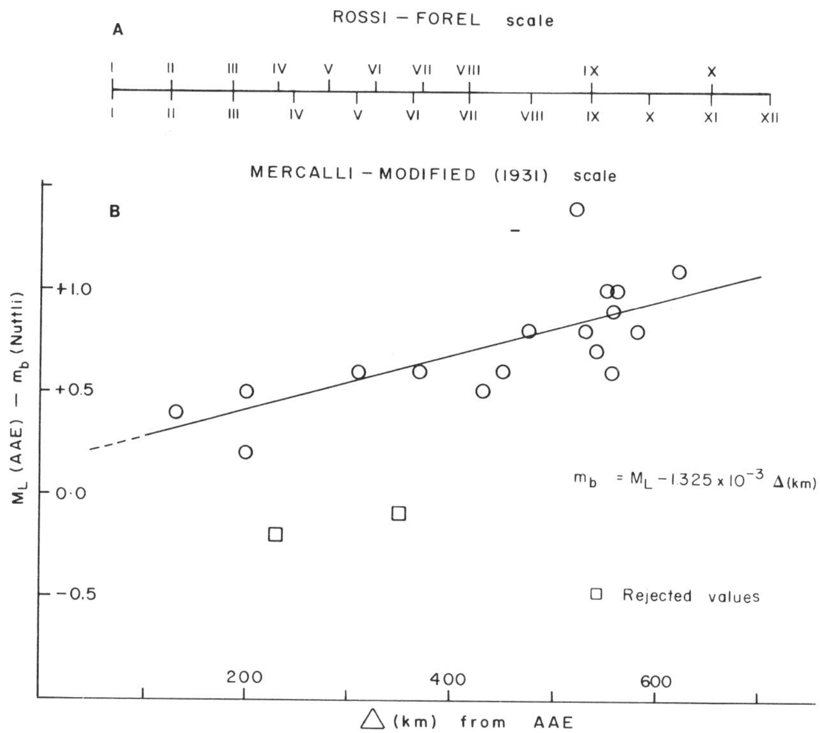

Fig. 9(A). Graphic comparison between the 10-grade Rossi-Forel intensity scale of earth shaking and the Mercalli-Modified (1931) scale. (B). Relation between M_L (AAE) and the empirical m_b values obtained by Nuttli (1973) for eastern North America.

Table 2. Rossi-Forel scale of intensities (1874) (an abridged version).

I	*Microseismic shock* — felt by an experienced observer
II	*Extremely feeble shock* — felt by a small number of persons at rest
III	*Very feeble shock* — felt by several persons at rest; strong enough for the direction and duration to be estimated
IV	*Feeble shock* — felt by persons in motion; disturbance of movable objects, i.e. doors, windows; cracking of ceilings
V	*Shock of moderate intensity* — felt generally by everyone; disturbance of furniture, beds, etc.; ringing of some bells
VI	*Fairly strong shock* — general awakening of those asleep; general ringing of bells; oscillation of suspended objects; visible agitation of trees; some people would leave their houses
VII	*Strong shock* — overthrow of movable objects; fall of plaster; ringing of church bells; general panic, without damage to good buildings
VIII	*Very strong shock* — cracks in the walls of good buildings
IX	*Extremely strong shock* — partial or total destruction of some buildings
X	*Shock of extreme intensity* — disaster; fissures in the ground; disturbance of geological strata; rock falls

Magnitude

Earthquake magnitude is the computed energy released at the source of an earthquake; it is expressed in this survey according to the following nomenclature:

m_b (or m) the body-wave magnitude calculated from short-period vertical P-wave amplitudes;

M_s (or M) the surface-wave magnitude using the 20-s Rayleigh-wave amplitudes from vertical long-period seismograms;

M_L the local earthquake magnitude — the techniques used for its determinations may differ from station to station;

M (estimated) in this report, it indicates a magnitude value inferred from the extent of the damage caused by an earthquake — such magnitudes are given only when no instrumental information is available.

Magnitude units in the text are usually followed by the code letters of the agency that determined them. Agencies and station codes are indexed on page 190.

From the very beginning, magnitudes of local events calculated at AAE (Addis Ababa, Ethiopia) have been based on the epicentral distance, the maximum trace amplitude and the period of the corresponding cycle recorded on horizontal short-period seismograms (Benioff, $T_o = 1.0$ s, $T_g = 0.8$ s). To speed up the calculations, we used the nomogram developed by Richter for Wood-Anderson equipment and later adapted by Nordquist (Gutenberg and Richter 1942, p. 164) to other types of seismic transducers.

The intermittent presence of Lg phases on AAE seismograms and the difficulty of differentiating Lg from Sg (Lg$\gg$Sg) on near-earthquake

Table 3. Mercalli-Modified (1931) intensity scale (an abridged version).

I	Not felt except by very few under favourable circumstances
II	Felt by only a few persons at rest, especially on upper floors of buildings; delicately suspended objects may swing
III	Felt quite noticeably indoors, especially on upper floors of buildings; vibrations like the passing of heavy vehicles; parked cars might rock slightly
IV	During the day, felt indoors by many; outdoors by a few, at night, some people are awakened; dishes, windows, and doors disturbed; walls make creaking sounds
V	Felt by nearly everyone; many awakened; instances of cracked plaster; unstable objects overturned
VI	Felt by all; many frightened and run outdoors; some heavy furniture displaced; a few instances of fallen plaster; damage slight
VII	Everybody runs outdoors; damage *negligible* in buildings of good design, *slight to moderate* in well built ordinary structures; *considerable* in poorly built or badly designed structures
VIII	Damage *slight* in specially designed structures; *considerable* in substantial buildings, with partial collapse; *great* in poorly build structures
IX	Damage *considerable* in specially designed structures; *great* in substantial buildings, with partial collapse; ground is cracked conspicuously
X	Most masonry structures destroyed as well as some well-built wooden structures; ground badly cracked; rails bent; landslides and rockslides
XI	Few, if any, masonry structures remain standing; bridges destroyed; broad fissures in ground; earth subsidence and landslides in soft ground
XII	Damage *total*

records caused an overestimation of the magnitude M_L(AAE) by 0.5–1.0 units when Lg was present. The error increased with epicentral distance, see Fig. 9. Two attempts were made to obtain more reliable values. At the suggestion of Anne Stevens (Energy, Mines, and Resources, Canada), as a first approximation, a comparison was made between M_L(AAE) and the m_b values obtained by Nuttli (1973) for epicentral distances smaller than 450 km ($0.5 < \Delta° < 4.0$) in eastern North America. For these distances, Nuttli's empirical formula is:

$$m_b = -0.1 + 1.66 (\log \Delta) + \log A/T$$

where: Δ is in kilometres; A is zero-to-peak ground motion in microns; and T in seconds is the period of the corresponding wave cycle.

Figure 9, based on 19 events, shows that the difference between M_L (AAE) and the body-wave magnitude obtained by Nuttli's formula is always positive and that it increases at a fairly steady rate from about 0.2 at $\Delta \simeq 50$ km to about 1.0 at 600 km. The root mean square (rms) slope of the curve is 1.325×10^{-3}.

The second attempt was to compare the m_b values published by the USCGS and NOAA with M_L(AAE) values for the same regional earthquakes. R. Searle obtained the following relationship:

$$M_L(AAE) = 0.24 + 1.05\ m_b\ (CGS)$$

with no reference this time to epicentral distances. Unless otherwise stated in the text, it is this empirical formula that I have been using to determine m_b(AAE).

Interpretation of the Magnitudes Published by Other African Agencies

BUL (Buluwayo, South Africa) uses a magnitude scale adjusted to USCGS body-wave values. The correlation, however, is poor (Bath, 1975).

LWI (Lwiro, IRSAC, Zaire) uses a local magnitude scale that exhibits for Ethiopian earthquakes a lower value than m_b (CGS). The ratio LWI/CGS ≈ 0.93. (Wollenberg 1966).

NAI (Nairobi, Kenya) magnitudes determined by Nairobi are values published by Rodriguez (PhD Thesis, University of Nairobi, 1969) and are higher than m_b (CGS) by a factor of 1.11.

Figures and Graphs

As far as is practically possible, uniformity has been kept in the selection and interpretation of symbols on maps, graphs, and other illustrations. To avoid overloading the figures, the meaning of each symbol has not always been repeated. Unless otherwise stated, the interpretation of the symbols is as follows:

✪ Addis Ababa, capital of Ethiopia. Symbol often used as a reference point on a map when only a few points are identified.

▲ Cities, towns, or population centres.

○ ● Instrumental epicentre directly related to an entry. The full symbol indicates an adopted location.

□ ■ Instrumental epicentre belonging to another entry. The full symbol refers to an adopted location.

★ Site of reported-felt tremors.

⬡ ⬢ Recomputed instrumental epicentre. The full symbol indicated an adopted recomputed location.

 Volcanoes, volcanic centres.

 Faults. When present, a stroke perpendicular to the fault strike indicates the downthrown side of the fault; stippling indicates the elevated side.

Earthquake History of the Ethiopian Western Plateau (Region A)

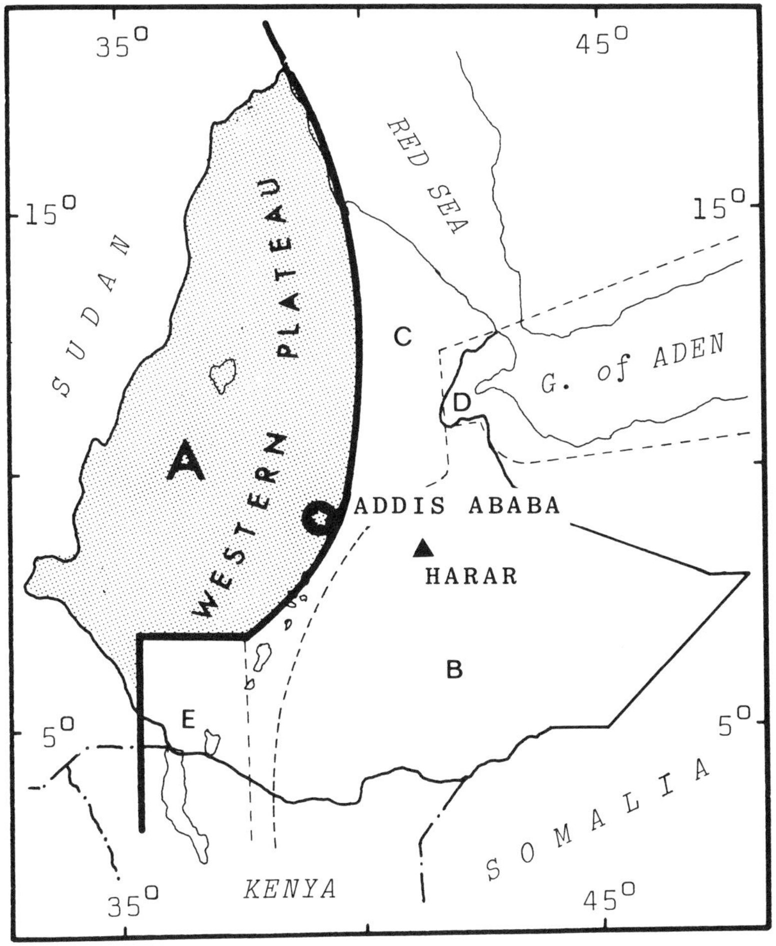

1431–32

An earthquake was felt in northern Ethiopia. The information is not sufficient to allow a good determination of the site where the shock was felt nor of its intensity. It is inferred, however, that the tremors originated along the eastern escarpment of the Plateau, slightly east of E 39.5° and north of N 10°.

Sources

Two manuscripts: one in Tigrigna, reproduced and translated in French by Kolmodin (1914, p. A31); the other in Ghe'ez, quoted and commented upon by Palazzo (1915, p. 298–301). It is my opinion that the original text was in Tigrigna and that the Ghe'ez text is a translation.

Comments

1. Dating the Event

Both manuscripts laconically state that *there was an earthquake during the year 85*. The Tigrigna version indicates that year 85 refers to the chronology *Amete Meheret* (Year of Mercy).

Amete Meheret, as a calendar identification, currently refers to the present Ethiopian Calendar that reckons its Day One of Year One, on the day corresponding to 29 August year 8 *Anno Domini*. In the manuscripts quoted above, however, *Amete Meheret* refers to a completely different chronology system, namely to the Third Cycle of the Diocletian Era in which Day One of Year One corresponds to 29 August of either A.D. 1346 or 1347 according to which school of thought the author belonged. In the 15th century, the designation *Amete Meheret* usually referred to the Diocletian Era. In the present survey, these ambiguities relative to the eras and to the corresponding dates in the Julian-Gregorian system *Anno Domini* are resolved by cross references to known astronomical phenomena — see comments in entry 1433–34.

To avoid any confusion as to the era to which the term *Amete Meheret* refers in this text, the following code has been adopted: A.S. when it refers to the Diocletian Era; and A.M. when it refers to the current Ethiopian calendar.

2. Probable Location of the Epicentre

It is inferred from the context and from recent seismological observations that the origin of the shock was most probably the eastern escarpment of the Western Plateau. The question is also discussed in entry 1433–34.

1433–34

During months following 29 August 1433, several earthquakes were reported from northern Ethiopia. As for the previous entry, the information supplied by the Ethiopian chronicler is insufficient to localize with any precision either the sites where the tremors were felt or their intensities. The epicentral region is presumed to have been the Plateau-Afar escarpment.

Sources

Same as in Entry 1431–32. The text reads:

In the Year 87, Zara Yagob became emperor and took the name of Constantinos I. That year, there was a total eclipse of the sun and many earth tremors were felt.

Comments

1. Date of the Events

The manuscript indicates that the tremors were felt during year 87; the Ethiopian chonology used by the chronicler has been identified as the Third Cycle of the Diocletian Era (entry 1431–32). In transferring this date to the present international calendar system *Anno Domini*, a 1-year ambiguity arises from the fact that there are two schools of thought concerning the year on which the Diocletian Era began and therefore each of its cycles. In the Coptic tradition year 87 ran from August 1433 to August 1434; whereas, the Ethiopian tradition puts it a calendar year later, from August 1434 to August 1435.

The ambiguity is resolved by the reference in the original text to a total solar eclipse, which is now identified as Eclipse 6286 in Oppolzer's classification (1887, edit. 1962, p. 252 and chart 126). This eclipse was visible in Ethiopia on 7 June 1434, that is during the Ethiopian year 87, which spanned from August 1433 to August 1434.

2. Estimated Epicentral Location

The text of the manuscript gives no direct indication concerning the sites where the tremors were felt; the context, however, leaves no doubt that the reports came from the north of the country. Moreover, it was an Ethiopian custom for a chronicle writer to be attached to and follow the Royal Court which, at the time, did not enjoy permanent residency but migrated across the country according to the whims of politics and wars. In 87 A.S., the Court was in Axum (N 14.1°, E 38.7°) at least for the Coronation festivities (Dillman 1884, p. 14). The earthquakes occurred during the Coronation Year. From the fact that the author of the Chronicles reported both the coronation and the earthquakes in the same paragraph of his text, it is inferred that the author witnessed both events at the same location, Axum, before the new Emperor moved his Court, southeast, to the vicinity of Lake

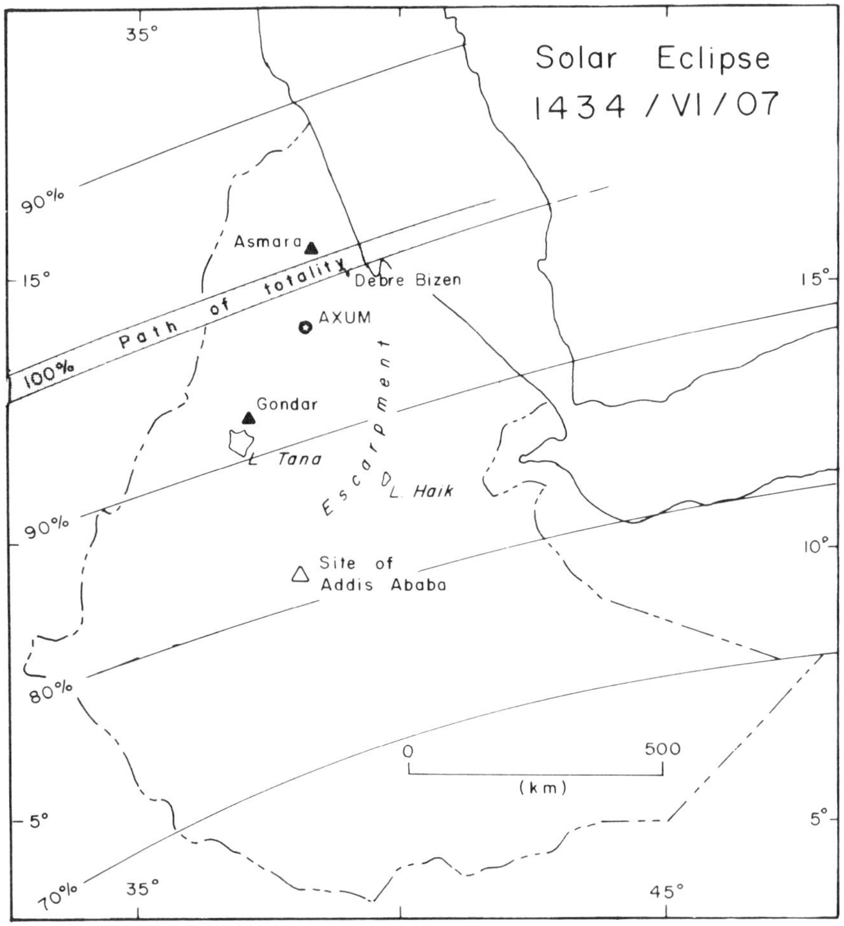

Fig. 10. Total solar eclipse of 7 June 1434, as observed over Ethiopia, and the relative location of its path of totality with respect to Axum and Debre Bizen. The maximum occultation at Axum occurred at about 07:00 Greenwich Mean Time, that is midmorning in Ethiopia at about 3½ hours Ethiopian Time as reckoned from sunrise. The elements of the eclipse were graciously computed by Jean Meeus.

Haik (N 11.5°) and later to Debre Berhan (N 09.7°). Computations show that about 95% of totality could be observed from Axum and almost 100% from the Debre Bizen monastery (see Fig. 10).

Later observations, both historical and instrumental, suggest that the epicentre(s) must have been located along the escarpment. See, for instance, entry 1845/II/12.

1518–19

Seismic activity was reported from northern Ethiopia during the years 1518–19.

Sources

Traditions of Tazzega and Hazzega edited by Kolmodin (1914, p. A33). Palazzo (1915, p. 298) quotes a Ghe'ez version of the same document kept at the Monastery of Debre Bizen. It reads as follows:

During the whole year 171 Amete Meheret, earth trembled very many times.

Comments

The chronology as in entries 1431–32 and 1433–34 is the Third Cycle of the Diocletian Era. As in these entries, the date was ascertained by cross-references with the eclipses of 7 June 1434 and 18 May 1528 (Oppolzer's numbers 6286 and 6497) mentioned in the same text. The paths of these eclipses are plotted on Fig. 8.

1553–54

Earth tremors were felt in Tigray. No exact location nor intensities are indicated.

Sources

Ghe'ez manuscript 160 at the Bibliothèque Nationale de Paris. (Parts of the manuscript were edited and translated to Italian by Conti Rossini, 1918, p. 279, 284–5, and 291. Palazzo (1915, p. 301–3) comments on the dating of the event).

This book was written in the year 206 Amete Meheret, during the 13th year of the reign of our king Galawdewos, after the earthquakes which occurred on a Monday and on a Tuesday.

Comments

Amete Meheret, in this context, refers to the chronology of the Third Cycle in the Diocletian Era (see comments in entry 1431–32). The correspondance of year 206 A.S. in Julian/Gregorian dates is made easier by two astronomical cross-references found in the same text, namely a solar eclipse in year 205 and a comet ("a star with a tail") in 208. The solar eclipse, number 6553 in Oppolzer's catalogue, was visible on 14 January 1553 during the Ethiopian calendar year running from August 1552 to August

1553, that is Year 205 A.M. The seismic activity followed during the period August 1553 to August 1554. The same reasoning applies to the observation of Comet Charles V, which was at its perihelion on 22 April 1556.

It is interesting to note that the author of the Ghe'ez manuscript 160 quoted in this entry used a different origin time in his A.S. chronology than the author of the Tigrigna manuscript quoted in entries 1431–32 and 1433–34; the difference is minus one year. This difference in dating the two manuscripts is not an error of computation by individual authors as is often mentioned in the literature; rather, it is a sign that an author belonged to one of the two traditional schools of thought concerning Ethiopian chronology: the first, which I have labeled "Ghe'ez Tradition," is usually found in the manuscripts originating from the monastery of Debre Bizen; the second, the 'Ethiopian Tradition' is usually found in the texts from the monastery of Debre Damo (see Explanatory Notes). Note that the remarks concerning the Ghe'ez and Ethiopian chronology traditions are restricted to the manuscripts consulted with reference to the present study. In short, Day One Year One of the Third Cycle of the Diocletian Era corresponds to: August 1347 in the Ghe'ez Tradition; and August 1348 in the Ethiopian Tradition.

1593/IX–XI

In September 1593, an earth tremor was felt in northern Ethiopia, presumably on the margin of the Plateau-Afar escarpment near Mai Chew (about N 13°, E 39½°). A second earthquake was reported from the same region during a month of December but the year is not indicated. It is presumed here that the second report referred to an aftershock of the September 1593 activity.

Sources

Ethiopian manuscript 142 of the Bibliothèque Nationale de Paris (edited by Basset 1882, p. 119); and manuscript EMML 5649 (property of Ato Johannis Afework, Addis Ababa).

1. The 31st year (of the reign of Malak-Segad) *began on a Wednesday, the time of the Gospel of Mark, golden number 23, epact 7. During that month, there was an earthquake.* (MS 142).

2. To-day, 13th of Tahsas, the earth trembled... (MS 5649).

Comments

Event No. 1. According to manuscript 142, the 31st year of the reign of Negus Malak-Segad started on 1 Meskerem 1586, the Year of St Mark, a Wednesday, that is 8 September 1593. The earthquake was felt, sometime, within the following 30 days.

This manuscript gives no information on the location of the tremors but from history it is known that Malak-Segad had, the previous year, moved his Court to Aiba (N 12.9°, E 39.6°), in Gojjam, near the present town of Mai Chew on the eastern margin of the Plateau. The region is seismically active (see, among others, entries 1853–54 and 1965/IX/30).

Event No. 2. The 13th of Tahsas fell during the first week of December. No year is indicated but the context refers to about the same period of history; it is likely that this second earthquake was an aftershock of the previous activity.

1632

An earthquake was felt, shortly before 7 June in the province of Beghemder. No exact location is given.

Sources

Private communication of C. Conti Rossini to Palazzo. Conti Rossini also mentioned that the earthquake occurred a few days before the battle of Dankaz, Beghemder, on 7 June 1632, but unfortunately, he did not indicate his sources (Palazzo 1915, p. 304).

Comments

Dankaz (N 12.4°, E 37.6°), now usually spelled Denches, is located 20 km SSE of Gondar. In 1632, Dankaz and Gondar were in the realm of Dembia; Beghemder, at the time, had its northern border on the eastern side of Lake Tana, about latitude N 12°. Conti Rossini, therefore, must have been referring to the present Administrative Region of Beghemder, not to the Beghemder of the 17th century (see Fig. 11).

In 1977, the Province of Beghemder was renamed Administrative Region of Gondar.

1667/X/22

On 22 October 1667 (15 Tekemt 1660) an earth tremor was felt in the region of Gondar (N 12½°, E 37½°). No intensity was reported.

Sources

Manuscript Bod. Eth., No. 30 (Oxford Univ.) and manuscript 167 of *Collection d'Abbadie: "Annales de Johannis I"* (a translation in French has been published by Guidi 1961a,b).

On Tekemt 15, the children of the deceased King ('Alam Sagad)... were brought to the mountain called Wahni. That very day, there was an earthquake.

Fig. 11. Northern and Central Ethiopia of the XVI and XVII centuries.

Comments

The date of the earthquake is ascertained by reference to the solar eclipse, Oppolzer's No. 6831, observed in Gondar on 4 November 1668 (Guidi 1961, note p. 56) and spectacularly described in the same manuscript as follows:

On Tekemt 28, the sun appeared some times as red as blood, some times green and other times covered with black smoke. It was surrounded by a dreadfully dark cloud. This phenomenon occurred at nine o'clock, a Sunday (Guidi 1961, p. 8).

NOTE: In Ethiopian time reckoning, sunrise is at 00 hours a.m.; nine o'clock Ethiopian time is therefore 3 p.m. The path and percentage of totality of this eclipse are plotted on Fig. 8.

1703/IX/10

Sections of the city of Gondar (population in 1650 was about 100 000) and the town of Gorgora on the north shore of Lake Tana were badly damaged. About 30 people were killed. Some historians have attributed the disaster to an earthquake. Contemporary documents, however, reveal that the panic, material damage, and casualties were not caused by earth tremors but rather by a strong whirlwind that tore off the roofs of some houses, dislocating the upper layers of the masonry walls. The falling stones caused casualties among those who had taken refuge in the tower.

Sources

A manuscript in Ghe'ez published by Guidi (1961, p. 227); a second manuscript, manuscript 142, Bibliothèque Nationale de Paris, published by Basset (1882, p. 54).

Comments

1. Nature of the Phenomenon that Caused havoc in Gondar

Two primary sources are very explicit on this subject:

The feast of St John fell on a Monday. On that day, the town was taken by panic... because a strong windstorm... uprooted the houses of the citizens and some of the King as well. The part of the King's palace called Jan Tekele fell down and crushed to death many people who had taken refuge in it (Guidi 1961, p. 227).

Meskerem 1 was a Monday. That day, there was panic amongst the population of Gondar because a violent tornado called Taro destroyed houses in the capital, those of the King and the tower Jan Tekele. Over 30 people were killed including Ato Wolde Berhan, a man of considerable prestige in Shoa (Basset 1882, p. 54).

A little further on, the same document describes a horseman who found himself suddenly sailing off the saddle of his mule when the wind caught his *jantela* (a sunshade in the form of an umbrella with a solid bamboo structure) as if it were a sail or a parachute!

These texts are explicit: the causative phenomenon was a strong whirlwind. Two questions come to mind: (1) How is it that a whirlwind could have been confused with an earthquake? and (2) Were there other instances in Ethiopia of whirlwinds strong enough to damage villages and towns? The answers are: (1) The error comes from the mistranslation of the Ghe'ez word *(delekeleke)* used in some of the documents relating the event. The word could mean earthquake if so qualified in a particular text but, in its first meaning, it is a generic expression for "disturbance, commotion or panic" (Dillman 1884). In a country where seismic activity is of common occurrence, the temptation is always strong to attribute disasters of uncertain origin to earthquakes under the pretext that "they happen all the time!" and (2) A clipping from the Ethiopian Herald of Addis Ababa, 23 Hamelie 1969 (Saturday, 8 July 1977) answers the question on the occurrence of devastating whirlwinds:

Dessie (ENA) — Thirty-two buidlings in the town of Mille on the Asseb road, including many government offices have been destroyed by a strong gale which hit the area but no loss of life was reported.

Among government buildings and offices which were destroyed by the winds last Wednesday afternoon were the town's clinic, the office of the Relief and Rehabilitation Commission, a school, a store belonging to the Desert Locust Control of East Africa and the Ethiopian Road Authority camp.

2. Date of the Event

In the literature, the event is dated 1704. The documents quoted above, however, specify that the winds occurred on 1 Meskerem, feast of St John, a Monday, that is 1 Meskerem 1696, i.e. 10 September 1703. In September 1704, 1 Meskerem 1697, fell on a Tuesday, feast of St Matthew.

1733/XI/30

During the 12 months following 29 November 1733, severe earthquakes shook northern Ethiopia; heavy damage and casualties were reported. The epicentres were probably located on the Eritrean escarpment at about N 15¾°, E 39°.

Sources

Manuscript in Tigrigna entitled: *Traditions of Tsazzega and Hazzega*, edited by Kolmodin (1912) and annotated in 1914.

In the year 386 Amete Meheret, during the night of Sunday to Monday, Hedar 23, the earth trembled. Men were killed during that earthquake, houses were destroyed, mountains and hills were visibly shaken, rocks and trees fell. Tremors were felt afterwards during the whole year (p. A44).

Comments

1. Dating the Seismic Activity

Year 386 of the Third Diocletian Cycle in this particular series of documents (see note on Ethiopian calendars in Explanatory Notes and entry 1431–32) began on the day corresponding to 9 September 1733; 23 Hedar corresponded therefore to 30 November 1733, a Monday. Note that Kol-

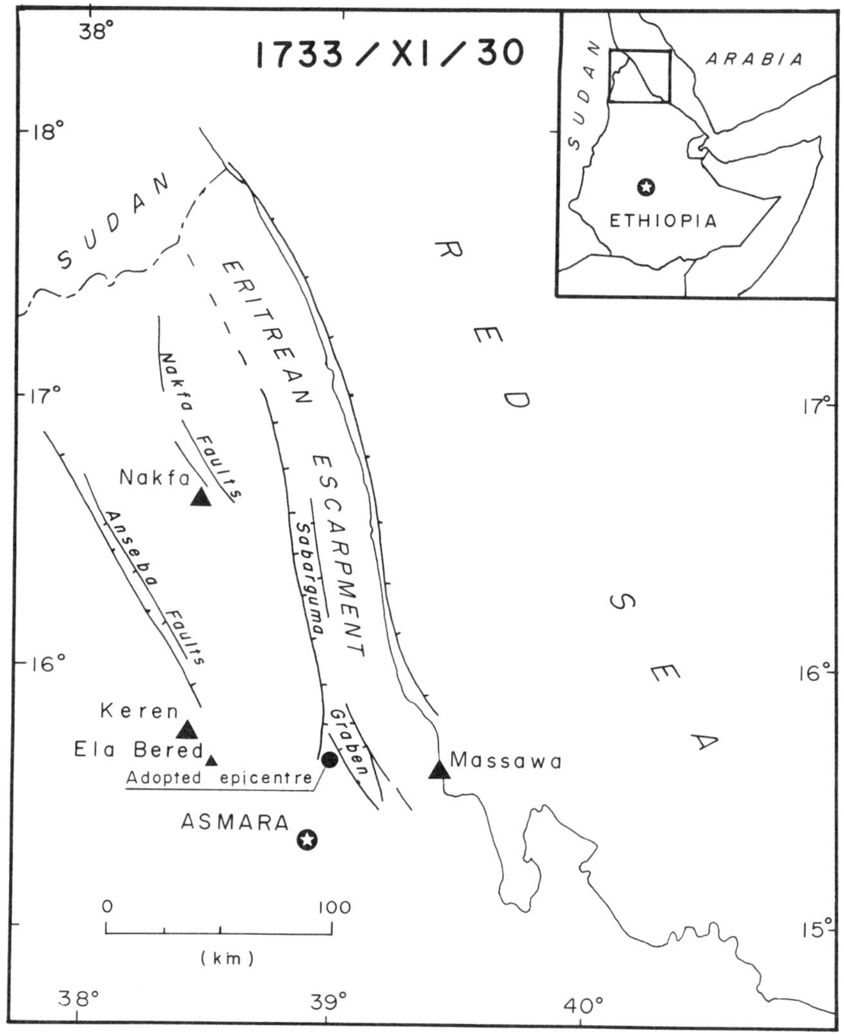

Fig. 12. Epicentral location of the earthquake of 30 November 1733 and the fault pattern in northeast Ethiopia.

modin (p. A44) dated the event 19 November instead of 30 November. Kolmodin's computation is correct for the Julian calendar, not for the Gregorian. Komoldin being a Russian was influenced by the fact that Russia had not accepted the Gregorian calendar before 1918.

2. Location of the 1733 Seismic Region

The manuscript unfortunately does not give any precise indication of the region where the disaster occurred. Inferences as to its probable location can be drawn from the following particulars.

(1) The manuscript was found in the Church of Tsazzega, near Asmara. Kolmodin (1914) established that it was either written there or at the Monastery of Debre Bizen, also located near Asmara. Considering both the language and the chronology used in the manuscript, it appears doubtful to me that it could have originated in Debre Bizen.

(2) The manuscript is primarily concerned with the history of the Deccatescim tribe, the rulers of the Hamasien region in which Asmara is located.

(3) If the disaster had occurred in the western sector of Tigray, it would most probably have been mentioned in the journal of the international religious community stationed in the Lake Tana region. There is no reference to an earthquake at that time in their 15-volume journal (Beccari 1903).

(4) The description of the 1733 disaster closely resembles that of 1875–1876 when earthquakes, located on the Eritrean escarpment within a 100-km epicentral distance from Asmara, caused havoc in the Keren region. Other earthquakes have subsequently occurred in the same region; that of 28 September 1915 had a magnitude of 6¾ (Gutenberg and Richter 1954).

As the Eritrean escarpment north of Asmara and slightly west of the 39th meridian is the most seismically active region of Eritrea, the probability is high that the 1733 activity was located there. See entries 1912–13, 1915/IX/28, 1963/VII/14, 1967/IX/15, and 1967/IX/23.

The adopted location is N 15.7°, E 39.0° (see Fig. 12).

3. The Magnitude of the Earthquake

Taking the text of the manuscript at face value, a shock that triggered landslides and rockslides and was followed by months of felt aftershocks could be confidently estimated at magnitude 6 or more. Subsequent earthquakes have occurred in the same area and that of 28 September 1915, reached a magnitude of 6¾.

1799/VIII/18

Before sunrise on 18 August 1799, an earth tremor was felt north of Lake Tana, in Dembia and in Tigray. People were frightened but no damage was reported.

Sources

Ethiopian manuscripts 118 and 225 in the *Collection d'Abbadie*. The first was edited by Conti Rossini (1954, p. 97) and the second by Blundell (1922, p. 463) (Palazzo 1915, p. 304).

In the year 7291 of Creation, Sunday the 13th day of the month of Nahassie, at the time the cock crows, the earth trembled (manuscript 118).

Having received the blessing of the priests, Dejesmatch Wolde Sellassie left (Axum) and spent the night on the slopes of Sarcé; from there he proceeded to Siré. Then the earth was shaken as if there was an earthquake and his enemies were so terrified that they did not dare to make a stand in front of his armies (manuscript 225).

Comments

1. No explicit indications are given that both documents referred to the same event and, moreover, the second (manuscript 225) is not dated except by reference to Dejesmatch Wolde Sellassie who ruled over Tigray as Ras from A.D. 1788 to 1816. The locations mentioned in the two documents being about the same and the year 1799 being within the reign of Ras Wolde Sellassie, it is assumed that both manuscripts relate the same earthquake experienced by two different authors.

2. From the context, Siré appears to be an older name for the present-day Enda Sellassie, 30 km west of Axum (N 14.1°, E 38.3°) (see Fig. 11).

3. Year 7291 *Amet'Alem* began on 9 September 1798; 13 Nahassie corresponded to 18 August 1799, a Sunday. Instead of 18 August 1799, Palazzo (1915, p. 304) indicated 18 July 1799 and Blundell (1922, p. 463) gave 1798 and 1801 as possible dates. None of these two solutions fall on the right day of the week, a Sunday, nor fit the astronomical cross-references.

1802/VI/30

On 30 June earth tremors were felt near Gondar in the province of Dembia, north of Lake Tana. No estimate of the intensity is given.

Sources

Ghe'ez manuscript classified at the British Museum as Orient 821 and translated by Blundell (1922) (the section quoted is on page 472). Palazzo (1915) quoted the same text from another source, manuscript 118 in the *Collection d'Abbadie*.

In the year of Creation 7294, year of Grace 1704, the month of Meskerem began a Thursday... The year was the year of Marc and the Negusa Negust Gualu was at Gondar... The month of Senie began on a Monday. On the 24th day of the month, there was an earthquake.

Comments

At the beginning of the 19th century, the capital of Tigray was Gondar, in Dembia. As the chronicler was attached to the Court, it is assumed that the earthquake was felt in the vicinity of the capital (N 12.6°, E 37.5°).

The year of Creation 7294 began on 10 September 1801; the first day of Senie was Monday, 7 June 1802 and therefore, 24 Senie corresponds to 30 June. (Palazzo indicates 28 May 1802; this is obviously erroneous).

1809/II/26

An earth tremor was felt in Gondar (N 12.6°, E 37.5°) during the battle of Samona Bar fought between Ras Gugsa and Dejesmatch Zawde. No intensity is reported.

Sources

D'Abbadie (1858, p. 89) mentions the event at the end of his report on the earthquakes of 12 February 1845. (Perrey 1859, p. 8; Palazzo 1915, p. 305–306).

Comments

Manuscript 118, *Collection d'Abbadie*, states that the battle at Samona Bar was fought on Sunday, 20 Yekatit 7301 A.A., that is 26 February 1809; this day is also the date of the earthquake.

1818/VI/30

A violent earth tremor was felt in Adua (N 14.2°, E 38.8°) on 30 June 1818.

Sources

Pearce (1831, p. 240). (Palazzo 1915, p. 306).

Comments

1. Intensity of the Tremor

Pearce was in Adua the day of the earthquake; he is therefore relating his own experience. What he means by *violent tremors* is difficult to ascertain because he does not give the criteria upon which he based his evaluations. As he does not report any material damage in the town, it is very unlikely that the intensity could have been any higher than grade V (MM).

2. Probable Epicentral Location

Recent seismological observations show that the region north of Lake Tana is almost aseismic; the state of conservation of historical monuments often built of poor masonry points to the same conclusion. On the other hand, tremors have continuously been reported from that area. They had, therefore, to be of low amplitude; their origin would have been the Plateau margin to the east. Figure 13 shows the most likely seismic region to produce tremors of intensity about IV from shocks of estimated magnitudes 5-5½.

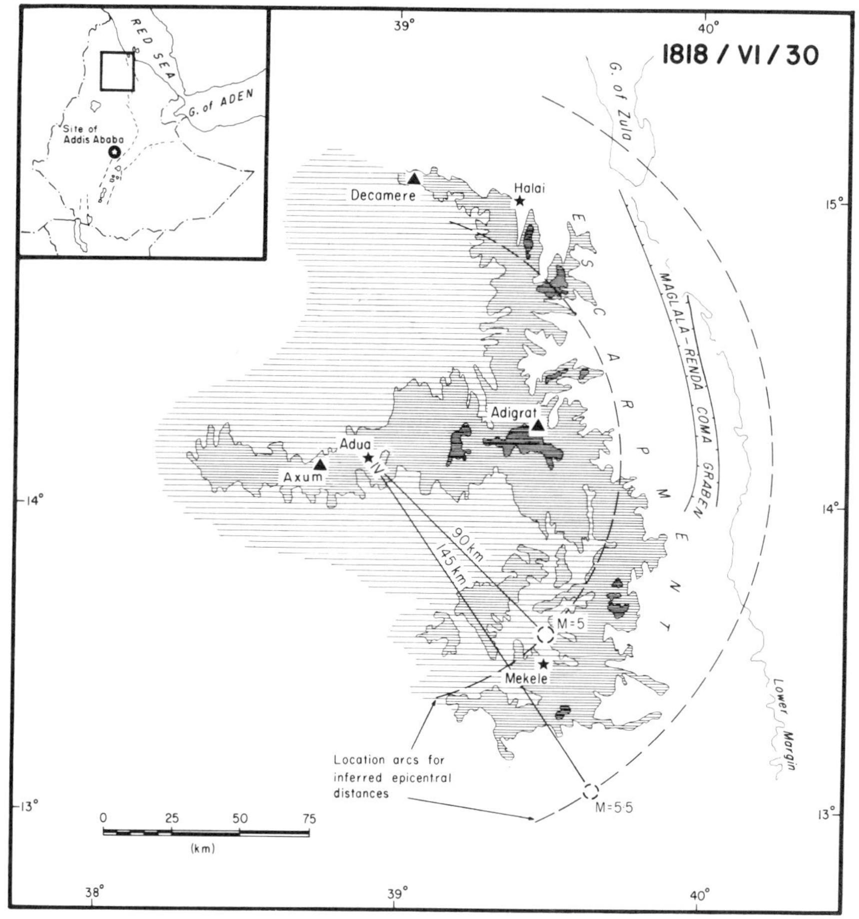

Fig. 13. Estimated location on the Plateau eastern escarpment of the seismic region from where shocks of magnitude 5 and 5½ would cause tremors of intensity IV–V in Axum and Adua. Later entries will localize the epicentres mostly along marginal graben(s).

3. The Seismicity of the Country

Pearce goes on and states, what seems to have been a common opinion, that *it is common in all parts of Ethiopia for the Earth to shake.*

1832/V/04

At about 1:30 p.m. on 4 May 1832, an earth tremor was reported about 10–15 km NE of Halai (N 15.0°, E 39.4°).

Sources

Ruppell (1838, p. 305). (Palazzo 1915, p. 306–307).

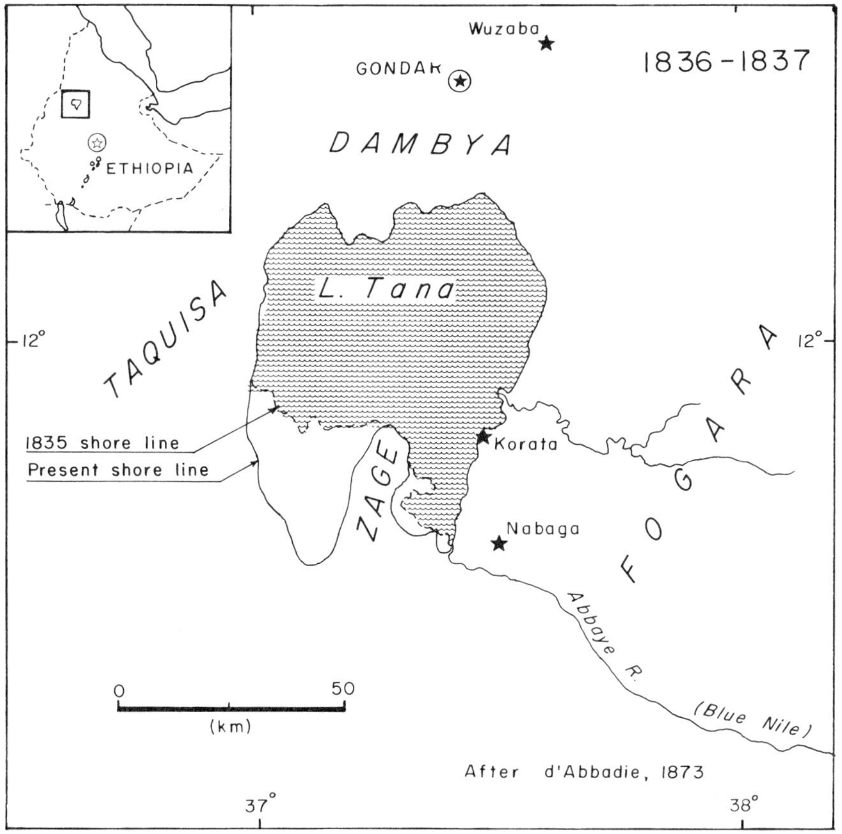

Fig. 14. Map of the Lake Tana region after d'Abbadie's "Géodésie d'Ethiopie 1873." The solid contour line indicates the shores of the lake as they were in the 1850s; the broken contour is the present shoreline. D'Abbadie's original geographic coordinates based on the Paris meridian have been converted to the Greenwich system. D'Abbadie's transliteration of Ethiopian locality names has been respected, e.g. Dambya.

Comments

Ruppell reports that all the members of his caravan following the historical trail from Adulis on the Gulf of Zula to Cohaito and Adi Caieh on the top of the escarpment felt the earth tremors while coming out of the valley of Assubo (Assaorta) in the direction of Halai (N 15.0°, E 39.4°). The position of Halai is indicated on Fig. 13.

To be felt by a caravan on the move, the tremors must have been of intensity IV–V.

1836–37

Earth tremors were reported around Lake Tana and throughout the eastern half of the province of Beghemder. Intensities in Fogara (about N 11½°, E 39¾°) were estimated as III–IV.

Sources

A private communication from some people in Gondar to d'Abbadie (1858, p. 89). (Perrey 1859, p. 8; Marinelli and Dainelli 1912, p. 111; Sieberg 1932, p. 886 — Palazzo lists these events in his catalogue but makes no comments).

The English translation of d'Abbadie's footnote in his report on the 1845 events reads as follows:

We were told on that occasion (earthquakes of 12 February 1845) *that an earth tremor had been felt eight years earlier in Gondar. The alaqa Habte Sellassie determined the time of occurrence at four feet of shadow before noon. At Wuzaba where the debteras of Kirkos were pleading court cases . . . three distinct tremors were felt. Three tremors were likewise felt at Nabaga in the district of Fogara. The shepherds of Fogara claim that there was another light tremor the following night.*

Note the meaning of some Ethiopian words: *alaqa* — a priest, head of a church; and *debtera* — a scholar in liturgy and Ghe'ez literature.

Comments

1. The location of the sites mentioned by d'Abbadie is indicated in Fig. 14. They are all situated in the eastern half of the province of Beghemder.

2. There has been confusion in the literature concerning the 1836–37 and the February 1845 earthquakes. The confusion stems from the fact that d'Abbadie related the two series of events together in one entry of his journal, that of 12 February 1845. His text is not perfectly clear as to whether the description of the seismic effects quoted above refers to the events "which happened eight years earlier" or to those of 1845 described in the first part of his entry (see text in entry 1845/II/12).

By comparing the time elements and the two lists of sites, the ambiguity is eliminated. First, Alaqa Hapte Sellassie clocked the tremors at 4 ft of shadow before noon, that is about 10 a.m.; the 1845 earthquakes occurred in the afternoon. (For details on time reckoning techniques based on the length of the shadow, see the pamphlet *Mengede Semaye* (Makarios 1967).

Secondly, the lists of locations given by d'Abbadie for both periods of seismic activity do not correspond. It is therefore evident that d'Abbadie's description quoted above refers to the 1836–37 tremors, not to those felt in 1845.

Sieberg's entry for these events (1932, p. 885, Table 182) is completely confused. It reads:

1837: strong earth tremors around Lake Tana, specially in Gojjam, Galla and Kaffa. Intensity VII–IX. Mountains crumbled, villages and people buried by landslides in Lasta and Wollo. Reported felt in Gondar.

This description applies to the 1845/II/12 events, not to those of 1836–37.

3. On the probable origin of the tremors felt in Lake Tana area, see comments in entry 1845/II/12.

1841–42

Between April 1841 and December 1842, several earthquakes shook the region of Debre Berhan, some 100–120 km NE of Addis Ababa. Along the Plateau's escarpment, they triggered landslides and rockfalls that destroyed the town of Ankober (N 09.6°, E 39.8°) then capital of Shoa, and some nearby villages. Negus Sahle Sellassie was then forced to relocate his capital on the summit of the escarpment at Debre Berhan (N 09.7°, E 39.6°).

The first shocks were reported on 23 and 24 April 1841. On 7 July 1842, two other earthquakes triggered a landslide that blocked the road from Debre Berhan to Ankober. On 8 December of the same year, stronger shocks caused substantial landslides — Ankober was destroyed. Surrounding villages were either destroyed or heavily damaged. There is no accurate account of the casualties.

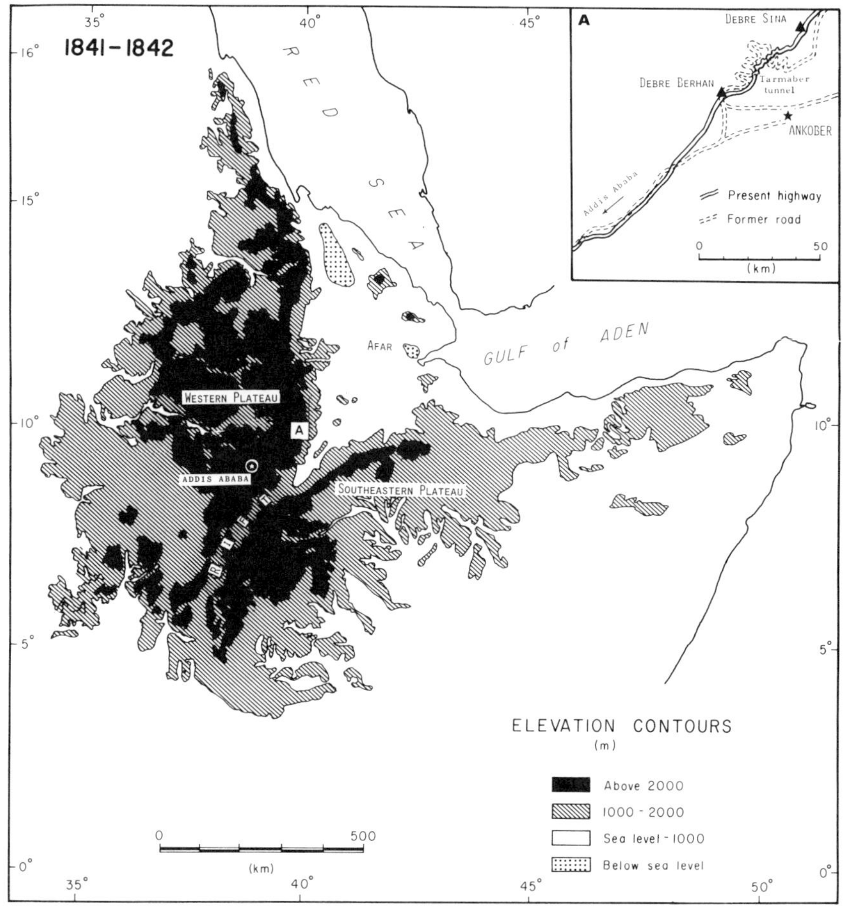

Fig. 15(A). Physiographic outline map of Ethiopia and of the Horn of Africa in 1000-m steps in elevation. The location of the Ankober region on the slope of the escarpment is marked by an open rectangle. The hairpins in the former road prior to the excavation of the Tarmaber tunnel (dashed lines in the inset) gives an idea of the ruggedness in the topography of the area. (B). A church, located on the site of the former Palace in Ankober. Notice the steepness of the escarpment and as well the high gradients over the fields (photo Chs Wood).

Sources

Graham (1841, p. 3), Harris (1843, 1844, II, p. 89), Beke (1844, p. 67), Johnston (1844, p. 193, 218, 279, 280). (These events are listed in the earthquakes catalogues of Marinelli and Dainelli 1912, p. 112 and of Pallazzo 1915, entry 23).

Comments

1. Seismicity of the Debre Berhan Area

Observations on (1) the magnitude and frequency distribution of earthquakes originating from the Debre Berhan region since 1958 and recorded at Addis Ababa as well as (2) the analysis of one particular earthquake sequence (Dakin 1975, p. 51–70, sequence 22 of August 1971) reveal that, in recent years, the seismicity of the region has mainly consisted of swarms of earthquakes, moderate in magnitude, among which the main shocks were seldom detectable. Although it would be imprudent to extrapolate this conclusion to the activity of over a century ago, one observes, for instance, that the description given of the 1841–42 sequence does not substantially differ from that of the 1971 swarm studied by Dakin, exception being made of the epicentral locations along the slope of the escarpment and of the amplitude of their secondary effects, namely hillside failures.

2. Extent of the Damage

Prior to the events of 1841–42, Ankober was the capital of Shoa, an important market centre for the caravans traveling from the Red Sea ports to the Ethiopian highlands. Ankober was a prosperous, peaceful, well-tended town. Isenberg and Krapf (June 1843) speak of the *ecstasy* they experienced after many weeks of travel through the heat of the Danakil desert when, at an altitude of 2800 m, they reached that cool oasis surrounded by rich vegetation and well-tilled fields. They described the Negus' Palace as a large house of stone and mortar; the other houses were made of wood, covered with thatched roofs and surrounded by gardens. The events of December 1842 changed that perspective.

A dramatic description of the disaster is extracted from a confidential letter written in January 1843 by Capt. Harris (Harris 1843), then British Envoy to Negus Sahle Sellassie:

On December 8, a deep rumbling noise was followed by the shock of an earthquake. The consequences were appalling. The soil, saturated with moisture, slipped from the steep rugged slopes and rocks, heaved from their resting places, persued a sweeping course of devastation to the glens below. Houses were engulfed and buried in the debris, or shattered to fragments by the huge masses bouncing on their course. Trees were torn up... and daylight presented to the eyes of the affrighted inhabitants, a strange scene of ruin.

Perched on the apex of the sloping peak, the palace had on the preceding evening frowned over the capital in all the security of its numerous encircling palisades, but now shorn of their bristling protection the buildings that has not been overthrown stood naked and exposed. Twenty open breaches, as though powerful batteries had been playing for a fortnight, swept up to the very steps of the banquetting hall, and palings and palisades hurried from their deep foundations lay broken and strewed over the entire face of the mountain. The roads along the scarp were completely effaced; tall green shrubs reclined with their roots reversed and not a vestige of fragile huts could be discerned in bare earthy tracts which marked the disastrous course of the treacherous slip.

Numbers of dead bodies were constantly being dug out, and placed in the market place... The sweeping desolation had spread for miles along the range. Houses and inmates' household gear had been buried away, and the ... (original text illegible) *from the green hill top and from the sheltered nook, announced the many victims that were immaturely buried... The destruction varied considerably according to situation and locality. Some villages were entirely obliterated by the descending tons of wet soil, and the inhabitants of others grieved only for the loss of cattle and farm stealing, but the destruction of life and property was altogether immense, and although shocks had often before been experienced, a similar calamity to the present had not befallen the country within the memory of man.*

It is to be noted that later, in his books *Highlands of Ethiopia*, Harris (1844) himself appreciably toned down the emotional content of his report written under the excitement of witnessing a seismic disaster at the onset of his first diplomatic mission.

To summarize the extent of the damage as presented in the sources quoted: (1) the Debre Berhan – Ankober road was blocked by an earthquake generated landslide during the spring of 1841; and (2) in December 1842, more important slope failures spread over many kilometres, destroyed the town of Ankober, destroyed or heavily damaged some villages, ruined the crops and caused a number of casualties (which cannot be estimated).

3. The Causes of the Damage

There is no doubt that the Ankober disaster was not primarily caused by the seismic activity, which was moderate, but by secondary effects, in this case landslides and rockfalls. The triggering of such slope failures was favoured by the heavy rains that had saturated the thin layer of clayish soil *(tchiqa)* precariously laid upon very steep slopes (see Fig. 15). The first signs of slope failures appeared at the end of the spring rainy season *(belghe)* 1841. To quote Johnston (1844): *A severe thunderstorm, attended by two distinct shocks of an earthquake at an interval of a few seconds, ushered in the first heavy fall of rain. No serious damage resulted; a few rocks were detached from the heights above, blocking the narrow road to Ankober in some places, and in another ploughing deep channels though the young green crops* (Vol. II, p. 193).

There is no reason to believe that between April 1841 and December 1842 the seismic activity had ceased, even if no tremors were reported in the literature. As indicated above, typical seismic activity in the Debre Berhan area consists of swarms of small or moderate magnitude earthquakes. The effects of continuous shaking, even if the shaking is light, on slopes of marginal stability is cumulative. No wonder that months later during a period when the soil was water saturated (Harris speaks of *tons of wet soil*), moderate seismic activity induced major failures on slopes whose stability had already been weakened during the previous months.

4. Slope Failure Occurrences in Ethiopia

Figure 15A broadly outlines the differences in elevation in Ethiopia; detailed topographic maps reveal local elevations ranging from about 5000 m above sea level to 100 m below, as well as individual features with sheer drops of thousands of metres. In such a disturbed topography, slope failures become major hazards.

1845/II/12

About noon, 12 February 1845, earth tremors were reported throughout western Ethiopia from the region north of Lake Tana to the southernmost provinces. In Gondar, the maximum reported intensity was V^+, a sign that the tremors were felt further north in the provinces of Tigray and Eritrea. In the south, the intensity-III isoline was located between the parallels N 05.5° and N 06°, suggesting an approximate epicentral radius of 650–700 km and a magnitude of the order of 6¼–6½.

Sources

D'Abbadie (1858, p. 86–94). (Perrey 1859, p. 8; Milne 1911; Marinelli and Dainelli 1912, p. 111; Palazzo 1915, p. 339; Sieberg 1932, p. 886).

Comments

1. Evaluation of the Sources of Information

The original documents relating the seismic events of February 1845 are d'Abbadie's accounts incorporated in his field notes. He and his crew were engaged in a geodetic survey of Ethiopia and on 12 February 1845, they were stationed in Gondar. Information from other localities was obtained in the course of the team's fieldwork during the following 2–3 months.

D'Abbadie was a professional earth scientist who also had mastered the language of the country; we feel confident that he critically evaluated the information received before recording it. The same degree of accuracy is not found in the "Earthquake Catalogues" of Milne and Sieberg relating the same events. Misinterpretation of d'Abbadie's reports was bound to occur because the toponymy and topography of Ethiopia were not so well-known abroad 100 years ago.

2. Reports on the Earth Tremors and Location of the Sites

Note: d'Abbadie's original longitudes were reckoned from the Paris meridian; an empirical correction of +2.3° obtained from statistical comparison of his values with more modern ones was applied to convert them to the Greenwich system.

From Gondar (N 12.6°, E 37.8°) — tremors were reported from different sectors of the city and a few dry-masonry walls collapsed. The maximum reported intensity is estimated at V^+. D'Abbadie remarks that, although he and his crew were stationed in Gondar at the time, they were so busy reducing geodetic field data that they did not notice anything unusual.

From the District of Lasta — 100–150 km east of Gondar: the tremors were stronger in Lasta than in Gondar and a village was destroyed by landslides. No name or precise location was obtained by d'Abbadie.

From the Southern Sector of Lake Tana — tremors were reported in Korata (N 11.6°, E 37.5°), but surprisingly enough, not in Bahr Dar barely 25 km away.

From Gojjam — in the region of Meca, near Mt Karni (≃ N 11.5°, E 37°), the church of Angata Kidane Meheret was destroyed. As no details are available on the type of structure and conditions of the damaged building, a conservative estimate of intensity VI is accepted.

From other parts of Gojjam, travelers reported that they felt tremors, not at noon as in Gondar, but around 3 p.m. If their time estimate is valid, they must have been referring to tremors caused by an aftershock with an epicentre possibly different from that of the main shock. (Note: the centre of Gojjam is about 200 km south of Gondar.)

From Central Wollo — a report from Wara Haimanot (N 11.3°, E 39.7°) states that three persons were trapped in a crevasse and crushed. No details were given.

From Gudru Region — (N 09.3°–10°, E 36.9°–37.3°), earth tremors were felt and the Oromo people reporting the event used the expression *the earth was dancing*, suggesting that they experienced long-period seismic effects.

D'Abbadie reports that local tradition states that no earthquake had been experienced in the region prior to 1845. In the light of more recent events, I strongly doubt the objectivity of this tradition.

From Southwest Ethiopia — it was reported to d'Abbadie that the earthquakes of 12 February were also felt in the region of Ennaria (N 07.8°–08.5°, E 36.5°) and in the province of Kaffa (about N 06–06.5°).

The location of the sites mentioned above are indicated by stars on Fig. 16. The meridional alignment in their geographical distribution comes

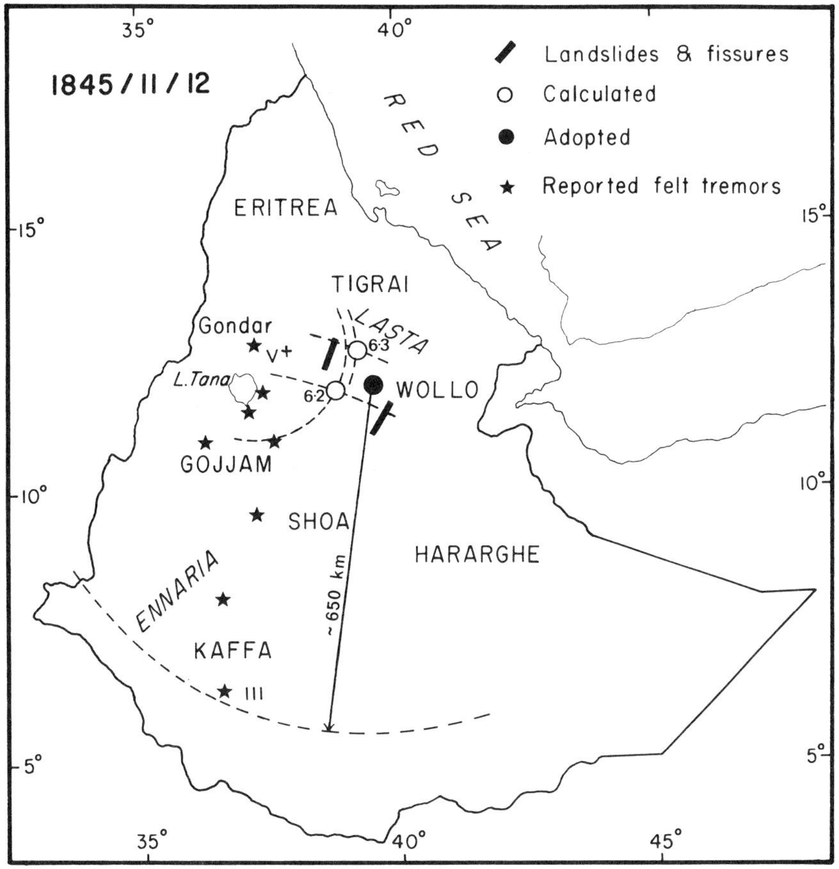

Fig. 16. Location map of the sites (★) from where tremors were reported, of the locations (O) inferred from the intensity distribution, and of the adopted centre (●) for the epicentral region. The dashed line arc-segments, centred on Gondar and S. W. Ethiopia, represent calculated epicentral distances for shocks of M = 6.2 and 6.3 to produce intensities V^+ and III, respectively, at the sites indicated.

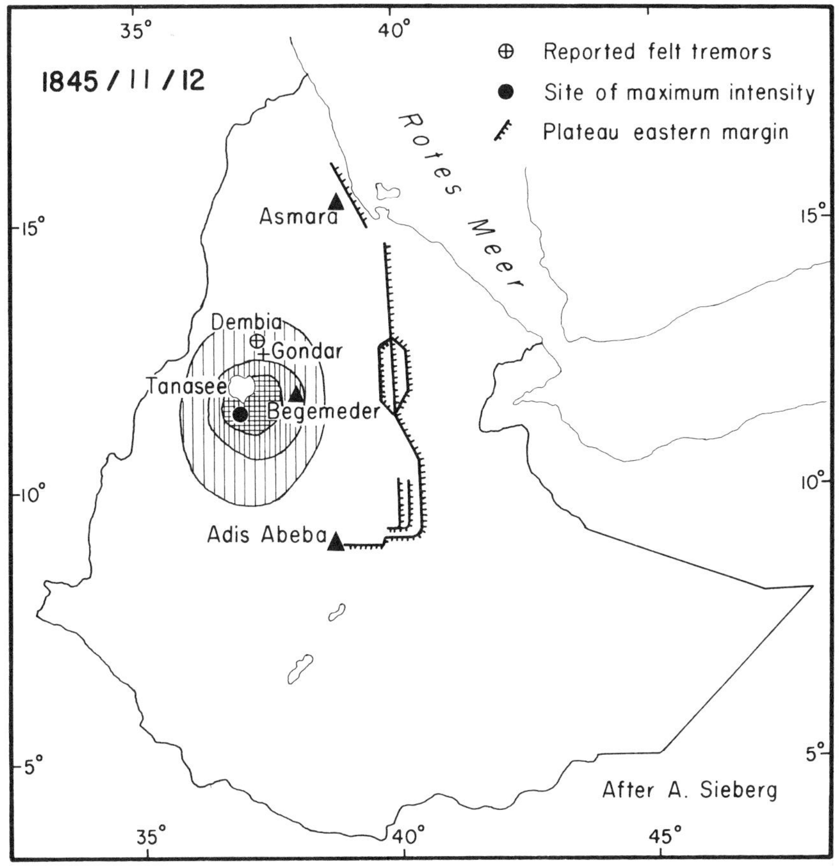

Fig. 17. Sieberg's solution for the epicentral region of February 1845 (note that Addis Ababa did not exist in 1845).

from the fact that d'Abbadie only interviewed people along his geodetic survey route.

3. Estimated Magnitudes

The reported intensities ranged in a north-south direction from V^+ in Gondar (N 12.6°) to about III some 750–775 km to the south (N 5¾°). The intensities are observed to increase eastward in the direction of the Plateau margin; a landslide and a fissure were reported in the vicinity of the E 39th meridian between N 12.5° and N 11.3°.

The graph of the seismic-energy attenuation rates observed in the southern sector of the Ethiopian Plateau and assumed to be valid for the northern sector of the same Plateau shows that an earthquake of magnitude 6¼–6½ is expected under normal soil conditions to cause tremors of intensity V–VI at epicentral distances of approximately 150–180 km and of intensity III between 750 and 850 km. Arc segments centred on Gondar and central Kaffa (boundaries of 1845) were drawn for magnitudes 6.0–6.5 by magnitude steps of 0.1. Only two arcs are reproduced on Fig. 18, those of M 6.2 and 6.3, showing that a magnitude $\simeq$ 6¼ would best correspond to d'Abbadie's accounts — assuming, naturally, that all the reports refer to one single shock, the main shock. If the time element of the report from

Gojjam is reliable, then one has to admit that the epicentres moved south from Lasta to Wollo, and that the fissuring of the ground reported in central Wollo was caused by a mid-afternoon shock. In this case, the magnitude of each of the two shocks would have to be reduced to about 6.

4. Estimated Location of the Epicentral Region

Inferences based on intensity reports and attenuation rates point toward an epicentral region centred about N 12¼°, E 39°, a region lacerated by deep ravines and steep fault-scarps that could easily be suspected of being seismically active. This region is located some 50–60 km west of the Guf Guf marginal graben, which is known to have been, and still be, seismically active. Earthquakes of magnitudes similar to those of 1845 occurred along this graben less than 10 years later, in 1853–54, and since then smaller ones have often been recorded.

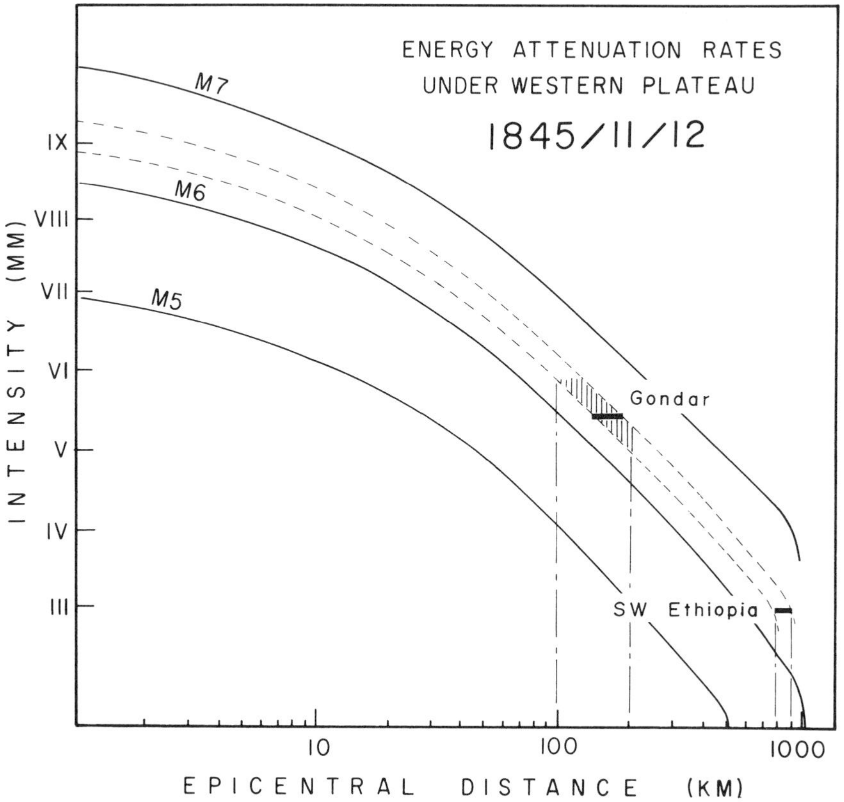

Fig. 18. Attenuation curves used in the determination of the epicentral distances shown in Fig. 16.

A realistic location (solid dot, Fig. 16) for the epicentres of 1845 would be some 60 km to the east of the location suggested by the reported intensities (open dots), near the Plateau-Afar margin and along the Guf Guf graben (for the tectonic map of the area, see entries 1853–54 and 1965/IX/30). The graben is situated in the vicinity of the E 39.5° meridian between latitudes N 12.0° and 12.5°. For computational purposes, location N 12.2°, E 39.6° has been adopted for the 1845 activity.

Sieberg (1932, p. 887, Fig. 466) attempted an epicentral location based on his own interpretation of d'Abbadie's accounts, an interpretation that proves to be inaccurate. He located the epicentral region at about N 11.5°, E 37.2°, south of Lake Tana (Fig. 17), a region that he suspected to be aseismic (p. 888). Unless Sieberg did not have access to d'Abbadie's original reports, but only to secondary sources, it is difficult to understand how he did not observe the tendency of the reported intensities to increase toward a causative centre east of Lake Tana.

1845/XII/07

On 7 December four earth tremors were reported from Rare (N 09.3°, E 37.2°), Gran Damot. Three occurred during the afternoon and were accompanied by noise that d'Abbadie described as *the roaring of the faraway artillery*. The fourth was felt around 7 p.m.

Sources

D'Abbadie (1859, p. 114). (Marinelli and Dainelli 1912, p. 113; Palazzo 1915, p. 339; Sieberg 1932, p. 885).

Comments

1. Location of Rare, in Damot

In 1845, Damot was the region located west of the 37th meridian, from Lake Tana in the north to the Gojeb river in the south (N 07½°). It was subdivided into two parts: Small Damot, north of the Abbai river; and Gran Damot, between the Abbai and the Gojeb. Some 200 km directly south of Lake Tana was Rare, a region that d'Abbadie identified by two conspicuous mountains, Mt Amara (N 09.2°, E 37.3°) and Mt Balballa (N 09.4°, E 37.1°). It is in this region that d'Abbadie felt earth tremors on 12 December 1845.

2. Probable Location of the Epicentres

D'Abbadie reports only the tremors and the accompanying noises; he does not mention any other site from which an intensity distribution could indicate the epicentral region. The loud noise suggests mountainous surroundings and proximity of the epicentre. This conclusion is supported by local tradition (entry 1846/I/26) and recent seismic records — both indicate that the region has been and still is seismically active. In January 1967 and 1968, for instance, instrumental epicentres were located at the intersection of Ambo-Ghedo and the Guder fault zones, 60–70 km southeast of Rare (open square on Fig. 19). Their location was N 08.7°, E 37.6°; details are given in entries 1967/I/23 and 1968/I/23. The 1845/XII epicentres were probably located along the same fault systems.

3. Sieberg's Epicentral Solution

On his seismicity map of Ethiopia, Sieberg (1932, Figure 466, p. 887) located the 1845/XII epicentres at N 06.8°, E 38.2°, on the floor of the main Ethiopian rift valley, about 250 km SE of Rare. The location is marked by a cross in a circle on Fig. 19; its description in Sieberg's *Erdbebengeographie* reads as follows (Table 182, p. 885);

1845, December 7: Moderate shocks in the mountains of Damot and at Lake Abbala (Queen Margherita Lake).

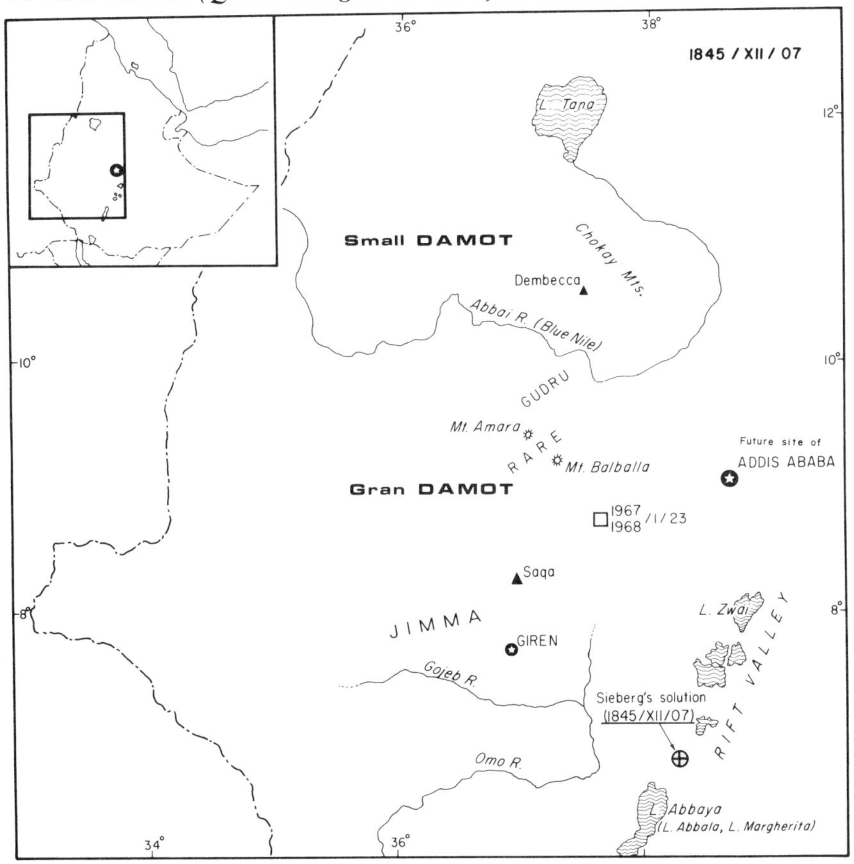

Fig. 19. Western Ethiopia after d'Abbadie (1859). The 1967 and 1968 epicentres are indicated by □ ; ⊕ is Sieberg's solution for the December 1845 epicentres.

D'Abbadie mentions no locality name except that of Rare from where his diary for 12 December 1845 was dated. The fact that locality names such as Damot and Lake Abbala (older name for Abbaya) crept into Sieberg's catalogue indicates that he did not have access to d'Abbadie's books *Sur le tonnerre en Ethiopie* and *Géodesie de la haute Ethiopie*, which are the primary sources for these events and which contain gazetteers of localities and 12 location maps.

1846/I/26

A light earth tremor was felt in Saqa, Kaffa, on 26 January 1846 at 5:06 p.m.

Sources

D'Abbadie (1858, p. 118). (Marinelli and Dainelli 1912, p. 114; Palazzo 1915, event 31).

Comments

1. Location of Saqa

There are two *Saqa* in the vicinity of Jimma, the present capital of Kaffa: Saqa (Jimma), 13 km to the southwest; and Saqa (Limmu Ennaria), 60 km to the north-northeast (Fig. 20). From the context, it appears that d'Abbadie was referring to the second site (Saqa-Limmu Ennaria), the historical capital city of the realm of Limmu. The coordinates of Saqa-Limmu are N 08° 12′, E 36° 57′. In d'Abbadie's time, the capital of the Jimma region was Giren.

2. Seismicity of Jimma Region

D'Abbadie (p. 64) reports that according to the local traditions, *the earth often shakes in Saqa*. This observation applies to the nearby region of Rare as well.

Comments on the probable location of the seismic activity often felt in the Jimma region are given in entries 1845/XII/07, 1967/I/23, and 1968/I/23.

1853-54

Severe earthquakes shook the region of Tambien in Tigray, from February until June or July 1854, fissuring the ground and triggering landslides over a large area. Tremors were reported over a minimum radius of 200 km.

Sources

Markham (1869, p. 244–247); Mgr de Jacobis' diary in Tekla Haimanot (1914, p. 123–234, 514–515); Milne (1911, p. 64); Palazzo (1915,

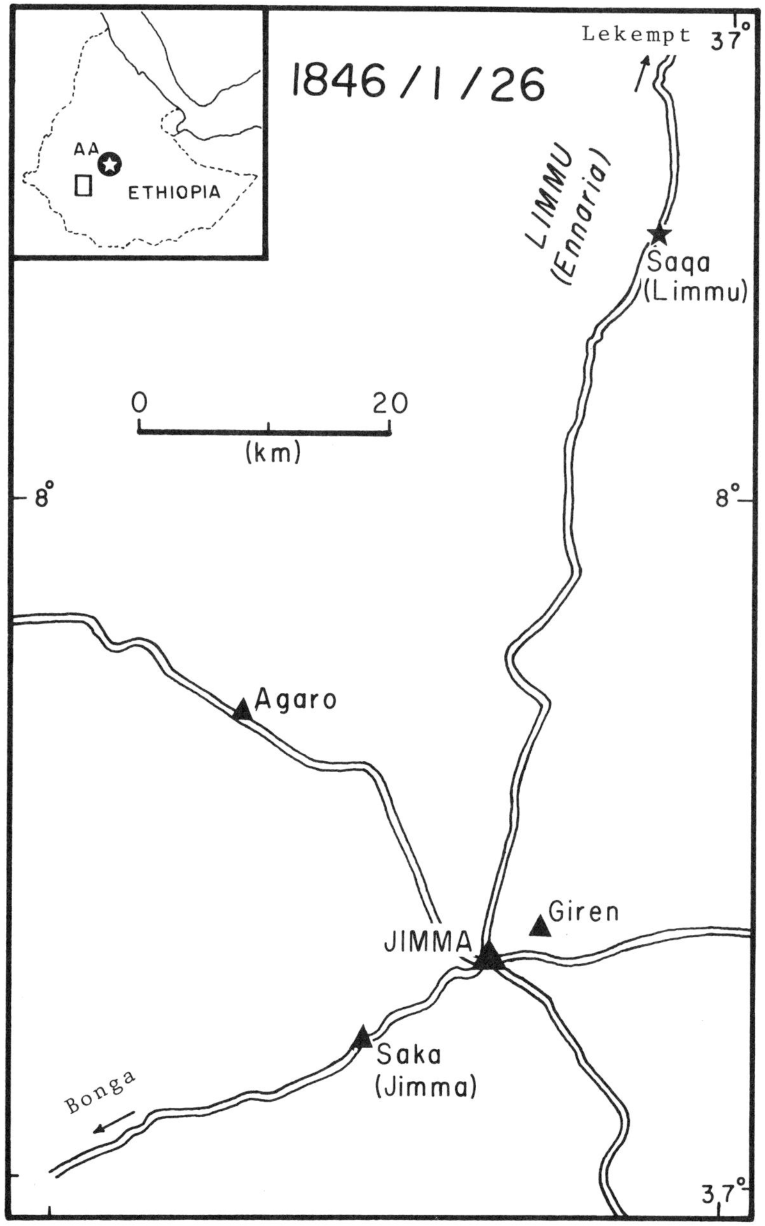

Fig. 20. Location of Giren, the former capital of Jimma, Saqa (Limmu), and Saka (Jimma).

p. 38–35); de Castro (1915, I, p. 84); Budge (1928, I, p. 142); Conti Rossini (1947, II, p. 310, 340).

Comments

1. Evaluation of the Sources

Diary of Bishop de Jacobis for 1853–54 — this diary is a firsthand report by a witness who personally experienced the seismic activity of 1853–54. The following excerpts are taken from Abba Tekla Haimanot (1914) who wrote a life of Mgr de Jacobis based on his writings.

October 1853, in Agula (N 13.7°, E 39.6°), near the river Inderta:

A very strong earthquake shook the hills. Army officers and soldiers were frightened and everyone ran away in search of solid ground. Men and animals were completely bewildered (p. 123–124).

Prior to June 1854, during a trip from Massawa to Gondar via Asmara and Mekelle:

Having left Celicot the party arrived at Samré (N 13.2°, E 39.3°) the market of Shoum Selous. Mgr. de Jacobis joined a caravan of merchants to continue his trip towards Gondar. One day during the trip, an earthquake happened which was much stronger than that of October 1853. Thunder was heard, mountains were shaken, in many places cracks appeared in the ground and some churches were destroyed. This happened many times until June 1854 (p. 124).

March–June 1854

In the year 1846 Amete Meheret, tremors shook the earth during the months of Megabit, Miazia, Genbot and Senie (roughly from March to the end of June 1854). *They happened quite often but not every day and were sometimes very strong. There was a noise like thunder. Mountains cracked, big trees were uprooted, and many churches, especially those built on top of the hills, were destroyed. In many places the sky became dark. Cracks appeared in the ground and new springs came out.* (Tekla Haimanot 1914, p. 514–515).

A Quote from Palazzo — an Ethiopian manuscript from Mrs Naretti's collection quoted by Palazzo, reads as follows:

During the Ethiopian Year 1846 Amete Meheret an earth tremor was felt mid-afternoon on a Thursday in February, a second one followed at sunrise on Saturday, and a third one occurred at about midnight on Tuesday. The tremors were very strong and the last one triggered a large landslide (Palazzo 1915, p. 310).

Amete Meheret 1846 started on 10 September 1853 and terminated on 9 September 1854. In 1854, 1 February was a Wednesday.

Conti Rossini (1947) — another Ethiopian manuscript refers to the tremors felt in Gondar (N 12.6°, E 37.5°) during February 1854:

During the month of Tir, 7347 Amet'Alem (i.e. February 1854), *the earth shook for many days* (Conti Rossini 1947, II, p. 408).

Sapeto (1857, p. 14–15) — in the report of his visit to northern Ethiopia during 1854–1855 Sapeto wrote:

In the provinces of Tambien and its vicinity, about year 1855 there was a subsidence of the terrain over an area of 12 to 15 miles in which a lake was formed. The site was not far from Lake Ashangi. According to a local legend, Lake Ashangi would also be the result of a subsidence at a place where long ago there was a flourishing city (Sapeto 1857, p. 14–15).

Markham (1869) — Markham was a geographer attached to the British Abyssinian Expedition of 1865–1866. His route took him through the province of Tambien. He reported:

There are deep cracks round the base of Amba Afaji which are said to have been caused by the earthquakes of 1854 and the natives assert that these earthquakes also caused great changes in the water-system of the Doba Valley, some springs dying out and others appearing (Markham 1869, p. 244).

The valley Markham refers to as Doba is today known as the Guf Guf Valley.

Round the north end of Lake Ashangi, there are deep fissures full of soft mud and quick-sands which are excessively dangerous. These fissures are said by the people to have been formed by the earthquake of 1854 (Markham 1869, p. 247).

The Technical Reports: Milne, Palazzo, and Perrey — From the different sources available, the following list of events felt at or observed from the town of Betalihem could be reconstructed: 2–16 February 1854: six tremors were reported.

On the 21st around 3 a.m., a first tremor with a strong vertical component was felt and lasted what appeared to be *a long time*. Many houses were damaged. On the slope of a nearby mountain, an important landslide was triggered producing a terrific noise. According to Milne and Palazzo, the first shock was the strongest. It was followed by four aftershocks on the same day and by many others on the 22nd.

A problem arises concerning the identification of the site of Betalihem: there are three Betalihems in virtually the same region: one at N 13.6°, E 39.4°, a second at N 12.7°, E 37.7° and a third at N 11.7°, E 37.9°. Perry (1854, p. 8) chose the third site probably because it was the only Betalihem indicated on the map at his disposal. Palazzo, following the same opinion, considered that an epicentral distance of 200 km was too high to relate the tremors felt in Betalihem to the Ashangi events and therefore listed the two sets of tremors separately. I am inclined to think that the first site near Mekele, less than 100 km from Lake Ashangi, is the relevant Betalihem. The Church there attracted many Ethiopian scholars, priests, and *debteras* (scholars in Ethiopian liturgy and Church history); they formed a scholarly group that could have observed and described the phenomena noted above (for a popular essay on Betalihem, see Packenham in *The Illustrated London News*, No. 6144, 9 March 1957). Another argument in favour of this choice is the fact that the events related by Mgr de Jacobis were located in the same vicinity and correspond to the facts described by other observers.

Budge (1928, I. p. 142) — Budge in his *History of Ethiopia, Nubia and Abyssinia* evidently quoted de Castro (1915, I, p. 84) who, in turn was quoting Sapeto and ended up with the fairy tale that earthquakes were rather frequent in the Province of Tambien around 1855, that they caused a major terrain subsidence underneath a large town, and that the town was simply obliterated by a newly formed lake, 10–12 miles long! (I took the liberty of paraphrasing his text).

2. Dating the Events

Although the dates of the seismic events listed above, as taken from different historical sources, do not prefectly correspond, it is historically certain that in 1853–54 the region of Tambien in Tigray was severely shaken by earthquakes. Reports of tremors and landslides came from the area between N 12.5°–N 14° and E 39°–40°; the felt area exceeded these

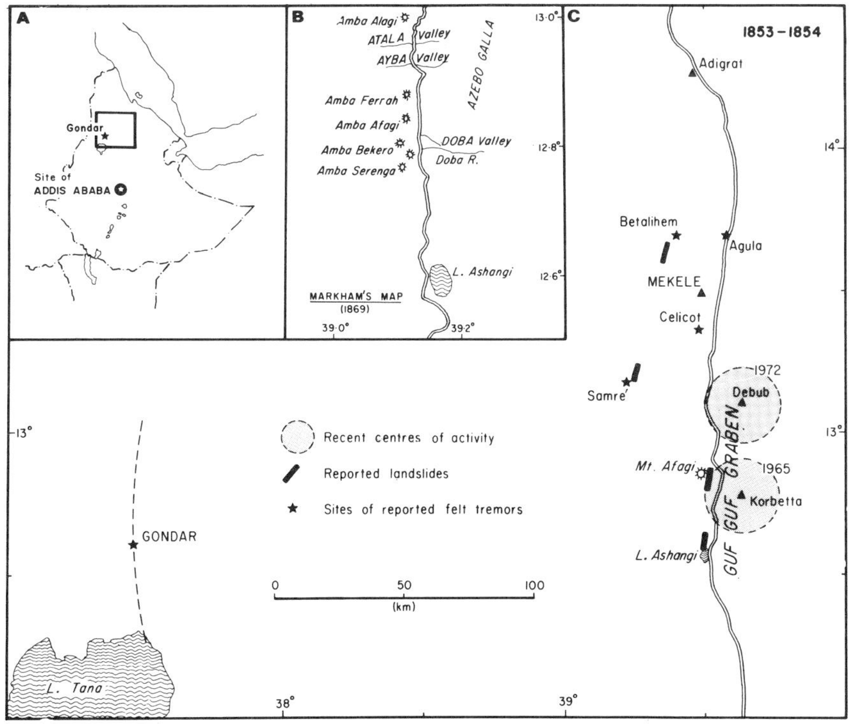

Fig. 21. Location map of the 1853–54 seismic region. Inset B is Markham's map dated 1869.

boundaries and probably reached Gondar, 225 km to the west. Note that a shock of magnitude 6 is expected to cause intensities IV–V at 200 km under the crustal conditions of the Plateau.

The earth tremors of October 1853, as reported by Abba Tekla Haimanot, could have been a foreshock of the 1854 activity, but in my opinion, it is more likely that an error was made in dating the events. Agula is situated only about 110 km north of Lake Ashangi (one of the centres of the 1854 activity, where the magnitude was estimated about 6 or 6¼) and is, therefore, located well within the felt area. As there are no other reports of tremors in Agula during 1854, the probability is that Tekla Haimanot made an error in dating the seismic events.

3. Estimated Location of the Seismically Active Area

From the reports, landslides occurred along and west of the escarpment of the Ethiopian Plateau between N 12.5° and E 13.7°; cracks and fissures seem to be restricted more to the south between N 12.5° and N 13.0°.

There is no doubt that in 1853–54 the whole length of the Guf Guf graben (see location and tectonic map, Fig. 21) was seismically active in a manner similar to the seismic activity along the Borkenna graben in 1961 (entry 1961/Kara Kore) where the epicentres were distributed over a very limited distance in longitude but over 1° in latitude.

There are recent reports of seismic activity in the Guf Guf graben near Mai Chew and Korbetta, e.g. 24 October 1965, 19 August 1972, etc.

4. Notes on the Location Map in Markham (1869) Included as an Inset in Fig. 21.

The modern Addis Ababa–Asmara highway passes east of Lake Ashangi, not west as in 1854.

Of the Ambas mentioned by Markham, only Alagi, Ferrah, and Sarenga could be identified on more recent maps. (Amba = a flat top mountain with very steep sides; it is equivalent to the geologic term "mesa".)

Insofar as the geographic coordinates are concerned Markham's latitudes need no correction but the longitudes have to be adjusted by +0.4° E.

1857/IV

During the spring of 1857, frequent tremors were felt in northern Ethiopia; a "town" was reported to have been destroyed.

Sources

Marinelli and Dainelli (1912, p. 115), quoting Munzinger (1857, p. 192; 1858, II, p. 33).

During the spring of 1857, frequent shocks were felt; they were not very strong in Massawa but a city was destroyed in Tigray.

Comments

The events reported here are indexed in Region C; they are recalled here only because it is mentioned that a town, in Tigray, was "destroyed." Comments on the interpretation of the report are to be found in entry 1857/IV, Region C.

1875/XI–1876/III

For 5 months, from November 1875 to March 1876, the provinces of Eritrea and Tigray were almost continuously shaken by earthquakes. The shocks were particularly severe during November 1875. Near Adua (N 14.2°, E 38.8°), a crevasse opened in the ground and a landslide was triggered on the slope of Amba Shelloda (2.5 km north of the centre of town). In and near Keren (N 15.8°, E 38.45°), a well-built church was partly destroyed, a brick building collapsed, and landslides were observed on the slopes of Mt Amba, 6 km west of Keren. Also at a distance of 5–6 h by foot or by mule (20–25 km from Keren) a large fissure opened; many people and animals were killed.

Some 40 years later, Palazzo questioned the *shimagillés* (elders) of the region about the earthquakes of 1875–76; they remembered the events but not the details.

Sources

Annales de la Congrégation de la Mission, 1876, 41, p. 127, 458-459; James (1884, p. 211); Palazzo (1915, p. 312–313); de Coursac (1926, p. 305).

Comments

1. Macroseismic Observations

In and around Keren (N 15.8°, E 38.45° —in older documents, Keren is also called Sanheit):

On November 2 (1875), a severe earthquake almost destroyed our church in Keren as well as the whole mission compound. The earth opened, mountains appeared to be oscillating... Rockslides destroyed everything along their paths. People thought the end of the world had arrived. Tremors and underground noises lasted more than one month. The epicenters must have been located about 5 or 6 hours from Keren where it is reported that a wide fissure open in the earth. Many people and animals were killed there... A new brick house... was entirely destroyed; 4 or 5 people were killed when it collapsed (Annales 1876, Vol. 41, p. 127).

... The beautiful church was seriously damaged; the dome is so badly cracked that it will have to be redone and one wall is in a very bad condition ... Day and night, Keren and its surroundings tremble five or six times a day (Letter dated Alitiena, Feb. 4, 1876, ibid, p. 458–459)...

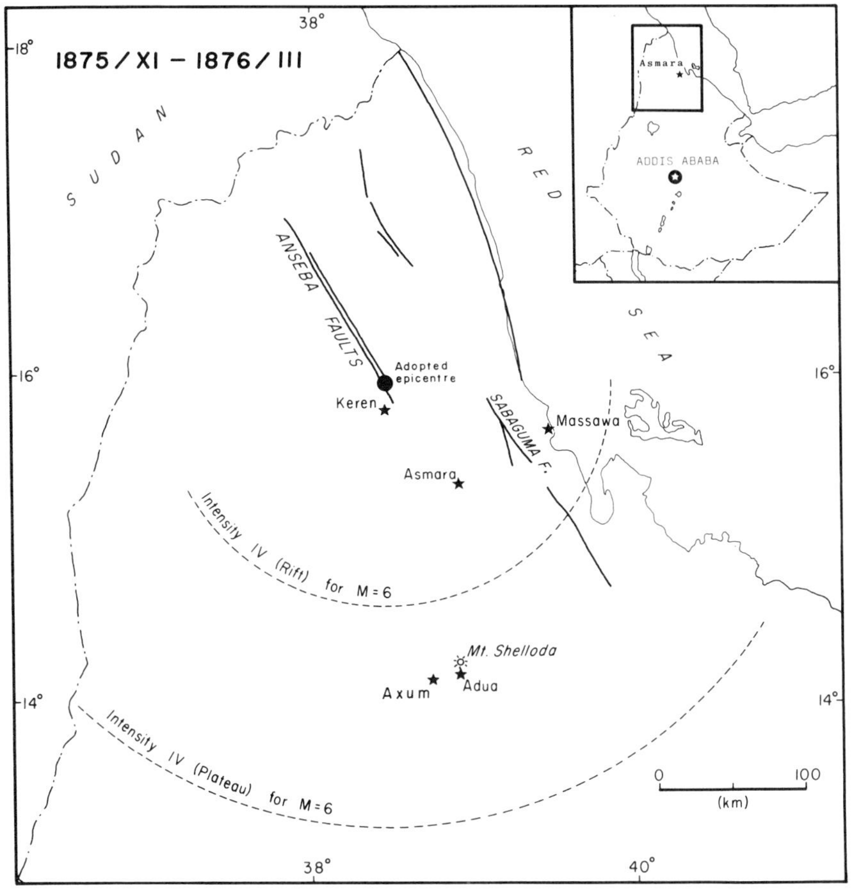

Fig. 22. The estimated location (•) of the 1875–76 seismic region and two intensity-IV ranges based on the attenuation rates observed in Ethiopia. Note that the Plateau rate applied better to Western Ethiopia.

We were told that two or three years ago, a great part of the roof had fallen in and that everything in the Church had been overturned except the image of the Virgin and Child over the high altar. (Communication of an Ethiopian priest to F. L. James, in Keren. James 1884, p. 211).

Near Adua (N 14.2°, E 38.8°):

King Johannes was leaving his palace (in Adua) *around eleven o'clock in the morning, November 2, 1875. A strong earthquake was felt and we saw an avalanche of rocks from Amba Shelloda, rolling near the King* (Papers from Mr de Sarzec, Vice-consul of France, reproduced in de Coursac 1926, p. 305).

2. Estimated Epicentre Location and Magnitude

Figure 22 shows the location of the two sites from which structural and surface effects were reported: Keren and Mt Shelloda near Adua. The intensities are estimated as IX at the epicentre, VII–VIII in Keren, and IV near Adua.

Assuming that the reported intensities were caused by a unique shock, in all probability the epicentre would have been located at the south end of the Anseba faults, some 20 km north of Keren, i.e. 5–6 h on foot as indicated in the documents for the site where the earth was fractured and some people killed. Assuming a magnitude 6 at the epicentre and an intensity IV at Adua, two arcs were drawn indicating the ranges of intensity IV based on the two rates of energy attenuation observed in Ethiopia. The diagram shows that, at Plateau attenuation rates, intensity IV at Adua could normally be expected from a shock of magnitude 6 or even 5¾, located at the southern end of the Anseba faults.

Intensities VII–VIII near Keren are above those expected for a magnitude 6, but the surface effects at or in the near-vicinity of a shallow focus epicentre are not valid criteria in the determination of a magnitude (Housner 1970).

The fact that the aftershocks lasted 3–5 months indicates in Ethiopia a main shock of $M \geqslant 6$.

For a description of the geology and tectonics of the northern Eritrea Plateau, see Mohr (1974a).

1878/XI

In November 1878 earth tremors were felt in Tigray during a period of about 6 days. No particular location or intensity is indicated.

Source

Palazzo (1915, p. 313), referring to an unpublished manuscript loaned to him by Mrs. Naretti.

1880/VI/30

An earthquake was reported in Tigray, on 30 June 1880. No exact location is given, but from the context it appears that the report originated in Axum (N 14.1°, E 38.7°).

Source

Manuscript 28 Casimir Mondon-Vidaihet, in Chaine (1913, p. 14).

During the 13th year of his reign, he (Johannes IV) *left his camp at*

Boru Mieda and moved to his territory in Tigray. He entered Axum, put up camp and constructed his palace. That year, on Hamlie 7th, the earth trembled. There was a great panic.

Comments

1. 7 Hamlie 1873 *Amete Meheret*, a Wednesday, corresponds to 30 June 1880, the 13th year of the reign of Johannes IV. He was crowned in A.D. 1868.

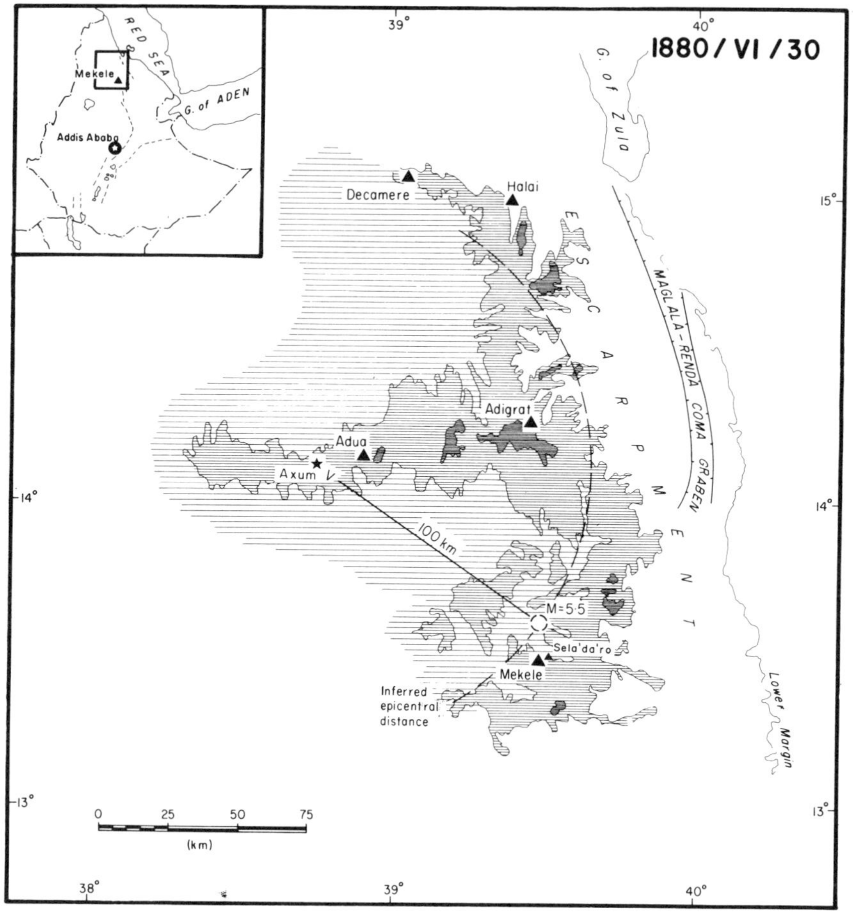

Fig. 23. Location of the sites in Eritrea and Tigray often mentioned in this catalogue. Also indicated as a reference is the inferred epicentral distance of a magnitude 5.5 earthquake located along the escarpment that would normally cause tremors of intensity V in Axum and Adua.

2. The town of Boru Mieda mentioned in the text is located a few kilometres north of Dessie, at N 11.2°, E 39.7° (Fig. 23).

3. To have caused panic, the intensity of the tremors must have reached grade V or VI (MM).

1884/X/14

At about 4 a.m. on 14 October 1884, a strong earth tremor was felt in Mekele. A subsidence, some 15-m wide (no length is given), appeared in the nearby village of Sela'da'ro and filled up with water.

Source

Palazzo (1915, p. 319), quoting an eyewitness.

Comments

There are three villages in northern Ethiopia that can be identified as Sela'da'ro: one in Shiré near Enda Sellassie (N 14.1°, E 38.3°), a second near Adi Ugri (N 14.9°, E 38.8°), and a third in Enderta near Mekele (N 13.5°, E 39.5°) (see Fig. 23). The third is the most likely location, 5 km NNE of Mekele, on the north shore of the river where a subsidence could easily occur and be filled by water seeping in from the river.

1886–87

Nine earth tremors were felt in Massawa from 8 May 1886 to 28 December 1887; some of these were also felt in Asmara. From the directions recorded on the seismoscope at Massawa and a greater intensity experienced at Massawa, it is presumed that the epicentres were off the coast in the vicinity of the 1884 centre of activity. See entry 1886–87 in Region C.

1889/X/04

Earth tremors felt in Keren (N 15.8°, E 38.4°) during the night of 4 October 1889.

Sources

De Zerbi (1891, p. 29). (Marinelli and Dainelli 1912, p. 116).

1894/V,VII

During the summer of 1894, the Meteorological Office in Keren reported two tremors, both accompanied by noise.

13 May at L.T. 06:45: earth tremor of about 6 s duration accompanied by noise.

4 July at L.T. 23:30: earth tremor accompanied by noise.

Source

Reports from the Meteorological Office in Keren reproduced in *Africa Italiana* 1894, 20 May, No. 228 and 8 July No. 235 (from Palazzo 1915, p. 322).

Comments

If no damage was reported, the intensities must not have been any higher than IV.

1894/XI/03

During the night of 3 November 1894, two strong tremors were reported in Halai (N 15.0°, E 39.4°), about 55 km SE of Asmara. They occurred at L.T. 21:30 and 23:00, respectively.

Source

Africa Italiana, 11 November 1894, No. 253 (from Palazzo 1915, p. 323).

Comments

Another instance of felt earth tremors in Halai was reported on 4 May 1832 (see Fig. 75).

1894/XI/08

At sunrise on 8 November 1894, an earth tremor was felt in Keren (N 15.8°, E 38.4°). No intensity was indicated.

Source

Africa Italiana, 11 November, No. 253 (from Palazzo 1915, p. 323).

1901/V

In May 1901, a rather severe earth tremor (qualified as *fortissima* by the officer-in-charge of the seismoscope station in Massawa) was felt in Amazi, a village about 14 km from Asmara. A jar fell down from one of the roofs and some wooden boards leaning on a wall tilted over. There was panic in the village and people rushed out of their houses. Estimated intensity V.

Source

Palazzo (1915, p. 324).

Comments

It is rather surprising, as Palazzo noted, that the tremor was not reported in Asmara. The epicentre must have been local and very shallow.

One of the workers on a construction site in Amazi remarked that *if the buildings (hŭdmó) in Amazi had not been built the Ethiopian way but in masonry as in Asmara, the damage would have been serious.*

1903/IX/26

The Meteorological Station in Asmara reported an earthquake at 5:42 a.m. on 26 September 1903.

Sources

Bullettino Ufficiale della Colonia Eritrea, XII, No. 41, Asmara, 10 October 1903 (Palazzo 1915, p. 327).

Comments

A SE-NW direction for the incoming wave is reported; it cannot be ascertained whether the information was based on subjective reaction or a seismoscope recording.

1904/V,VI

Two light earth tremors were reported from Addis Ababa, one at the end of May and the other at 1:55 p.m., 30 June 1904.

Source

Palazzo (1915, p. 328).

Comments

For both of these events, isoseismal maps were drawn by de Castro and attached to his monthly meteorological reports. Efforts to obtain these original isoseismal maps, both in Ethiopia and in Rome, were unsuccessful.

1904/VI/01

Sieberg (1932) reported that a tremor of intensity V was felt in Eritrea on 1 June 1904. The fact could not be verified.

Sources

Sieberg (1932).

Comments

Sieberg indicates no precise location where the tremor was felt; moreover, the event is not reported in any local nor international report.

1905/III/09

At 5:10 p.m. on 9 March 1905, an earth tremor was felt at Adi Ugri and Asmara, Eritrea (N 14.9°, E 38.8° and N 15.3°, E 38.9°, respectively).

Sources

Tancredi (1906, p. 1236–37), Marinelli and Dainelli (1912, p. 116), Palazzo (1915, p. 328).

Comments

Tancredi noted in his report that this was the sole seismic event felt in Adi Ugri during the last seven years; yet he himself, as director of the station at Adi Ugri, had reported one on 2 April 1900! For the location of Adi Ugri and Asmara see Fig. 75.

1905/XII/23

A few light tremors were felt near Adua (N 14.2°, E 38.8°) on 23 December 1905.

Source

Marinelli and Dainelli (1912, p. 116), Mercalli (1913, p. 3).

Comments

Palazzo does not confirm this statement. Mercalli adds that the tremors were also felt in Asmara.

Marinelli and Dainelli identified the site at which the tremors were felt as Mai Enda Maruglo; the name has been changed to Adi Malgundy. It is located near Mt Damo Galila, 12 km south of Adua. The coordinates of Damo Galila are N 14.05°, E 38.9°; the distance from Damo Galila to Asmara is 140 km.

1906/V/26

On 26 May at 9 p.m. an earth tremor was reported from Boru Mieda (N 11.2°, E. 39.7°), 15 km north of Dessie. The village is located in a downfaulted basin near the edge of the eastern escarpment of the Ethiopian Plateau.

Source

Il Messaggiero and *La Tribuna*, Rome, 30 May 1906 (from Palazzo 1915, p. 330).

Comment

On more recent maps, the village Boru Mieda is called Boru Debre Berhan.

1906/Shoa

From 25 August to 8 November or later in 1906, the Province of Shoa was subjected to severe earthquakes. Gutenberg and Richter (1954) rated the magnitude of the main shock on 25 August at 6¾; the estimated epicentral location obtained by de Castro from his isoseismal intensity map is N 08°, E 38.5°, 100 km south of Addis Ababa.

Source

Bulletin de la Commission Sismique Centrale de Russie (1906, p. 30–35); Bushman (1906, p. 117); de Castro and Oddone (1909, p. 442); Szirtes (1910, p. 48-50); Marinelli and Dainelli (1912, p. 117); Palazzo (1915, p. 330-331); Mérab (1922, II, p. 192); Gutenberg and Richter (1954); Goëtz (1965, personal communication).

A description of the events described in some of these sources follows.

1. In Addis Ababa

Mérab states that the August tremors *were exceptionally violent.* Marinelli and Dainelli mention oscillation of suspended objects, cracking of roofs, doors, and windows. Bushman, who was at the time sitting in a rocking chair in his garden, compared his impressions to *those shocks made by sea waves breaking from time to time on the sides of a vessel and making it suddenly vibrate.* It is also reported that the shock of 28 October was strong enough to cause the bells of the Church in the Finfini district to ring spontaneously. (Finfini in 1906 was located on the old Municipality Hall premises).

The population of Addis Ababa was greatly afraid. Damage, however, was slight because: (1) the town, being barely 10-years-old, had not yet fully developed; and (2) the indigenous constructions (the Ethiopian *tukuls*) could, according to experts, withstand any shock short of a wide rending of the earth directly underneath them. This remark has been made by many observers and was spectacularly confirmed at Kara Kore in 1969 (see entry 1961/Kara Kore).

2. In Adamitulu (N 07.85°, E 38.70°)

Adamitulu is 125 km south of Addis Ababa. This report was given to me by an old friend, Herr Herman Goëtz, who had settled in Adamitulu on the shores of Lake Zwai in 1902. He was over 80-years-old when I asked him about the details of the 1906 earthquakes. He recalled that about five shocks were felt during the afternoon *siesta*; he saw through the window of his room Mount Alutu *dancing* on the south-eastern horizon; to the south-east along the canyon of the Bulbula River, he first noticed *vertical motions in the surface of the earth.* Landslides were then triggered along the steep banks of the river; he noticed columns of dust rising in the air and progressing toward the SSW, indicating therefore that from Adamitulu, the direction of the epicentre had a N or NE component.

Herr Goëtz reported no damage to structures in Adamitulu, which was a very small village in 1906; he observed no cracks in the ground.

3. At Lake Langano

Herr Goëtz also reported that during the 1906 period of seismic activity, a geyser started on the island O'a (N 07.5°, E 38.6°) in the northeastern bay of Lake Langano (see Fig. 24). At its birth, in 1906, the height of the hot water column was 25–30 m and the rate of the pulsation 2/min. A report dated 1926 mentions that the height of the water column was about 1 m and that the period of pulsation was about 30 min.; another report claims that in 1963–64, it was 2-m high. In August 1965, Mohr and Dean reported that the original geyser, which they classified "Spring 10," was then a 2-m diameter circular pool lying within a circular basin 10–15 m in diameter. Its water level fluctuated by 5–10 cm in a complete 9–10 min cycle of fill, steam-surge, and withdrawal. In 1970, Lloyd (UNDP 1973, p. 89) observed that the surge periodicity was no longer noticeable.

4. In Silte

In Silte (N 08°, E 38.3°, 120 km SSW of Addis Ababa), tremors were also felt.

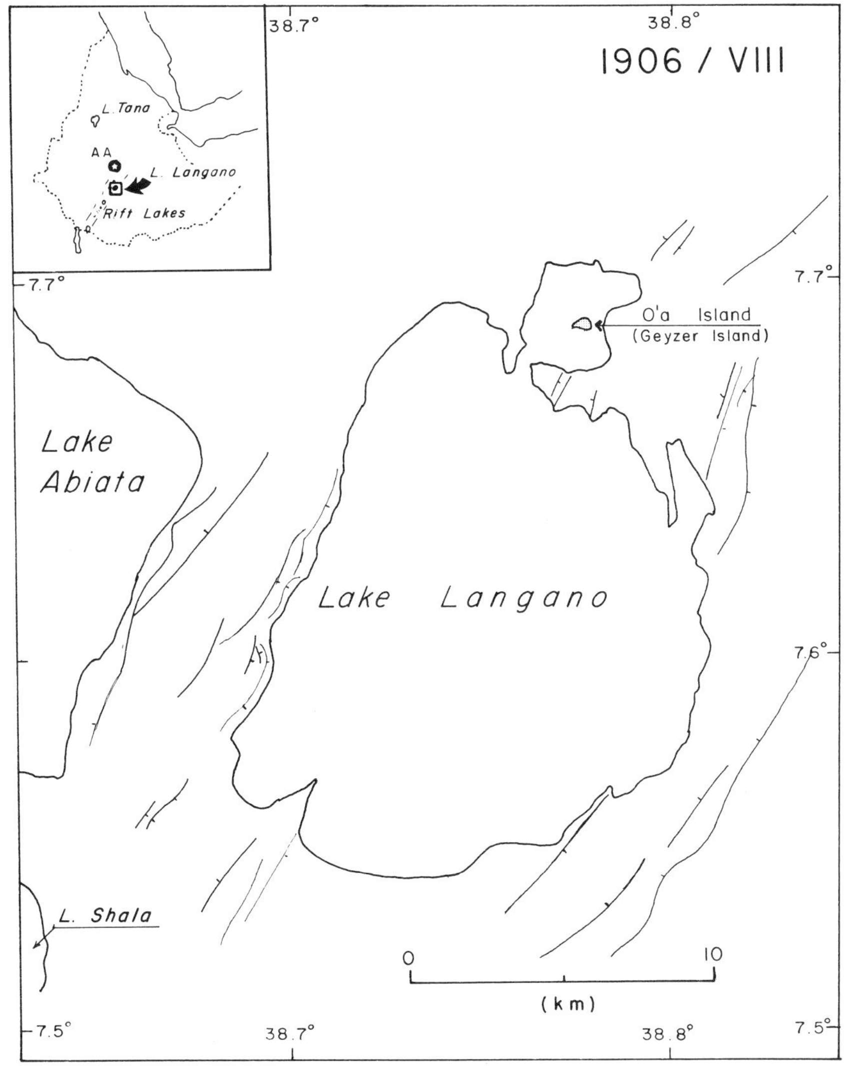

Fig. 24. Location of O'a Island in the rift lake Langano on which a geyser started during the seismic activity of August 1906.

5. Gurage Country

In the Gurage country on the northwest shoulder of the Ethiopian rift valley even today many elderly people still give their age as so many years after the great earthquake that was thought to *mark the end of the world*, meaning the earthquakes of August 1906.

Comments

1. Instrumental Data

Teleseismic recordings show two important shocks on 26 August 1906: the first at U.T. 11:54:48 and the second 2 h later at U.T. 13:47:36. The magnitudes were 6.6 and 6.8 and the number of reporting stations 30 and 40, respectively.

Gutenberg and Richter (1954) located the epicentres of the two shocks at the same site N 09°, E 39°, barely 13 km east of Addis Ababa. Szirtes (1910, p. 48–50) put it at N 09.1°, E 38.7°. I attempted a recomputation of the original data; the results were not coherent due to inaccuracies in the recorded arrival times at individual stations.

2. Field Data

De Castro mentioned at the end of a meteorological report that an isoseismal map of the 1906 activity was attached. That map has been lost. On the strength of his intensity distribution map, de Castro concluded that the centre of the epicentral region was located at N 08°, E 38.5°, about 100 km slightly W of S from Addis Ababa. His epicentral location is very realistic in terms of intensity distribution and tectonic structure. It lies at the foot of the western escarpment of the Ethiopian Rift Valley in a zone of positive Bouguer anomaly associated with recent mantle intrusion under the rift (Searle and Gouin 1972, p. 41–52), therefore a potentially active area.

Figure 25 shows a postulated isotropic intensity distribution for an earthquake of magnitude 6¾ based on the composite energy distribution rates adopted by Gouin for Ethiopia (1976, p. 20–21); the computed intensity VI–VII for Addis Ababa corresponds to the observed values cited above.

Besides the shocks at U.T. 11:55 and 13:48 on 25 August, five others were felt in Addis Ababa; details are given in Part II (p. 189).

3. Regional Seismicity

Sieberg (1932, p. 887, Fig. 466) published a map on the seismicity of Ethiopia on which the activity of the Ethiopian rift is based mainly on the events of August 1906. The relevant part of this map is reproduced in the inset of Fig. 25. As one notices, the lowest intensity envelop is elongated toward the east and southwest due to the inclusion in the map of the events of May 1832, April 1842, and December 1845, all of which were of moderate amplitude.

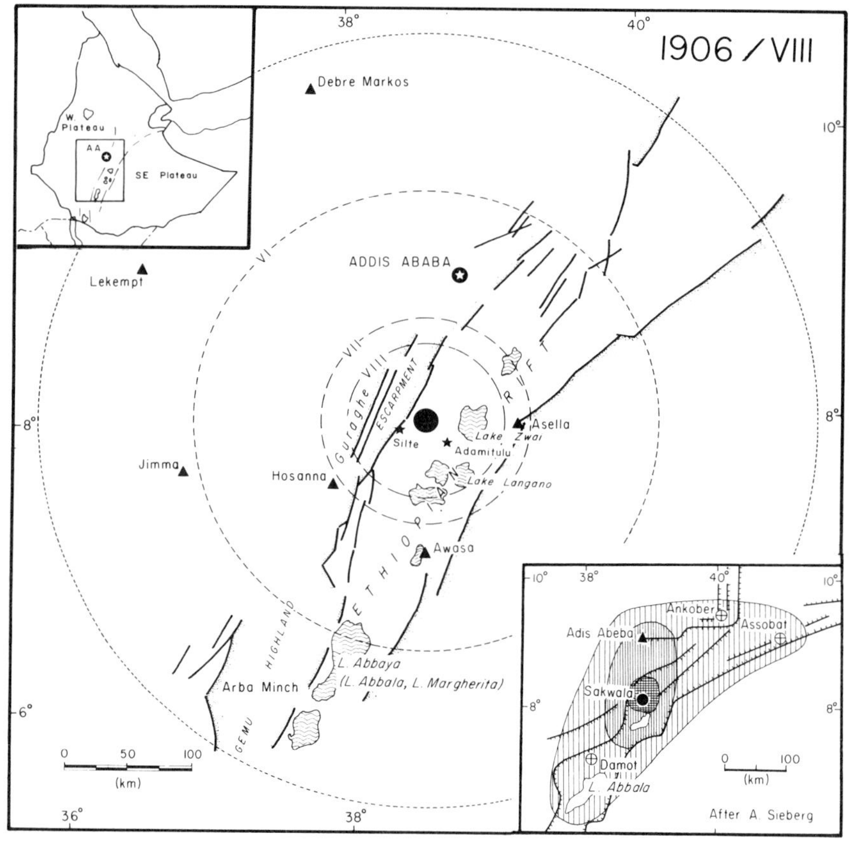

Fig. 25. Location of the 1906 centre of seismic activity and of the provincial capital cities where intensities larger than V were presumed felt. In the plotting of circular isolines, the different attenuation rates observed under the plateaus and across the rift were not taken into consideration; in reality, one would expect an elliptical distribution pattern with its major axis oriented NE-SW and a NW-SE semiminor axis compressed by the higher attenuation rates across rift structures. The inset reproduces the isoseismals of the same events as interpreted by Sieberg (1932). The three highest intensity contours are those of the 1906 earthquakes; the outer contour line is arced by the presence of two other events indicated by the ⊕ symbols. Sieberg's spelling of locality names has been respected.

The entries in *Erdbebengeographie* from which Sieberg built his seismicity map are:

1832, May 4: One tremor felt in the Assabot valley.

1841, April 23 and 24: Small tremors felt in Ankober.

1845, December 7: Moderate shock in the Damot mountains and at Lake Abba (Queen Marguerita Lake).

1906, August to December: Several shocks, often strong, reported felt at Addis Ababa; a few felt in Gibbi.

Note: There is no locality known as "Gibbi," but there is a river in the Guraghe country by the name of Gibbié. On older maps, such confusion in site identification is easily explained.

Identification of the sites mentioned by Sieberg and a discussion on the accuracy of each entry are found at the indicated dates in this survey.

Note that Fig. 25 of this survey presents isoseismals for a homogeneous crust with no reference to any actual felt reports except those from Addis Ababa and Adamitulu. As mentioned above, the original reports summarized by de Castro have been lost but it is not unrealistic to believe that Sieberg had them in 1932.

1907/IV/12–17

Three light tremors were felt in Addis Ababa on 12 and 17 April 1907. People were frightened, remembering the 1906 experience, but no damage was reported.

Sources

De Castro (1909, p. 442), (Marinelli and Dainelli 1912, p. 118; Palazzo 1915, p. 333).

Comments

Here again, the reporters observed that Ethiopian *tukuls* could withstand shocks of any amplitude.

1908/XII/28

At sunset, on 28 December 1908, an earth tremor accompanied by a rumbling noise was reported from Ela Bered (N 15.7°, E. 38.5°), approximately 60 km northwest of Asmara (see Fig. 26).

Source

Palazzo (1915, p. 334–335).

1910/II/04

After midnight on 4 February 1910, an earth tremor was felt in Addis Ababa. No damage was reported.

Source

Private letter of Dr de Castro to Palazzo (1915, p. 335). (Sieberg 1932, p. 887).

1912/III

In March, an earth tremor was felt in Addis Ababa. No intensity is given.

Source

Mérab (1922, p. 192).

1912–13

Dainelli describes as follows the seismic activity that, for several months in 1912, shook northern Ethiopia: *In more recent times, we must recall the seismic period which began before the end of 1912 and continued without interruption until the end of Spring 1913. The earthquakes were felt throughout Eritrea and Tigray but more severely in Massawa and Adua* (Dainelli 1935, II, p. 127).

Sources

Palazzo (1913, p. 110–127, 1915, p. 335–337), Dainelli (1935, II, p. 127–128, 1943, III, p. 713), Gutenberg and Richter (1949, p. 305), ISS (Turner 1917–34).

Comments

1. Reported Times of Occurrence, Magnitudes, and Intensities

In the region of Asmara, the more intense activity spanned from the end of January to May 1913. From that region, 457 shocks were reported to have been felt at a mean rate of 3.7 per day during four consecutive months (Dainelli 1943, III, p. 713). The frequency of the shocks felt in Asmara and Massawa was distributed as follows: from 24 January to 8 April, 208; 9 April to 9 May, 117; and 10 May to 28 May, 132.

The two most violent shocks occurred on 27 February and 27 March 1913. Gutenberg and Richter estimated their magnitudes at 5.8 and 5.5,

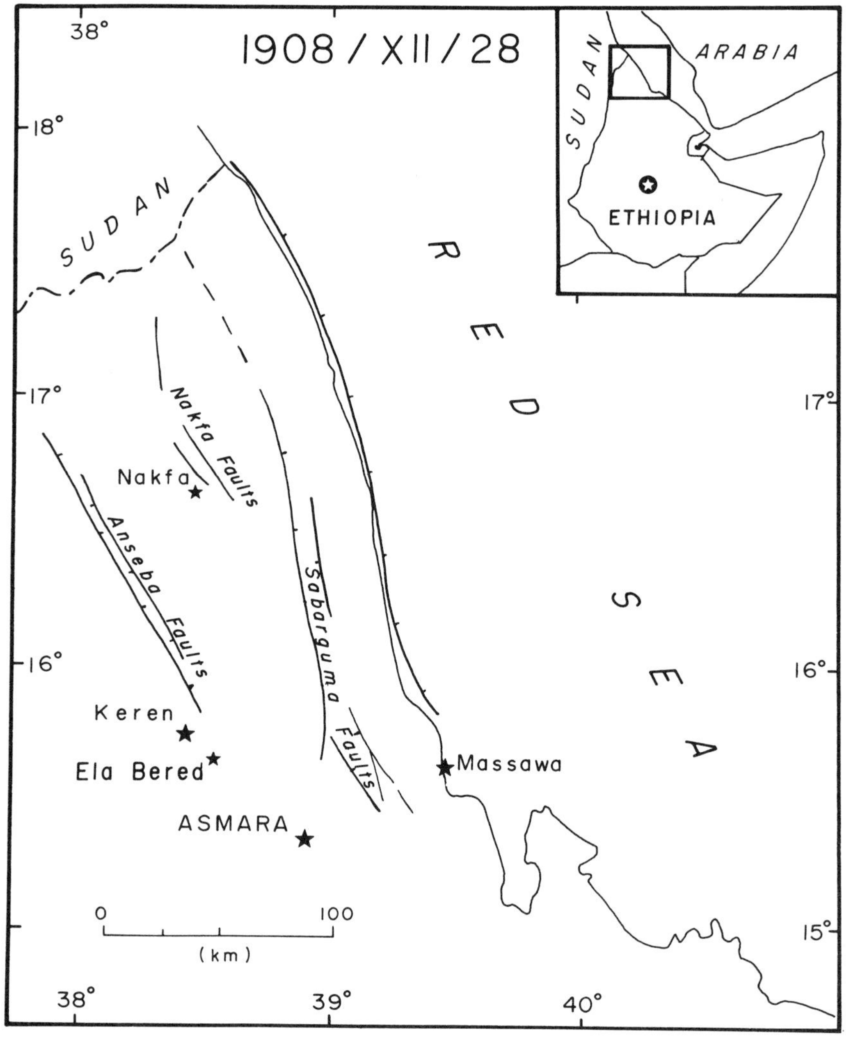

Fig. 26. Location map of potentially active faults and of sites from where earth tremors have been regularly reported in northeastern Ethiopia.

respectively. The social impact was such that the Government of Eritrea decided to install a seismic station in Asmara. A description of the station and of its equipment is given in Palazzo (1913, p. 110–127). Suffice to state here, that from 6 June to 16 July 1913 the new seismic station (N 15.298°, E 38.830°) recorded 141 local shocks on its Agamemnone seismographs of static magnification ×30. Palazzo's report on the Asmara station is the only one available today; all efforts to trace the original seismograms or their analysis were of no avail. It is not known, either, how long the station stayed in operation.

Some intensities were reported from Asmara, Massawa, Adi Caieh, and Ela Bered; the maximum values being VII–VIII in Asmara. No reports are available from Adua despite the explicit quotation by Dainelli that the city had "severely suffered." Sieberg (1932) estimated the intensities in Asmara at VI–VII on the Mercalli-Sieberg scale; details are given in Part II (p. 189). The Asmara station files must have been lost, otherwise it would be inconceivable that Palazzo, the chief-seismologist at the time, would not have left any report on the amplitude of the first seismic crisis recorded by his station, on the structural damage and ground effects, or on the intensities of the tremors felt throughout Eritrea.

2. Location of the Seismic Region Inferred from Reported Magnitudes and Intensities

During the period of seismic activity only two events were of magnitude high enough to be recorded on the low sensitivity seismographs used in 1912–13 at distances greater than 2000 km. For these two events, the parameters are:

	H	Coordinates		M
27 February 1913				
G & R	U.T. 16:22:54	N 17.5°	E 39.0°	5¾
ISS	U.T. 16:22:34	N 14.0°	E 39.0°	
PUL	U.T. 16:22:34	N 16.0°	E 39.0°	
27 March 1913				
G & R	U.T. 03:13	N 16.5°	E 39.0°	5½
ISS	U.T. 03:12:45	N 15.9°	E 39.0°	
TIF	U.T. 03:11:45	N 15.9°	E 39.5°	

In these solutions, the longitude values agree perfectly, whereas the latitude estimates differ by as much as 3.5°. This is to be expected of older data when station time standards were not radio-controlled and when the stations were concentrated in a geographic sector north and northeast of Ethiopia.

The longitude coordinates (≃E 39°) are accepted at their face value. In reality, they correspond to unstable tectonic features (Nakfa and Sabarguma graben(s)), heading 010° from N 15.5° to N 17.2° and along which seismic activity is known to have occurred on other occasions. It is fortunate that in the location of epicentres on the Ethiopian Plateau or along the Rift escarpment, the errors are usually maximum in latitude and minimum in longitude. As most of the regional tectonic structures are N–S trending, even a large location error in latitude still keeps the epicentre on the same tectonic structure.

Two elements are used to restrict the error in latitude: reported intensities and magnitudes. The highest intensities (VII–VIII) having occurred in Asmara and the magnitudes at the origin being ≃ 5¾, it follows that the epicentres must have been within a radius of 30–40 km from Asmara. A circle of such radius encompasses the Sabarguma faults (or quasi-graben as opposing scarps are not parallel) which are known to be seismically active (Fig. 27).

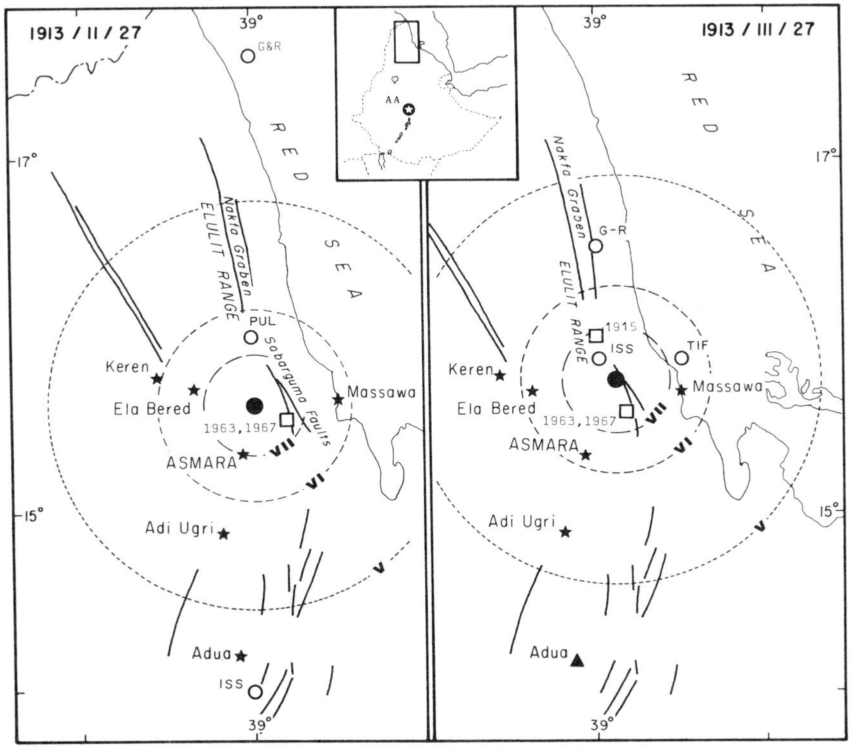

Fig. 27. Location maps of the two main shocks during the 1912–13 seismic crisis. Open circles (○) are the published locations; solid circles (●) are the adopted locations for the 1912–13 activity. The open squares (□) refer to earthquake epicentres of other dates.

The intensity circles show that all of Eritrea and Tigray should have felt tremors from these two earthquakes; Massawa could have been damaged, but not Adua, where the intensity was no higher than IV.

It is my personal opinion that: (1) during 1912–13, the whole length of the Sabarguma fault system was active; (2) the most probable location of the two main shocks was that of the 1963 and 1967 epicentres marked by an empty square symbol on Fig. 27; and (3) no serious damage could have been inflicted beyond the intensity–V limits. When Dainelli wrote his note in 1935, it seems that for greater journalistic impact he exaggerated to a certain extent; he did not substantiate these earlier statements in his *Geologia dell' Africa Orientale* published in 1943. The reason for two slightly different epicentral locations on the accompanying maps is to justify the reports of equal intensities in Asmara originating from shocks of different magnitudes.

1915/IX/23

At U.T. 08:14:43 there was an earthquake of magnitude 6 ¾ (G & R) in Eritrea. Shide (Turner 1917, p. 8) from 31 station reports located the epicentre at N 16.0°, E 38.5°; Gutenberg and Richter (1949) located it at N 16°, E 39°; and the Pulkovo Observatory at N 15.7°, E 34.8° (Fig. 28).

Tremors of intensity VI were felt in Asmara (N 15.5°, E 39.0°); people ran out of their houses and cracks formed in the front wall of the hotel Albergo Italia near Corse de Re, in the centre of town.

Sources

Turner (1917, p. 8); Bellamy (1936, p. 3); Gutenberg and Richter (1949, p. 205); Sr. Vaccaro Salvatore, Asmara (personal communication 1974).

Comments

The tectonic structure nearest to two of the calculated epicentres (N 16.0°, E 38.5–39.0°) and susceptible to seismic activity is the Sabarguma graben on the margin of the Eritrean escarpment. Subsequent activity along the same graben is described in entries 1963/VII/14, 1967/IX/15–18, and 1967/IX/26. Pulkovo's solution has been neglected.

The intensity at 50 km from a magnitude 6¾ is expected to be about VII. This means that the epicentre could have been located slightly north of 16° along the escarpment, in the direction of the Nakfa graben (see Fig. 27).

1922/I/23

On 23 January 1922 the tremor of an earthquake was felt in Wejja, near Waldia (N 11.8°, E 39.6°), some 75 km due north of Dessie, capital of Wollo (see Fig. 29). No intensity was reported.

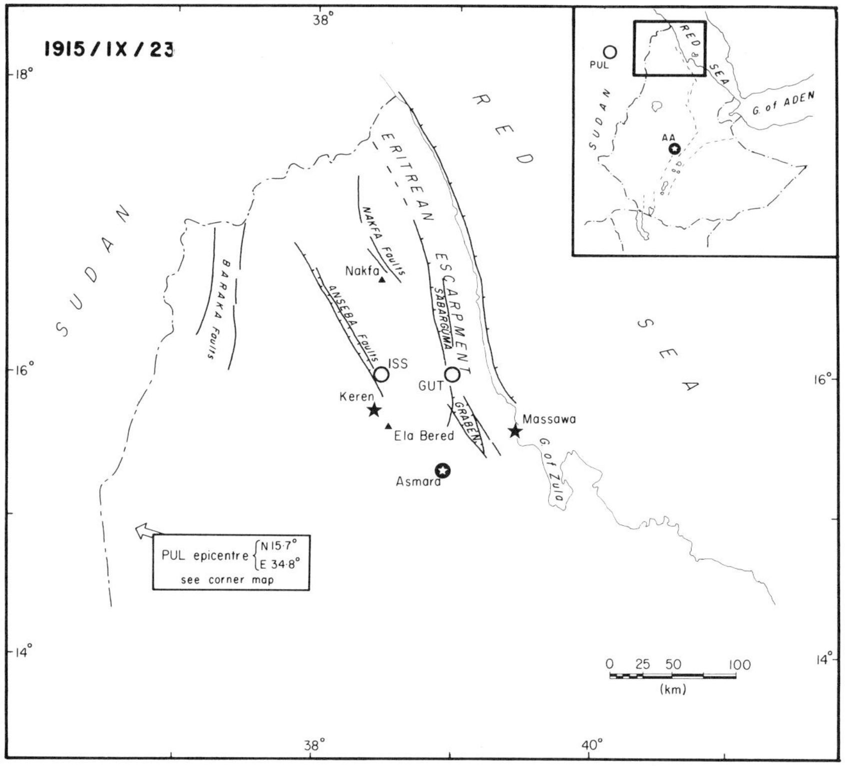

Fig. 28. Instrumental epicentre locations for the magnitude 6¾ earthquake of 23 September 1915.

Source

A marginal note written in a hymnbook (*degua*) of the Wonzai Abbo Church (manuscript 3822 at the Microfilm Library of the Orthodox Church, Addis Ababa).

In 1914 A.M., Year of Marc, Ter 15, during the reign of Dejasmatch Habte Mariam, there was an earthquake.

Comment

The date, 15 Ter 1914 A.M., corresponds to 23 January 1922.

1924/III/13

On 13 March 1924, there was an earthquake of unknown magnitude along the Eritrean escarpment at N 16.0° ± 0.5°, E 38.8 ± 0.5°.

Source

Bellamy (1936).

Comments

The original ISS location was the same as for the epicentre of 23 September 1915, i.e. N 16°, E 38.5° (see comments in entry 1915/IX/23).

1930/IX/23

On 23 September 1930, an earth tremor was felt in Waldia (N 11.8°, E 39.6°), a village located on the margin of the Plateau-Afar escarpment about 75 km in horizontal distance due north of Dessie, capital of Wollo. No intensity is indicated.

Source

A marginal handwritten note found in a hymnbook of the Wonzai Abbo Orthodox Church, near Waldia (manuscript 3822, Ethiopian Microfilm Library of Orthodox Church, Addis Ababa).

In 1923 A.M., Year of Luke, Meskerem 13, during the time of Dejesmatch Admassu, there was an earthquake.

Comments

The date 13 Meskerem 1923 A.M. corresponds to 23 September 1930 (see Fig. 29 for location).

1931/V/01

On 1 May 1931, ISS located an epicentre at the extreme north tip of the Eritrean Plateau, slightly north of the Ethiopian-Sudanese border. A recomputation (PG3) of the ISS data relocated the epicentre across the Red Sea, on the coast of Yemen. Details are indexed in Region C, entry 1931/V/01.

1934/II/01

An earth tremor of intensity III was felt in Asmara at 11:50 p.m.

Source

Fantoli (1966, Appendix L, event 149).

1935/VI/22

On 22 June 1935, an earth tremor was felt in Wejja near Waldia (N 11.8°, E 39.6°), some 75 km due north of Dessie. No intensity was reported.

Source

A marginal note written in a hymnbook of the Wonzai Abbo Church (manuscript EMML 3822, kept at the Microfilm Library of the Orthodox Church, Addis Ababa).

In 1928 Amete Meheret, Year of John, Senie 15, during the Italian occupation there was an earthquake.

Comments

No mention of the event could be found in contemporary newspapers. For location, see Fig. 29.

1938/X/20,23

Two earthquakes of magnitude 5.6 and 4.5, respectively, occurred at the foot of the Plateau-Afar escarpment, 35–40 km NE of Debre Berhan. The epicentre coordinates obtained by Gutenberg and Richter, by ISS, and by recomputation of the original ISS data (1974) are as follows:

	H	Coordinates		M
20 October 1938				
G&R	U.T.13:14:58	N 10.0°	E 39.5°	5.6
ISS	U.T.13:14:55	N 09.5	E 40.3	
Recomputed	U.T.13:15:05	N 10.11±1.2	E 40.23±0.8	
23 October 1938				
ISS	U.T.02:28:47	N 09.5	E 40.3	
Recomputed	U.T.02:28:567	N 10.01±1.0	E 40.33±0.3	

Sources

Gutenberg and Richter (1949); ISS for 1938.

Comments

1. These two events belong to a group of five that occurred between north latitudes 10° and 11° and east longitudes 39° and 40° where the SW corner of the Afar triangle funnels into the main Ethiopian Rift Valley (see Fig. 30). Despite the standard errors attached to the recomputed solutions,

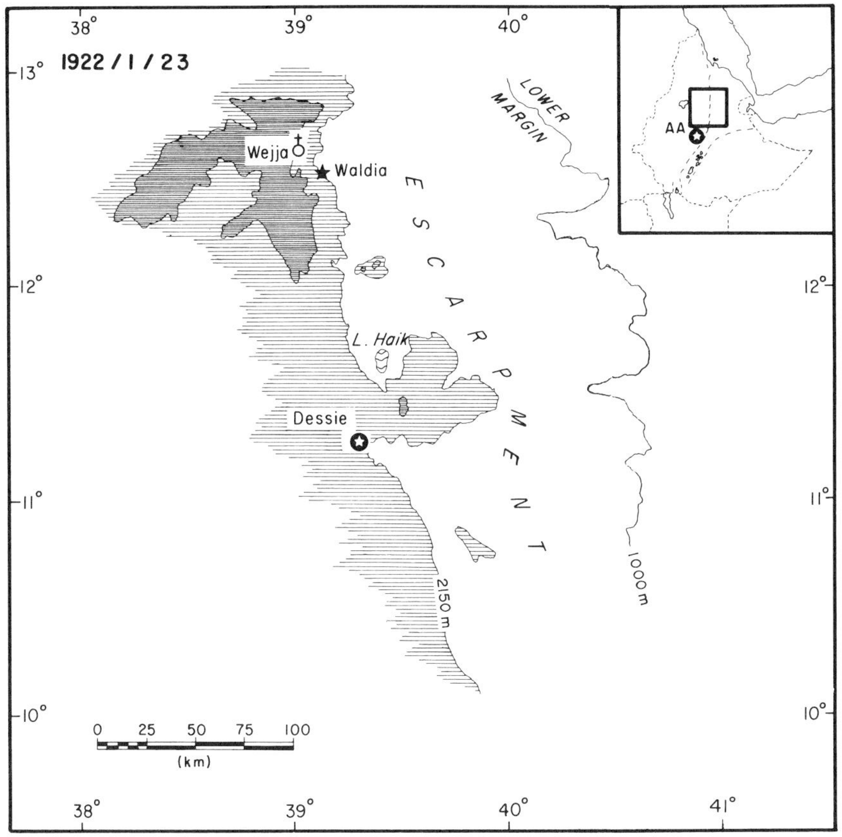

Fig. 29. Location of Wejja and Waldia on the upper margin of the Western Plateau escarpment.

the probability based on recent observations remains that these epicentres are related to the escarpment at a latitude where the general fault trend, which was N25°E, suddenly changes direction by −50° and heads NNW (Mohr 1974a, p. 34).

These two earthquakes are also indexed in Region C, entry 1938/IX–X.

1942/XI/18

At U.T.12:01, 18 November 1942, an earthquake of magnitude 5.3–5.9 occurred on the Ethiopian Plateau. Gutenberg and Richter (1949) located

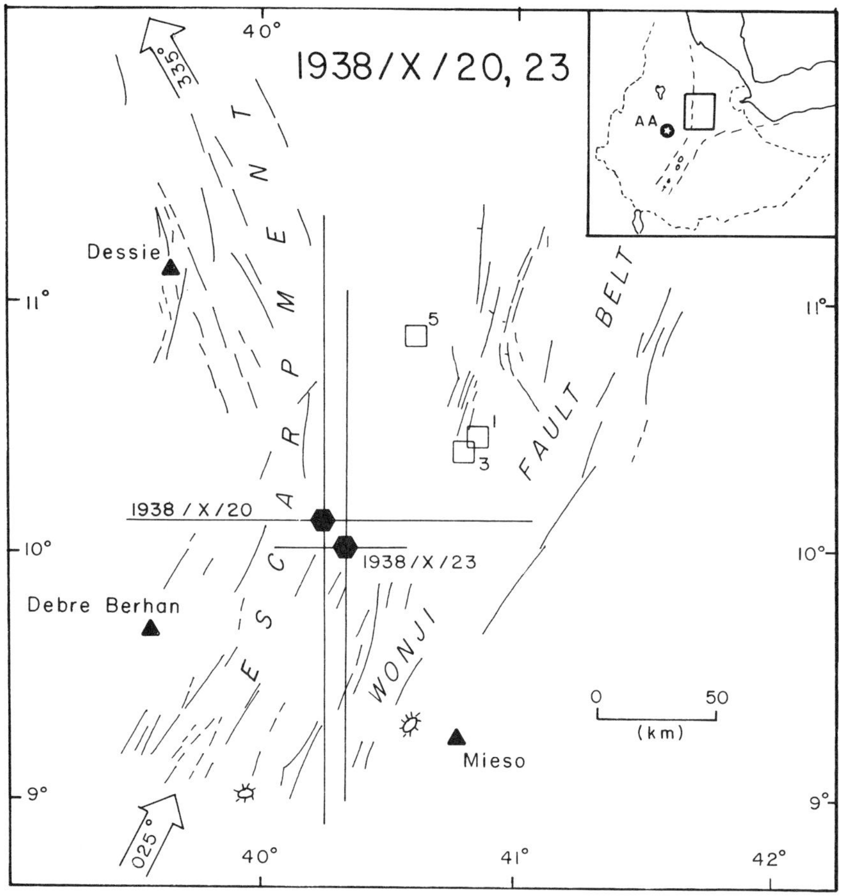

Fig. 30. Location map of the October 1938 epicentres at the foot of the Plateau-Afar escarpment near the N 10° parallel (⬢). Three other earthquakes occurred at the same time; they are discussed in Region C, entry 1938/IX,X; their adopted locations are indicated by empty square symbols (□).

the epicentre at N 12.0°, E 40.5°; ISS at N 10.0°, E 38.0°. The recomputation of the ISS original data using the SPEEDY computer program yielded, after 7 iterations, the following parameters: H U.T.12:01:20.1 ± 02.2; N 11.345 ± 0.334°, E 39.156 ± 0.153°.

The recomputation shifted the original ISS determined epicentre by 195 km N040E.

Tremors were reported from Wonzai Abbo monastery (N 11.8°, E 39.5°); no intensity was indicated (see Fig. 31).

Sources

Gutenberg and Richter (1949); ISS (1942); Ethiopian manuscript EMML 3882 kept at the Microfilm Library of the Orthodox Church, Addis Ababa. The last document reads as follows:

In 1935 Amete Meheret, Year of Luke, Hedar 9, during the life of Lig Makuas Abagas in Waldia, there was an earthquake.

Comments

The recalculated instrumental epicentre for the earthquake of 18 November 1942 is in a region highly dissected by strong faults trending NE–SW and NW–SE respectively. In that region, the Wonzai Abbo Church (Wejja), where the manuscript was written, is situated on the upper margin of the Plateau escarpment (see Fig. 29). The intensity at Wejja must have been about V.

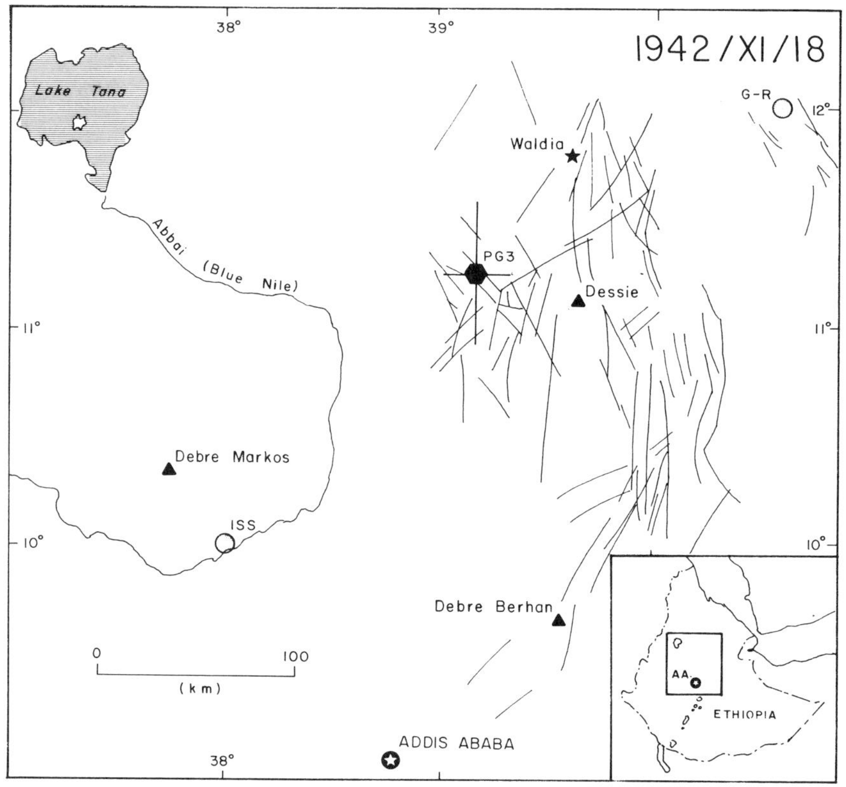

Fig. 31. Location map of the three teleseismic solutions available (G & R, ISS, and PG3) for the earthquake of 18 November 1942.

1945/III/31

At U.T.23:07:48, 31 March 1945, ISS reported an earthquake of unspecified magnitude in Eritrea at N 14.5°, E 39.5°. A recomputation (PG-3, 1974) of the original ISS data file yielded a new location at N 14.87 ± 0.42°, E 39.54 ± 0.16°. The magnitude must have been of the order of 5 because Calcutta recorded it at a distance of 5200 km.

Source

International Seismological Summary for 1945.

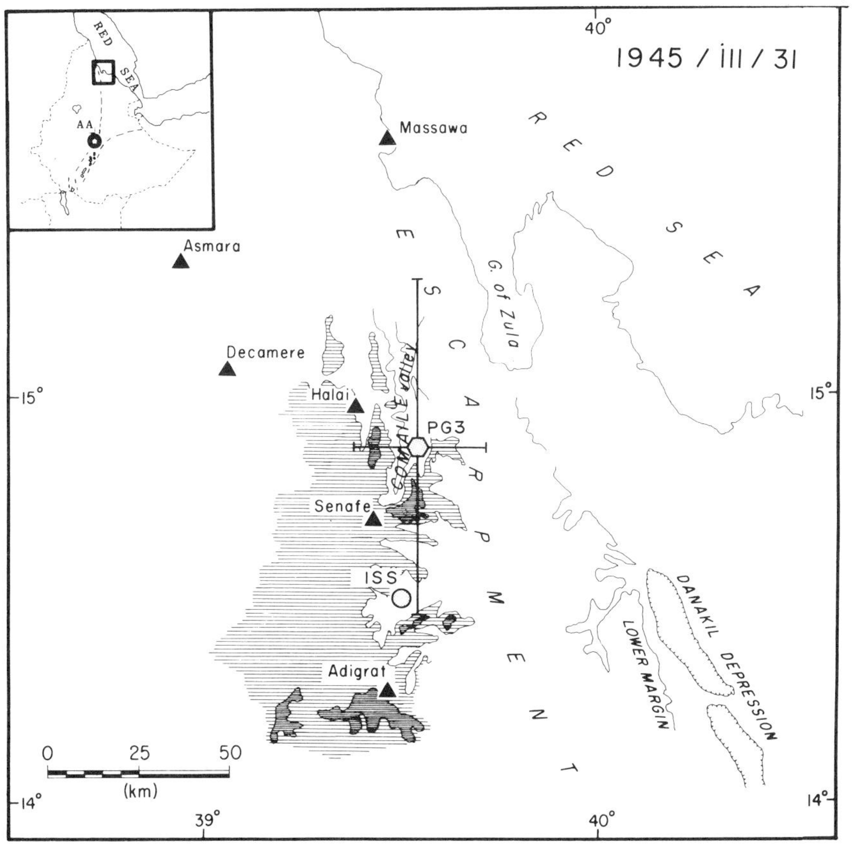

Fig. 32. Location of the two instrumental epicentres available for the earthquake of 31 March 1945.

Comments

The recomputation based on 26 station reports did not substantially alter the original ISS epicentre; it shifted it by 40 km northward along the same tectonic structure, the Plateau-Afar escarpment. The area is known to be seismically active (see Fig. 32).

1948/II/17

On 17 February 1948, an earthquake of magnitude about 5½ occurred at N 13.7 ± 0.5°, E 39.8 ± 0.2°, on the margin of the Plateau-Afar escarpment, northeast of Mekele (Fig. 33). Thirty stations reported it to ISS. No local report is available.

Sources

ISS (1948); Rothé (1954).

Comments

ISS (1948, p. 94) located the epicentre at N 14.0°, E 40.2°; Rothé (1954) at N 13.3°, E 40.0°. Recomputation of the original ISS data using the SPEEDY computer program, after 3 iterations, yielded the following results: H U.T.22:11:06.2 ± 3.5; N 13.71 ± 0.48°, E 39.85 ± 0.25°.

These results perfectly match the mean value of the solutions obtained by ISS and Rothé (N 13.7°, E39.9° versus N 13.7°, 40.1°).

1950/III/26

At U.T. 16:53, 26 March 1950, an earthquake of magnitude 5.5 (BCIS) was reported in Tigray, near Adigrat. The instrumental location is N 14.3 ± 0.2°, E 39.3 ± 0.2° (Fig. 34). No local reports are available.

Sources

BCIS; ISS for 1950, p. 182.

Comments

Basing their computations on 26 and 47 stations, respectively, BCIS and ISS fixed the epicentre at N 14.5°, E 39.5°. Recomputation (using SPEEDY program) of the original ISS data yielded the following coordinates: N 14.318 ± 0.244°, E 39.316 ± 0.196°, a location barely 25 km away from that of the other solutions.

The ISS data file shows three other events with the same coordinates N 14.5°, E 39.5°: 1945/III/31; 1950/VIII/02; and 1952/IX/10.

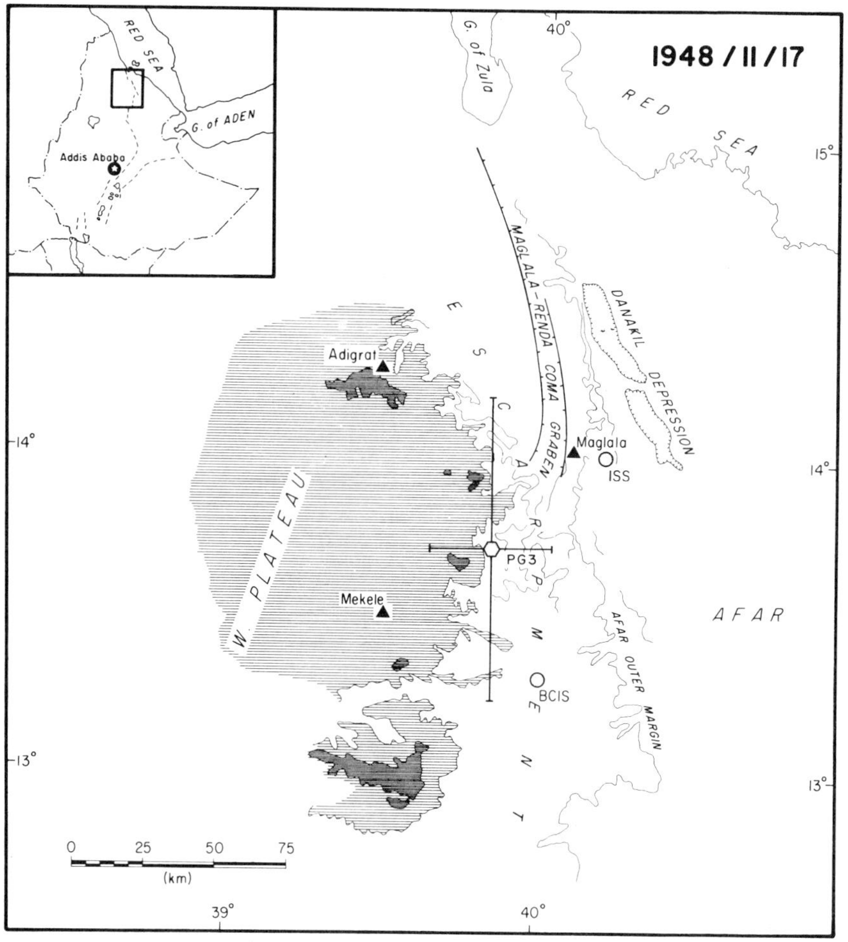

Fig. 33. Location map of the teleseismic solutions for the epicentre of 17 February 1948.

1950/VIII/02

On 2 August 1950, at U.T. 13:50, an earthquake of magnitude 6.2 occurred at N 14.5 ± 0.1°, E 39.7± 0.1° in northern Ethiopia (Fig. 35). No local reports were found on this earthquake although 191 international stations reported it to ISS.

Sources

BCIS; ISS; USCGS; Rothé (1954).

Comments

Six agencies published epicentral coordinates for this shock: four (BCIS, CGS, ISS, Rothé) located it between N 14.5–15.0° and E 39.5–40.0°; Poona (POO) put it about 150–175 km to the east at N 15.0°, E 38.0°; and Prague (PRA), much more to the southeast, at N 12.0°, E 43.0° in the Gulf of Aden. The mean value of the six solutions is: N 14.3°; E 39.9°. The recomputation (SPEEDY, 3 iterations, 115 stations) of the information contained in the ISS file yielded new coordinates (N 14.522 ± 0.101°; E 39.715 ± 0.076°), which differ only by 0.2° from the mean of the previous six solutions. For the seismicity of the same region, see entries 1945/III/31, 1950/III/26, and 1952/IX/10.

Poona (POO) and Strasbourg (BCIS) indicated focal depths at 150 and 500 km, respectively. These figures appear to be much too large for Ethiopia and Africa in general.

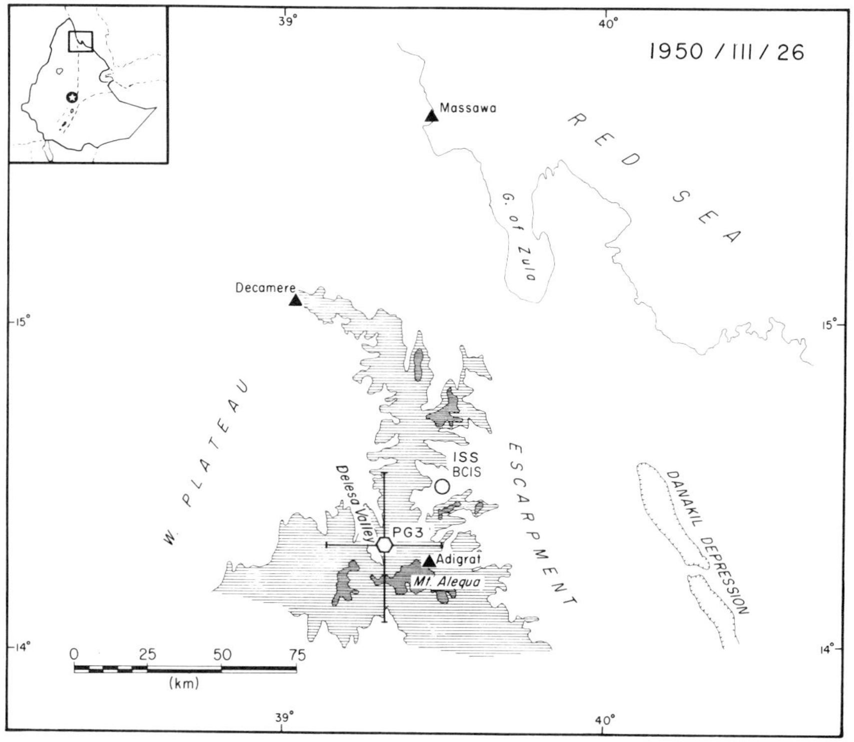

Fig. 34. Epicentre of 26 March 1950.

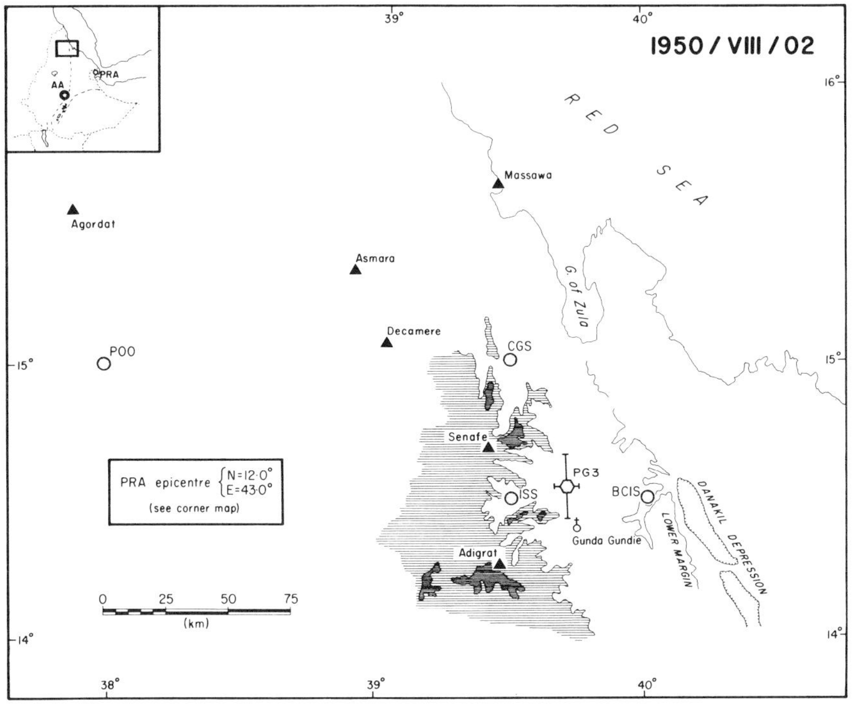

Fig. 35. Location of some instrumental epicentres for the earthquake of 2 August 1950.

1952/II-VIII

Onnis (1957, p. 74–77) reports that 22 earth tremors were felt in Asmara between 26 February and 18 August 1952.

Source

Onnis (1957, p. 74–77).

Comments

No precise dates, no exact times, and no primary sources are given in the text. Moreover, the contemporary newspapers published in Asmara make no mention of the events. Onnis's information is, therefore, disregarded in the present survey.

1954/I/17

On 17 January 1954, BCIS reported an epicentre at N 16.5°, E 36.0°. The location is in Eastern Sudan, but so near the Ethiopian border that it had to be included in this survey to avoid fringe effects in contouring seismicity parameters for Ethiopia.

The location is indicated by an open square on Fig. 80.

1957/X-XI

Four earth tremors were felt in Mai Chew (N 12.5°, E 39.5°) between 1 October and 6 November 1957. The maximum intensity was estimated as IV.

Source

A letter from H. Wilson to the Climatological Services in Addis Ababa (Civil Aviation File, Addis Ababa):

There were distinct shocks which occurred between IMT 17:30 and 20:30 on October 1, 1957. The first and the third shock waves were much more severe than the second and the fourth. They lasted several seconds and were, as the first, accompanied by rumbling noises... Articles fell from cupboards, shelves and residents saw their roofs shake and move (Wilson, H., personal communication).

Comment

For location map and analysis, consult among others, entries 1853–54, 1972/VIII/18–19.

1961/V/IX

Seismc Crisis in Wollo/Kara Kore — Majete

In 1961, from the end of May until September, over 3500 earthquakes of magnitude M_L (AAE) $\geqslant$ 3.5 shook Central Ethiopia. The village of Majete (N 10.5°, E 39.9°) was completely destroyed; in the nearby town of Kara Kore (N 10.4°, E 39.9°) most masonry houses collapsed. Cracks, fissures, and subsidences of up to 1-m deep developed on the Addis Ababa – Asmara highway; many culverts and retaining walls along the road had to be rebuilt. Rockslides and landslides were observed on steep escarpment slopes and a 15–20 km long fissure, in places 6–7 m deep, formed in unconsolidated soil along the eastern scarp of the Borkenna Graben (see Fig. 36, 37). There were no casualties. Thirty-four epicentres have been located from teleseismic reports; the maximum magnitude reported was 6.6 (PAS).

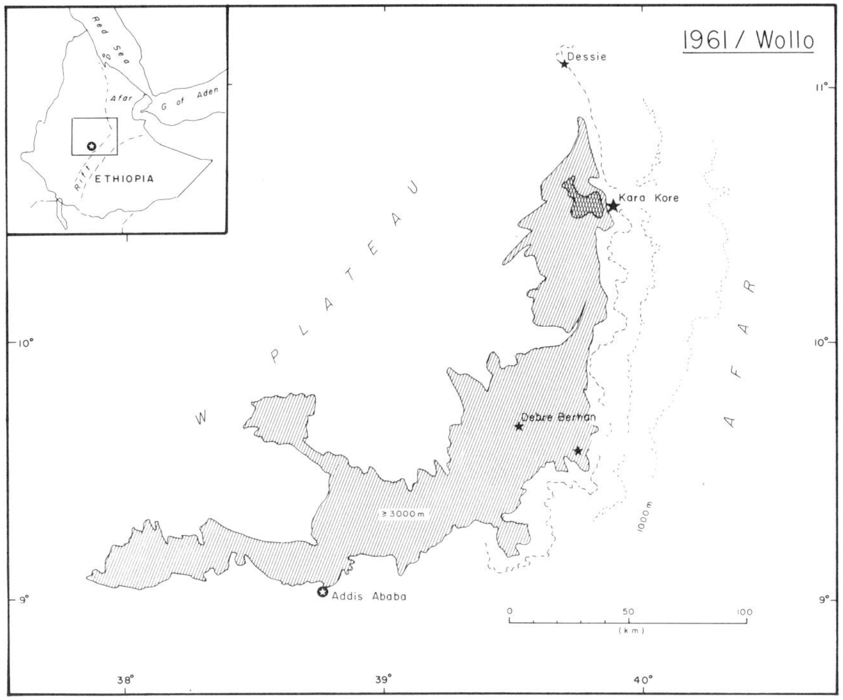

Fig. 36. Location of Serdo on the upper margin of the Plateau-Afar escarpment.

Fig. 37. Location of Kara Kore in a pass along the upper margin of the escarpment. Kara Kore's site is an example of the critical position of many Ethiopian towns located along or at the foot of steep escarpments.

Sources

AAE Data File and personal field notes; BCIS; Fairhead and Girdler (1970); Gouin (1970, 1975); Herrin et al. (1968, p. 1276); ISS; MOS; Rothé (1969); Sykes and Landisman (1964); USCGS.

Comments

1. Damage

Zone of Maximum Damage — Geographically, the area of observed maximum damage was situated along the Addis Ababa – Asmara highway between latitude N 10° and 11° and longitude E 39.7° and 39.9°. At these latitudes, Hwy. No. 1 runs along the upper margin of the eastern escarpment of the Ethiopian Western Plateau, which is dissected longitudinally by the Borkenna and Robi marginal graben(s) and transversely by strong NE-SW faulting curving in from Afar.

Description of the Damage in the Epicentral Area:

(1) To Man-Made Structures — The village of Majete (N 10.27°, E 39.56°) was a UNESCO Training Center with single-story houses built of cinder blocks joined by rather poor-quality cement (class D in Richter's classification of structures; see Explanatory Notes p. 22); Kara Kore, on the contrary, was an Ethiopian town where houses of crude masonry design (stones, mud, or poor cement, and thatch roofs) alternated with typical

Ethiopian *tukuls* built of clay (locally called *tchika*) reinforced with eucalyptus framing. Majete was totally destroyed; whereas, it was revealing to observe that in Kara Kore all the masonry houses collapsed while the *tukuls* withstood the shocks very well.

Along the highway, damage was spectacular. Large boulders from rockslides, some estimated to weigh 12–15 tonnes, blocked the road. Rubble from landslides covered the pavement in many places. Bridge pillars were fissured and parapets destroyed; cracks as wide as 60 cm and as deep as 100–150 cm were opened in the road surface; heavy slumping and subsidences with a resulting difference of some 100 cm in the surface level rendered the road impassible (see Fig. 38). All the bridges and culverts between kilometre posts 240 and 255 from Addis Ababa had to be rebuilt. (An estimation of the repair cost could not be obtained as the Highway Authority, at the time, calculated road construction and maintenance costs on a yearly basis only).

(2) Alterations in the Landscape — Numerous rockslides and landslides have been mentioned above. In addition, a piedmont scarp in unconsolidated materials opened along the escarpment of the Borkenna graben. This scarp could be followed over 12–15 km until it became obscured by rubble. In some places, the vertical differential displacement reached 2 m, the depth 5–7 m, and the width at the surface over 1 m.

Reports from Outside the Epicentral Zone — In Combolcia and Dessie, some 60 km north of the actual seismic zone, the tremors were very strong but no damage was reported.

In Asseb (N 13.0°, E 42.7°) on the Red Sea shore, tremors of intensity III were felt.

Along Hwy. No. 1 from Kara Kore to Addis Ababa, strong tremors were reported. No damage was observed except in Robi where a tobacco drying plant was destroyed and near Debre Berhan (N 09.7°, E 39.5°) where a power transformer was brought down from its supporting poles. At the Army camp near Debre Berhan, vehicles and other supplies were kept out of their shelters for many weeks.

Boulders were dislodged and rolled down the slopes of volcanoes Fantale (N 09.0°, E 39.9°) and Boseti Guda (N 08.5°, E 39.5°).

At Addis Ababa, some 200 km south of Kara Kore, the first earth tremors were felt at breakfast time on 29 May; many others were felt off and on for at least 2 weeks. Their occurrence was characterized by an abrupt, obvious, and surprising silence from birds and dogs (a rare event in Addis Ababa!) followed by a wave of rustling noises generated by the twisting of the corrugated iron roofing sheets covering most houses. Some masonry structures cracked, partition walls in reinforced concrete frame-buildings were dislocated by shearing motion, etc . . . especially along the Filohawa fault zone in the southern sector of the city. At Africa Hall, a 7-story building still under construction but whose frame-structure was completed, the steel flagpoles on the roof were seen crisscrossing during the tremors (Constructor's report).

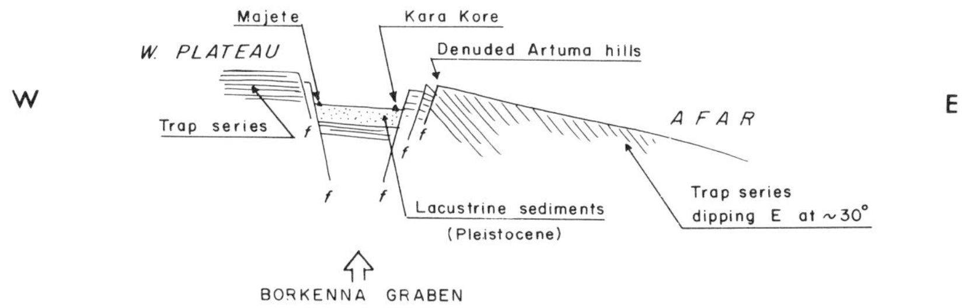

Fig. 39. Geological east-west cross section of the Plateau-Afar escarpment and upper margin at the latitude of Kara Kore, the region that was seismically active in May–August 1961.

Some inhabitants in Addis Ababa were really frightened. During the night of 1 June, all students boarding at the University College Arat Kilos campus rushed out of the residences, and students in other boarding schools behaved similarly. The manager of the Gheon Hotel queried the prudence of keeping his guests indoors; many schools requested their students to sleep out-of-doors for a while. So did many families. At the request of the Geophysical Observatory, the Ministry of Education closed the schools in Addis Ababa for 3–4 days.

At the Observatory, the seismograph units were displaced on their piers.

Figure 39 shows the geological east–west cross section of the Plateau–Afar escarpment and upper margin at the latitude of Kara Kore.

2. Location of Epicentres

Instrumental Epicentres Based on Teleseismic Data Alone — Originally, the USCGS located 34 epicentres for the 1961 seismic period of activity, which is classified in Ethiopia as the "Kara Kore Earthquakes." These 34 epicentres have been recalculated by Sykes and Landisman (1964) and by Fairhead and Girdler (1970) using two different computing techniques: Sykes and Landisman analyzed each epicentre as an independent event; whereas Fairhead and Girdler used the group location technique developed by Douglas (1967) and coded JED (for Joint Epicentre Determination) in this report.

The coordinates of the 34 epicentres are listed in Part II; their epicentre plots are illustrated in Section 2 of Fig. 40. In each plot, the centre of observed maximum intensity is indicated by an open star. Considering that the zone of maximum intensity should logically be located in the vicinity of the maximum magnitude epicentres, it appears at first sight that both the USCGS and Sykes-Landisman locations are realistic although their respective distribution patterns partially differ. The JED plot, on the other

Fig. 38. Plates 1–3. Sections of the fissure that opened in May 1961 along the escarpment of the Borkenna graben between kilometre posts 240 and 255 (old road survey) north of Addis Ababa on highway no. 1. The vertical displacement between the tips of the fissure reached 2 m in some places, the horizontal width over 1 m, and the depth 5–7 m. Plates 4–6. Rockslides and landslides at the same site along highway no. 1. Some of the boulders reached and blocked the highway (plate 6).

Plates 7 and 8. Road conditions south of Kara Kore. In plate 7, to the west of the observer, the lower section of the subsidence was about 100 cm below the normal level of the highway. Cracks as wide as 50–60 cm often ran across the pavement. Wherever the road ran along a slope, heavy slumping was observed on the shoulders. Plates 9 and 10. Condition of the parapets along the retaining walls.

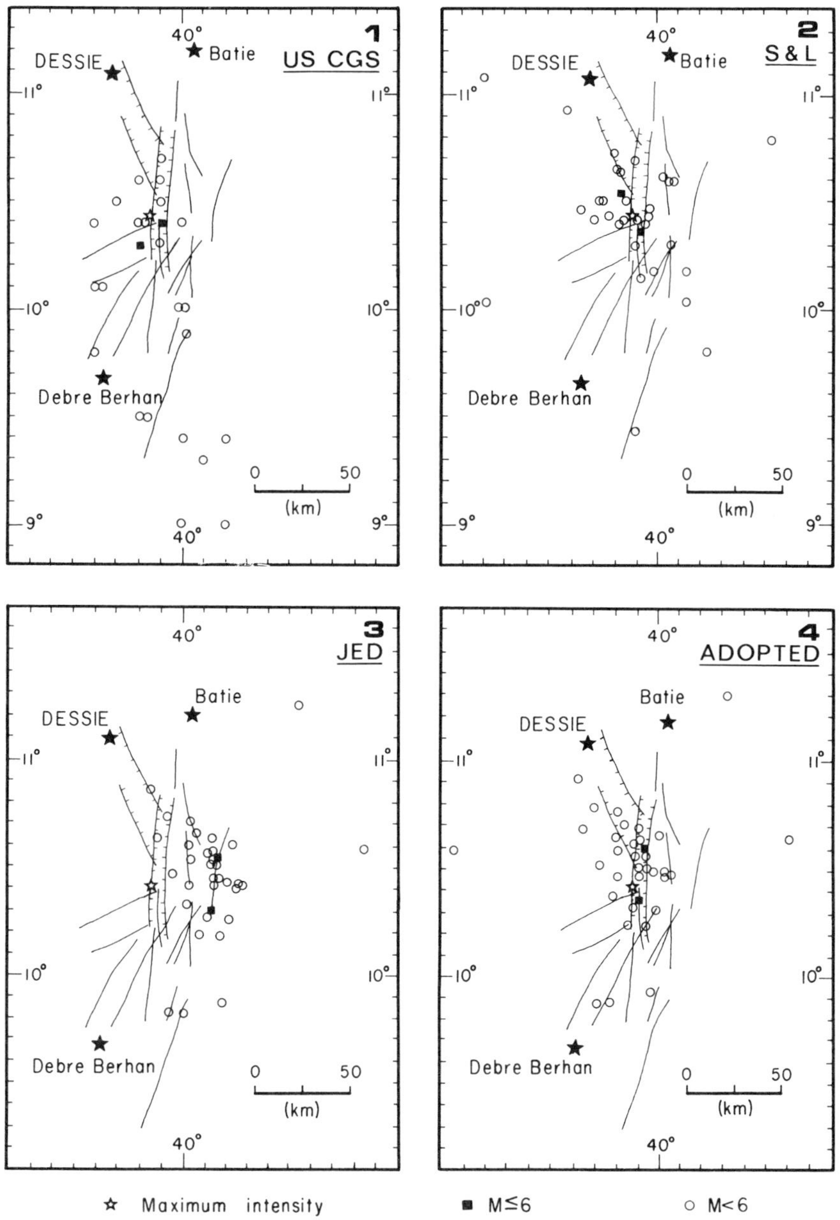

Fig. 40. Location maps of the 1961 Wollo epicentres based on teleseismic data (maps 1, 2, and 3) and the adopted locations corrected by field observations.

hand, is obviously biased toward the east with respect to the centre of observed maximum intensity. Such a bias in the absolute geographical location of a JED epicentre cluster can easily happen as its absolute location is dependent on the accuracy of a master or submaster reference event and on the tectonic regional anomalies separating the reference event from the actual zone of activity. In this case, the "restraint" reference is the epicentre of 18 April 1966, located in the Gulf of Aden, about 1000 km east of Kara Kore. Its parameters determined by USCGS and ISC are given below; they suggest the margin of incertitude that is attached to the absolute geographic location of the JED cluster.

	(1) USCGS	(2) ISC	(1–2)
H (U.T.)	08:14:18.8 ± 1.8s	08:14:22.0 ± 1.8s	
Lat. (N)	12.92 ± 0.04°	12.97° ± 0.03°	−0.05°
Long. (E)	48.31 ± 0.05°	48.40° ± 0.03°	−0.09°
h	57 km ± 17.7	84 km ± 17	−27
N	34 stations	128 stations	
Magnitude	5.4	5.1	

Readjustment of the Instrumental Epicentres — The field observations on which the readjustments in epicentral location are based were collected in three ground surveys made during and shortly after the period of seismic activity. Visual observations were complemented by extensive interviews with eyewitnesses. An air survey was twice attempted but failed due to heavy rains on the Plateau.

Taking into consideration (1) the site of observed maximum intensity at the joining of the Borkenna and Robi marginal graben(s); (2) the tectonic structures susceptible to seismic activity, such as the steep scarps of the graben(s); and (3) the macroseismic evidence provided by ground failures, an "educated" relocation compromise suggests a readjustment of the JED values by -0.06° in latitude and -0.64° in longitude. The larger correction in longitude is, in my opinion, due to the disturbed crustal conditions in Afar separating the 1961 seismic region on the Western Plateau from the Gulf of Aden where the reference epicentre of August 1966 was located.

The epicentre plot in section 4 of Fig. 40 illustrates the epicentre relocations adopted in this survey. The open star represents the observed centre of maximum intensity, the full squares the epicentres of maximum magnitude, and the empty circles 32 out of more than 3500 shocks and aftershocks.

Frequency, Magnitude, and Intensity — From 25 May to the end of September, over 3500 shocks were recorded from the Kara Kore area at the Geophysical Observatory in Addis Ababa, 200 km from the centre of activity. (Note that in 1961, the seismic station AAE was not yet equipped

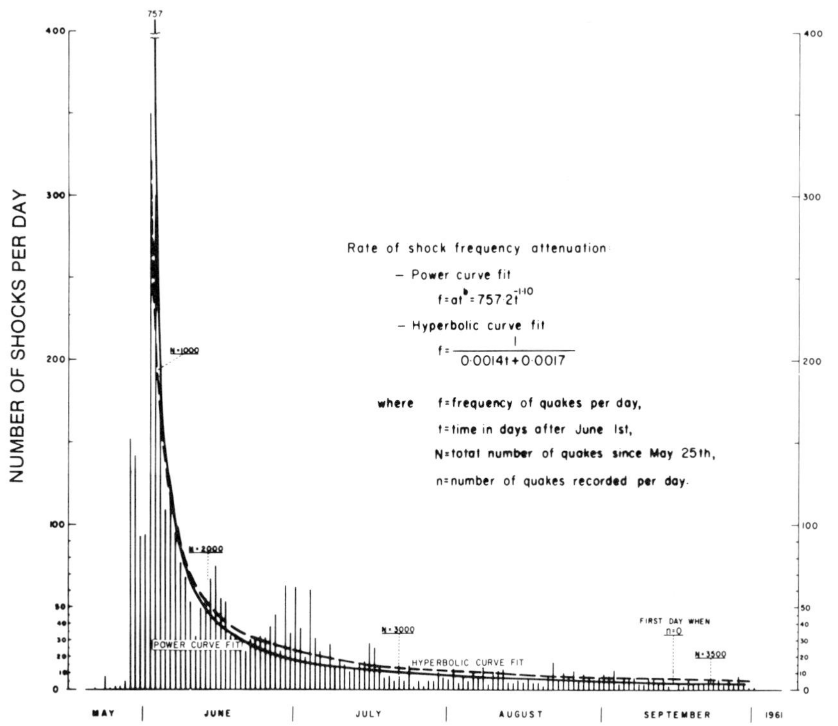

Fig. 41. Attentuation graph of the number of aftershocks in Wollo during the 1961 seismic activity at Kara Kore (recorded at Addis Ababa, summer 1961).

with calibrated WWSS equipment.) In the first series of earthquakes, which began on 29 May, the frequency reached a maximum of 150 per day; a second series began on 1 June with a peak frequency of 350 per day. If the shocks prior to 1 June are considered as foreshocks, the mean rate of decrease in the number of recorded shocks per day was about: $n = 1/(0.0014t + 0.0017)$, where t = number of days after 1 June (Fig. 41).

Out of a total of over 3500, two shocks had a magnitude $m_b \geqslant 6.4$ and seven a magnitude $m_b \geqslant 5.0$. The felt area was estimated as about 300 000 km²; relatively higher intensities were observed in the southeastern sector of the zone. The maximum intensity at the centre of the epicentral zone was estimated as VIII–IX on the Mercalli-Modified scale.

In a comparative study of earthquake sequences in Ethiopia, Dakin (1975, p. 51–70) concluded that during the 1961 Kara Kore sequence: (1) "each renewal of activity was associated with a main shock due to movement along a different, but coupled, tectonic line"; and (2) there was "indication of a 20° shift in the epicentral alignment following the second peak of activity." Such statements favour the epicentral relocation along the Borkenna and Robi graben(s) because their central axes meet at an angle of approximately 16–18°.

3. Other Geophysical Observations: Variations in the Earth's Magnetic Field

The onset of the earthquake on 29 May (iPZ 10:51:59.5; magnitude 5.5 (MOS)), was marked on the Addis Ababa magnetograms by a decrease of 20 gammas in the H-component.

Because the magnetic station is located at almost zero dip-equator, a change in the H component is equivalent to an almost equal change in the total (F) magnetic field. On 2 June at U.T. 05:44:54, under the seismic shock of M 5¾, the three magnetic traces (H, D, and Z) disappeared for $\simeq 6$ min but no permanent variation of the magnetic field was observed. On 14 June, 40 min ahead of the earthquake of M 5.7 at U.T. 20:32:18, a small $\Delta H \simeq 2$–3 gammas was observed for no good reason.

These observations are given here for future reference. For other geomagnetic observations coinciding with seismic activity in Ethiopia, see Gouin (1965, p. 541–543).

1961/VIII/25

An earth tremor strong enough (IV) to awake the students of Menelik II School in Addis Ababa at 45 min past midnight (local day 26 August) was felt in Addis Ababa. It was recorded by the Observatory at U.T. 21:43:48.9 on the 25th. The (S-P) time was 24.1 s corresponding to an epicentral distance of about 200 km. No azimuth could be determined.

Sources

AAE Data File.

1962/IX/09

An earth tremor was reported from Agordat (N 15.5°, E 37.8°) in northern Ethiopia on either 9 or 10 September. No damage was reported.

Source

Personal communication.

Comments

Agordat is not located in a seismic zone that is presently active. In the absence of other felt reports in the region, not too much importance should be given to this observation. For location, see Fig. 35.

1963/VII/14

During the evening of 14 July 1963, a shock of magnitude 4.5 (MOS, AAE), immediately followed by an aftershock, jolted the city of Asmara (N 15.5°, E 39.0°). The cinema hall at Comboni College was evacuated. The maximum intensity in Asmara was IV–V; no serious damage was reported.

Sources

Report from Comboni College; AAE Data File; BCIS for July 1963, p. 1598.

Comments

1. Two solutions based on teleseismic data are available for the main shock:

	Origin Time	Coordinates		M	Distance from AAE
BCIS	17:18:10	N 15.6°	E 39.0°		731 km
MOS	17:18:06	N 15.1°	E 38.7°	4.5	675 km

Figure 42 shows both computed epicentres at almost equal distances north and south of Asmara.

2. The two teleseismic instrumental epicentres are 56 km apart; which one is nearer to reality? Local seismograms should provide the answer. For the first earthquake, the seismic station in Addis Ababa recorded the Pn phase at U.T. 17:19:47.0 and a clear Lg at 17:21:35. The Sn phase was practically absent. Basing the computations on: (1) the arrival times of Pn and Lg; (2) the travel times for Ethiopia determined by Searle and Gouin (1971a,b), where $T_{Pn} = (4.6 \pm 0.5) + (\Delta/(7.95 \pm 0.03))$s; (3) the apparent Lg phase velocity of 3.64 ± 0.02 km/s observed during the Djebbel Dumberi (Sudan, N 12.6°, E 30.8°) earthquake of 9 October 1966, whose path to Addis Ababa was approximately the same as the present event; and (4) the origin times H_1 and H_2 calculated by Strasbourg and Moscow, respectively, we obtained the following epicentral distances (Δ) to Addis Ababa:

(1) for Pn – H_i, where i = 1,2: $\Delta_1 = 735 \pm 4$ km and $\Delta_2 = 766 \pm 3$ km;
(2) for Lg - H_i: $\Delta_1 = 746$ and $\Delta_2 = 761$ km;
(3) for Lg - Pn: $\Delta = 756 \pm 5$ km.

Note: It is fully realized that epicentral distances based on Lg arrival times are not accurate. These are used here only to ascertain some approximate distances.

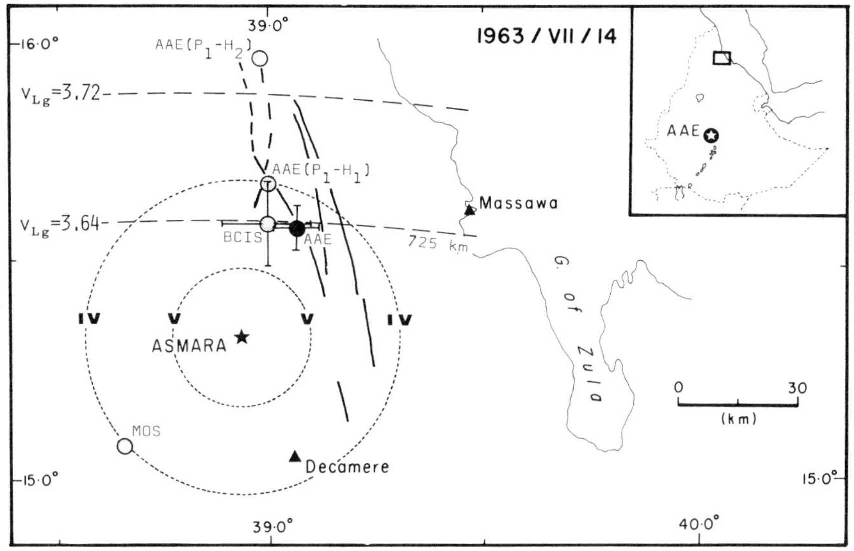

Fig. 42. Instrumental solutions and generalized felt intensities for the earthquake of 14 July 1963.

When the values given above, which range between 735 and 766 km, are compared with the teleseismic epicentral distances (731 and 675 km) knowing that the location of Asmara is about 720 km from Addis Ababa, it is obvious that all epicentre locations based on the AAE seismograms lie north of Asmara. Therefore, the Moscow solution, southwest of Asmara, is to be discarded. Had the MOS epicentre determination been valid, it would have been surprising not to have received any felt report from Decamere, which would be at the same MOS epicentral distance as Asmara. It is worth noting that, in general, the Russian solutions of epicentres in Ethiopia and in the Horn of Africa, when compared to the determinations by other international agencies, are usually biased toward the west and the south.

How does the BCIS solution, north of Asmara, compare with the AAE epicentral distances? The BCIS did not publish the standard errors attached to its solutions; it is inferred from other observations that for a BCIS solution based on 29 station reports the accuracy is certainly not any better than ± 10 km, and that therefore the epicentre is located between 715 and 735 km from Addis Ababa. The AAE solution based on Pn-H_i, gave a distance range of 731–739 km. The confidence limits of both solutions overlap between 731 and 735 km. Strasbourg's epicentre is therefore adopted, but it is moved slightly eastward toward the Sabarguma faults, a fault system that has often manifested seismic activity during the last four centuries.

1964/VII/03

Seismic activity, 25–30 km WSW of Dessie, capital of Wollo. The main shock was of magnitude m_b 5 (CGS), M_L (AAE) = 4.6. Damage of intensity V–VI was reported from Dessie. The tremors were extensively felt in Addis Ababa.

Sources

AAE Information Cards and Data File; BCIS; ISC (1964, I(2), p. 201); USCGS (PDE 56–64, p. 10); Fairhead and Girdler (1970); Rothé (personal communication, 1961).

Comments

1. Epicentral Location of the Main Shock

One local and five teleseismic solutions are available for the main shock of 3 July:

	H	Coordinates		Δ°	Az°	h(km)
AAE		N 11.0°±0.1°	E 39.4° ±0.1°	2.1	17	
BCIS	19:18:32	N 11.2°	E 39.8°	2.4	25	
JED	19:18:33.0	N 11.32°	E 39.57°	2.4	19	
ISC	19:18:32.7±1.8	N 11.05°±0.18°	E 39.7°±0.15°	2.2	24	54±19
Rothé	19:18:34	N 11.00°	E 39.30°	2.0	15	
USCGS	19:18:34.0±1.5	N 11.01°±0.08°	E 39.28°±0.08°	2.0	14	60

The above values are plotted on Fig. 43.

The AAE epicentral location is based on the (S-P) time interval of 29.0 recorded at Addis Ababa and on the empirical formula, in which $\Delta° \simeq (t_{(s-p)} - 4.0)/12$, deduced from 86 recent local shocks recorded at AAE and at 25–50 other stations. The results indicate an epicentral distance of 2.1° or about 230 km at an azimuth 017° from Addis Ababa, and a location about 25–30 km WSW of Dessie where intensities of IV–V were observed (assessment of the damage in Dessie was made by the Engineering Department of the Commercial Bank of Ethiopia).

Concerning the distance from Addis Ababa, the teleseismic ISC, Rothé, and USCGS solutions are more realistic than the BCIS and JED (Fairhead and Girdler) solutions. For computations, the AAE epicentral solution has been adopted. The M_L (AAE) = 4.6 corresponding to m_b 4.2.

Tectonically, the epicentre is located in a zone of transition where the marginal grabens, the Menebay-Hayk to the north and the Borkenna to the south, are dextrally displaced by 25 km with respect to one another.

2. Pattern of Seismic Activity on 3–4 July 1964

The main shock, which occurred at U.T. 19:18.5 on 3 July, was preceded by insignificant foreshock activity and followed within 26 h by 8 aftershocks ranging in magnitude M_L (AAE) between 1.9 and 3.2. The (S-P) time intervals were restricted to between 28.5 and 29.0 s indicating a

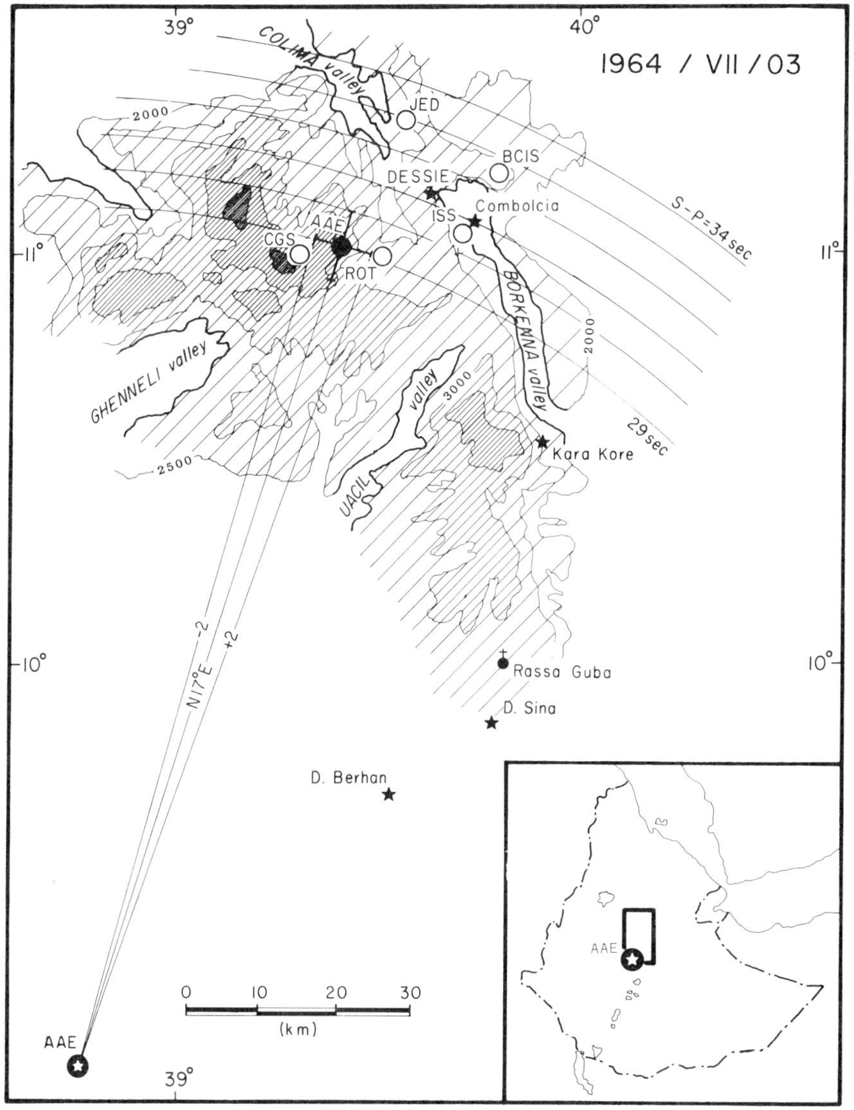

Fig. 43. Instrumental locations for the epicentre of 3 July 1964 in Wollo. The arcs indicate epicentral distances from AAE based on S-P time intervals. The elevation contours are 500 m apart.

seismically active area of about 0.02° (≤2.5 km) in radius.

The energy released by the eight aftershocks corresponded to the energy released by a single shock M_L= 3.4.

1964/X/17

Earth tremors of intensity III were felt in Addis Ababa around eight o'clock in the evening of 17 October 1964.

Source
AAE Data File.

Comments
1. Two isolated shocks, from an epicentral distance 1.4–1.5° and azimuth ≃ N40°E, were recorded by AAE at U.T. 17:41:45.9 and 18:03:28.0, respectively. Their M_L (AAE) magnitudes were about three. Figure 44 shows a common epicentral location at N 10.20°, E 39.75° (±0.2°). The region is situated on the top of the Plateau-Afar escarpment; it is heavily faulted and appears to be the focus of important cross-rift structures.

2. Apparently, the intensities observed at Addis Ababa were anomalous. For an epicentral distance of 165–170 km, an intensity III under the best propagation conditions suggests a magnitude ⩾4.5, not 3.0 as registered at AAE. I do not want to discuss this point because of the scarcity of the data, but experience has shown that earth tremors caused by earthquakes located along the upper margin of the Plateau-Afar escarpment have often been, in the southern sector of Addis Ababa, of higher intensities than expected.

1965/IV/14

At midmorning on 14 April 1965, earth tremors were felt in Dessie (N 11.1°, E 39.6°). No damage was reported.

Source
AAE Data File.

Comments
AAE recorded the iPn at U.T. 07:53:10.8 and the Sn 30.2 seconds later. The apparent epicentral distance from Addis Ababa was about 260 km in a NE direction. M_L (AAE) was estimated as 4.3; m_b 3.9. On 3 June 1964, USCGS located an epicentre of m_b 5.0 at N 11.0°, E 39.3°; Fairhead and Girdler (1970) relocated it at N 11.32°, E 39.57°. For that event of 1964, the S-P time interval at AAE was 29.0 s as compared with the present S-P = 30.2 s. Both events certainly originated from the same region, slightly west of Dessie. The adopted epicentre for the 1965/IV/14 earthquake is N 11.3°, E 39.6°.

1965/VI/29

An earth tremor was felt in Asmara (N 15.5°, E 39.0°) during the night of 29 June 1965.

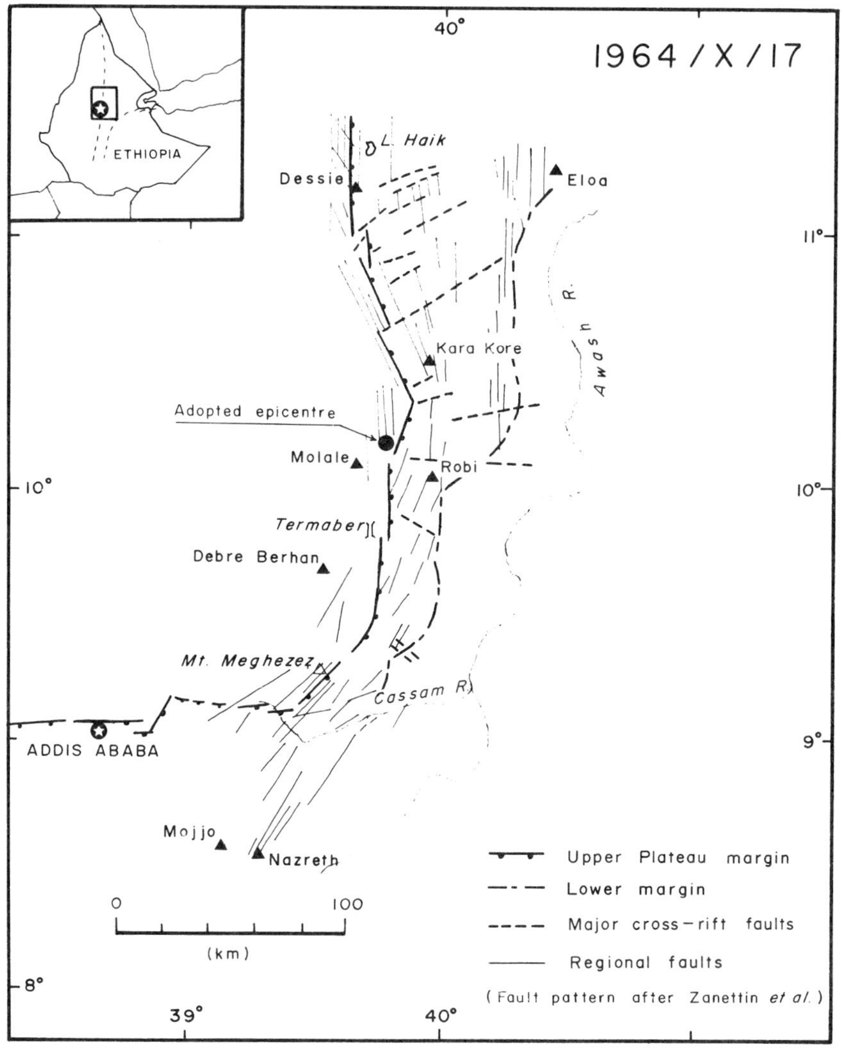

Fig. 44. Location of the two epicentres of 17 October 1964. (The schematic tectonic outline is after Zanettin and Justin-Visentin 1975 and from an unpublished photo-geological map by P. Tacconi.) A topographic outline of the region is presented in Fig. 43.

Source

AAE Data File.

Comments

AAE recorded the event at U.T. 22:04:33.3. The magnitude M_L (AAE) is indicated as 4.2, m_b 3.8, for an epicentral distance of 780–790 km almost due north of Addis Ababa. A second shock followed at (eP) U.T. 22:30:23; its m_b was 3.4.

1965/IX/30

Three earthquakes of M_L (AAE) ranging between 4.0 and 4.5 occurred along the edge of the Ethiopian Plateau-Afar escarpment at an epicentral distance of 420 ± 10 km from Addis Ababa. Tremors were reported in Mai Chew, Korbetta, and Mekele; slight damage occurred in Mai Chew and Korbetta (N 12¾°, E 39¾°).

Sources

Provincial Police reports; AAE Data File.

Comments

The three events were recorded at AAE, at U.T. 19:00:19, 21:23:55.6, and 22:18:45, respectively. The computation of the three epicentres gave the same location N 12.7°, E 39.7°, a few kilometres SE of Mai Chew. The epicentral distance from AAE is 420 ± 10 km on the western scarp of the Guf Guf marginal graben along the Plateau-Afar escarpment.

As a check on attenuation rates, two circles are drawn around the epicentre indicating intensity III for rift and plateau attenuation rates (see Fig. 45). If rift attenuation rates are used and intensity III considered as the threshold of perceptibility, then Mekele should not have felt the tremors, but it did. Therefore, at least for the sector east of the escarpment, plateau attenuation rates are more appropriate.

1966/II/05

During the early morning of 5 February 1966, an earth tremor was felt in Asmara (N 15.5°, E 39.0°). Windows rattled and plaster fell from ceilings in some of the older houses. The estimated intensity was IV.

Sources

Comboni College and U.S. Mapping Mission personnel in Asmara.

Comments

The P-onset of the earthquake was recorded in AAE at U.T. 04:20:10. The magnitude was m_b (AAE) 3.8 and the apparent epicentral distance 785 km, that is, about 65 km further away from Addis Ababa than Asmara. No precise azimuth could be estimated.

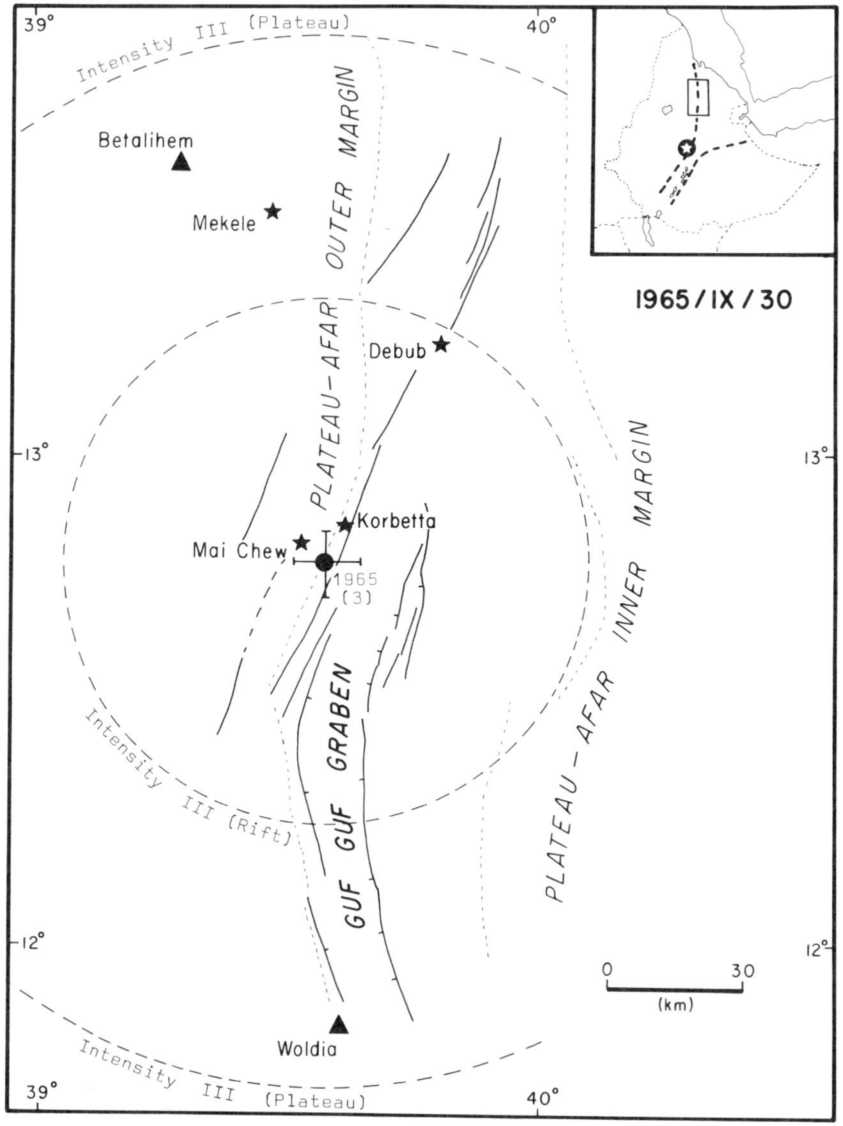

Fig. 45. Location of the three epicentres of 30 September 1965 near Mai Chew. The dashed lines indicate the threshold of perceptibility (intensity III) for the two rates of energy attentuation observed in Ethiopia.

1967/I/23

An earth tremor of intensity III-IV was felt in the Ambo-Lekempt region around 08:30 a.m. on 23 January 1967. No damage was reported.

Sources

AAE Data File and Information Cards; BCIS; IRSAC/LWIRO and LUANDA Bulletins.

Comments

For this event, the following station reports are available:

	Arrival times		
AAE	iP 05:33:19.1	iS 05:33:35	
NAI	eP 05:35:38.0	eS 05:37:54.5	L 05:38:14.5
AST	iP 05:36:10.4	05:38:59.9	05:39:14.6
LWI	iP 05:36:14.4	05:36:29.1	05:39:05.5
BNG	iP 05:37:24.7	S 05:41:29.2	
Delcommune	05:38:01.3		
SDB	L 05:52.00		

The AAE solution based on the crustal model of Berckhemer et al. (1975) yielded the following parameters: H:05:32:59.6; N 08.67°, E 37.65°; $\Delta°$ 1.17°, 130 km; Az 254°; M_L (AAE) 4.1. It is identified by a solid square on Fig. 48. A solution based on the teleseismic data listed above has not been attempted.

For comments on regional travel times and geologic structures in the Ambo-Guder region, see entry 1968/I/23.

1967/I/30

An earth tremor was reported at Asmara during the evening of 30 January. The Cinema Hall at Comboni College had to be evacuated. The estimated intensity was IV.

Source

AAE Data File; Report from Comboni College.

Comments

The shock was recorded at AAE at U.T. 16:28:35.7. The apparent epicentral distance from Addis Ababa is between 725 and 775 km almost due north of AAE; the estimated M_L (AAE) 4.2; m_b 3.8. Phases are not recorded clearly enough to justify a more accurate estimate of the epicentre.

1967/IX/15–18

There was seismic activity along the Sabarguma faults in Eritrea, north of Asmara. Two shocks were reported from Asmara (N 15.5°, E 39.0°) and Decamere (N 15.1°, E 39.0°): the first at 01:00 during the night of 15–16 September; the second at 05:00 on 18 September. None of the tremors caused any damage.

Sources

AAE Data File and Information Cards; Fairhead and Girdler (JED) 1970; ISC; USCGS (EDR 67-67); Mobil Oil Office in Asmara (personal communication from Mahdi M. Shibru).

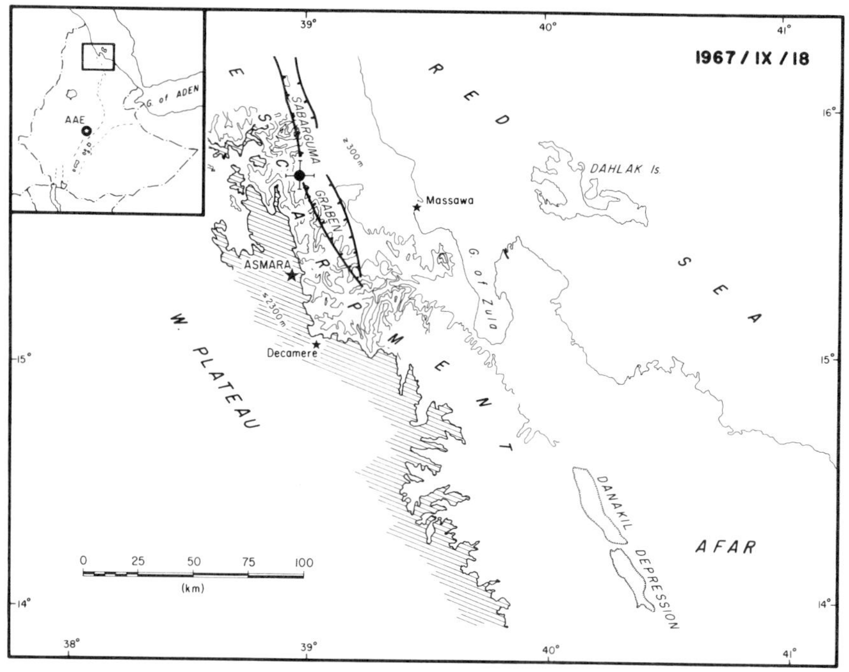

Fig. 46. Location of the mean instrumental location (●) of the earthquake of 18 September 1968 along the Sabarguma graben at the foot of the Eritrean escarpment.

Comments

1. Time of Occurrence and Epicentral Locations

On 15 September the following phases were recorded at AAE: iPn U.T. 22:04:07.6; Sn U.T. 05:21.0; and Sg/Lg U.T. 05:41.6. Using the same computation formula as in the following event, we obtained an epicentral distance of 735 ± 5 km north of Addis Ababa (see Fig. 46).

For the earthquake of 18 September the USCGS indicated an origin time of H = 02:02:59.8 ± 0.3 and a magnitude m_b 4.8. Three teleseismic solutions are available for the epicentre:

	Coordinates		From AAE			
			Δ°	Δ(km)	Az°	h(km)
ISC	N 15.78	E 38.94	6.8	751	1.4	58
JED	N 15.80	E 38.93	6.8	753	1.3	
USCGS	N 15.69 ± 0.08	E 39.03 ± 0.10	6.7	741	2.2	33
Mean	N 15.76 ± 0.06	E 38.97 ± 0.06	6.8	748	1.7	

The epicentral locations of both earthquakes indicate that the western scarp of the Sabarguma graben was active over a length of some 15 km. For earlier activity, see entry 1912–13 (27 February 1913).

2. Transit Times

The three teleseismic solutions given above agree so well (within ± 6 km) that they are used here to determine a regional transit-time equation for seismic waves originating in Eritrea and following the Plateau-Afar escarpment path where the crust-mantle interface steeply dips westward and along which a travel-time delay has often been observed. From the phases recorded at AAE and the mean distance (748 km) from the teleseismic data, we obtained the following empirical formula for one set of arrivals (Pn and Sn):

$$\Delta(\text{km}) = \frac{(\text{Sn}-\text{Pn})-1.3}{0.0984}$$

as compared with the following equation:

$$\Delta(\text{km}) = \frac{(\text{Sn}-\text{Pn})-3.636}{0.0984}$$

obtained for Ethiopia in general (Searle and Gouin 1971a,b).

3. Simultaneous Occurrences

The 1967 seismic activity along the Sabarguma graben coincided with strong activity in the Gulf of Zula, in the NE sector of the Danakil Depression, and at the north end of the Danakil Alps. There is no way of knowing if there was any seismic coupling between the three zones.

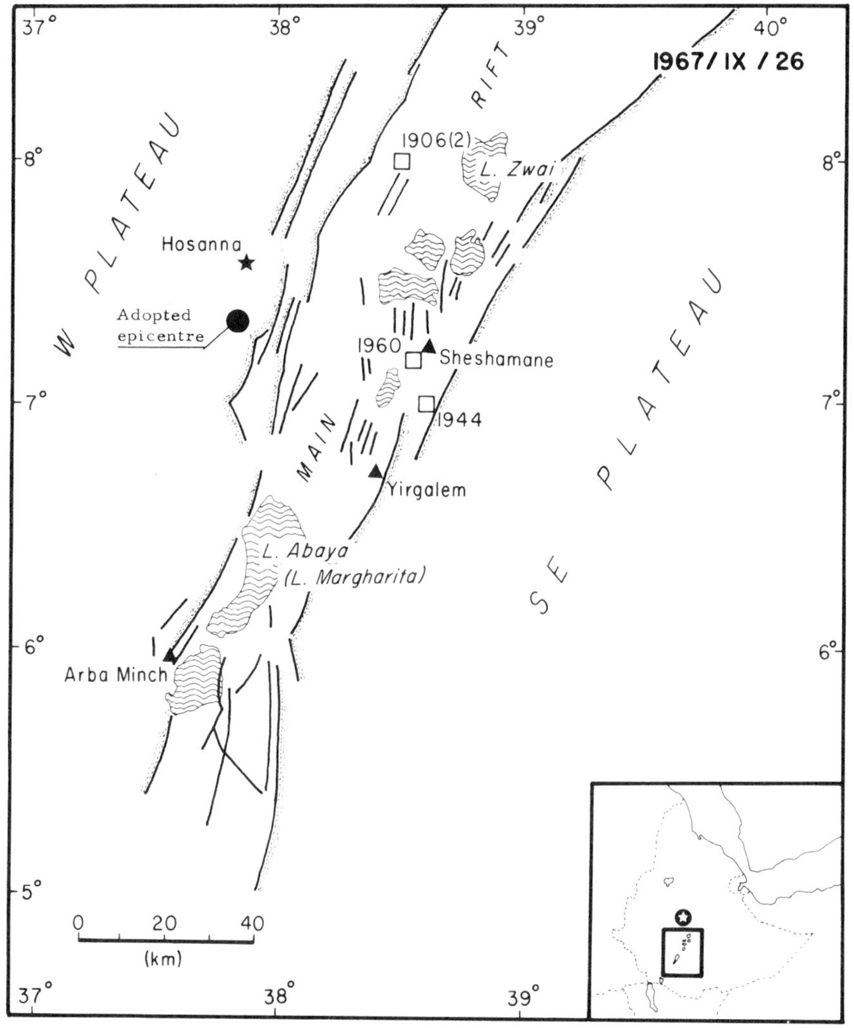

Fig. 47. Instrumental location of the 16 September 1967 epicentre.

1967/IX/26

During the evening of 26 September 1967, an earth tremor was felt in Hosanna (N 07.6°, E 37.9°), a town located on the western margin of the rift-valley escarpment.

Sources

AAE Data File and Information Cards

Comments

The readings at AAE show an iP phase at U.T. 17:21:58.9 and an iS at 17:22:23.4 giving an apparent epicentre at N 07.34°, E 37.84°, epicentre 208 ±10 km at 210° from Addis Ababa with M_L(AAE) 2.7.

The azimuth (210°) was controlled by the intensity III reported in Hosanna. The range of intensity III for a shock of magnitude 2.7 is about 10–15 km. Both the 208-km distance from Addis Ababa and the 15-km intensity radius centred on Hosanna intercept at N 07.34°, E 38.84°. The adopted epicentre location is N 07.3°, E 38.8° (Fig. 47).

1968/I/23

At about 10:20 p.m., 23 January 1968, an earthquake was reported from almost every locality in southwestern Ethiopia between N 06.9–11.0° and E 34.8–38.8°. The instrumental epicentre was located at a distance of 125–130 km WSW (250°) of Addis Ababa. The magnitude m_b of the shock was 5.1 (CGS), 4.9 (ISC). Some typical reported intensities were:

Intensity	Site	Epicentral distance
V	Ambo	45 km
IV	Addis Ababa	125–130 km
III		Radius of 250–275 km

Sources

AAE Data File and Information Cards; BCIS (1968, p. 401–402); ISC (1968, no I/736, p. 306); USCGS (EDR 6-68, p. 39); Fairhead and Girdler (1970).

Comments

1. Epicentral Location Based on Teleseismic Records

Three teleseismic solutions are available:

	H	Coordinates	
CGS	19:18:13.0 ± 0.3	N 08.71 ± 0.04°	E 37.66 ± 0.05°
ISC	19:18:14.4 ± 0.4	N 08.69 ± 0.05°	E 37.41 ± 0.07°
JED	19:18:14.7	N 08.74 ± 0.03°	E 37.56 ± 0.04°

The standard errors indicated for JED coordinates are those of the master event used in the computations; they should be considered as minimum s.e. values. The ellipses of the standard errors for the three solutions practically overlap and define a maximum probability epicentral region of 35 km², centred at N 08.71 ± 0.03°, E 37.53 ± 0.12°, slightly southwest of the point where the Guder transverse lineaments (Mohr 1971c, p. 33) crisscross the Ambo faults at almost a right angle.

Note that, as should be expected, the axis of maximum probable error (0.12° versus 0.03°) in epicentral determination is E-W in the Ambo region, i.e. parallel to the axis of the Ambo faults; whereas, it is observed to be N-S along the Wonji Fault Belt and the Plateau-Afar escarpment.

2. Energy Attenuation Rates

The two large circles around the epicentral region (Fig. 48) indicate the ranges of perceptibility computed for rift (solid line) and plateau (dashed line) crustal structure conditions. The formulae used are those adopted for

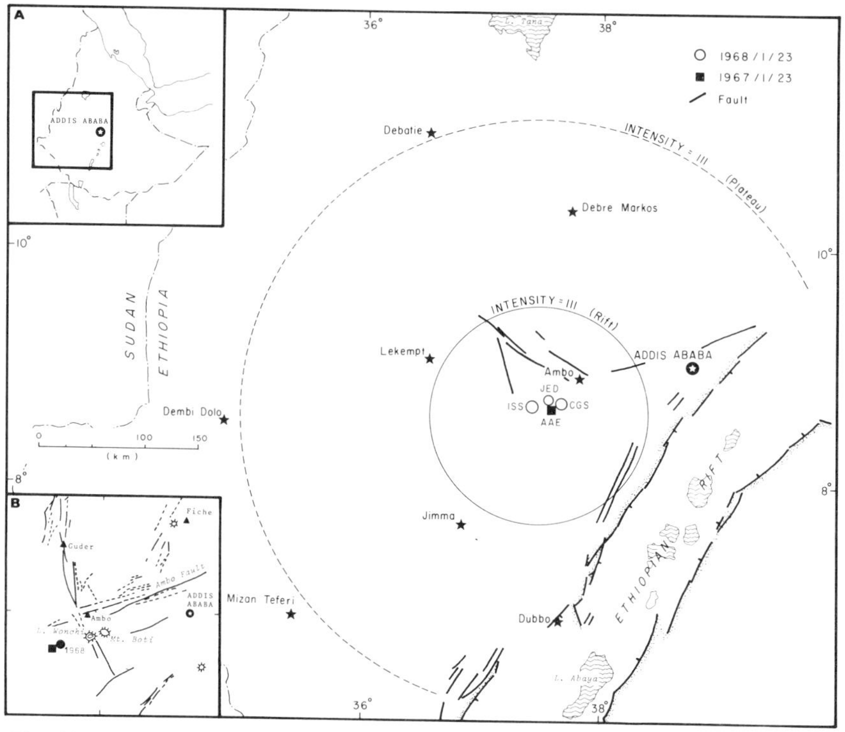

Fig. 48. Epicentral location of 23 January 1967 (○) and of 23 January 1968 (■) near Ambo. The star symbols (★) indicate the sites of reported intensities. Inset B outlines the tectonic setting of the Ambo region. The intensity III range (dotted line) is used in the computation of the energy attenuation rates under the Plateau.

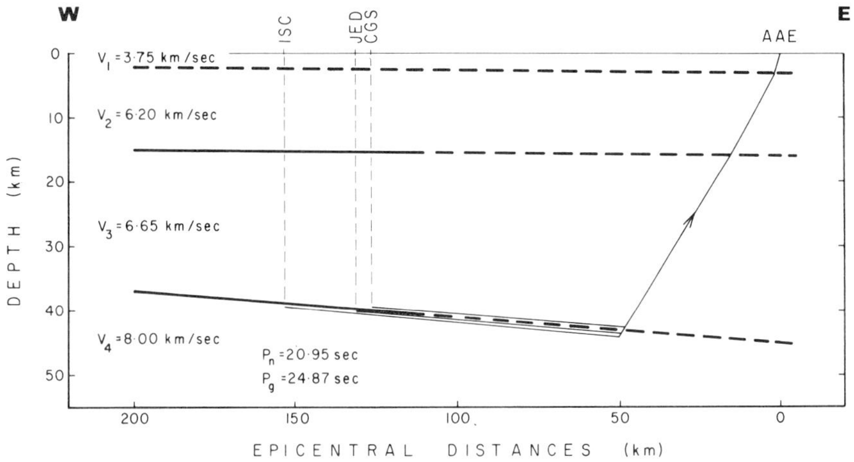

Fig. 49. Crustal structure model from Addis Ababa westward through the 1967 and 1968 seismic region near Ambo. Thick lines marking interfaces are documented data; dashed lines are interpolations (adapted from Berckhemer et al. 1975).

Ethiopia (Gouin 1976, p. 19–22). The intensities reported from nine sites well distributed in azimuth leave little doubt that the energy attenuation rates observed under the southern part of the Ethiopian Plateau are almost identical to the attenuations observed under Eastern Canada (Milne and Davenport 1969) from where, as a first approximation, a family of curves was adopted.

3. Crustal Structure, Empirical Travel Times, and AAE Seismograms of 23 January 1967 and 1968.

The high accuracy in location of the present epicentre computed from teleseismic reports and from the local AAE seismograms offers the opportunity of checking the validity of the regional travel time equation observed by Searle and Gouin (1971): $T_{Pn} = (4.6 \pm 0.5) + (\Delta/(7.95 \pm 0.03))$seconds (Equation 1). For this test, we used the regional crustal model obtained by Berckhemer et al. (1975, Profile I) from deep soundings conducted in 1972.

Such a validity test will allow an assessment of which earth model, among those used by international agencies, best fits the Ethiopian records.

The elements considered are the following: (1) the original times ($H_{1,2,3}$) supplied by the respective agencies; (2) the iPn (AAE) onset time at U.T. 19:18:32.5; (3) the epicentral distances ($\Delta_{1,2,3}$) derived from (1) and (2); (4) the Pn travel times: Pn (AAE) – $H_{1,2,3}$ given in the table below; (5) the Pn travel times obtained by equation 1 for the respective $\Delta_{1,2,3}$; (6) the Pn travel time and epicentral distance fitting Berckhemer's crustal structure (Fig. 49); and (7) the depth h = 38 km as per Fig. 49. Some relevant parameters among these are listed below.

Agencies	Distances Δ (km)	Pn travel times (s) Pn-H	Equation 1	Berckhemer's model
CGS	126.5 ± 5.6	19.5 ± 0.3	20.5 ± 0.5	19.7
ISC	153.6 ± 6	18.1 ± 0.4	23.9 ± 1.2	23.0
JED	136.3	17.8	21.7 ± 1.4	21.0
AAE	125 ± 5	—	20.4 ± 1.2	19.7

At first sight, it is apparent that the epicentral distance (Δ_1), origin time (H_1), and T_P travel time (iPn-H_1) given by the USCGS best fit Berckhemer's regional crustal model. This was to be expected as equation (1) was almost exclusively derived from USCGS information. It follows, however, that up to now, and until more regional information is available, the USCGS earth model is the one that best agrees with the AAE seismograms for earthquakes originating on the Plateau. The agreement is not so good for the events originating in the Gulf of Tadjoura (see Region D).

4. Tectonic Setting in Epicentral Region

The two important features of the Ambo region are the Addis-Ambo-Ghedo fault zone and the Guder lineaments crossing one another almost at a right angle near the town of Ambo (Hagere Hiwot). The major fault of the Addis-Ambo-Ghedo zone, often referred to as the Ambo faults, runs NNE-SSW (070°-072°), is downthrown to the south, and is lined with thermal springs. The Ambo-Ghedo faults have been described as the border fault zone between the Amhara and the Kaffa highlands. To the south of Ambo town, on or near the Guder zone, are the young Wonchi crater lake and the Bati volcano.

As shown on the map (open circle for the 1968 earthquake; solid square for 1967), the two epicentres are located some 20 km SW of the intersection of the fault zones.

Available tectonic maps of the Ambo region do not agree on the exact location and orientation of the Guder geological structures; as reference, inset B on Fig. 48 shows them as they appear on Landsat imagery (Mohr 1974a). Despite secondary discrepancies in the available descriptions of the fault pattern, the fact remains that the area is tectonically disturbed and at present, seismically active. For more details on the geology of the Ambo region see Smeds (1964), Baker et al. (1972), and Burke and Whiteman (1973).

5. Recent Seismicity of the Ambo Region

During the 6 h that followed the main shock of 23 January 1968, three aftershocks of lesser magnitude were recorded at AAE. Their phases were not recorded clearly enough to warrant any accurate location.

Exactly one year earlier, on 23 January 1967, an earthquake of M_L(AAE) 4.9 occurred in the same region. Its location is identified by a solid square on Fig. 48 (see entry 1967/I/23).

1968/V/11

An earth tremor was reported from Decamere (N 15.1°, E 39.1°) during the night of 11 May 1968. Some were awakened by the shock; no damage was reported. The estimated intensity was IV.

Source

AAE Data File

Comments

AAE recorded two earthquakes of M_L(AAE) 4.6, m_b 4.1, on 11 May at U.T. iP 18:17:17.6 and eP 20:17:38 (9:17 and 11:17 p.m.) with S-P time intervals of 65.8 and 70.0 s, respectively. The apparent epicentral distances were 625 and 670 km almost due north of Addis Ababa; the geocentric distance from AAE to Decamere is 676 km.

Adopted epicentres for computations: N 14.4°, E 39.8° and N 15.0°, E 39.8° along the Plateau escarpment.

1969/V/16

An earth tremor was reported in Adigrat (N 14.3°, E 39.5°) at sunset on 16 May 1969. No damage was reported.

Source

AAE Information Cards.

Comments

No trace of this tremor was recorded at AAE.

1969/XII/20

Lasa Center (LAO) reported an epicentre at (N 10.8°, E 37.6°) on 20 December 1969, at U.T. 07:51:33. This is no doubt a misinterpretation of data originating from a single autonomous array. No trace of such an earthquake appears on the AAE seismograms.

Sources

ISC 1969(2), No. 12/690, p. 279, quoting LAO.

1970/VII/16

Lasa Center erroneously reported an earthquake in Wollo on 16 July 1970 at U.T. 07:13:47, 85 km ENE of Addis Ababa. The published coordinates were N 08.8°, E 39.5°. This event is not confirmed by AAE records.

Sources

ISC 1970(2), No. 7/423, p. 230.

1971/VI-VIII

From 5 June to 4 August 1971, swarms of microearthquakes of M_L (AAE) $\leqslant$ 3.8 occurred almost continuously along the upper margin of the Western Plateau in the region of Debre Berhan (N 09.7°, E 39.5°). The radius of the epicentral region was estimated to about 30 km. Slight damage was caused in Debre Berhan on 4 August.

Sources

AAE Data File; Dakin (1975, p. 51–70) and personal notes.

Comments

1. For 60 days during the summer of 1971, 151 microearthquakes of M_L (AAE) $\leqslant$ 3.8, $m_b \leqslant$ 3.4, were identified by Dakin (1975) as originating at an epicentral distance of about 1° and azimuths varying between 052° and 063° from Addis Ababa. One of the last events of M_L (AAE) 3.8 caused light damage in the town of Debre Berhan; Miss Dakin personally surveyed the seismic region and plotted the isoseismic map reproduced in Fig. 50. The SW-NE elongated intensity distribution pattern illustrates the positive influence of the Plateau marginal fault zone on the amplitude of ground shaking.

2. The seismic activity near Debre Berhan during June–July 1971 is a typical example of an "earthquake swarm": (1) when the number of shocks increases and then decreases with time (Fig. 51); and (2) when no event is of magnitude predominant enough to be singled out as the main shock of the sequence. Mogi (1963, p. 644) adds that the number of events in that type of sequence should exceed 10 and that the ratio of the maximum daily number of earthquakes to the square root of the time duration should exceed two. During the 1971 Debre Berhan swarm, the number of events recorded at epicentral distances of 100–120 km from Addis Ababa was 151, the maximum daily occurrence of shocks with $M_L \geqslant 1.5$ reached 3.1, and $N_{max}/\sqrt{t} =$ 4.0.

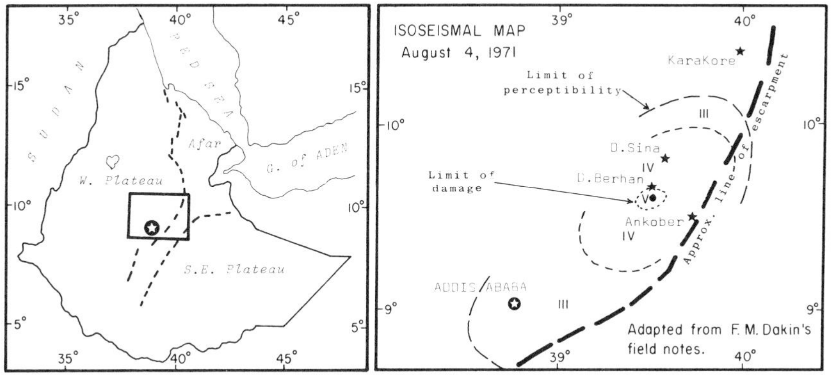

Fig. 50. Location and intensity distribution of the m_b (AAE) 3.4 or M_L 3.8 earthquake on 4 August 1971.

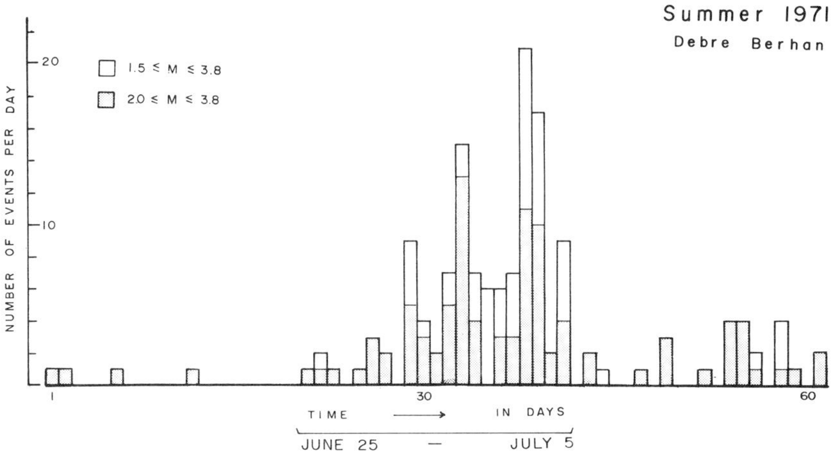

Fig. 51. Daily frequency of occurrence of the shocks recorded at Addis Ababa during the Debre Berhan swarm of the summer of 1971.

Figure 51 gives the frequency versus time histograms for the events of $M_L \geqslant 1.5$ and $M_L \geqslant 2.0$.

Mogi explains that this type of earthquake sequence occurs in highly fractured regions. He reasons that as the applied stress gradually increases, a high stress concentration appears around numerous cracks and faults; local fractures begin to occur but as they are already under a low stress this prevents any single predominantly large fracture from developing (Mogi 1963, p. 649). Such a geological setting of highly fractured and heterogeneous structures is of common occurrence in the Plateau marginal fault zone (see Fig. 44).

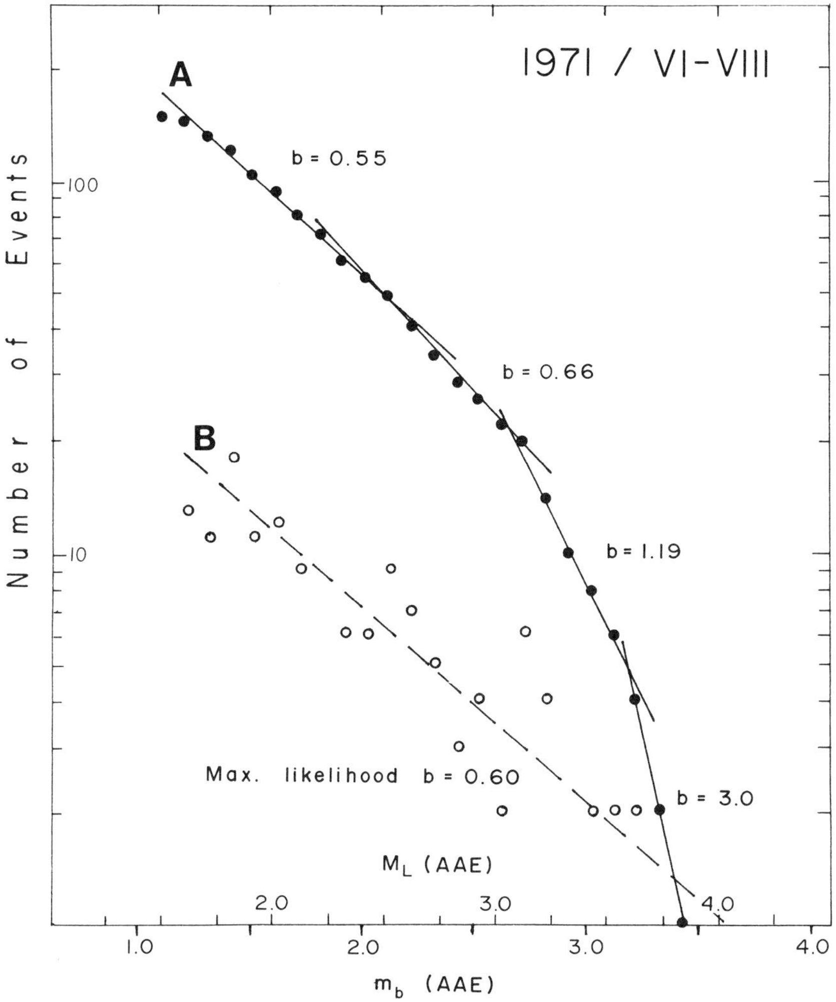

Fig. 52. Frequency-magnitude relations for the 1971 swarm in Debre Berhan. The four segments of curve A indicate the least-squares best fit to the data; curve B was obtained by the maximum-likelihood method.

One interesting observation that comes out of the cumulative frequency-magnitude distribution graph (Fig. 52) is the marked discrete increase of the b-values with increasing magnitudes. Four legs are clearly discernible in the frequency-magnitude curve. The regression slopes of the curves defined by the classic frequency-magnitude relationship of Gutenberg and Richter (1954) ($\log_{10} N = a - bM$) were determined by two different techniques: (1) by the conventional least squares best fit technique; and (2) by the method of maximum likelihood, for which Utsu (1965) suggests that the b-coefficient should be considered as: $b = (\log_{10} e)/(M - M_0)$, where: $M = \sum_{i=1}^{n'} M_i / n'$ is the mean magnitude in a class of earthquakes; n' is the number of events larger than M_0; and M_0 is the lower limit of the lowest magnitude class considered. If the magnitude values are given at intervals ΔM and the central value of the smallest magnitude class is M_0 then $M_0 = M_0 - \Delta M/2$.

Merits and shortcomings of both techniques for determining the b-coefficient in Gutenberg-Richter frequency-magnitude relationships are not discussed here. (A relevant reference on the subject is the controversy between Duda, Rashidi, and Gibowicz in Acta Geophysica Polonica, XXII(2), 1974, p. 201–210.) Because the least square method assumes a normal distribution of the events about the mean and the maximum likelihood method assumes various distributions, it follows that the curve slopes should be identical only in the case of a normal distribution; they exhibit substantial differences in other cases. In the present report, it is always specified by which technique b-values were calculated. During the 1971 Debre Berhan swarm, the regression line for the cumulative frequency curve brakes up into four segments with respective b-values of 0.55, 0.66, 1.19, and 3.0. The maximum likelihood curve has a mean slope of 0.60. What does the observed variation in b represent? Bath (1973, p. 146) suggests a variation in the mechanism of energy release at different levels.

3. Dakin (personal communication) described the characteristics of the seismic traces from the Debre Berhan swarms as follows: first P motions Z↑, NS↓, EW↓; three distinct body waves phases Pn, Sn, and Sg with time intervals of Sn – Pn of about 17 s and Sg–Pn of about 21 s; Sg is the largest horizontal sharp motion; and the maximum trace amplitude is larger on the NS component than on the EW.

1971/VIII/14

During the night of 14 August 1971, a tremor of intensity III was reported from Mizan Teferi (N 06.9°, E 35.5°).

Sources

AAE Information Cards and Data File.

Comments

The event was poorly recorded at AAE. No clear P onset; an S phase at U.T. 19:19:07.5. No epicentral location was attempted. For the location of Mizan Teferi, see among others Fig. 48.

1971/XI/13–23

There was a period of seismic activity in Wollo, south of Dessie. The main shock was of magnitude m_b 5.5 (mean of eight determinations). Intensities up to IV–V were reported from Wollo and Shoa but they should not be considered as the maximum intensities. The instrumental epicentre determination is questionable.

The seismic activity above magnitude M_L (AAE) 1.7 lasted until the 29th; two foreshocks and 50 aftershocks were identified on AAE seismograms.

Sources

AAE Information Cards and Data File; BCIS (1971); ISC (1971, p. 99); USGS (EDR 83-71, p. 42–43); Seismological Bulletin 1971 of F.R. Germany.

Comments

1. Epicentral Location of the Main Shock

Teleseismic Solutions — Four teleseismic solutions are available for the main shock of 13 November:

	Coordinates		△ (km)	A_z°	h(km)
BCIS	N 11.0°	E 39.8°	247	27	
ISC	N 11.03 ± 0.03°	E 39.71 ± 0.04°	245	25	39 ± 13
MOS	N 11.0°	E 39.5°	233	20	15
USGS	N 10.97 ± 0.03°	E 39.70 ± 0.04°	238	25	24 ± 19

The four agencies located the epicentre in the transition zone between the southern end of the Menebay-Hayk marginal graben that runs between latitudes N 11.75° and 11.17° and the northern end of the Borkenna graben at N 10.9° (Fig. 53). The axes of both grabens are dextrally displaced 25 km relative to one another and the offset is marked by large transverse faults, upthrown to the SSE (Mohr 1975, p. 86).

Local Anomalies — Despite perfect agreement in the teleseismic solutions (Δ = 241 ± 6 km; A_z = 24 ± 3°), local observations do not completely concur with the computed location.

AAE Seismograms: The local seismic station AAE recorded an iP_n (Z) at U.T. 15:48:16.2, and iS_n (EW) at U.T. 15:48:40:2. The mean epicentral distance obtained from Pn – H_i (see equation in comments 3, 1968/I/23) agrees perfectly with the mean teleseismic location ($\overline{\Delta}$ = 238 ± 14 versus 241 ± 6 km); on the other hand, a discrepancy appears in the S-P travel

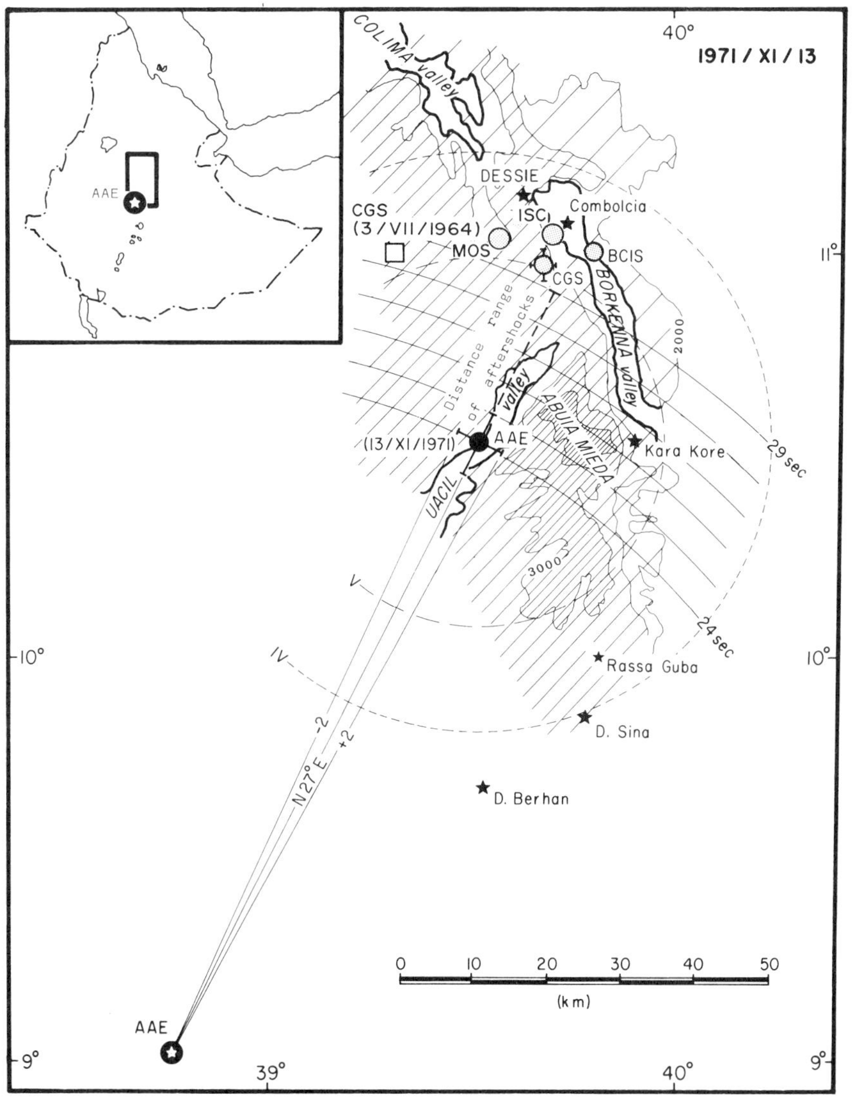

Fig. 53. Location map of instrumental epicentral solutions for the main shock of 13 November 1971 in Wollo province (○, ●). Also indicated, by a square symbol (□), is the epicentre of 3 July 1964; by dashed isolines, the reported intensities; and by solid curves, the epicentral distances from Addis Ababa. The topographic elevation contours are 500 m apart.

times of two shocks originating at apparently the same site. On 3 July 1964, five agencies located a shock of magnitude m_b 5 at N 11.10 ± 0.13°, E 39.51 ± 0.22°; the S-P time difference was 29. The present earthquake with a mean instrumental epicentre at N 11.00 ± 0.02°, E 39.68 ± 0.13° recorded an S-P of only 24 s. A 5 s difference in S-P suggests an epicentral distance about 0.5° (about 50 km) nearer to Addis Ababa.

Macroseismic Evidence: Two intensity reports, both between IV and V, were filed from sites at 12 km N and 105 km S of the mean instrumental epicentre location: Combolcia (N 11.9°, E 39.7°) and Rassa Guba (N 10.0°, E 39.8°) (see Fig. 53). Unless these intensity reports refer to two different shocks about 100 km apart, such an intensity distribution is not easily explained.

The observed S-P time difference at AAE and the intensity anomalies would not exist if the epicentre was located about N 10.5°, E 39.5°, some 0.5° SSW of the mean instrumental epicentre. It should be noted, however, that the S-P time differences for the aftershocks ranged between 24 and 29 s; therefore, some of the aftershocks did occur at the original teleseismic location (see Fig. 54).

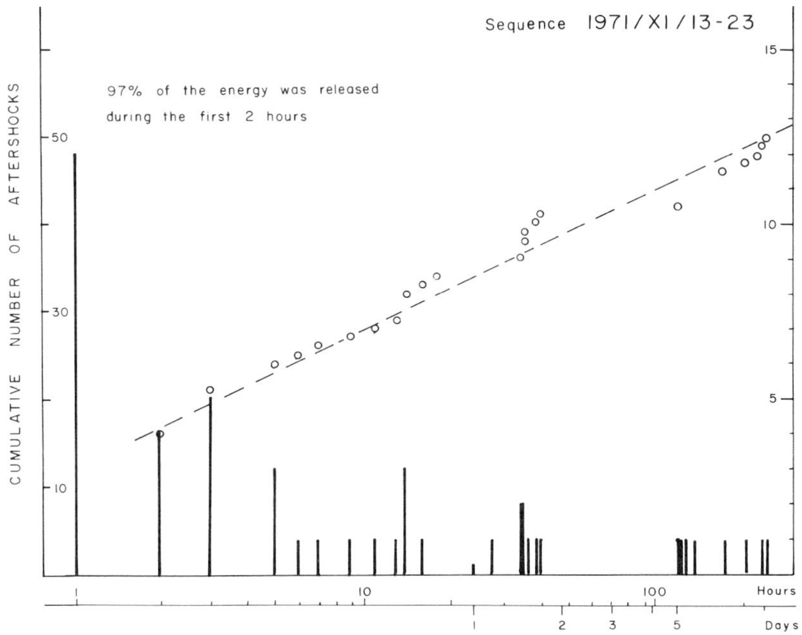

Fig. 54. Aftershocks from 13 to 23 November 1971; their cumulative number and hourly frequency distributions.

2. November 1971 Seismic Sequence

The earthquake sequence of November 1971 in Wollo was comprised of two foreshocks of magnitude M_L(AAE) 2.2 or 2.3 preceding the main shocks at intervals of 4 h; one main shock of magnitude m_b(ERL/GS) 5.3, M_L(AAE) 5.4, on 13 November at U.T. 15:48; and 50 aftershocks of magnitude M_L(AAE) ranging from 1.7 to 3.8 distributed in time from 13 to 23 November. They migrated northward during the period of activity.

For more information, see Dakin (1975, sequence 23), but note that some of the data have been revised and differ somewhat from hers.

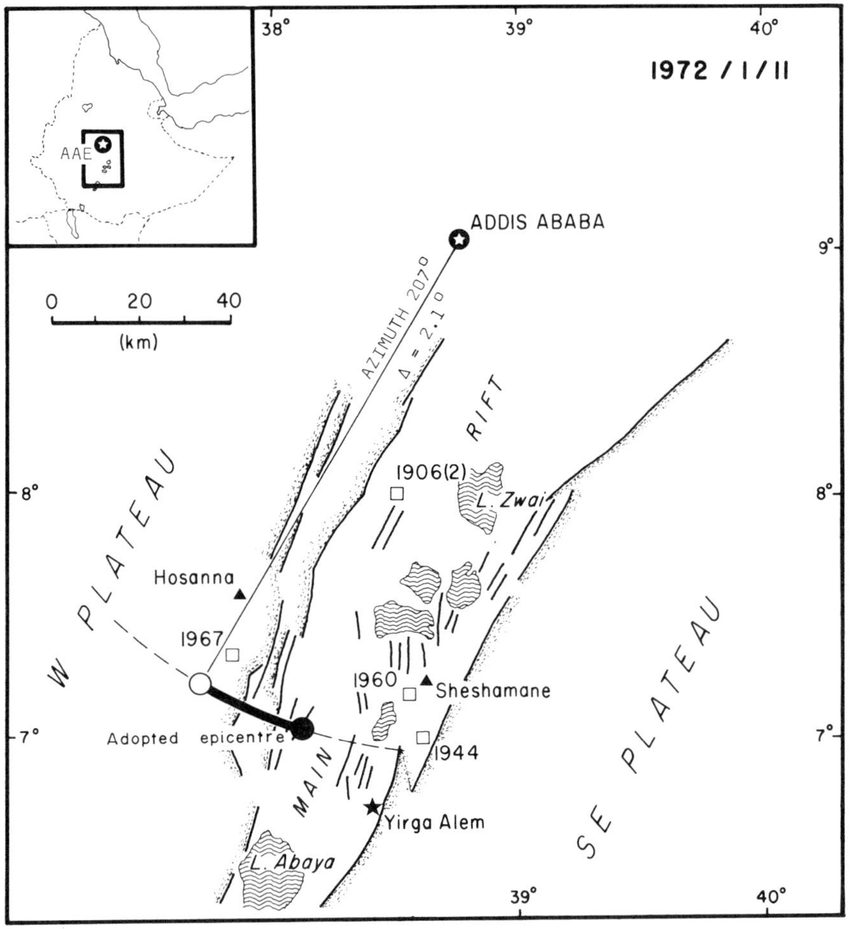

Fig. 55. Instrumental epicentral location of the main shock on 11 January 1972 and its possible SE relocation if the "felt report" from Yirga Alem refers to this shock, not to an aftershock.

1972/I/11

The tremors of an earthquake were felt at Yirga Alem (N 06.8°, E 38.4°) in the late afternoon of 11 January 1972. No damage was reported.

Sources

AAE Data File and Information Cards.

Comments

The source of the tremors was a shock of magnitude M_L(AAE) $\leqslant 4$ apparently located 235 km SSW (207°) of Addis Ababa on the upper margin of the Ethiopian Rift Valley. Its iP-onset was recorded at U.T. 15:11:14.2 (local time 18:11:14.2) in Addis Ababa. The main shock was followed by five aftershocks of lesser amplitudes.

AAE seismograms suggest an epicentral location at about N 07.2°, E 37.7°. On the other hand, there is a report that tremors were felt at Yirga Alem, 90 km away to the SE, on the floor of the Rift. Under normal circumstances, a shock of magnitude $\leqslant 4$ would not cause tremors of intensity III at such a distance; 50–60 km would be more conceivable. There is also the possibility that one of the aftershocks could have been located along the slope or at the foot of the escarpment and that the felt report referred to it. As there is no way of finding to which event the felt report was referring, the location map indicates both the instrumental location and the prolongation of the epicentral distance arc across the escarpment toward Yirga Alem (Fig. 55).

1972/VIII/18–19

Six earthquakes of magnitude m_b(AAE) $\leqslant$ 4.0 originated in the Guf Guf graben or its vicinity on 18 and 19 August 1972. Their epicentral distances ranged between 350 and 550 km almost due north of Addis Ababa and most of the shocks were felt at Debub (N 13.2°, E 39.8°). No damage was caused.

Source

AAE Data File.

Comments

The tectonic features affected by the seismic activity of August 1972 were the Guf Guf graben and probably the Dergaha-Sheket graben, which run northward between the latitudes north 12° and 14°, slightly west of the 40th meridian. The first one abuts the upper margin of the Plateau-Afar escarpment; the second, displaced eastward with respect to the axis of the Guf Guf, runs at midslope of the escarpment (for a detailed description,

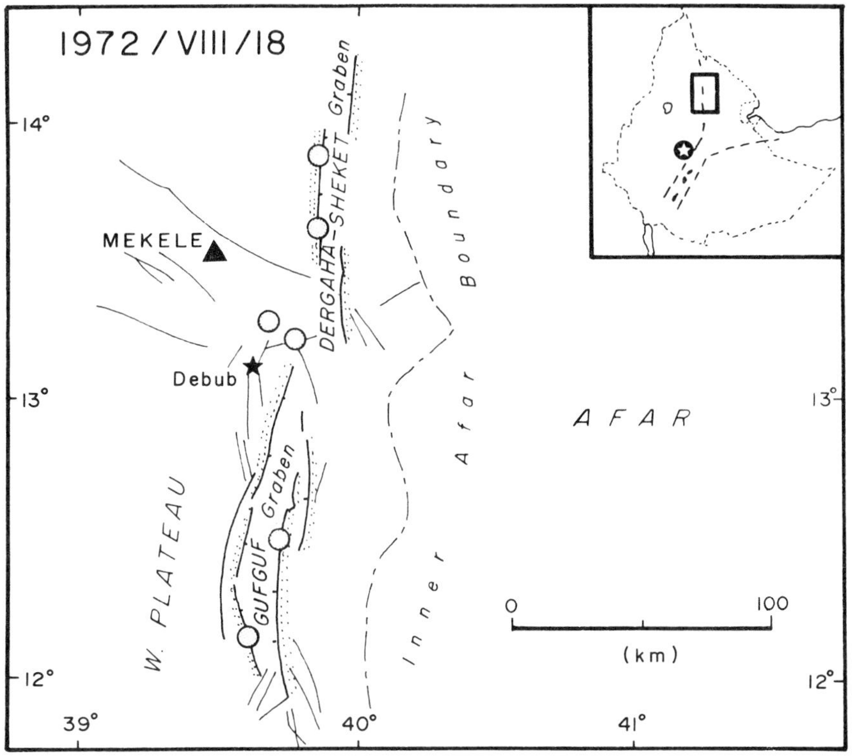

Fig. 56. Adopted locations for six of the epicentres of 18 and 19 August 1972, along the Plateau-Afar escarpment.

see Mohr 1975, p. 83). Both grabens are known to have been seismically active.

The adopted epicentres are plotted on Fig. 56; their relative distances from Addis Ababa are estimated to ± 10 km. The same accuracy cannot be claimed of their azimuths. The fact that the shocks, which were of low magnitude, were felt in Debub restricts the possibility of error. However, no felt report was received from Mekele.

1972/XI/13

An earthquake of M_L(AAE) 4.3 occurred in northern Shoa, north of the N 10° parallel. The tremors were felt as far south as Addis Ababa at an epicentral distance larger than 150 km.

Sources

AAE Information Cards and Data File.

Comments

The AAE seismographs recorded Pn and Sn arrivals at U.T. 07:08:21.1 and 07:08:43.0, respectively, indicating an epicentral distance of about 1.5° in a NE direction.

The epicentre was located south of the Robi graben, the last of the graben structures that characterize the Plateau-Afar margin north of N 10.17°. The region is disturbed mainly by NNE and NE trending faults crossing the escarpment, which at these latitudes heads almost due north.

Despite the hour of the day (10 a.m.) when people are busy at work, tremors were felt in Addis Ababa, especially in high rise buildings located near the Filoha fault. Curiously enough, no reports were filed from the towns situated to the north of the epicentre.

1974/II/24–25

During February 1974, tremors were reported in the Debre Berhan – Debre Sina region (about 100 km NE of Addis Ababa); some of these tremors were also felt in Addis Ababa. No damage was reported.

Sources

AAE Information Cards and Data File; ISC (1974, p. 118; USGS (EDR 25-74); Dakin (1975).

Comments

The tremors that were reported were caused by a sequence of earthquakes with magnitudes less than 4.5 originating from the region to the north and northwest of Debre Sina.

The epicentre of the main shock on the 25th was reported by ISC, USGS and by the Tendaho local network TLN:

	H	Coordinates		m_b	N
ISC	16:05:18.0 ± 1.3	N 10.20 ± 0.13°	E 39.80 ± 0.23°	4.6	9
CGS	16:05:15.7 ± 1.4	N 09.84 ± 0.17°	E 40.03 ± 0.19°	4.5	6
TLN	16:05:12.5	N 10.00	E 39.78		

For the same earthquake, the AAE data file gives the following parameters: iPn at 16:05:40.6; M_L 4.2; Δ= 140 ± 10 km at 047°; location at N 09.9°, E 39.7°. The adopted site for this epicentre is the arithmetic mean of the four original values: N 09.99°, E 39.83°.

Of the 36 shocks of the swarm identified on AAE seismograms and ranging between magnitudes M_L(AAE) 3.6 and 4.2, six epicentres were determined by the Tendaho network. These are plotted on Fig. 58A and

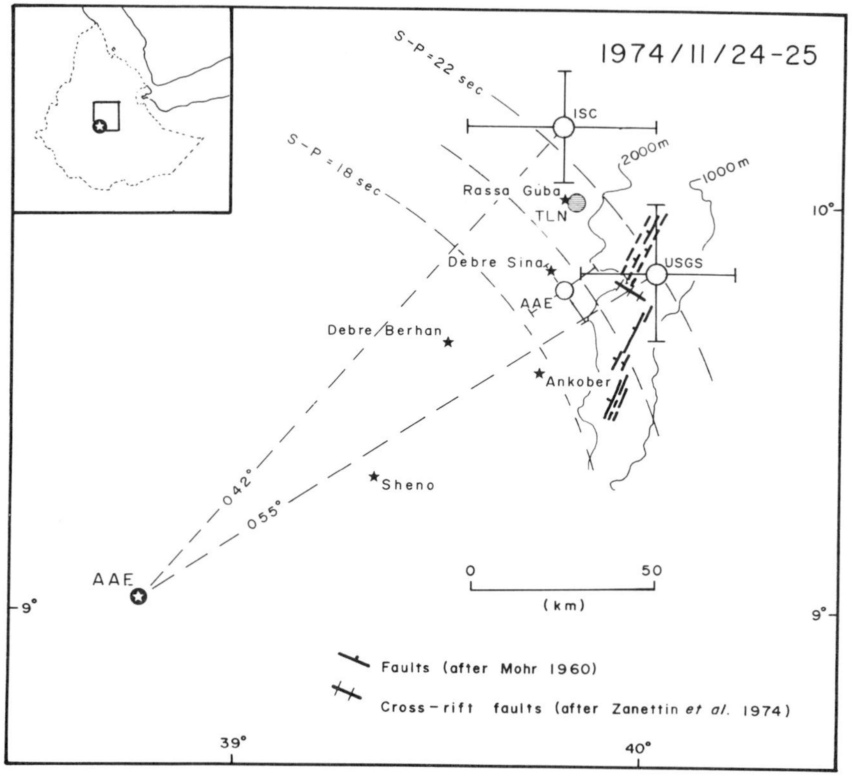

Fig. 57. Seismic region in Wollo during February 1974. The TLN epicentre is a location obtained by the local network operated by the University of Durham near Tendaho in central Afar.

identified by a solid square or circle within a larger open square (see next entry). Note that their alignment suggests an active geological structure heading NW from Debre Sina.

1974/II–IV

During the 124 days of intermittent operation of the temporary seismic network that Durham University installed near Tendaho in central Afar from 24 February to 23 September 1974 (information on this network is to be found in entry 1969/Serdo, Region C), 106 epicentres of low magnitude earthquakes could be determined on the Western Plateau between the 39.5th and 40th east meridians and the parallels north 10th and 13th. Exception being made of: (1) a cluster of about 50 events at the south end of the Guf Guf graben (south of N 12°); and (2) an alignment of epicentres northwest from Debre Sina, the other epicentres were distributed very regularly along the upper margin of the escarpment. The distribution pattern confirms that there is no seismicity gap along the upper eastern margin of the Plateau between N 10° and N 13° (see Fig. 58A).

Source

W. Ridgen, Durham University, 1978 (personal communication).

1974/XII/17–20

Earth tremors were felt in Combolcia (N 11.07°, E 39.7°) and Dessie (N 11.1°, E 39.6°) during the night of 17–18 December 1974. People ran out of their houses: intensity was estimated as V. No damage was reported.

Sources

AAE Data File; Ethiopian News Agency.

Comments

Six earthquakes were identified on the AAE seismograms as having their origin in the Dessie region:

Date	Arrival times		Apparent magnitudes	
XII	P	S	Δ(km)	M_L(AAE)
17	19:19:15.0	19:19:43	220±10	3.6
17	23:14:33.3	23:15:02.5	230	2.9
18	00:44:49.7	00:45:19.0	230	3.8
18	09:21:36.5	09:22:03.7	215	2.5
20	05:55:08.7	05:55:36.4	215	2.9
20	10:11:25.0	10:11:52.2	215	1.9

Apparent distances and azimuths are not accurate enough to pinpoint the epicentres. However, intensity V from a shock of M_L(AAE) 3.8, equivalent to m_b about 3.4, located in a highly rifted region suggests an epicentral distance of about 15–20 km from Dessie (Fig. 58B).

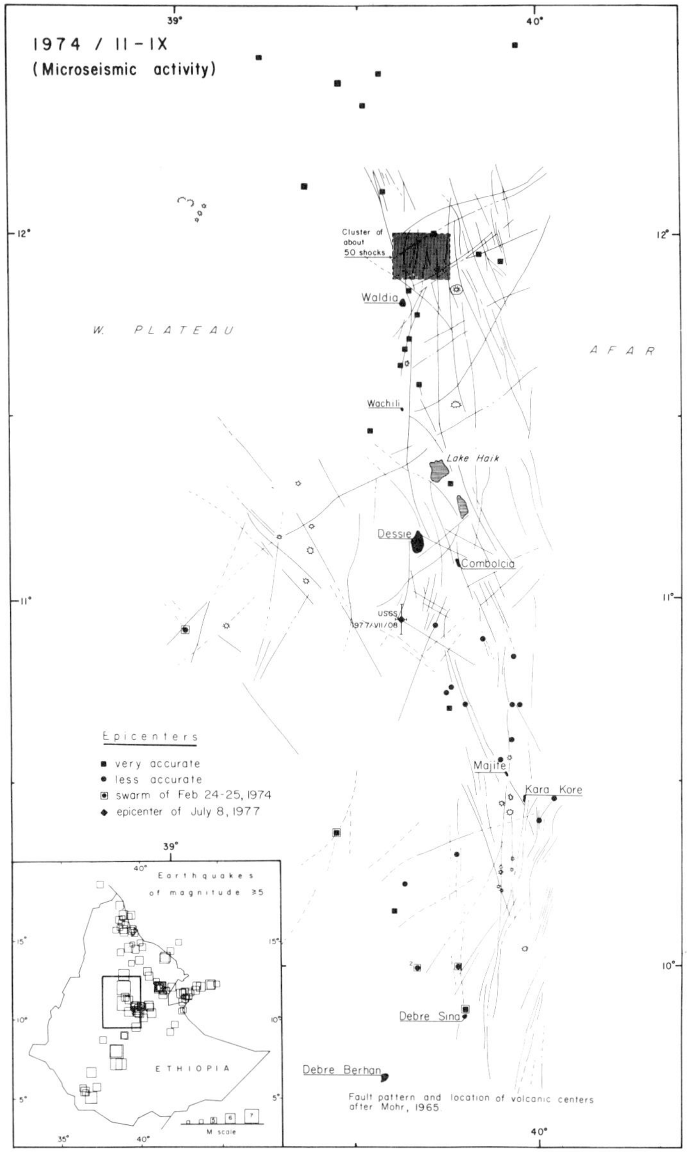

Fig. 58A. Location of two low magnitude epicentres along the upper margin of the Plateau-Afar escarpment determined by the Durham University seismic network (TLN). For comparison sake, an epicentre map of magnitude ⩾ 5 is included in the inset.

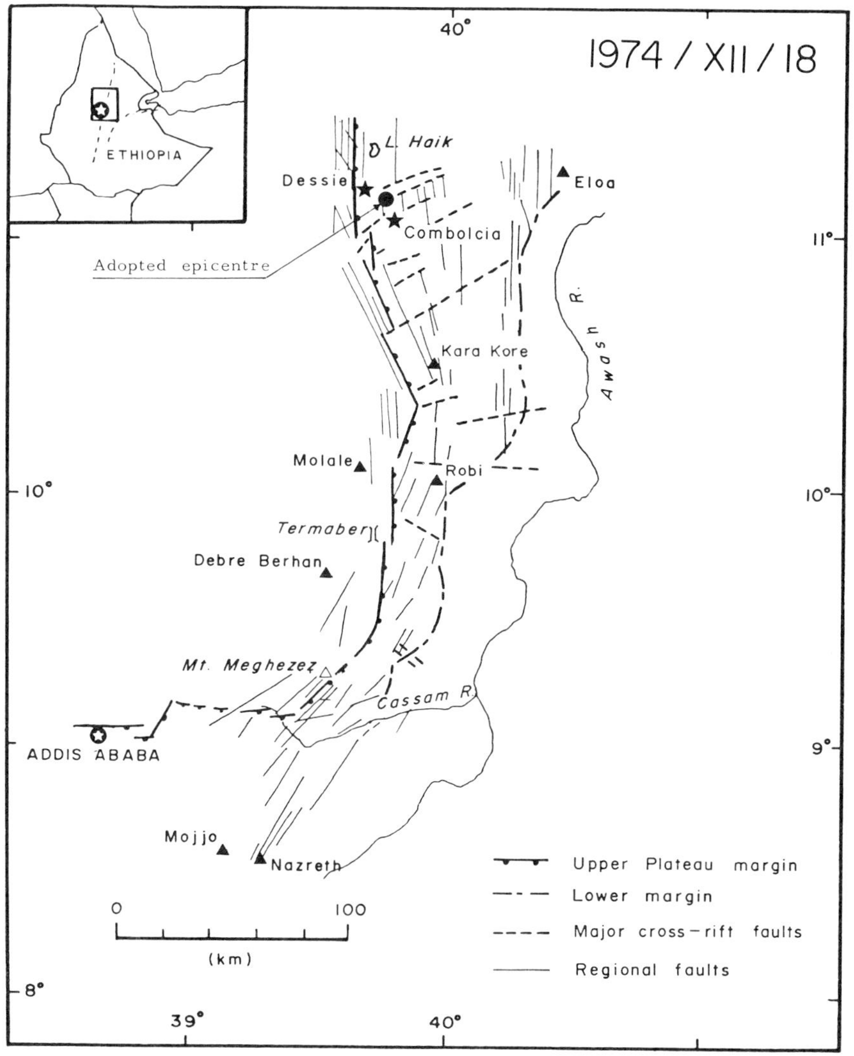

Fig. 58B. Tectonic map of the Plateau-Afar escarpment, north of Addis Ababa. The December 1974 seismic region is located in the upper part of the map (fault pattern after Zanettin and Justin-Visentin 1975 and from an unpublished photogeological map by P. Tacconi).

1975/VIII/23

On 23 August 1975, two earthquakes of magnitude m_b 5.2 and 4.5, respectively, occurred during the night in the Kara Kore–Majete region

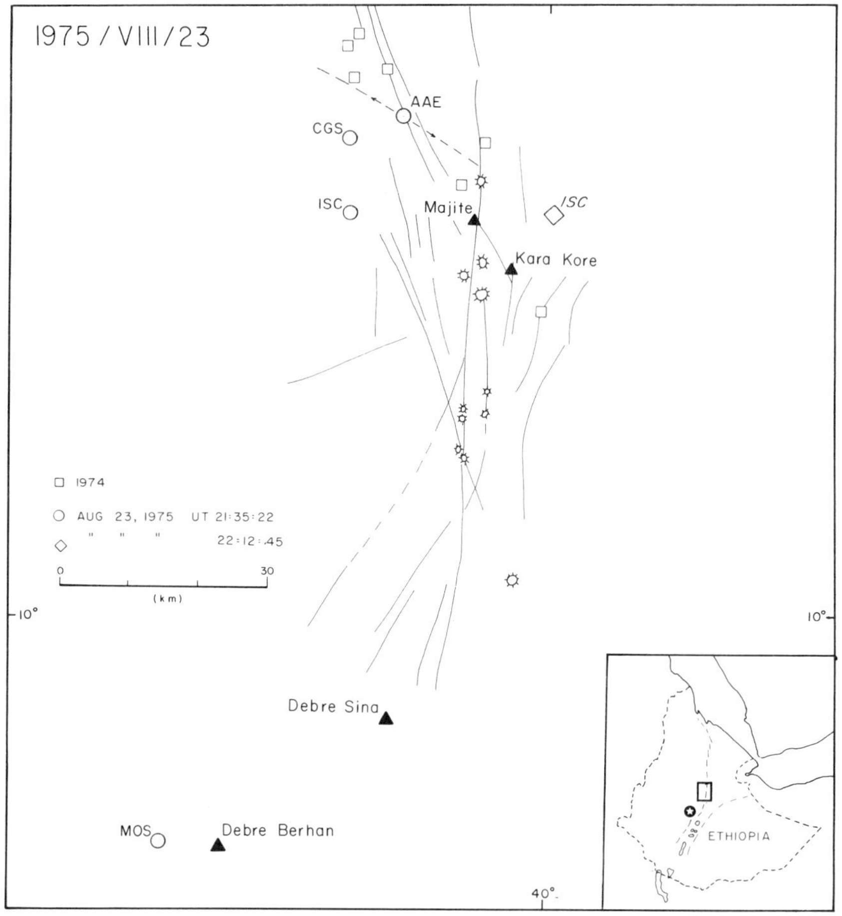

Fig. 59. Epicentral locations for the two epicentres of 23 August 1975 in the Kara Kore area.

along the upper margin of the Plateau escarpment (Fig. 59). Tremors were reported from Derek Amha (Dry Mountain). Four instrumental solutions are available for the first epicentre, one for the second. Their parameters are listed in Part 2.

Sources

ISC; USGS (EDR 33-75).

1977/VII/08

At U.T. 06:23:02 on 8 July 1977 an earthquake of magnitude m_b (USGS) 5.0 occured near the city of Dessie, capital of Wollo. The USGS located it at N 10.940 ± 0.042°, E 39.629 ± 0.028° and indicated a focal depth of 37.6 ± 5 km. Ninety eight stations reported the event, which was the main shock of a sequence of 12 recorded at AAE (one foreshock and 10 aftershocks).

No structural damage to houses was observed in Dessie. However, the Governor's Office reported that the main shock triggered a landslide along the slope on which Dessie was built and that two women working in the fields were killed. Tremors were felt with different degrees of intensity throughout Addis Ababa.

Sources

AAE Data File; USGS (EDR 12-77, 2).

Comments

The main shock of 8 July 1977 provided an example of intensity distribution during earthquake tremors in Addis Ababa. At ground level, the intensities were much higher downtown, south of the Filohawa zone where the houses are built on thick less consolidated geological formations, than in the northern part where solid rock is near the surface. It was also observed that the direction of the main axis of a structure greatly influenced the amplitude of the tremors.

The casualties in Dessie exemplify the need for planning the selection of a site, especially the site of a capital. Dessie is in a location similar to Ankober, which was destroyed in 1842 by an earthquake-triggered landslide!

The epicentre for the earthquake of 8 July 1977 is plotted on Fig. 58A. For other epicentres in the same region see entries 1964/VII/03 and 1971/XI/13.

Earthquake History of the Southeastern Plateau (Region B)

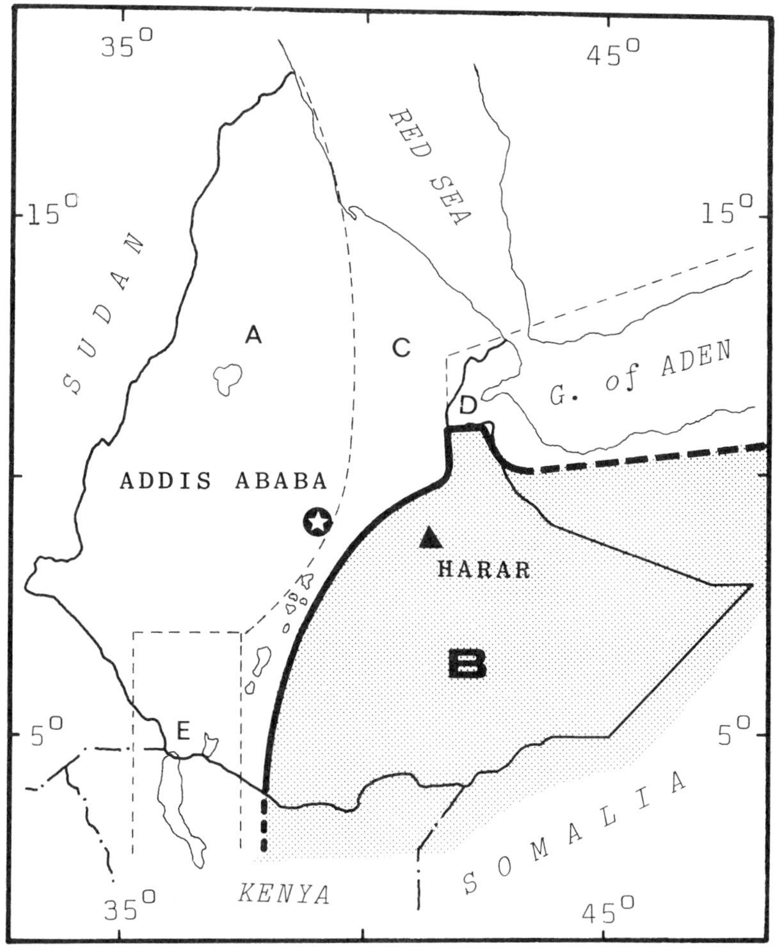

The data file on Region A is comprised of some 90 entries, many of which are complemented by topographic and tectonic sketch maps covering practically the whole Western Plateau. Fortunately for the security of its inhabitants, the seismic file for Region B is comprised of only thirteen entries, and therefore very few local maps are presented. To compensate for the paucity of maps and to place the few that are available within the context of the whole Southeastern Plateau, a general location map (Fig. 60) is offered. It is a sketch map on which the differences in elevation are determined by only two isolines (2000–3000 m; and above 3000 m); the sites that are indicated are those mentioned in the text or in international reports on earthquakes.

1884/I/2–3

Light earth tremors were felt in Harar (N 09.3°, E 42.2°) during the night of 2–3 January 1884.

Source

Diary of Mgr Taurin Cahagne (Capucins Fathers' archives in Toulouse, France).

1911/III/02

On 2 March 1911, at 8:20 a.m., a light earth tremor was felt in Dire Dawa (N 09.6°, E 41.9°). No damage was reported.

Source

Semeur d'Ethiopie, Dire Dawa, March–April 1911.

1930/X/24–31

From 24 to 31 October 1930, the Horn of Africa was shaken by a series of earthquakes that apparently had their origin on the Aisha Horst, identified on the location map of Region B by a spur protruding northward from its northern border (see Fig. 63A). Tremors were reported from Zeila and Djibouti on the coast to as far inland as Awash Station (?) in central Ethiopia. The first important shock occurred on 24 October at U.T. 10:47; its magnitude was about 5½. Others followed on 25 and 27 October. Aftershocks were reported until 31 October.

Sources

ISS (Bellamy 1936, p. 3); Dreyfuss (1931, p. 342); Gutenberg and Richter (1949); Taylor (1931, p. 34). (Macfayden 1933, p. 38).

Dreyfuss (1931) gives the following report:

The swarm of earthquakes started on October 17, at L. T. 08:18. Three shocks were felt on the 17th, one on the following day, none on the 19th and one on the 20th. On October 24th, the shaking started again at L. T. 13:44; six or seven shocks were felt every day until October 29th.

Comments

1. Reliability of Intensity Reports

There are inconsistencies in the evaluation of the intensities reported from Djibouti. The newspaper *Réveil de Djibouti* in its edition of 12 October 1963 stated damage in Djibouti and Tadjoura easily rated VII (M.M.). I quote (the translation into English is mine): *In Djibouti, the earthquake of October 1930 heavily cracked the walls of the houses that had been left intact by the earthquakes of 1929. In Tadjoura, two mosques collapsed. The tremors were also felt at Ali Sabieh, and in Ethiopia as far as Dire Dawa and Awash station; curiously enough, they were not reported felt in Dikhil.* On the other hand, the meteorologist-in-charge at Pointe-au-Serpent, Djibouti, in his official report for the same period (17–31 October 1930) mentioned 50 tremors but none with an intensity any higher than III–IV. (The intensities from Djibouti are always rated according to the Rossi-Forel scale; they are transposed here in Mercalli-Modified units).

It is unfortunate that the editor of the *Réveil* did not quote his references because the report from Awash Station becomes questionable and with it the evaluation of the radius of the felt area.

Taylor (1931, p. 34; Macfayden 1933, p. 38) reported 18 tremors from Zeila (N 113°, E 43.5°) on 24 and 25 October. As no intensities are indicated, it is surmised that they did not rank any higher than III–IV.

2. The Probable Epicentres

In the literature, four earthquakes of magnitude class "d" (5.3–5.9) are said to have occurred by the end of October at the locations listed below:

Date	H	Coordinates		Agencies
(1) 24 Oct	U.T. 10:47	N 10.5°	E 43°	G & R
		N 10.3	E 42.7°	ISS
(2) 25 Oct	U.T. 16:29	N 11.5°	E 44°	G & R
		N 11.5°	E 42°	ISS
(3) 25 Oct	U.T. 17:42	N 11.5°	E 44°	G & R
		N 11.5°	E 42°	ISS
(4) 27 Oct	U.T. 23.29	N 11.5°	E 43.5°	ISS

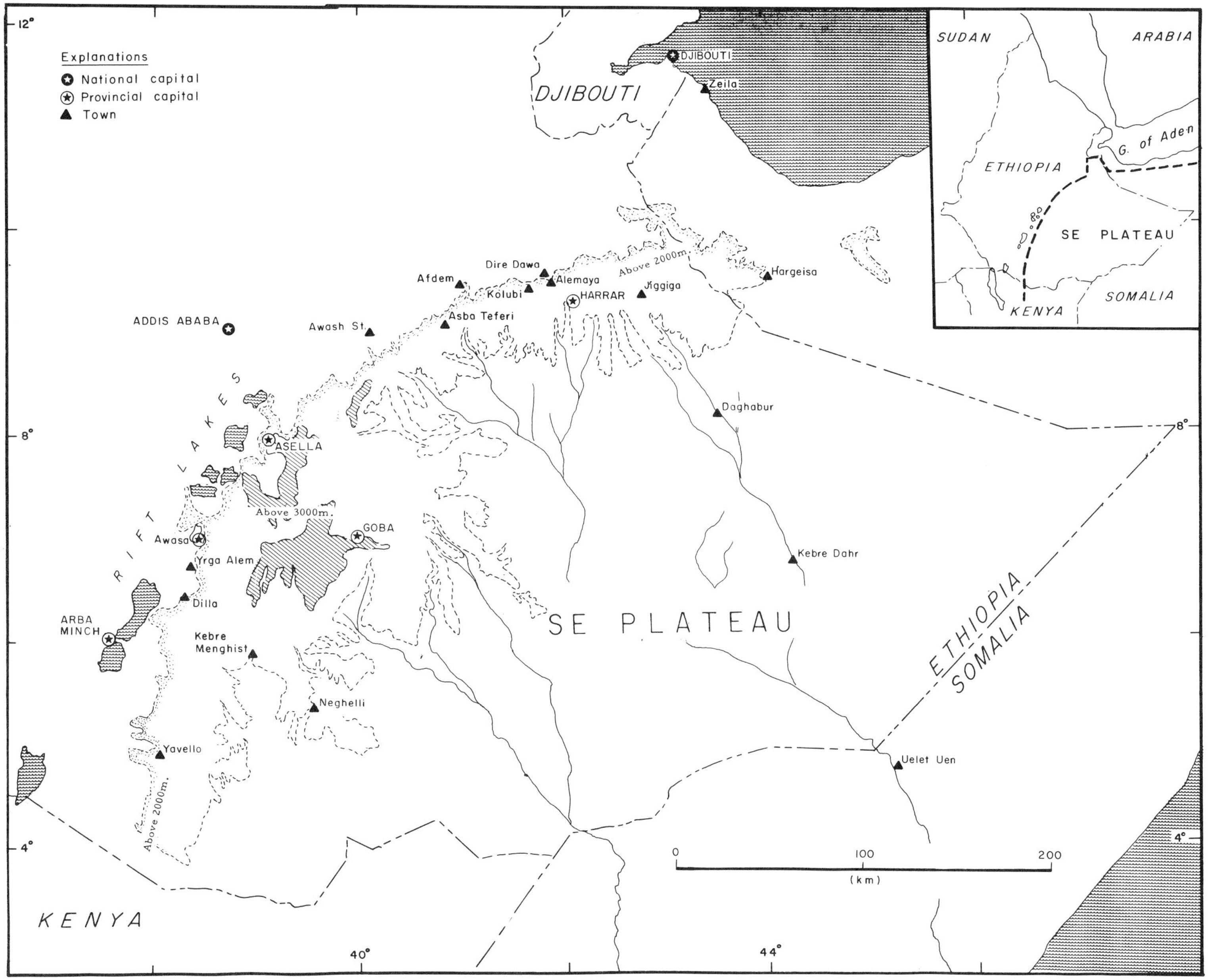

Fig. 60. Outline of the Southeastern Plateau.

If these 1930 solutions are taken at face value and the four shocks are assumed to have originated from only one active area, then this area would be more or less defined by a radius of about 2° centred around N 11°, E 43°, inland, some 60 km south of Djibouti.

A recomputation of the four epicentres based on the information contained in the ISS and CGS data files was attempted. The results are given below; the last column identifies the data files.

Date	H	Coordinates		File
(1) 24 Oct	U.T. 10:47	N 09.1 ± 3.0°	E 42.3 ± 0.8°	ISS
		N 10.7 ± 1.0°	E 43.0 ± 1.0°	CGS
(2) 25 Oct	U.T. 16:29	N 08.2 ± 10.2°	E 41.4 ± 3.7°	ISS
(3) 25 Oct	U.T. 17:42	N 08.4 ± 12.8°	E 42.3 ± 3.4°	ISS
(4) 27 Oct	U.T. 23.29	N 10.6 ± 1.0°	E 42.6 ± 0.5°	ISS

Solutions (2) and (3) show unusually large standard deviations (± 10.2° and ± 12.8°). These deviations are due to the limited number of reporting stations, to station errors in clocking P arrival times, and to an extremely bad spatial distribution of the reporting stations. In both cases, the stations were concentrated due north of the epicentral region over azimuth angles of only 47° (345–032°) and 60° (332–032°), respectively. These two solutions are to be rejected, but they undoubtedly suggest epicentral locations south of N 11.5°.

For the earthquake of 24 October there are two independent solutions. A cursory analysis of the time residuals accepted in the computation of the ISS data reveals a residual of +9.8 s, which is much higher than any other (the nearest = 2.1 s) and is attached to a station (Malaga, Spain) located northwest of the epicentral region (see Fig. 61). If this station report had been rejected, the epicentral location would have been higher in latitude than N 09.1°. These observations suggest that the original latitudes N 11.5° are most probably too high, that N 09.1° is too low, but that the original N 10.5° and 10.3° found by Gutenberg and Richter and ISS for the event of 24 October and the 10.7° and 10.6° found in two recomputations are most probably the best approximations. Such a latitude (N 10.6°) corresponds at these longitudes to the vicinity of a major wrench fault (Black et al. 1972), the Bia Anot fault, which cuts through the Aisha 'Horst' in a northeasterly direction between approximately E 42.6° and 43° (Fig. 62).

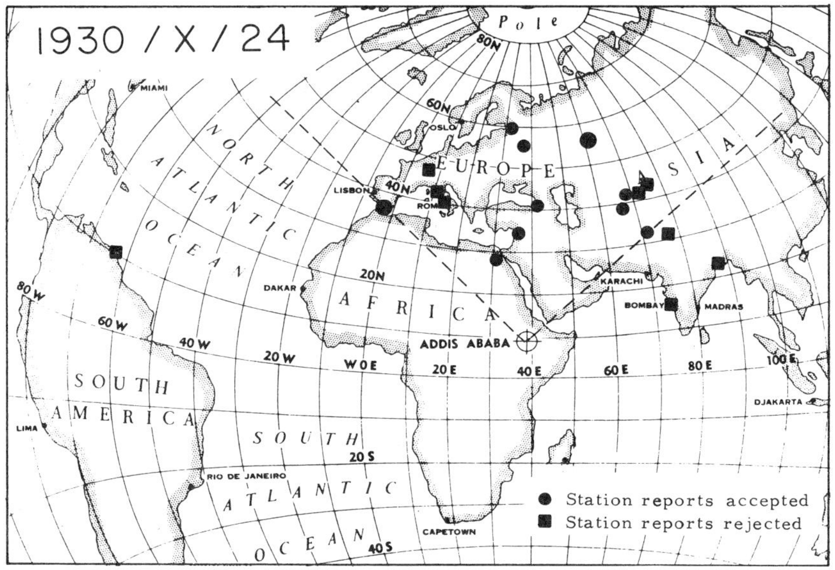

Fig. 61. Distribution of the seismic stations that reported to ISS the Ethiopian earthquake of 24 October 1930. Note that all the stations whose reports were accurate enough (O-C ⩽ 10 s) to be accepted in the epicentre determination are clustered in a 90° sector north of Ethiopia.

Provisionally, the following epicentral parameters have been adopted:

Date	Time	Coordinates		Magnitude
(1) 24 Oct	U.T. 10:47	N 10.4°	E 42.8°	5.8
(2) 25 Oct	U.T. 16:29	N 10.3°	E 42.7°	5.6
(3) 25 Oct	U.T. 17:42	N 10.4°	E 42.7°	5.4
(4) 27 Oct	U.T. 23:29	N 10.6°	E 42.6°	5.0

For reference to later seismic activity in the same region, see entry 1932/II/01. As three out of four of these instrumental epicentres were originally located in Somalia by Gutenberg and Richter and by ISS, these events are also indexed under Region D.

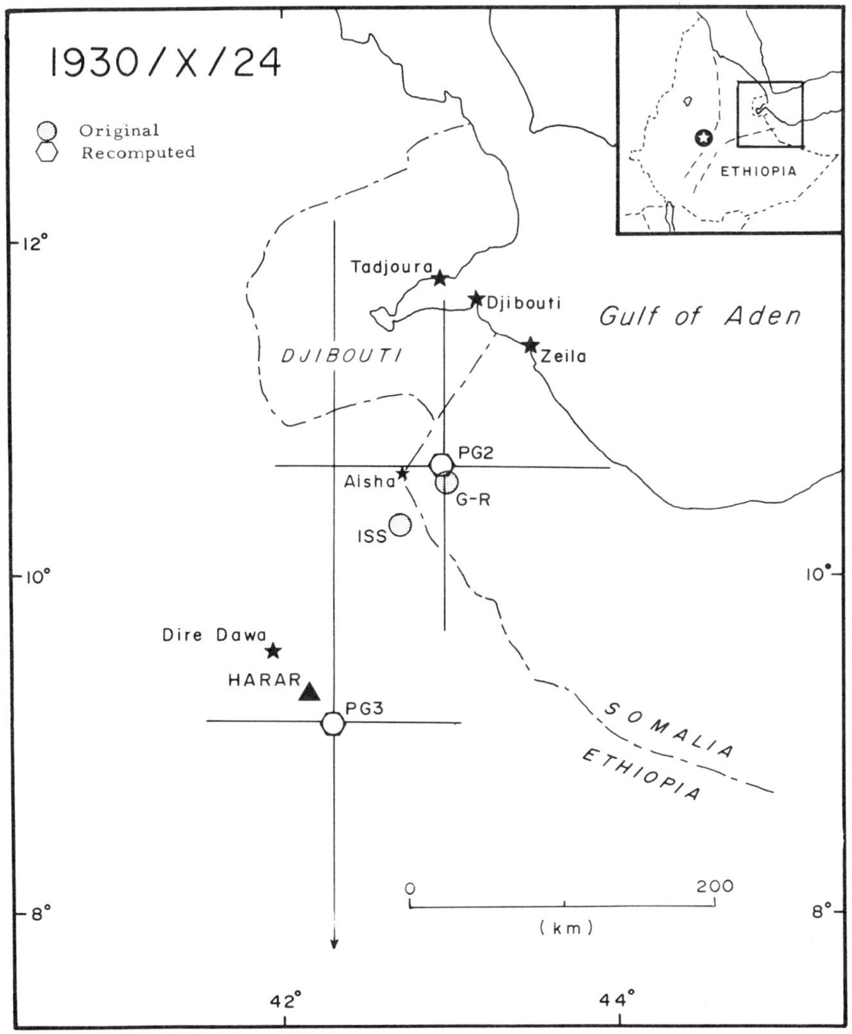

Fig. 62. Instrumental epicentre location of the earthquake of 24 October 1930. The other epicentres on 25 and 27 October have been relocated in the same region.

1932/II/01

On 1 February 1932 at U.T. 07.6, ISS reported an epicentre of undetermined magnitude at N 10.3°, E 42.7°, on the western scarp of the Aisha horst, near the Ethiopia-Somalia border.

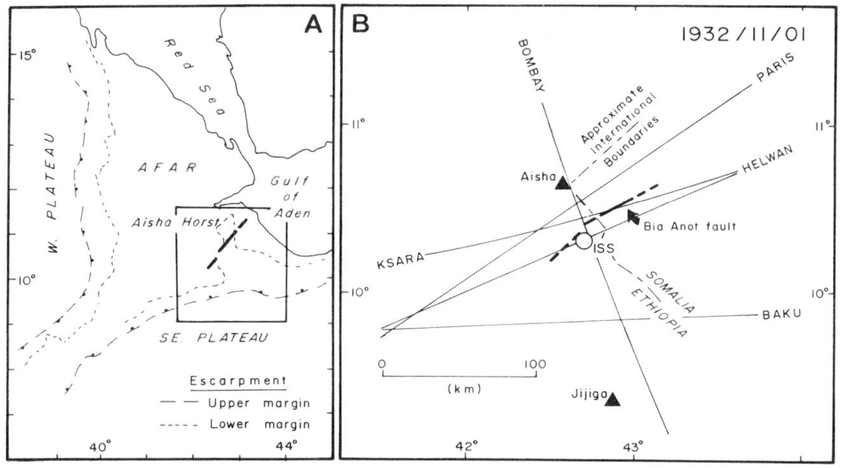

Fig. 63(A). Location of the Aisha horst with respect to Afar and the Southeastern Plateau. (B) Distance arcs centred on the stations that reported the earthquake of 1 February 1932.

Sources

Bellamy (1936).

Comments

Only six stations reported the event to ISS and their geographic distribution is concentrated in the quadrant NE of the epicentre. A recomputation of the original data has been attempted, but no solution of any better reliability than the original ISS epicentre could be obtained. Epicentral distance-arcs centred on five stations (Helwan, Ksara, Bombay, Baku, and Paris) are reproduced in Fig. 63; the arc centred on Calcutta was neglected because of its location some 5° west of the epicentre.

The site is very realistic; it is in the vicinity of the important Bia Anot fault, recognized as a left-lateral transcurrent fault by Black et al. (1972, p. 170). The Bia Anot fault cuts through the Aisha horst in a northeasterly (060°) direction. Some geologists have suggested that it could indicate a prolongation inland of some of the transform faults recognized in the Gulf of Aden and Tadjoura (Black et al. 1972, p. 171). See also Region D, entries 1930/X/24 and following.

The Meteorological Office at Djibouti reported tremors of intensity II-III (Rossi-Forel intensity scale) on 30 January. Could there be a slight error in the date? Djibouti is about 130–140 km from the adopted epicentre location.

1944/IX/06

An earthquake of magnitude larger than 6 occurred in southern Ethiopia on 6 September 1944. No local report was found; 1944 was during the war.

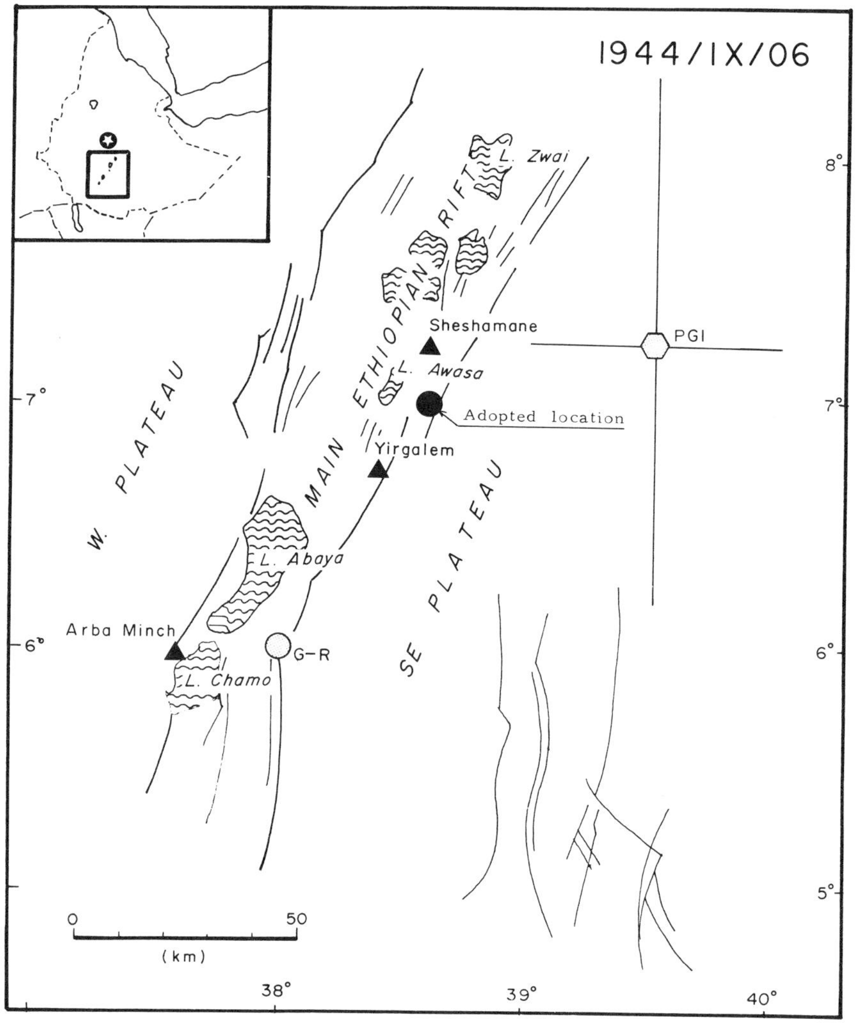

Fig. 64. Location map of the two instrumental epicentres and of the adopted location for the earthquake of 6 September 1944.

Gutenberg and Richter located the epicentre at N 06°, E 38°; recomputation (courtesy of Prof Rothé) relocated it on the Southeastern Plateau at N 07.24 ± 1.08°, E 39.29 ± 0.52°, a region that, as far as is known, is presently aseismic. As there are good reasons to believe that the earthquake belongs to the Rift, the particulars are indexed under Region C. Figure 64 locates the two instrumental epicentres based exclusively on teleseismic data and the adopted location in the Rift.

1953/V/28

On 28 May at 4.30 a.m. an earthquake of magnitude 5.2 occurred a few kilometres southwest of Harar city, in the northern sector of the Southeastern Plateau. Five aftershocks were reported the same day. Damage of intensity VI and V was observed in Harar and Dire Dawa, respectively.

Sources

BCIS; *Ethiopie d'Aujourd'hui*, 29 May and 5 June 1953; *Ethiopian Herald*, 5 and 6 June 1953.

Comments

1. Descriptions of the Damage

Reports from contemporary newspapers and from eyewitnesses interviewed 10 years later do not agree entirely. The events can, however, be summarized as follows:

Harar City (N 09.3°, E 42.2°) — Some reports claim that the tremors were preceded and followed by roaring noises; walls were fissured in the residence of the Duke of Harar and of the Deputy-Governor, in the administrative buildings of the Army, and in some private houses. According to the *Ethiopian Herald*, 75% of the city was more or less "affected." What the news reporter meant by the term 'affected' is in no way defined.

It is known that the students boarding at the Teachers Training School slept a few nights in the open. The estimated intensity in Harar city must have been about VI.

Dire Dawa (N 09.6°, E 41.9°) — The tremor at 04.25 a.m. (U.T. 01:25) was the first and the strongest felt in Dire Dawa; it was accompanied by a *roar like that of artillery*. Damage was less severe than in Harar; the heaviest being done to the Airport Terminal and to the Mohajan Indian School.

2. Epicentral Location

The following parameters for the main event of 28 August 1953 are those of BCIS; USCGS accepted them as such: time, 01:24:46; coordinates N 09.2°, E 41.9°; magnitude 5.2. The location is at a horizontal distance of about 30 km from Harar city and 45 km from Dire Dawa (see Fig. 65). A shock of M_s 5.2 is expected, under normal conditions, to cause intensities IV and V at these distances.

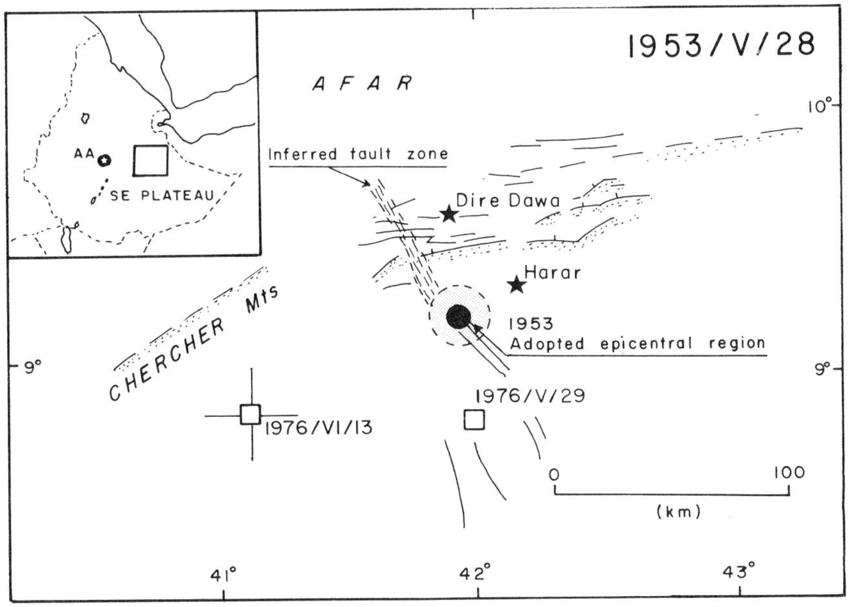

Fig. 65. Location of the seismic region near Harar on the Southeastern Plateau.

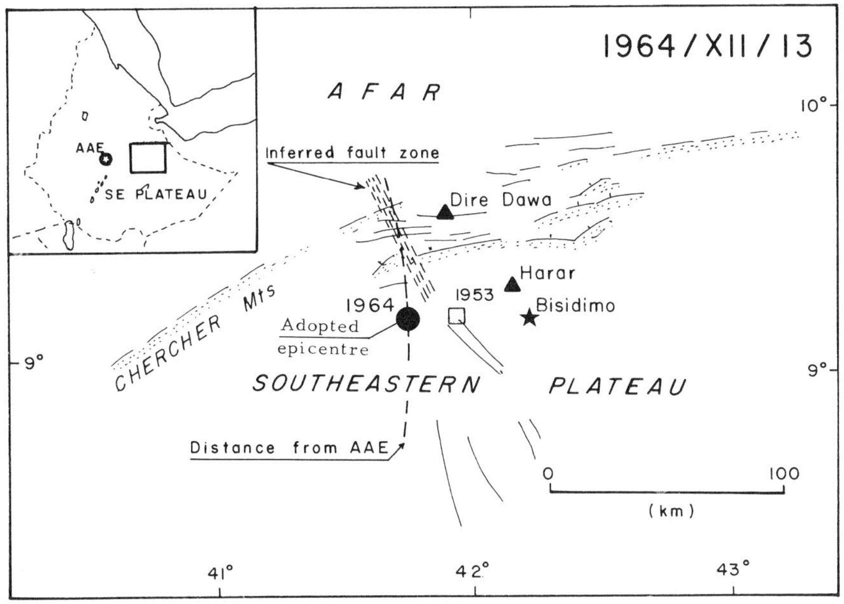

Fig. 66. Location map of the 13 December 1964 epicentre in the northern sector of the Southeastern Plateau.

For a well-determined epicentre in the same region, see next entry (1964/XII/13).

1964/XII/13

On 13 December 1964 at 9.40 p.m. a tremor with strong vertical motion was reported from Bisidimo (N 09.2°, E 42.2°) in the Administrative Region of Hararghe.

Sources
AAE Data File; Bisidimo Leprosarium Administration report.

Comments
For this earthquake, AAE recorded the following phases: iP(Z) U.T. 18:42:44.2; iS (EW) U.T. 18:43:24.5, (NS) U.T. 18:43:25.6; Sg/Lg 18:43:34.9.

The AAE seismograms indicated an epicentre almost due east of Addis Ababa at a distance of 3°, about 330 km. The epicentral arc-distance passes through the village of Kolubi, and at a minimum distance of 50–55 km from Bisidimo where tremors of intensity III–IV were reported. The estimated magnitudes at AAE were M_L = 4, m_b = 3.6.

The adopted epicentre (N 09.2°, E 41.7°) lies on the SSE projection of a postulated buried tectonic zone (Fig. 66), which is revealed at the foot of the Plateau northern escarpment, a few tens of kilometres west of Dire Dawa, by a sudden reversal in the dipping of the tilted blocks that form the escarpment east and west of that zone. The escarpment is made of blocks, cut by synthetic faults and tilting to the south; in this particular zone, at the foot of the escarpment, there are rows of similar tilting blocks separated this time, by antithetic faults and tilted to the north (Black et al. 1972; Schacknai 1972; Kasmin, personal communication). As most of the tectonic trends in that part of the Plateau are NNW-SSE, it is thought a valid assumption to postulate a SSE extension of that zone underneath the Plateau.

1973/VI/03

An earth tremor awakened people in Jiggiga (N 09.3°, E 42.7°) during the early hours (2.45 a.m.) of 4 June (3 June U.T. 23:39) 1973. No damage was reported.

Sources
AAE Information Cards and Data File.

Comments

The shock was recorded by AAE at U.T. 23:39:41.0; its epicentral distance was 435–450 km almost due east of Addis Ababa (Fig. 67). The estimated magnitudes were M_L = 4.2, m_b = 3.8.

For information on the Marda fault zone see Morton (1974, p. 24–25) and Purcell (1975, p. 133–140; 1976, p. 569–571).

1973/XI/16

A sharp tremor of intensity IV⁺ was extensively felt throughout Dire Dawa (N 09.6°, E 41.9°) in the early afternoon of 16 November 1973. The reports came from the railway station, Notre-Dame school, and the Somali sector of the town. No damage was observed.

Sources

AAE Information Cards.

Comments

Field investigations confirmed that the information was genuine and reliable. The time given for the tremor was "siesta time," that is after 1 p.m. The survey revealed that the felt area was restricted to about 1 km in diameter. As the tremor was felt over so restricted an area, and not recorded by AAE, a distance of 350 km, the focus must have been very shallow and underneath the centre of Dire Dawa.

There is a report of a tremor felt in Jiggiga at about noon, the same day, a distance of 1° (110 km) ESE (099°) of Dire Dawa. Both reports could refer to the same event. If so, the epicentre would have to be located ESE of the town, along the slope of the escarpment.

1975/IV/27

During the night of 27–28 April 1975, an earth tremor of intensity III was felt in Harar; it was also felt as a sharp shock in Bisidimo, which awakened many people and was accompanied by thunderlike noises. The Geophysical Observatory at Addis Ababa recorded the shock's Pn phase at U.T. 23:33:52.5 on the 27th and located the epicentre at N 09.4°, E 42.8°, in the vicinity of the Marda Fault Zone (see Fig. 67).

Sources

AAE and TFAI Data Files.

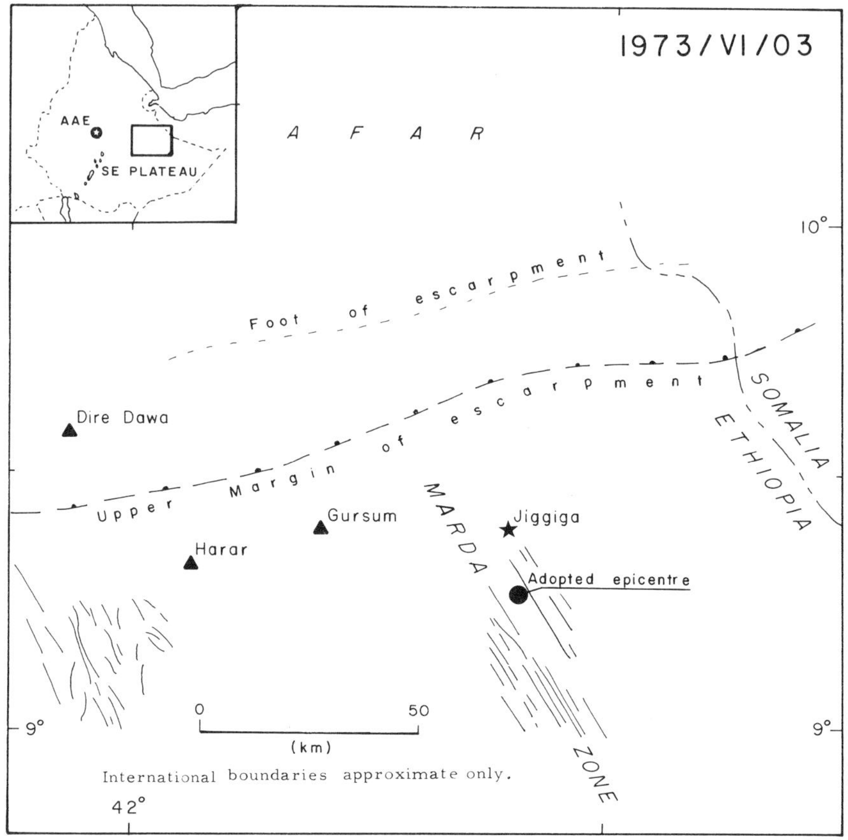

Fig. 67. Location map of the 03 June 1973 epicentre on the Marda Fault Zone.

1976/V/29

Around 2 a.m. on 30 May 1976, many inhabitants of Harrar were awakened by a sharp earth tremor. No damage was reported.

Sources

AAE and Arta Data Files.

Comments

The seismograms of Addis Ababa and Arta (Djibouti) indicated an epicentre at about N 08.7°, E 42.0°, 65 km south of Harar, an origin time at U.T. 22:59:01 on 29 May, and a magnitude M_L (AAE) 4.4. The location is plotted on Fig. 65.

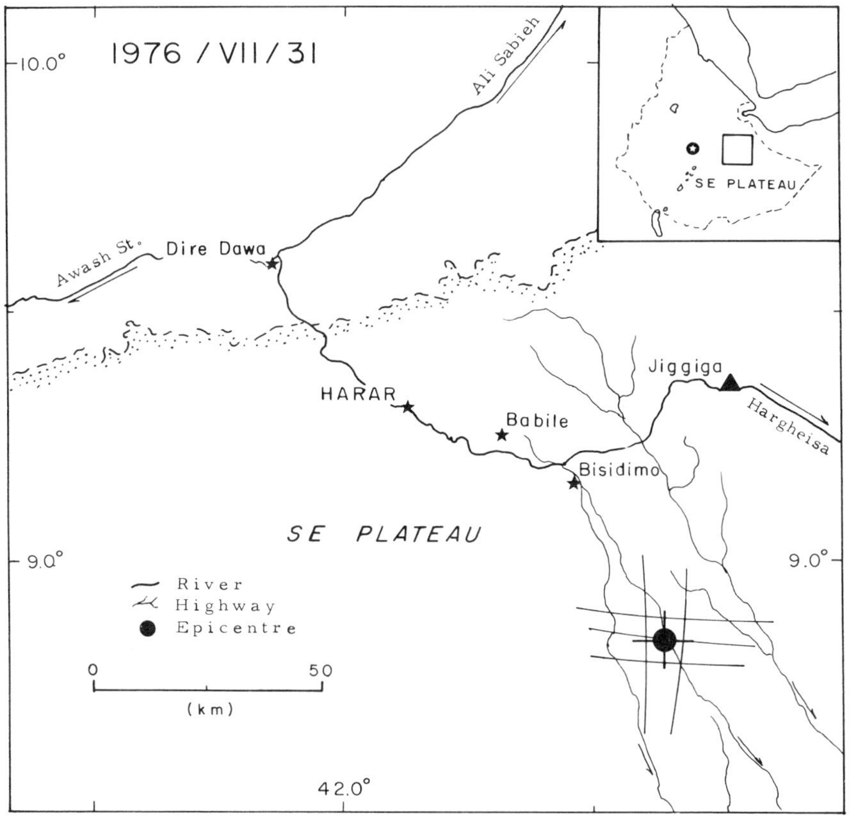

Fig. 68. Epicentre location for the earthquake of 31 July 1976. The star symbols indicate the sites from where tremors were reported felt; the arc-segments surrounding the epicentre are centred on the stations at Addis Ababa and Debre Zeit to the west and the IPG network of stations (Djibouti) to the north.

1976/VI/13

At U.T. 01:05.0 on 13 June 1976, a magnitude M_L (AAE) 3.2 epicentre was located at about N 08.8°, E 41.1°, southeast of Asba Teferi on the SE Plateau. The location, based on two station reports only (Addis Ababa and Arta) might not be more accurate than ±0.2°. The information is given because of the unusual location of the epicentre (see Fig. 65).

Sources

AAE and Arta Data Files.

1976/VII/31

Two tremors shook the region of Harar on 31 July 1976. The first occurred at 5:30 a.m. and the second at 2:30 p.m. By order of amplitude, the reported intensities — which by no means represent the total number of sites at which tremors were felt during the second shock — were:

V in Harar city (N 09.4°, E 42.1°) — people rushed out of their houses; part of the plaster ceiling in Kedous Michael Church fell and cracks opened in many walls in the town.

IV^+ in Babile (N 09.3°, E 42.3°) — noises were heard; bottles fell from shelves at the mineral water bottling plant; people were highly frightened.

IV^+ in Bisidimo (N 09.2°, E 42.4°) — no material damage.

IV in Dire Dawa (N 09.6°, E 41.9°) — some people left their houses. No material damage.

Sources

AAE Data File; *Ethiopian Herald* for 1 August 1976.

Comments

The epicentre based on the AAE and Debre Zeit records (Ethiopia) and on those of the Djibouti network was located at N 08.8 ± 0.1°, E 42.6 ± 0.1°, (Fig. 68).

The Earthquake History of the Ethiopian Main Rift, Afar, and the Southern Red Sea (Region C)

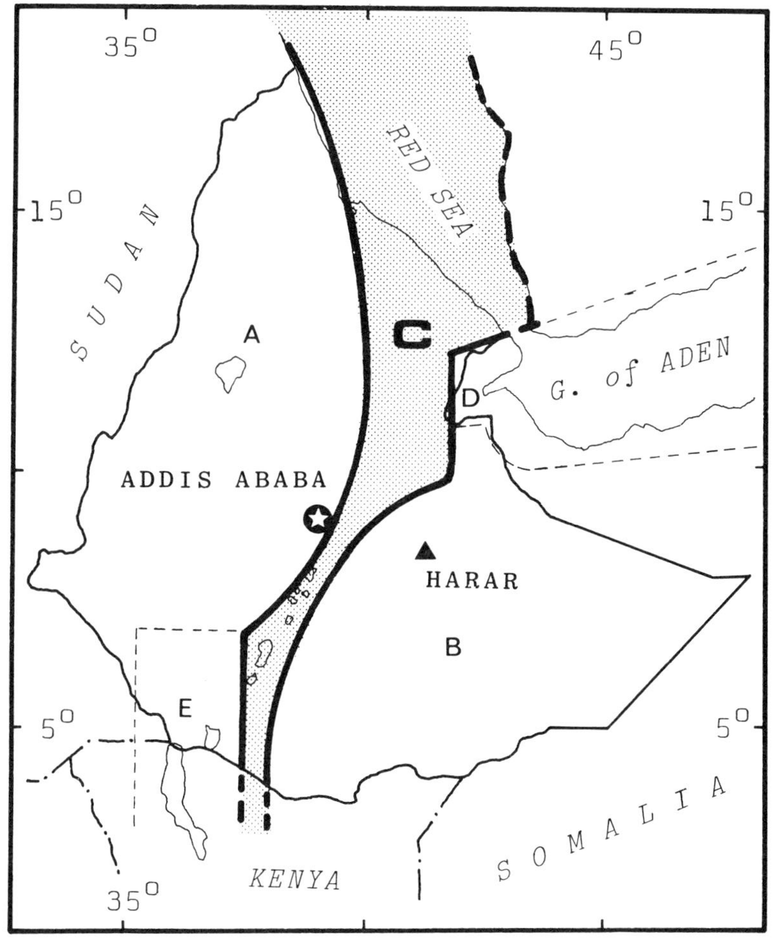

It is stated in the Introduction that volcanism in Ethiopia and in the northern sector of the Horn of Africa has always been intimately linked with crustal and lithospheric zones of weakness and therefore with seismic zones. Volcanic activity dates as far back as early Miocene. What is of prime interest to this survey is: (1) the volcanic activity that presently increases the degree of natural hazards; and (2) the location of volcanic centres that could give us a clue to potentially active tectonic structures. Figure 3 maps the most important volcanic units that can still be identified. Those located on the plateaus cluster either east and southeast of Lake Tana on the Western Plateau or near the western shoulder of the Southeastern Plateau at the latitudes of lakes Shala and Awasa. No distribution pattern is discernible among these units. From the viewpoint of potential hazards, these volcanoes are presumed dead.

Such is not the case for the rift volcanoes (see Fig. 3). They line up along active tectonic structures. Three alignments are identified by numerals on this map. Alignment 1 coincides with the rift axial fault zone, the Wonji Fault Belt, from the south end of the rift valley to lake Gamarri in central Afar; a second alignment, comprised of many active units, tops the

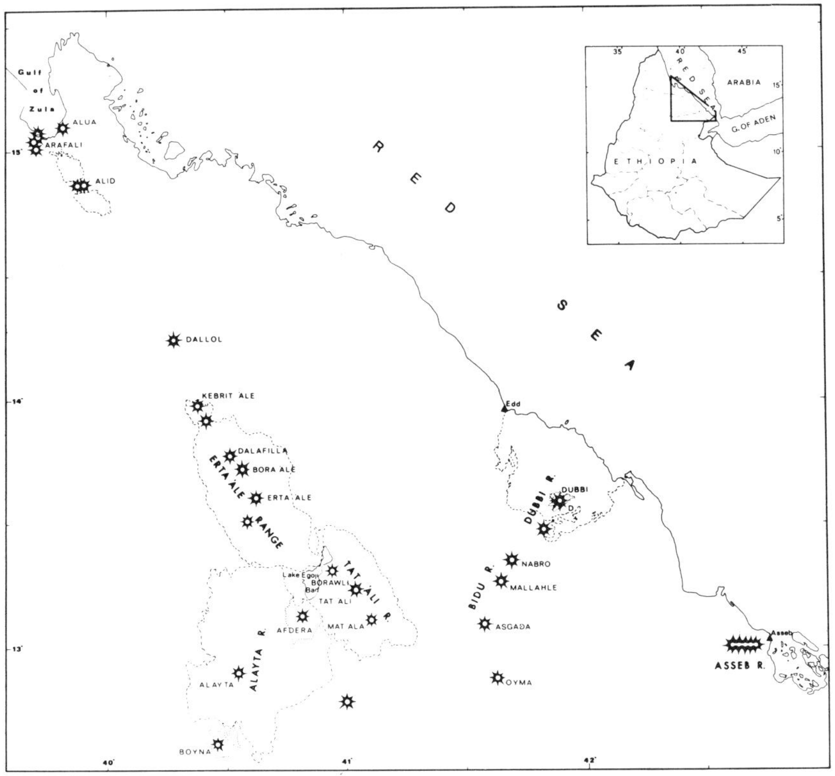

Fig. 69. Identification map of important volcanic centres in northern Afar.

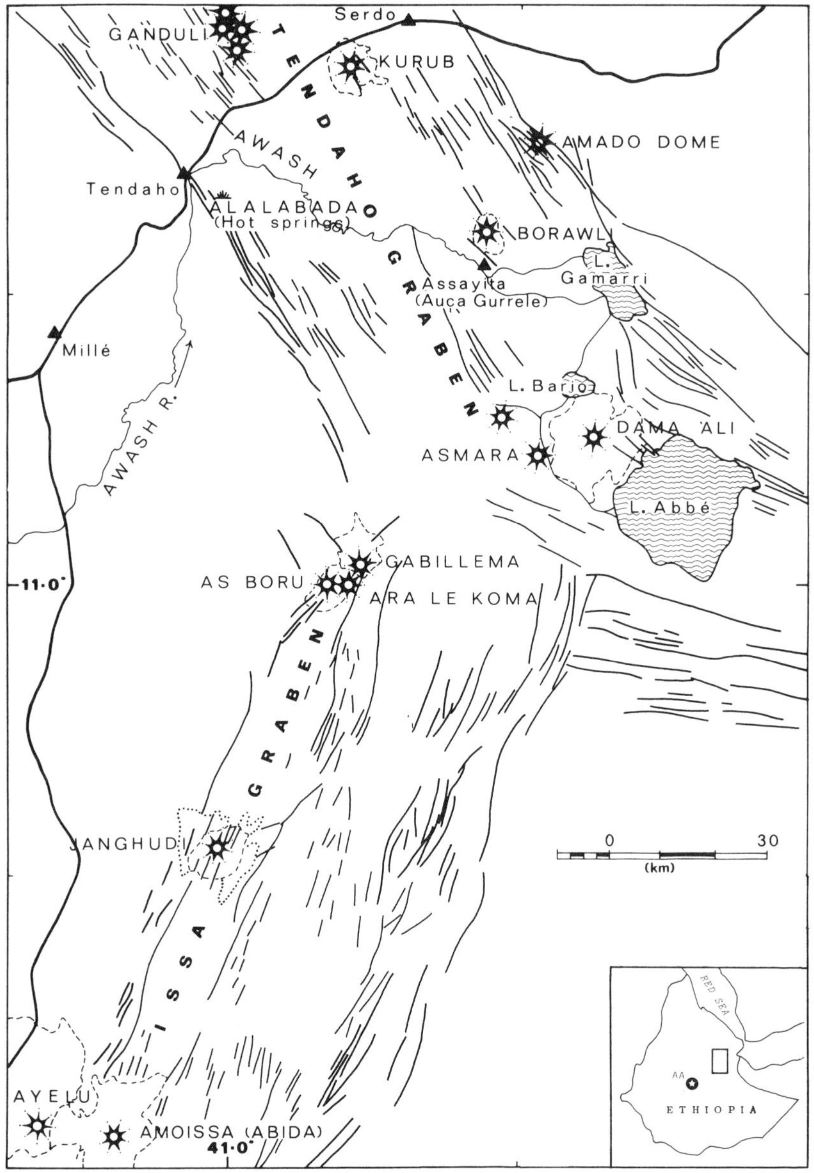

Fig. 70. Identification map of important volcanic centres in southern Afar.

axial graben of the Afar depression from central Afar north-northwestward to the Gulf of Zula. This zone is recognized as an area of crustal spreading (Tazieff and Varet 1969). Finally, the map identifies the Bidu-Dubbi volcanic range heading NNEward toward the Rea Sea (3). The units of group (4) could easily be considered as tail volcanoes of the first two alignments. Sections of these alignments might be artificial. What is important is to realize that younger volcanoes and potentially active fissures are located within seismic zones and that their presence increases the hazards to life and property. Fortunately the most active volcanoes are located in sparcely populated areas of Ethiopia such as the Danakil desert. Even so, entry 1631/II/10 describes the destruction of Assayita (Auçâ Gurelle) and entry 1861/V-IX relates the eruption of volcano Dubbi, which presumably destroyed the villages of Moobda and Ramlo, caused over 100 casualties, and killed herds of domesticated animals. Figures 69 and 70 identify by name the important volcanoes of Afar.

On the main rift floor there is also evidence of recent fissural and central activity; as for instance, the lava flows on the southwestern flank of the volcano Fantale (N 08° 58′, E 39° 54′). Perhaps not later than 150 years ago, the Sabober — a satellite centre of the Fantale — spilled extensive flows of scoriaceous olivine basalts that reached the shores of lake Metahara (Mohr 1962b; Gibson 1967). Tradition has it that these flows destroyed in their path a village and its monastery (Azaïs and Chambord 1931). This surge of activity is still remembered, at least by the threat that parents in the locality use to scold their naughty children: *You'll melt like the Fantale.*

The flows are physical evidence of the recent activity, but the destruction of the village might more objectively be attributed to war during the reign of negus Dawit (1382–1411) than to a volcanic eruption (Gebre Kaustos manuscript, Ethiopian Hagiographical Sources, National Library, Addis Ababa).

The most recent instance of fissural activity dates from the early 1900s; it is a flow of pitchstone ejected along the Wonji Fault Belt a few kilometres southeast of lake Galilee (Koka). Eyewitnesses are still alive and were interviewed by Dakin and Morton in 1975 (personal communication). Since then many instances of volcanic activity have been reported to the Geophysical Observatory; none have been substantiated.

1400/Dubbi

During the summer of A.D. 1400, sailors on the Red Sea observed a violent volcanic explosion on the coast of Ethiopia. They reported that the seismic activity that accompanied the eruption lasted more than a week.

Source

An Arabic manuscript on the history of the Resuliyy dynasty in the Yemen written before 1414 by the 'Imam el-Khazrejiyy. (The English translation given below is from Redhouse 1704, volume II, p. 285. A translation in Italian and some commentaries are given by Palazzo 1915, p. 296–298).

One of those voyaging by sea told me that there happened in the 'Land of the Blacks' (Sudan) a tremendous earthquake that lasted continuously for a number of days less than ten, from which a number of places and many hills fell down. Thereafter broke out in a canton thereof an immense fire with a great volume of smoke and the inhabitants fled from the region. The fire continued from some days; the smoke became massed. Later, the smoke took solid form and became a group of hills in that locality where, up to that event, no hills of any kind were known. The whole of this occurred in the course of the latter half of the year Hegira 802 aforesaid. Inch Allah.

Comments

1. Meaning of the Term Sudan

Redhouse renders the Arabic expression *Biladu's Sudan* (literally "the Land of the Blacks") by the word Sudan, which should not be considered to refer to the present Republic of Sudan. *Biladu's Sudan* was in the 15th century a generic term applied by medieval Arab writers to a territorial belt crossing Africa from the western coast of the Red Sea to Cape Verde in the Atlantic. In this context, el-Khazrejiyy refers to the African shore of the Red Sea (see also Redhouse, III, annotation 1634, p. 226).

2. Location of the Observed Volcanic Activity

From Suez at the north end of the Red Sea to the Gulf of Zula in Ethiopia, there is no evidence of recent volcanic activity. Nor is there any along the southernmost portion of the coast between Asseb and the Gulf of Tadjoura. Therefore, the volcanic and seismic activity described by the sailors could have occurred only on the Ethiopian coast between N 15° and N 13°, that is between the Gulf of Zula and Asseb.

Figure 71 is a location map of the volcanic centres situated between N 15° and N 13° whose activity could have been directly spotted at sea. Near the 15th parallel there are two volcanoes, Alid and Asa Ali. Both are at a late stage of development (see entry 1901–02/Alid–Zula) and have certainly not been spectacularly active during the last 500 years although fumaroles have been reported on their flanks.

To the south between N 14° and N 13° two geologically recent volcanic groups can be seen from the sea: the Dubbi complex and the Asseb range. At these latitudes, there are other dormant and active volcanoes deeper inland; however, their activity would not likely be observed from the sea.

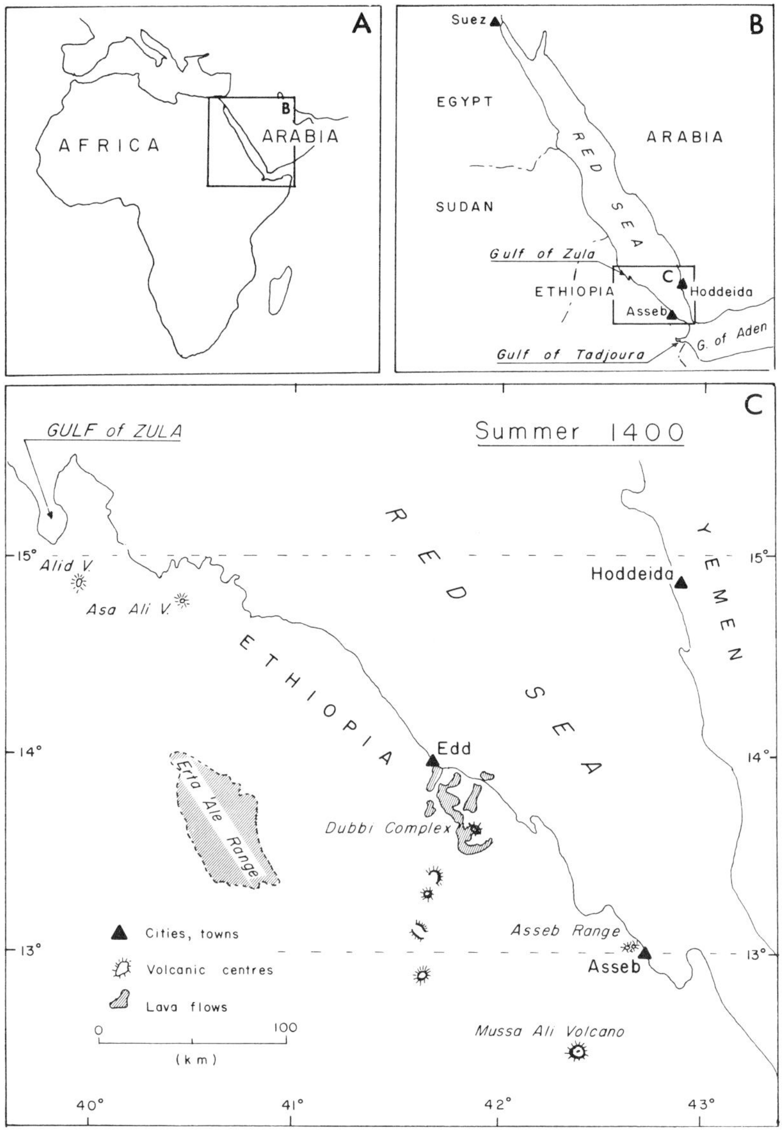

Fig. 71. Location map of the volcanic eruption and seismic activity in A.D. 1400. Section A shows the relative position of the Red Sea with respect to Africa, Arabia, and in particular, to Ethiopia. Section B is a location map of the sites mentioned in the text. Section C indicates the volcanoes in northeastern Ethiopia whose activity could have been observed from the sea.

The Asseb range is characterized by older fissure activity and more recent flows, along with well preserved spatter cones and two younger explosive craters (de Fino et al. 1973). The explosion of these two craters could have been observed at sea, but they probably predate 1400.

The phenomenon the sailors observed in 1400 was, in all probability, an upsurge of activity in the Dubbi *(Djebel Dubbeh)* stratovolcanic complex that towers to 1625 m above sea level. The explosion of one or many of its craters must have been, for the sailors, a spectacular sight. The description of the 1400 eruption given by el-Khazrejiyy matches that of 1861 when the lava flows were reported to have buried two villages and caused over 100 casualties (see entry 1861/V–IX).

3. Magnitude of the Largest Shocks

The probable peak amplitude of the seismic shocks that accompanied the volcanic eruption can be inferred from the intensity of the tremors and the distance at which they were reported. The present evaluation is based on Richter's empirical relations between the observed intensities, the radius of perception of the tremors, and the energy released at the origin (1958, p. 352–353). Richter's observations were made in California under geological conditions that are somewhat similar to those of the Red Sea coast of Ethiopia, i.e. the Danakil Alps and the Red Sea rift.

El-Khazrejiyy's text is not clear as to where the tremors were felt. It could be heresay from visitors to the Ethiopian coast, in which case no reliable epicentral distance could be estimated. On the other hand, the sailor-narrator was from the Yemen and the nearest port-of-call from Dubbi was Hoddeida *(al Hudayah)*, some 200 km to the northeast across the sea. One can assume, therefore, that the sailors heard about the tremors at Hoddeida for the very reason that they were felt there.

The accepted perceptibility level of earth tremors is between grades III and IV on the Mercalli-Modified intensity scale. At a radius of 200 km from an epicentre, earth tremors of intensity III–IV are generated, according to Richter, by earthquakes of magnitude about 5¾ (on the Richter scale). Therefore, an M of 5¾ is a minimum value as a minimum intensity was assumed. As a matter of fact, and as a basis for comparison, the earth tremors that accompanied the eruption of the same volcano in May 1861 were reported not only at Hoddeida but as far as Aden, a distance of 375 km from Dubbi (see entry 1861/V–IX). The magnitude of the main shock in 1861 was estimated as 5¾–6.

4. Changes in the Topography

The smoke that *took solid form and became a group of hills* would refer to the lava fields that reached the Red Sea shore; these can be distinguished from those of 1861 and are shown by shaded areas on Fig. 71.

5. Date

The year Hegira 802 began on 3 September 1399 (Julian calendar). The latter half of the Islamic year corresponds to the summer of 1400.

1608/XII/23

Secondary volcanic activity was observed in Aussa on 23 December 1608. *Smoke* came out of the earth at the foot of Mt Waraba (Doma 'Ali volcano ?), on the western shore of Lake Abhé. Estimated coordinates of the site: N 11.4°, E 41.7°.

Sources

An Arabic manuscript published with an Italian translation and annotations by Cerulli 1931, p. 87–88. The dotted line in the translation below corresponds to part of the original text that is missing.

The date when the smoke came out of the earth in Waraba near the lake east of Mount Waraba... was a Tuesday evening during the month of Ramadan, 15 days after the beginning of the fast, in the year Hegira 1017. Iman Umar-din al-Madayti lo Safi'ita was reigning.

Comments

Discussions on the origin of the document and on the probable location of Mt Waraba in Aussa are presented in entry 1631/II/10, which relates the surge of activity of the volcano and the destruction it caused.

1631/II/10

Seismic tremors were reported almost continuously over a period of several months in Aussa, southeastern Afar. The town of Waraba was destroyed. The seismic tremors coincided with volcanic activity.

Sources

An Arabic manuscript published by Cerulli 1931, p. 87–88. (The dotted lines indicate the parts of the original manuscript that are either missing or illegible.)

A great earthquake was felt in the city of Aussa on a Monday night during the unique month of Ragab. A second tremor was felt on Thursday morning on the 11th of the same month. During the night of Friday 12th, a tremor was felt and at about midnight a great fire was seen accompanied by a loud noise... On the 23rd of the same month at the time of the evening prayer, a strong earthquake destroyed all the huts in Waraba... about 50 people... The earthquakes lasted for six months. Inch Allah.

Comments

1. Location

The region of Aussa (spelled Auça on contemporary XVII century maps) was not always as well-defined territorially as the present *awradja* of Aussa, but it always had its centre in the fertile lower Awash Valley. The capital city in 1625–31 was Aussa or Auça Gurrelle in the vicinity of present-day Assayita (N 11°33′, E 41°27′).

The site *Waraba*, mentioned in the original Arabic document, refers to a mountain and a town not known by that name to contemporary geographers such as d'Almeida or Paëz. Cerulli (1931, p. 74) acknowledged that he could not identify the place. From the context, Waraba is different from the capital of Aussa, the present Assayita. (Note that the words *waraba* in the Somali language and *warabesa* in Galligna means "hyena." In that region, one finds names like "Hyena Hill" and "Hyena Well"; so it is probable that *waraba* was a popular nickname given to particular sites by the inhabitants of the lower Awash Valley.)

In another paragraph of the same Arabic document, the town called *waraba* is said to have been located at the foot of a volcano, west of a lake. Among the volcanic centres in that region, three could fit the description of the original text and were investigated:

(1) The Dama'Ali (or Doma Ale) volcanic complex on the northwestern side of Lake Abhe. Its coordinates are N 11°17′, E 41°43′. There are recent laval flows both inside its crater and on the outer flanks; the latter flows could well have destroyed a village in their path. Activity both in the crater and on the flanks of the mountain would explain the "great fire" observed by the inhabitants (Dakin et al. 1971, p. 24; G. Marinelli, personal communication).

(2) The volcano Gabillema (N 11.4°, E 41.3°) where recent lava has flowed through a densely inhabited plain surrounded by swamps (Marinelli, personal communication).

(3) The Amado dome, 12 km northeast of Assayita (Lloyd 1971, p. 203–4), which is the youngest rhyolitic cone in the area and where fumarolic and hot spring activities are still present (Fig. 72). Lacustrine sediments and preserved terraces to the east indicate the presence of a former lake. Marinelli, however (personal communication), is of the opinion that the Amado dome extruded gradually and never had activity violent enough to destroy villages.

Of the three volcanic centres, the Dama'Ali volcano remains the most likely site. It should also be noted that recently at least two earthquakes have been instrumentally recorded from the immediate vicinity of the Dama'Ali:

Date	H	Coordinates	
1938 27 Sept	U.T. 02:31:51	N 11.13°	E 41.74°
1965 07 June	U.T. 13:43:58	N 11.35°	E 41.53°

Fig. 72. Amado dome. Shekayto fumaroles can be seen at the right front of the dome; to the left of the dome are the extensive sinter deposits from old hot springs. The Gamari fault is in the background (photo L.L. Lloyd, November 1970).

For the location of the three volcanic centres mentioned above, see Fig. 70.

2. Damage

The manuscript mentions "huts destroyed" and "50 people . . . ". This last part of the text is illegible; it is presumed that the missing word was "killed." It is extremely unlikely that 50 people could be exterminated by a slowly progressing lava flow; there had to be an explosion.

3. Date

The date of the events is not clear. As the text is a marginal note adjacent to a paragraph on Imam Umar-din ibn Ahmaddin who reigned between Hegira 1036 and 1046 (A.D. 1626 and 1636), it is logical to assume that the event occurred during his reign. The note actually mentions a Thursday, the 11th day in the month of Ragab; the only Hegira year during the reign of Umar-din when the 11th day of Ragab fell on a Thursday was Hegira 1040 (A.D. 1631). If the assumption that the marginal note referred to the reign of Imam Umar-din is correct, it follows that the seismic and volcanic activity at Mount Waraba started on 10 February 1631, a Monday, corresponding to 8 Ragab A.H. 1040. The partial destruction of Auça Gurrele by earthquakes and volcanic activity in 1631 is substantiated by a contemporary document in Portuguese dated from Gorgora (Tigray, Ethiopia) in 1662: . . . *Afterwards in 1631, during the month of December, fire rained from the sky upon the city of Auça Gurrele and almost entirely burned it. This was known by letters which many people wrote from there to Ethiopia.* (d'Almeida 1662, III, Ch. 9). The author mentions December instead of February; he probably meant "winter 1631."

Cerulli (1931, p. 88) dated the seismovolcanic activity in Aussa A.D. 1627, the year corresponding to Hegira 1036. That year, A.H. 1036, 11 Ragab did not fall on a Thursday as indicated in the original text but on Sunday, 28 March. Cerulli's dating is therefore erroneous.

1838/II–III

From 25 February to the end of March 1838, a series of earth tremors were felt at Massawa (N 15.8°, E 39.6°) and at Muncullo/Imakullu, 12 km inland from Massawa. No damage was mentioned (d'Abbadie's spelling of Massawa is Muc'awwa).

Sources

D'Abbadie (1873, p. 107). (Marinelli and Dainelli 1912, p. 111–112; Palazzo 1915, events 13–21; Sieberg 1932. The original reports, part of which are given below, are from d'Abbadie, a geodesist who was at that time stationed in Massawa. The times of occurrence of the shocks are to be considered accurate as d'Abbadie used his astronomical chronometer for his observations. The time indicated by d'Abbadie is probably the Solar Apparent Time).

23 February at L.T. 23:18. *An earthquake was felt in Muc'awwa. It was accompanied by a loud noise from the sea, similar to that of a school of fish leaping out of the water.*

25 February at L.T. 23:20. *A light tremor was felt in Muc'awwa. I heard a sound like distant thunder accompanied by other noise like the breaking of mighty waves on the shore.*

27 February at L.T. 22:25. *Strong tremor felt in Muc'awwa. There were lead bullets upon a shelf in my house which fell down. From the direction of their fall, the direction of the seismic waves must have been east to west, that means at right angle to the main axis of the Ethiopian mountains.*

27 February at L.T. 22:48. *New tremor but weaker.*

27 February at L.T. 23:02. *Small tremor, weaker than the previous one.*

1 March at about 03.00 a.m. *Very strong earth tremor.*

22 March at L.T. 13:38:53. *In Muc'awwa a light tremor was felt. Its direction seemed to me to be from south to north; it lasted 2–3 seconds and was accompanied by a noise like distant thunder.*

Comments

D'Abaddie's reports on *loud noises from the sea* suggests nearby epicentres along the Massawa Channel and in the vicinity of the 1921 seismic region (Fig. 73). (For information on noises accompanying seismic activity, see Davidson 1938, p. 147–161; Hill et al. 1976, p. 1159–72 with a comprehensive bibliography on p. 1171–72.) Such an epicentral location would confirm d'Abaddie's observation of 27 February that the seismic waves — if they were P waves — came from the east. It is fully realized, however, that noninstrumentally detected directions of seismic waves are not to be given much credibility.

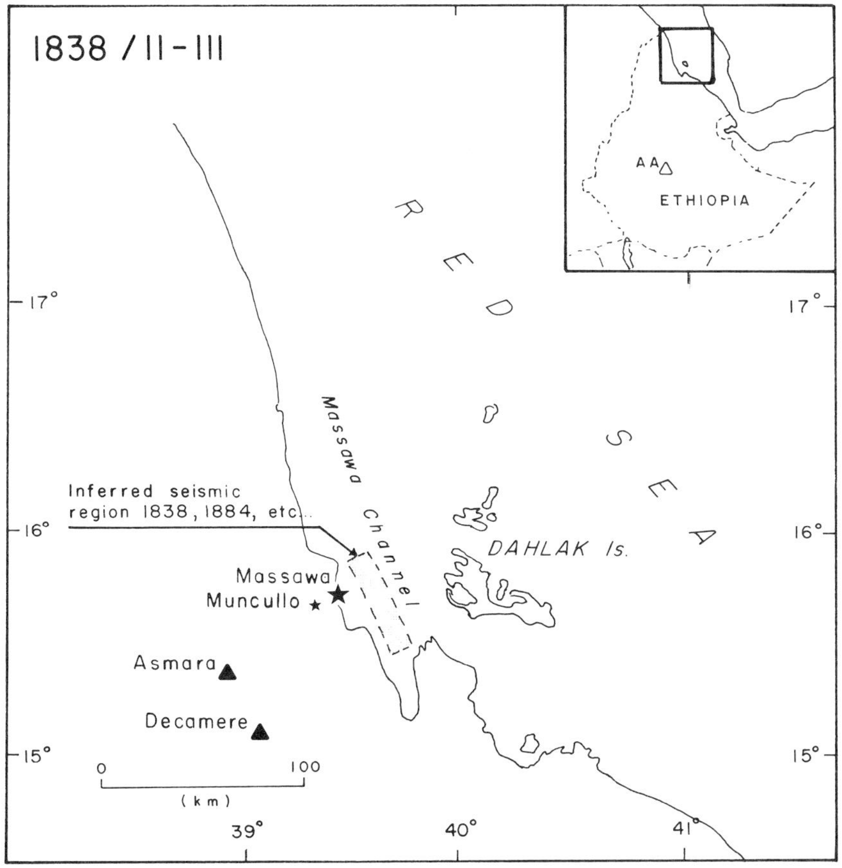

Fig. 73. Location of the seismic region presumed to have been active in February–March 1838.

If the maximum intensity in Massawa (Mac'awwa) was about IV$^+$and the epicentral region at about 20–25 km from the city, it follows that the magnitude of the main shock may not have exceeded 4.

1844/X/23

There are three reports of earth tremors having been felt near Massawa during October 1844. Two of these reports did not specify any date; the third one, signed by d'Abbadie is dated 23 October. The tremors were reported in Muncullo/Imakullu (N 15.5°, E 39.4°), 12 km southeast of Massawa.

Sources

D'Abbadie (1858, p. 86–87); Perrey (1859, p. 7). (Marinelli and Dainelli 1912, p. 112; Palazzo 1915, events 24–26. D'Abbadie makes no other comment than to quote his sources.)

Until historical evidence is found to the contrary, I believe the three reports refer to the same event, which d'Abbadie describes as follows:

On that occasion, on the sea coast north-east of Imakullu, a noise like that of canon-fire was heard; there were no battleships around at the time. The natives affirmed that a similar noise was heard 20 years ago.

Comments

I am not aware of any documents related to earthquakes occurring in the same area during the 1820s.

The adopted location for this event is in the Massawa Channel around N 15½°, E 39½° (see Fig. 73 and entry 1921/Massawa Channel).

1848/VII–VIII

During July and August 1848, swarms of earth tremors were reported in Muncullo, Massawa, and the Samhar region of Eritrea (see Fig. 73).

Sources

D'Abbadie (1858, p. 114–115 and 1873, p. 11), (also in Perrey 1859, p. 12; Marinelli and Dainelli 1912, p. 114; Palazzo 1915, events 32–37).

Comments

The list of reported tremors reads as follows:

13 July, noon: light tremor in Samhar.

1 August, midnight: strong tremor in Muncullo. Apparent direction of the waves: N-S

2 August, about 02 a.m.: second shock in Muncullo, weaker than first one.

3 August, 09–10 p.m.: very light tremor in Muncullo.

6 August, about 3 p.m.: light tremor in Massawa and Muncullo.

10 August, about 3:30 p.m.: tremor of intensity IV (MM) in Muncullo. *Plates were rattling on their shelves.*

Tremors of intensity IV in Muncullo suggest a magnitude 4 for a shock originating in the Massawa Channel.

1857/Spring

Marinelli and Dainelli (1912, p. 115) quoting Munzinger reported that frequent shocks were felt in Tigray and Eritrea. They were light in Massawa; damage to an unidentified town was reported from Tigray.

Source

Marinelli and Dainelli (1912, p. 115).

1861/V–IX

Eruption of the volcano Dubbi *(Djebbel Dubbeh:* N 13.5°, E 41.8°) on the Red Sea coast of Ethiopia. The eruption was accompanied by strong earthquakes that were felt over a minimum radius of 200 km. Two villages were destroyed, animals killed, and 106 people lost their lives. The tremors were felt during four months.

The first phase of the volcanic and seismic activity started simultaneously on 7–8 May 1861. Earth tremors were reported at Edd (N 13.9°, E 41.6°) and Massawa (N 15.8°, E 39.6°) on the African coast of the Red Sea, and at Hoddeida (N 14.8°, E 42.9°) and Mukka (13.3°, E 43.3°) on the coast of Yemen. Gutenberg and Richter (1954) determined the epicentral coordinates of the main shock (7 May U.T. 22:20) as N 13.74°, E 41.55°.

A renewal of volcanic activity probably occurred during September; no further earth tremors were reported (Dainelli 1943, II, p. 649).

Sources Concerning the Seismic Activity

Perrey (1864a, p. 95–99). (Palazzo 1915, p. 311; Sieberg 1932, p. 885; Dainelli 1943, III, p. 648–649).

Sources Concerning the Volcanic Activity

Times, London, 20, 21 June and 24 September 1861; *Opinion Nationale*, Paris, 29 October 1861; Steudner (1864, p. 115); von Heuglin (1868, p. 329); Dainelli (1943, III, p. 631, 649); CNR-CNRS (1973, p. 465); Richard and van Padang (1957, IV, p. 15–16, including a bibliography); Sapeto (1879).

Comments

A description and location of volcano Dubbi has been given in entry 1400/Dubbi; incidentally, the 1400 eruption is not mentioned in the current literature on volcanoes. The present description of the 1861 eruption is based on reports from eyewitnesses and from investigators who visited the site shortly after the eruption.

1. Sequence of Events

8 May around 02:00 a.m.: Earth tremors were felt in Edd, a small village located on the coast at the northern margin of lava fields, some 50 km from the centre of the Dubbi complex.

8 May after sunrise: Fine whitish ash started falling. It completely darkened the sky and covered the southern Red Sea and the coast of the Yemen. At noon, the colour of the ash changed from white to red. At the end of the day, in Edd, the ashfall was knee-deep. From the Dahlak Islands, some 300 km to the north of the volcano, one seaman reported that he could not keep the deck of his ship clean, even by sweeping it continuously.

8–9 May at night: For the first time people in Edd realized that a volcanic eruption was going on. When the opacity of the dust filling the air had sufficiently decreased, they noticed fire and smoke over the Dubbi mountain to the southeast.

17 July: Steudner (1864, p. 115) and von Heuglin (1868, p. 329) reported that rain mixed with ash fell on the Western Plateau. Dainelli (1943, III, p. 631) admitted the possibility that such a phenomenon could have been connected with the Dubbi eruption in May but refused to comment on the grounds that he could not obtain any sample of this "mudfall" for analysis. It must be noted here that from October to May, the direction of upper air currents over NE Ethiopia is normally NE-SW due to a stable barometric high over the Arabian peninsula; during that period, the dust could have been transported over the Plateau. Later on, during the rainy season in late June (Tato 1964, p. 29), the air currents reverse direction by 180°; the winds then become SW-NE. Such a meteorological phenomenon may not favour the assumption that the "mud and rain" fall reported over the Plateau was linked with the Dubbi eruption, unless the ash, having been carried away to the SW during the dry season, was still suspended in the air and fell during the rainy season (?).

September 1861: Dainelli (1943, III, p. 649) indicated that a second eruption took place in September of the same year. Noises similar to those of May were again heard as far as Massawa; no earth tremors, however, were felt. There was a new fall of ash and lapilli over Edd. This time lava was ejected and the flow took the direction of the sea shore. Unfortunately,

Dainelli does not quote his primary sources. To my mind, this report by Dainelli might be questionable.

Other reports on subsequent activity:

Summer 1863: Many authors interpret Sapeto's report (1879) about some ash fall over the southern Red Sea as an indication that Dubbi had a spell of activity in 1863. I do not think that this interpretation is valid. In fact, Sapeto is suspected of having written his notes not while in the "field" but after his expedition was over. Memory can be tricky and an error on dates can easily happen, for instance 1863 for 1861!

About 1900: The CNR-CNRS group (1973, p. 465) reports lava ejection "some 70 years ago." No documents that I know of confirm this statement.

2. Earth Tremors and Noises during the Volcanic Activity

Tremors were also felt on the eastern side of the Red Sea along the coast of Yemen, particularly at Hoddeida (N 14.8°, E 42.9°) and Mukka (N 13.3°, E 43.3°) and as far as the mountain range, 50 km inland. The tremors were accompanied by sharp loud noises very similar to cannon shots. They were heard distinctively enough in Massawa (330 km away) to make the authorities suspect that the city of Dessie was being bombarded and send an investigation party. In Perim Island (200 km), people also had the same impression of a cannonade along the coast.

3. Type of Volcanic Eruption

The volcanic activity observed at Djebbel Dubbeh in 1861 (Fig. 74) can easily be compared to the type observed at the Laimawa volcano in Hawaii in 1960 (Kuiper 1965, Comm. No. 49).

4. Loss of Life and Damage to Property

During the very first days, there were an estimated 106 casualties. Other damage included the loss of large herds (in terms of damage to property, it has to be remembered that for desert people the herds are their whole wealth) and the destruction of two villages, Moobda and Ramlo.

5. Excerpts from Primary Sources

The excerpts are first-hand detailed descriptions on the 1861 volcanic and seismic activity of the Dubbi volcano. Even if lengthy, they are introduced here because the original documents are not readily available and their existence is often not known.

Report from a Resident of Edd — Edd is a village located along the coast, 50 km north of Dubbi. Edd, at the time, was a refuelling port-of-call for shipping companies. The report was submitted to Capt. Playfair, British Political Resident in Aden:

On the night of the 7th or the morning of the 8th of May, the people of Edd were awakened by the shock of an earthquake followed by others which continued with little intermission for about one hour. At sunrise, a quantity of fine white dust fell over the village, like rain. About noon, the character of this dust appeared to change and then resembled red earth.

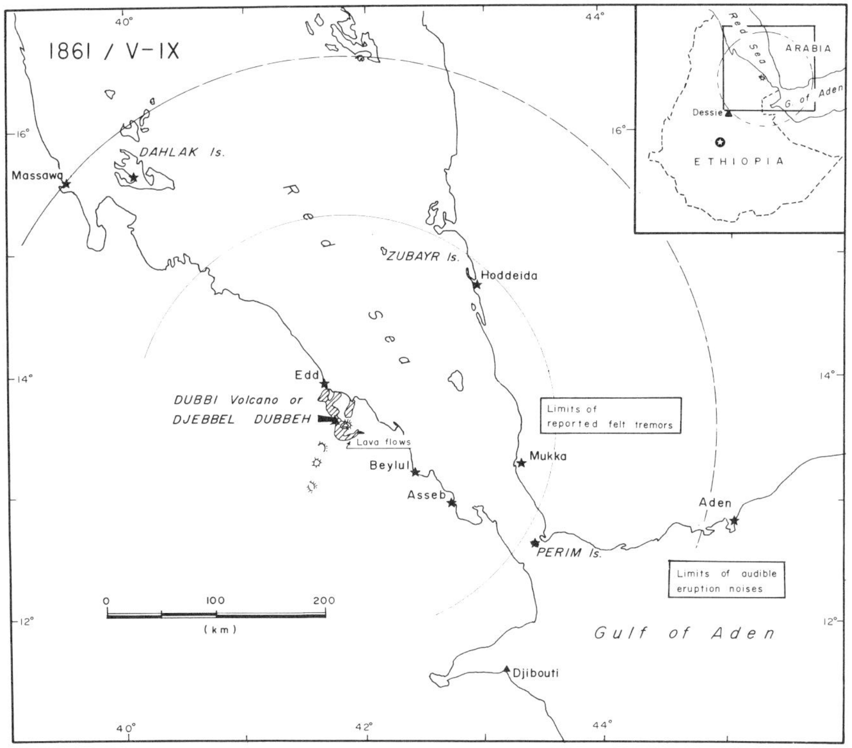

Fig. 74. Location of volcano Dubbi (Djebbel Dubbeh) and vicinity from where the phenomena that accompanied the 1861 eruption were observed.

Shortly afterwards, it increased to such an extent that the air was perfectly darkened and we had to light lamps in our houses. It was darker than the darkest night, and the whole place was covered with dust, nearly knee-deep.

On the 9th, the dust somewhat abated, and we were able to see a little in our houses without light. At night, we saw fire and dense smoke issuing from a mountain called Djebbel Dubbeh, situated a day's journey inland; and this continued all the time I remained in Edd. The ashes only fell for two days. Sounds like the firing of guns issued from the mountain. This mountain is inhabited, but no one had reached Edd thence when I left. Nothing of the kind had ever been heard of before, and the people were exceedingly frightened (Times, London, 21 June 1861).

Report from Captain Playfair — Playfair visited the volcano shortly after the eruption (Playfair's original text, was published in French in the *Opinion Nationale*, Paris, 29 October 1861).

The summit of the mountain looked whitish, but it had been blackened by fire. We dug in the ash 1½ feet before reaching the earth. On the mountain we observed 19 craters, 18 of which were smoking day and night, and were bright as a lamp.

A very large crater, about 100 arm-lengths long and 100 wide, burns night and day and ejects enormous stones. When a stone was projected, it went straight up until it appeared not any greater than a crow, and then it fell straight back and was crushed into atoms. When fire and stones were ejected from the crater, their ejection was accompanied by noises like the roaring of cannons.

During the time the stone was in the air, we heard a muffled noise; when it came down, it exploded and another one was ejected in its place.

We saw another marvellous thing on the volcano. At about 15 arm-lengths from the fire, water was abundantly flowing from the earth like sea waves. When fire was coming out, the water was ejected upwards to a man's height. Then and after the explosion, the water returned from where it came. The place where the fire is now produced is called Arooma. The eruption started during the night of Wednesday, Shewal 27, 1277 (May 8, 1861).

The name of the mountain is Dubbeh and is at a distance from Edd that a man can walk from sunrise to 5 p.m. The names of the villages burned which were located near the mountain, are Moobda and Ramlo. One hundred and six men and women were killed and their bodies were not recovered. The number of animals which perished is unknown. The distance from the foot of the mountain to the top is 2 hours by foot.

Note: The whitish colour reported by Playfair probably refers to a layer of sublimates, soluble in water, as has also been observed in other instances, for instance at Nyiragongo in Zaire.

Conditions in the Red Sea — A report to the editor of the London Times filed by Charles Beke, a former British Envoy to Ethiopia (*Times,* 20 June, p. 10; 26 June, p. 6; and 24 September, p. 10, London, 1861).

This acount (referring to the report of Capt Playfair) *has been amply confirmed from other sources, and the most remarkable feature of the case is the immense extent affected by the disturbance.*

At Massawa and Dahlak Islands — the supposed firing of guns was heard as coming from Ansley Bay, and so exact was the resemblance that the whole town was in a state of consternation. Note: Ansley Bay is suspected to be the Bay of Anfile, 175 km southeast of Massawa. Its direction with respect to Massawa is SE and that of Dessie is SW. There is therefore an apparent contradiction in the interpretation of the direction from where the "sound of bombing" was coming.) *It was thought that the French were bombarding Dessie, and the authorities dispatched special messagers to ascertain the cause of such an unaccountable proceeding. The nacoda* (the captain or master) *of the boat which brought the news from Massawa to Aden was detained 10 days in the Dahlak Archipelago, unable to continue his voyage owing to the dense cloud of dust which darkened the air. Many other nacodas reported the same thing, and one brought a specimen of the dust. He said it fell in such quantities that he could not keep his poop clean by continual sweeping. The dust appeared like very fine powdered pumice, containing minute particles of mica. Although the greater part of the shores of the Red Sea are of igneous origin, no active volcanoes have been known in modern times, save in the Zeberi Islands (Zubayr?) one of which was observed in a state of activity by the commander of the Indian Navy Steamer Victoria in, I think, August 1846.*

On the Coast of Yemen — On May 8th, shocks of an earthquake were felt at Mukka and Hoddeida, and there, as along the entire coast of Yemen, and inland as far as the mountain range the dust described as the "white ashes" fell for several days; the noises were also heard and, as usual, were attributed to artillery.

Southern Red Sea — Both the steamers Candia and Ottawa reported having had two very hot days in the lower part of the Red Sea, and on the 10th they encountered what appeared like "London Fog", which continued for several hours. The captain of the latter vessel described this fog as consisting of very fine dust, so that he could not see the length of the ship, and during its continuance the weather was perfectly calm.

In the Straits of Bab el Mandeb at the Entrance of the Red Sea into the Gulf of Aden — At Perim, the sounds emitted by the volcano were distinctly heard and they were attributed to a bombardment. The firing (as it was supposed to be) commenced at about 2 a.m. on May 8th, and continued with long intervals up to the 10th and 11th. The general idea at Perim was that the sound proceeded from the African Coast. The 'firing' on the 8th was heavy, and continued for 9 or 10 hours.

6. Regional Seismic Activity

During the 1400 eruption, earth tremors were reported by sailors on the Red Sea and presumably on the coast of Yemen. The minimum distance from the volcano to the inhabited coast of Yemen is about 150 km. Tremors strong enough to be felt at such an epicentral distance under rift crustal conditions could not be caused by shocks with magnitudes any lower than 5½ or 6 (see entry 1400/Dubbi).

For the 1861 eruption, Gutenberg and Richter (1954) calculated that the main shock was of "class d," that is of magnitude 5.3–5.9. Since then, at least three instrumental epicentres were located to the NE of Dubbi (entries 1950/IX/18, 1958/II/13, and 1960/XII/16). The location of the Dubbi complex on a major fracture transversely cutting through the Danakil mountains suggests that seismic activity is expected to continue in the future for the following reasons: (1) the fracture behaves as a hinge zone that already allows a 15° change in the axes of the Danakil Alps to the NNW with respect to the Danakil Horst to the SW; (2) the fracture's length extends from the SE end of the Bidu volanic group (CNR-CNRS 1973, p. 474) to, most probably, the eastern shore of the Red Sea (entries 1958/II/13 and 1960/XII/16); and (3) the zone has been sporadically active since prerift times.

Fortunately, Dubbi is located in a very sparcely populated area. The nearest village of any size is Edd (or Idi), at present a hamlet of some 400 households located at the very edge of the volcano's lava field, 50 km WNW of the Dubbi centre. In 1861, Edd was covered by some 50 cm of ash and lapilli. No other serious damage to Edd was reported. Another village in the vicinity is Beylul, which is at approximately the same distance as Edd, to the southeast of Dubbi. And then, there is the port of Asseb at a distance of 120 km. It should be remembered that there is an oil refinery in Asseb!

1864/III/05

On 5 March 1864, at about 8 p.m. an earth tremor was reported from Massawa; no indication of the intensity was given (see Fig. 73).

Sources

Blanc (1865, p. 78), (Marinelli and Dainelli 1912, p. 115; Palazzo 1915, event 49).

1864/IX/14–15

Two shocks were reported from Massawa (N 15.6°, E 39.5°); the first at 11:15 p.m. on 14 September and the second on the following day at 1:45 p.m. No damage was reported.

Sources

Blanc (1865, p. 80); Perrey (1868, p. 75–76); Rassam (1869, I, p. 30). (Palazzo 1915, p. 311–312).

Comments

There are reports of four different earth tremors having been felt between 12 and 15 September. One reporter is the anonymous author of the meteorological report from Massawa for September, 1864 (Blanc, p. 78). In a similar report for the same month, Blanc mentions three tremors: two at 11:00 hours on the 13th and one at 11:00 (a.m. or p.m.) on the 15th (Blanc, p. 79–80). The fact that four tremors were reported by different persons at exactly the same hour (11:00 a.m. or p.m.) on consecutive days is suspicious. Palazzo (events 50–53) accepts these reports as referring to different events. Perrey is of the opinion (cat. for 1864, p. 75–76 and addendum for 1868, p. 22) that the four reports refer to one single event (cf. also Rassam 1869, I, p. 30). Personally, I prefer to give credit to Rassam who was living at Massawa at that time and recorded in his diary one shock at 11:15 a.m. on the 14th and one at 1:45 p.m. on the 15th.

On these events, we also have the entries in the logbook of the SS Dalhousie, which was anchored in Massawa harbour; they are reported by Rudolph (1894, II, p. 572–573) in his study on submarine earthquakes and volcanic eruptions. Rudolph dated the first on 13 September (instead of 14) at 11 p.m. and the second on 15 September at 11 a.m. It is not clear from Rudolph's text whether he quoted the ship's log directly or through Perrey's version (M. A. Br. 18, 1866).

1864/X/21

On 21 October at 8:45 a.m., an earth tremor of estimated intensity V was felt in Massawa. No damage was reported.

Sources

Rassam (1869, I, p. 43). (Perry, suppl. for 1868, p. 32; Palazzo 1915, event 54).

Reports Rassam:

There was a severe shock of an earthquake at 08:45 this morning. The vibrations were so strong that Dr. Blanc and I rushed out of the house for fear that it might fall upon us.

D'Abbadie adds:

It was so strong that all the people had to rush out of the houses (in Perrey, p. 32).

1883A

Prior to 16 April 1883, seven tremors were felt near Mount Janghudi (N 10.5°, E 41.1°) in southern Afar.

Sources

Antonelli (1889, II, p. 541). (Martinelli and Dainelli 1912, p. 115; Palazzo 1915, event 58).

Comments

Antonelli does not indicate the exact dates on which the tremors were felt, but his diary is dated from Gawani, 16 April 1883. The earthquakes, therefore, occurred prior to this date. See comments and maps in entry 1883B; 1938/IX, X, and 1971/XII/05–19.

1883B

Earth tremors were reported in Asseb (N 13.0°, E 34.5°) during the summer of 1883.

Comments

Marinelli and Dainelli (1912, p. 115) include in their catalogue a report from Licata that earth tremors were felt in Asseb during the summer of 1883. Licata's text is so vague that one may strongly question the scientific value of his report. Asseb, however, is a site where earthquakes, even of moderate magnitude, originating either from the Red Sea central trough or from central Afar could easily be felt.

In the previous entry (1883A), Antonelli reported having felt earth tremors while at camp near Mt Janghudi (N 10.5°, E 41.1°). The distance between Asseb and Mt Janghudi is barely 300 km. If the epicentres of 1883 were located in the Aussa territory, where earthquakes have often occurred (among other entries see 1631/II/10 and 1969/III–V (Serdo-Central Afar)) the tremors could easily have been felt at both places. Entries 1883A and 1883B could refer to the same events; in such a case, if the epicentres were in Aussa, the estimated magnitude of some of the shocks would be above 5.5 or 6. The difference in time between the two reports, that is April and Summer, is not significant. On the other hand, epicentres are known to have been located near Gawani and Mt Janghudi (see entries 1938/IX, X, 1971/XII/05–19), and the tremors of 1883 could have originated there as well.

1884/VII–X

Between July and October 1884, a series of earthquakes, whose epicentres were mostly probably located between the mainland and the Dahlak Archipelago, devastated the coastal region near Massawa. The shock on 20 July, of estimated magnitude 6.2, triggered a small "tidal" wave. High water waves (seiche type) built up in the harbour at Massawa, especially in the bay between Taulud and Edaga Barai, and swept over the causeway; ships in the harbour were seen to rock violently. Several times the sea drained off so rapidly that fish were seen stranded on the exposed seabed.

Sources

London *Times*, 25 July 1884; Winquist (1884, p. 90–94); Detaille (1885, p. 185–187); Fuchs (1886, p. 153–154); Milne (1912, p. 83); Palazzo (1915, p. 313–319); Sieberg (1932); Gutenberg (1929, IV, p. 886); Gutenberg and Richter (1954).

Reports from Massawa

(1) A dispatch concerning these earthquakes was cabled from Suakim (Sudan) to the London *Times* on 25 July 1884. The original text of the cable was kindly supplied to me by the Editor of the *Times:*

Further particulars regarding the earthquake in Massawa state that nearly all the houses in the town have been destroyed or damaged by the shock. All ships in the harbour were seen to rock violently. The inhabitants were greatly alarmed and fled to the mainland.

(2) The following report is taken from the diary of a Swedish missionary, Miss E. Winquist, who was in Keren, about 100 km inland from Massawa, at the moment of the first destructive shocks and who visited Massawa and Muncullo a few days later:

In Massawa the water had risen so high it was feared that it would take the whole town back into the depth of the sea. The destruction was dreadful (Winquist 1884, p. 93).

(3) Almost 30 years later, Palazzo collected the following information from the inhabitants of Massawa (Palazzo 1915, p. 314–315):

One Sunday morning in June or July 1884, at about 09:30, a light seismic tremor was felt. It was accompanied by a noise similar to that of a distant explosion. About one hour later, there was a much more violent shock. At that moment, between the Taulud-Edaga Berai causeway and Scheik Saïd Island and also in Gherar Bay, the waters suddenly rose to a height of 20 cm above the causeway and then rapidly receded far beyond the shore leaving fish wriggling on the dry sea bottom. This was repeated two or three times in the space of a few minutes.

Two or three masonry houses were heavily damaged, large fissures appeared in the walls, but none collapsed. Cooking pots on the fire in preparation for the feast marking the end of Ramadan, a few days later, toppled over. (In 1884, the end of Ramadan was on 26 July.)

On that day, more than 20 tremors were felt in Massawa but were of less intensity than that of 09:30. At the end of August, other tremors were also felt.

During the period of seismic activity, the citizens of Massawa living in masonry houses moved out to live in Otumlo in improvised branch huts or slept in the open; the foreigners took refuge in the Swedish mission. After two to three months, they moved back to Massawa.

It is thought that more shocks were felt in the Dahlak Islands but were less destructive.

Report from Keren (N 15.8°, E 38.5°, 115 km WNW of Massawa)
One evening (the main shocks occurred in the morning) the walls of

the room in which we were sitting suddenly started to rock to and fro. The table dropped; and the petroleum lamp standing on the table luckily went out. This was the beginning of the earthquakes which came when one least expected them. I lived in permanent suspense and anxiety (Winquist, p. 90; Keren is not explicitely mentioned — it is inferred from the context, as is the date 12 July.)

Report from Muncullo (about 15 km west of Massawa)

When we put up our tents the night before reaching Muncullo (the party was coming from Keren), *we heard the well known rumbling in the earth announcing an earthquake. Suddenly, the earth under us started to go up and down like the waves on the sea. The army camp next to us became dead silent ... when we rode into the mission station early next morning an unexpected view met us: the roof of missionary Samuel Bengysson's house had collapsed and everywhere big cracks appeared in the walls. We built grass huts not too far away in the desert because we did not dare to pass the night in the large stone house* (Winquist, p. 94).

Comments

1. Damage to the City of Massawa

The city of Massawa is comprised of two small islands, Taulud and Scheik Saïd; two peninsulas, Gherar and Abdul-Cadir; and a section on the mainland, Edaga Berai (see Fig. 75). The dam or causeway between Taulud and Edaga Berai is at present about 150 cm above mean sea level. The mean sea level in 1884 is not known but, according to the Harbour Authority, it should not have been much different from what it is now.

The damage to the houses in Massawa, as described to Palazzo some 30 years after the earthquakes, does not seem to correspond exactly to the content of the cable to the *Times* nor to what Winquist witnessed in 1884. Thirty years is likely to render reminiscence rather hazy. Sieberg in *Erdbebengeographie* (1932) estimated the damage in Massawa at VIII–IX° on the Mercalli-Sieberg intensity scale. Milne (1912, p. 83) in his *Catalogue of Destructive Earthquakes* classified it of grade III, that is the highest grade in his scale of three units. We do not know on which documents Sieberg and Milne based their estimates of intensities; they correspond to the description given by Winquist and by the *Times'* correspondent. It must also be noted that Massawa is built on an uplifted reef and that lightly consolidated geology amplifies seismic tremors.

2. Date of Larger Shocks

There are certain ambiguities in the different historical sources as to the exact dates of occurrence of these events. One source of confusion is the fact that some authors mistook the date of the telegram from Suakim to London for the date of the main shock. The instrumental data used by Gutenberg and Richter are the best evidence that the main shock occurred on 20 July; this date is confirmed by the information collected by Palazzo that the shock hit Massawa at the end of Ramadan marked by the feast of Baïram, 26 July in 1884.

The sequence of the major events as given by Palazzo has been adopted. It is reproduced in Part II, p. 189.

3. Estimated Magnitude and Location of the Main Shock

Gutenberg and Richter evaluated the magnitude of the main shock on 20 July at 5.3–5.9 (class d) and the epicentre at N 16°, E 41°, in the central trough of the Red Sea, east of the Dahlak Archipelago. Sieberg and Milne estimated the damage intensity at Massawa at VIII–IX. To me, VIII is more realistic.

Gutenberg and Richter's epicentre is about 170 km away from the city of Massawa. Under normal conditions no earthquake of magnitude less than 8 would cause damage of intensity VIII at a distance of 170 km from its epicentre even on poorly consolidated grounds. Therefore, Δ = 170 km is too high. Because: (1) the felt intensity was reported less in the Dahlak Islands than in Massawa; and (2) the damage of intensity VIII was apparently restricted to the city of Massawa, it is concluded that the focus was very shallow and the epicentral distance from Massawa much smaller than 170 km. It is therefore suggested that the 1884 epicentre was located not in the central trough of the Red Sea as calculated by Gutenberg and Richter but in the Massawa Channel, somewhere in the vicinity of the 1921 seismic region. The adopted coordinates for the central region of activity in 1884 are N 15.7°, E 39.6°.

Gutenberg and Richter rated the instrumental magnitude at 5.3–5.9; such a magnitude would cause tremors of intensity VIII at 20–25 km from the epicentre and of intensity V at 130–140 km as reported from Keren.

4. On a Report from Seiberg

Without quoting his sources of information, Sieberg (1932) added that the tremors were felt in Arabia and as far north as Cairo in Egypt. Cairo is 15° (1665 km) from Massawa. A felt area, 15° in radius, presupposes a release of energy larger than ever experienced! Lions (1907) published a list of earthquakes felt in Egypt; he made no mention in it of tremors having been felt in Cairo during July 1884.

1886–87

The tremors listed below were reported from Massawa; some of these were also felt in Asmara. The direction of the incoming seismic waves were obtained from the seismiscope that was in operation at the Meterological Station in Massawa (see Fig. 75). The times quoted are L.T.

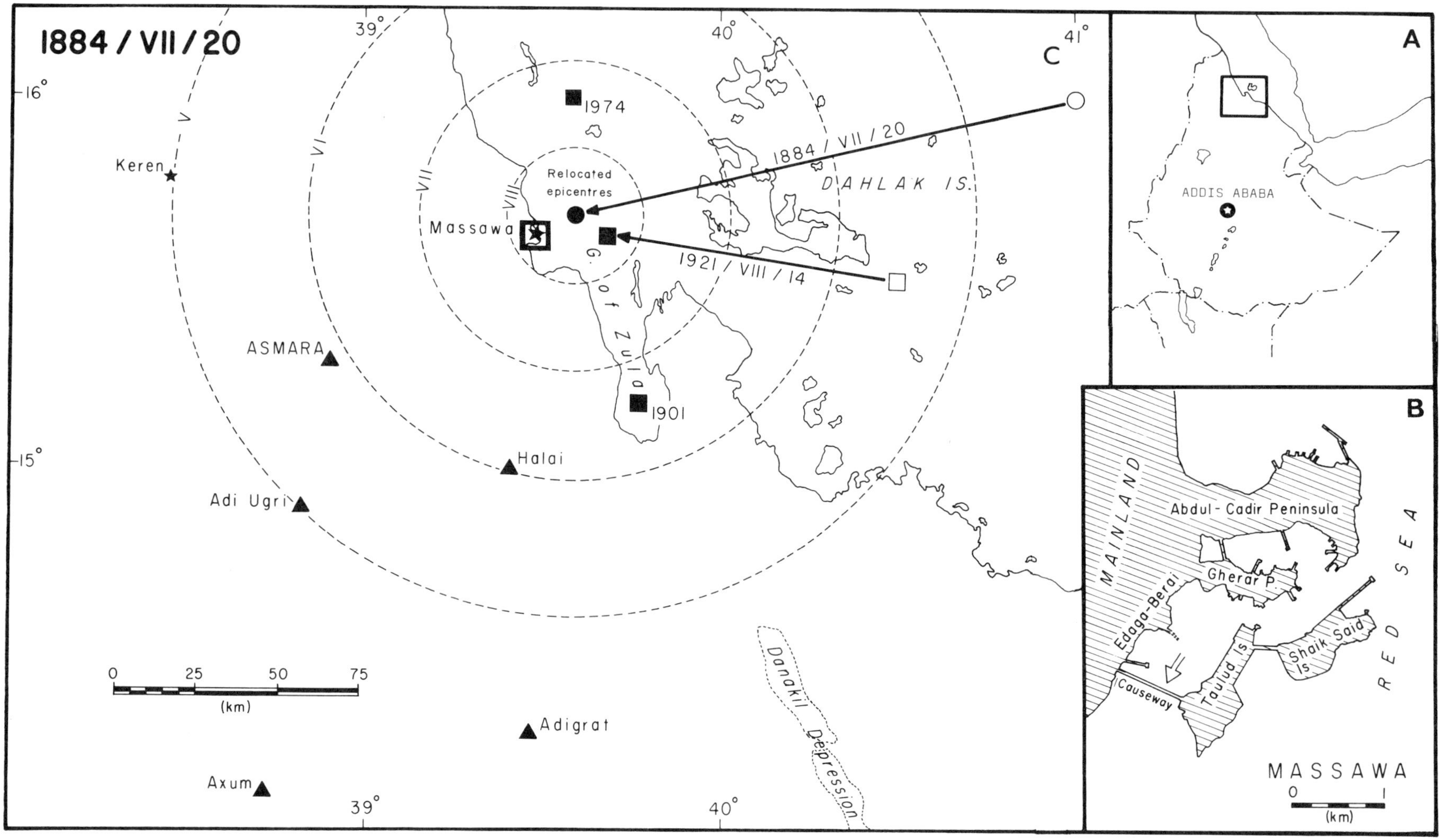

Fig. 75. Relocation of the 1884 earthquakes in relation to the harbour of Massawa and subsequent seismic events. The circles indicate the theoretical intensity values under rift crustal conditions. The arrow (inset B) indicates the causeway that the seiche waves overflowed.

8 May 1886	at 17:30, 17:51, 17:53, and 21:38	
15 June	at 09:45	
8 August	at 17:48 & 21:17	(direction SW-NE)
7 December	at 06:25	(direction SE-NW)
23 December	at 18:10	
30 January 1887	at 23:44	(direction N-S)
1 February	at about 0200	(two light tremors)
14 March	at 10:00	
28 December	at 02:45	(direction NW-SE)

Source

Palazzo (1915, events 67–78).

1891/II, IV

The Meterological Station in Massawa reported the three tremors listed below. The direction given is presumed to have been read on the seismoscope. The time is L.T.

12 February at 12:40 — light tremor followed by a second one about 1 h later. The direction of the seismic wave was W-E.

27 April at 04:40

Source

Report from the Meteorological Station at Massawa (from Palazzo 1915, p. 321).

1892/XI/23

On 23 November 1892, light tremors were felt in Massawa (N 15.6°, E 39.5°).

Source

Report from the Meteorological Station in Massawa (from Palazzo 1915, p. 321).

Comments

The newspaper *Corriere Eritreo* of 27 November 1892 states that the tremors were felt on 25 November instead of 23. This seems to be a typographical error on the part of the newspaper, an error which was not corrected by Marinelli and Dainelli (1912, p. 116).

1894/IX/12

At L.T. 12:50 on 12 September 1894, a strong earth tremor which lasted 3 s was reported by the Meteorological Station in Massawa. The direction of the seismic wave was N-S.

Sources

Report quoted in *L'Africa Italiana*, 16 September 1894, No. 245 (from Palazzo 1915, p. 323).

1896/XII/11

A small tremor was felt in Massawa at 2:20 a.m. on 11 December 1896. Despite the fact that the tremor was light, people rushed out of their houses and gathered in the streets, recalled the reporter. They remembered the destructive earthquakes of 1884 and dreaded the coming of larger ones.

Sources

L'Africa Italiana, No. 362, 13 December 1896 (quoted by Dainelli and Marinelli 1912, p. 116).

1897/IX/30

On 30 September 1897, the Massawa Meteorological Station recorded a shock in an E-W direction. It was also felt by people in the surrounding area. Sieberg estimated the intensity at V–VI.

Source

Bollettino Mensuale della Societa Meteorologica Italiana, Series II, Vol. XVII, 1897, p. 73 (from Palazzo 1915, p. 323).

1900

About 1900 or so, pitchstone was ejected from a fissure on the rift floor, southeast of the Koka reservoir (N 08.3°, E 39.4°). The "rain of hot grains" as witnesses described the eruption, destroyed crops many kilometres away. The fact was investigated by Dakin and Morton (1975). The witnesses' narration is backed up by the presence of very recent pitchstone flows.

Source

Dakin and Morton (Dept. Rep., Geology Dept., Addis Ababa University, 1975, personal communication).

1900/IV/02

An earth tremor was felt in Massawa (N 15.6°, E 39.5°) and Adi Ugri (N 14.9°, E 38.8°) during the night of 2 April 1900 (Fig. 75).

Source

Meteorological reports from the stations in Massawa and Adi Ugri (from Palazzo 1915, p. 324).

Comments

The two reporting meteorological stations were situated about 110 km apart. The station at Massawa reported the time of the earthquake at L.T. 22:20 and qualified it as "strong"; the station at Adi Ugri put it at L.T. 22:00. Although the reports came from two reliable sources, the probability is that they both refer to the same event and that there was an error in recording the time at one of the stations. Palazzo shares this opinion.

1901–02/Alid–Zula

Between 11 November 1901 and March 1902, tremors were reported from the region surrounding the volcano Alid, a few tens of kilometres SSE of the Gulf of Zula. During approximately the same period, although the recorded times of the individual events do not correspond, tremors were also felt in Massawa, some 40 km to the northwest of Alid. It is believed that these tremors were caused by a common centre of activity near the volcano (Fig. 76). They are listed below in chronological order:

11 November 1901 — Strong tremor accompanied by noise felt in Buia, a waterhole at the crossroads of caravan routes, 12–13 km SW of Alid volcano. Two aftershocks followed.

29 January 1902 — at L.T. 01:15 light tremors felt in Massawa.

24 February — at L.T. 19:00 light tremors felt in Babala Maderto, 7 km SSE of the volcano.

4 March — at L.T. 20:30 two sharp tremors felt at Ueten, a few kilometres SSW of Alid.

5 March — at L.T. 02:20 light tremor felt in Ueten.

3 May — at L.T. 02:00 light tremor felt in Massawa.

26 May — at L.T. 09:25 light tremor registered in Massawa. The seismoscope indicated a S-N direction.

29 September — at L.T. 22:30 light tremors recorded in Massawa: direction SW-NE.

Source

Palazzo (1915, p. 325–327).

Comments

Marini, in private communication to Palazzo, expressed the opinion that the 1901 seismic activity around Alid was an obvious manifestation of its volcanic activity. This relationship might not, in fact, be so obvious. Although there are fumaroles in Alid's caldera and on its flanks, the volcano is at a late stage of its development and it seems unlikely that its vitality remained sufficient to *shake the earth.*

To the northwest and southeast of the volcano there are two large fields of basalts broken by spectacular open fissures. These lava fields predate 1900. It is conceivable that some of these fissures opened during the release of seismic energy in 1901.

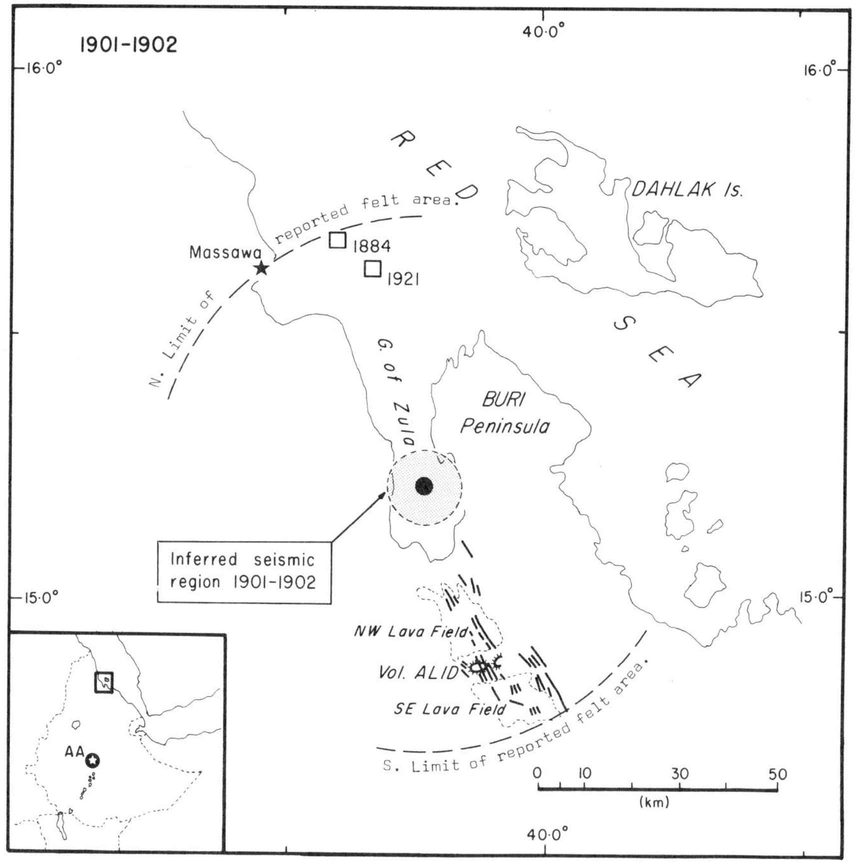

Fig. 76. Location map showing the suggested centre of the 1901–02 region of seismic activity, as well as the "reported felt" area. Included in the same map are relevant epicentres from other entries.

The epicentral region could either be ESE of Alid, or simply in the Gulf of Zula. The latter is a reasonable assumption if the Gulf of Zula is considered as the northwestward prolongation of the tectonic zone so well identified further to the SSE by the Danakil depression and the Erta'Ale range (see also Fig. 77), and to the NNW, by recent epicentres along the Massawa Channel. The adopted coordinates for the centre of the 1901–02 active region are N 15¼°, E 39¼°.

1906–07/Afar Depression

For a period of almost 2 years, observations of renewed seismic and volcanic activity in and around the Erta'Ale and Alayta volcanic ranges in the Danakil depression were reported. The centre of volcanic and seismic activity was the Alayta range located at about N 13°, E 40.5° (Fig. 77).

Sources

Mainly reports from government officials who were on duty patrol in the Danakil desert.

The first report on the activity of 1906–07 came from Sr. Odorizzi, Commissioner for the Danakil area, who was patrolling the Salt Plain in March 1906. He reports having experienced strong earth tremors and having observed volcanic activity in the distance. He explicitly mentions in his text the volcano Erta'Ale (Palazzo 1915, p. 329), but it is far from certain that he could identify it from among the other volcanoes in the Depression.

During his second tour of inspection in May 1907, Odorizzi observed similar seismic and volcanic phenomena in the same region (Palazzo 1915, p. 333).

In June 1907, Raggi (1907, p. 1271–72) and Tancredi (1907, p. 490–91) reported that a volcano named Afdera was in eruption in the Erta'Ale range and that the volcanic activity was accompanied by severe earthquakes. Tancredi mentioned that the number of felt tremors reached 20 per day. It is not known were Raggi obtained his information but it is known that Tancredi was stationed in the Danakil Alps. At that distance, he erroneously located the fissure eruption that he was observing; he thought the Afdera volcano was in action; whereas, in fact, it was the Alayta. Recent investigations by the CNR-CNRS scientists leave no doubt on this point (CNR-CNRS Afar team 1973, p. 470).

On 4 August 1907 there was still evidence of activity.

On November 1907, the Ministry of Foreign Affairs reported that the activity had ceased.

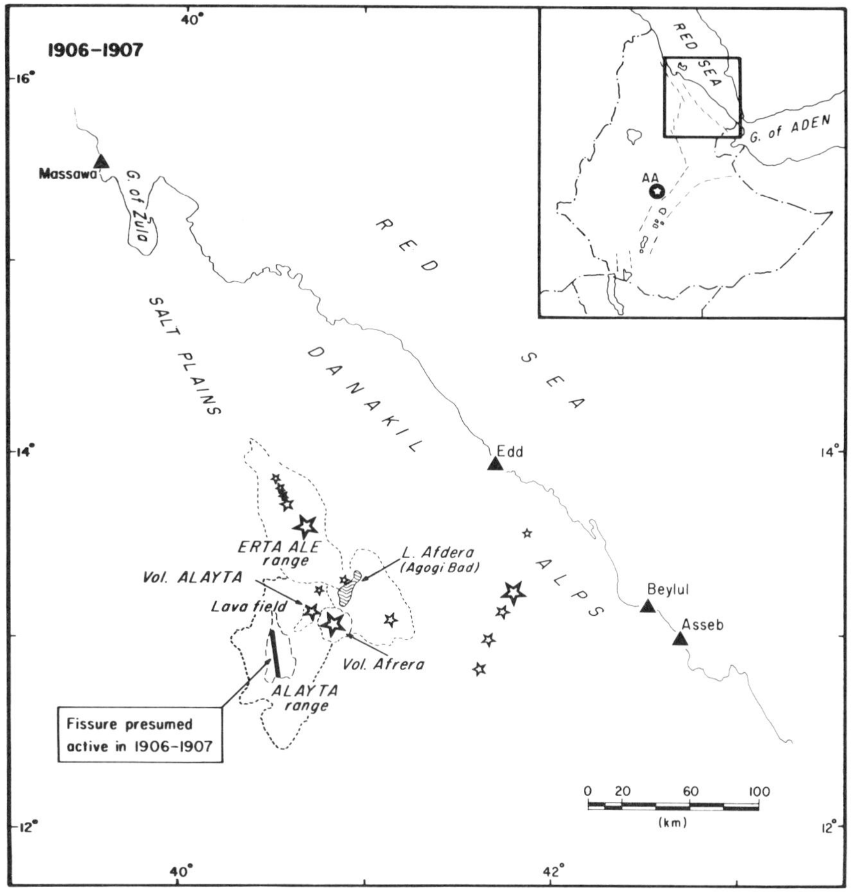

Fig. 77. Schematic outline of the axial volcanic ranges in northern Afar showing the fissure in the Alayta lava field presumed active in 1906–07 and observed by Tancredi, probably from Beylul.

Comments

1. Accuracy of the Reports

The government officials who filed these reports were neither volcanologists nor seismologists so that their reports are necessarily limited to: (1) visual observations of some volcanic phenomena occurring at rather large distances; and (2) subjective impressions of earth tremors. The first report is dated March 1906, the last November 1907. These dates should not be interpreted as the absolute beginning or end of the seismovolcanic activity.

2. Identification of the Volcanoes

In these reports there are errors and confusion about the identification of the volcanic centres. One source of possible error comes from the fact that the best-known volcano in the area is called Erta'Ale (the "smoking mountain," in the Afar language) and that the same name Erta'Ale is also given to a whole range of about 10 active volcanoes. (At the time, the Alayta range was not identified as a unit separate from the Erta'Ale range.) It is difficult to decipher in these early reports whether the authors were referring to a particular volcano or to a whole volcanic range.

A second source of confusion comes from the absolute lack of contemporary topographic information on the area. The Salt Plain in the depression of the Danakil Desert is one of the most hostile and forbidding regions of the world (see Englebert 1970) with air temperatures reaching over 50 °C. Small wonder that we have had to wait for the age of aerial photography and airborne logistics to map the Erta'Ale range and identify its individual units. A good example of the consequences of the lack of topographic information is the fact that de Zerbi (1891) organized an expedition to scale the Erta'Ale and climbed the Dallafilla instead! Tancredi (1907) mistook Alayta volcano for Afdera.

3. Location of the 1906–1907 Seismic Area

There is no doubt that earthquakes accompanied the volcanic activity in the Salt Plain during 1906 and 1907. The centre of seismic activity was the volcano Alayta, or rather the Alayta lava field (N 12.7°, E 40.6°). This does not mean that the seismic activity was restricted to the Alayta complex. Abundance of fresh fissural lava flows, cauldron subsidences, and recently opened fissures not only in the Alayta complex but also along the Erta'Ale range, suggest that some of them formed during the 1906–07 period.

Although the seismic activity was severe in the Salt Plain, it was not reported in Asseb, nor in Asmara or Massawa. From these observations, it is inferred that the focus was rather shallow (some focal depths in the same region near Serdo have been observed to be as shallow as 4 km) and that the maximum magnitude might not have exceeded 5.

1921/Massawa Channel

During the whole year of 1921, earth tremors originating in the Massawa Channel jolted the Province of Eritrea. The shocks were particularly severe on 14 August and 21 September when two shocks reached magnitude class "d," that is, between 5.3 and 5.9. Reports of tremors came from all over the province. Massawa was the site that suffered most; it was so badly damaged that all buildings still standing were ordered to be demolished as a security measure. Eritreans remember 1921 in their folklore as *Zemene delekeleke,* the era of the earthquakes.

Sources

Bulletino Ufficiale della Colonia Eritrea, XXX, Decree 3845, Asmara, 10 December 1921; Cavasino (1922, p. 131–3); Diary of O.F.M. Cap. in Massawa for 1921; Rihani (1942, p. 219); Gutenberg and Richter (1954).

Comments

1. Evaluation of the Damage in Massawa

The official Gazetta of the Ministry of Public Works, dated 10 December 1921, sums up the situation as follows:

Considering that the most severe damage was inflicted by the earthquake of August 14, 1921, and its aftershocks, and that the city of Massawa was particularly afflicted by the disaster:

a) it will not be permitted to restore most of the houses still standing, but they must be demolished . . . ;

b) it becomes necessary to draw a new rational plan of the city . . . taking into consideration public security, in view of the fact that other earthquakes will eventually occur . . .

These statements were followed by the rulings of a new Security Building Code that was to be enforced in Massawa and Ras Murdur *(Bull. Uff. della Colonia Eritrea, XXX,* Suppl. to No. 22, Decree 3845, Asmara, 10 December 1921). For the relative location of Massawa and Ras Murdur, see Fig. 75.

A description of the destruction of Massawa is also found in the diary of the Capucin Fathers who had a monastery and a church in the city. Cavasino estimated the intensity of the damage in Massawa as grade VIII–IX on the Mercalli scale; this would correspond to IX on the Mercalli-Modified scale.

2. Discussion Concerning the Location of the Epicentral Region

Gutenberg and Richter (1954) placed the epicentre of the main shock on 14 August U.T. 13:15:28, at N 15.5°, E 40.5° in the Red Sea southeast of Dahlak Island. ISS located it on the mainland, at N 15.5°, E 39.0°, 50 km WSW of Massawa, almost 35 km due south of Asmara, near Decamere (Fig. 78).

Gutenberg and Richter's location, at first sight, appears more realistic than the ISS one because the maximum intensity was observed in Massawa; Asmara and Decamere were not damaged. Considering that the epicentral locations of 1912 earthquakes are far from reliable, a reevaluation based on instrumental data and macroseismic evidence was necessary.

Three options could be considered in the appraisal of the 1921 epicentres:

(1) To accept the original epicentre as determined by Gutenberg and Richter (G-R1) at 100 km east of Massawa and alter its 5.3–5.9 magnitude. Under normal conditions, an earthquake of magnitude below 6 does not

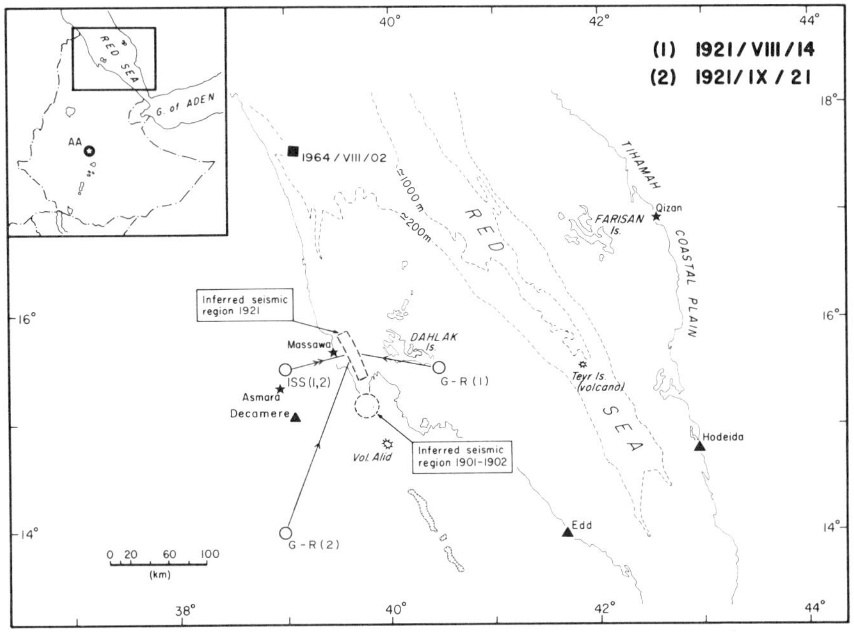

Fig. 78. Relocation of the 1921 epicentral region.

cause damage of intensity IX (Cavasino 1922) at 100 km from its epicentre even if soil conditions are bad. To make the original G & R location acceptable, a magnitude higher than 6.5 or 7.0 is required.

(2) To move the epicentral location further east or northeast toward the Red Sea central trough where scores of more recent earthquakes have been located. Such a relocation would mean increasing the epicentral distance from Massawa and therefore would require a still higher magnitude than in option (1).

(3) To relocate the epicentre in the Massawa Channel, 25–30 km east of Massawa (site on Fig. 75). A shallow depth earthquake of magnitude "class d," at an epicentral distance of 25–30 km, could have produced the destruction of the city of Massawa, when one considers both the geology of the coast (uplifted coral reef) and the type of masonry used in Massawa in 1921; it could also explain why the area of damage was restricted to a few tens of square kilometres.

The presumption that the northern part of the Massawa Channel might be a seismic line of crustal weakness is supported by the fact that USCGS located an epicentre on 30 June 1974 at N 16.0°, E 39.6°, in the middle of the Channel. The steep slope of the bathymetric drop between the 100 and 200 fathom contour lines north of 16.6° at the northern end of the channel and the presence of the Gulf of Zula at its southern end suggest the presence of a tectonic fault system parallel to the Eritrean escarpment. (Note: For mariners, the Massawa Channel does not enter the Gulf of Zula but veers to the east and passes between the Dahlak Islands and the Buri Peninsula.) The fact that damage was restricted to the coastal area during the 20 July 1884 earthquake (entry 1884/VII/20) could also be explained by a shallow earthquake near the coast.

Rihani's observations, even if the date indicated does not correspond exactly to that of the main shock, imply some sort of volcanic activity coincidental with or triggered by the seismic events. If *millions* of fish were poisoned by sulfur hydroxide, the source of the gas must have been submarine. Such a source could either be the floor of the Red Sea trough, where sulfide deposits have accumulated, or the vicinity of the volcanic island Djebel Teyr (N 15°42′, E 41°45′). As a matter of fact, Djebel Teyr is about halfway between Qizan and Hoddeida, the two sites between which the phenomenon of the dead fish was observed.

A question now comes up: could the recent epicentres of 25 August 1962 and 30 June 1976, both located east of the Dahlak Islands, added to Rihani's observations give greater weight to option (2)?

1929/V/18

A magnitude 6 earthquake occurred near the town of Serdo, in Central Afar.

Sources

Bellamy (ISS, 1936); Gutenberg and Richter (1954).

Comments

The original epicentral locations given by Gutenberg and Richter and by ISS were N 11.5°, E 41.5° and N 11.5°, E 42.0°, respectively. The recomputation of this epicentre (courtesy of Sykes 1965) shifted the location to N 12.03 ± 0.10°, E 41.36 ± 0.10°, in the NW-SE tectonic zone that was seismically active in 1969 (Fig. 79) (also see entry 1969/III–V/Serdo–Central Afar)

1931/V/01

On 1 May, ISS indicated an epicentre at the extreme northern tip of the Eritrean Plateau, slightly north of the Ethiopian border with Sudan. A recomputation of the original station reports relocated the epicentre in Yemen, across the Red Sea.

Source

Bellamy (1947, p. 2).

Comments

The epicentral location published by Bellamy and based on seven stations was N 18.0°, E 37.5°, on the Sudanese side of the border with Ethiopia. A recomputation in 1972 of the ISS original data using the SPEEDY computer program at the Institute of Geological Sciences in Edinburgh (Scotland) after 10 iterations yielded the following parameters for each reporting station:

Station	Residual (s)	Δ (from AAE)	Az (from AAE)	Epicentre	
HLW	− 0.102	11.15°	142.3°	N 20.84°	E 38.61°
KSA	+ 0.042	12.91°	167.7°	N 21.18°	E 38.81°
BAK	− 0.125	21.21°	203.2°	N 20.53°	E 32.15°
FEO	+ 0.185	23.97°	173.4°	N 21.16°	E 38.25°
TAS*	+30.075	32.41°	224.9°	N 15.80°	E 46.14°
ANR*	+ 1.217	34.22°	227.7°	N 14.66°	E 46.90°
SVE*	− 2.187	42.45°	198.7°	N 15.52°	E 47.66°

* Neglected in the final epicentre determination

The parameters for the original epicentre were: H, U.T. 09:47:55; N 18.0°, E 37.5°; recalculation yielded H, U.T. 09:48:29.5; N 21.19 ± 1.13°, E 39.16 ± 0.33°

The discrepancy between the original ISS and the recomputed epicentres comes from the fact that ISS used all seven station readings; whereas, in the recomputation, the last three, which apparently belong to another event, were discarded.

The recomputed epicentre represents a shift of the original ISS location by 392 km in latitude and 172 km in longitude, that is, a total NNEward (026°) shift of 393 km (Fig. 80).

The original location was in Africa, the relocated epicentre is on the Euro-Asian continent.

1934/VII/02

A light earth tremor was felt in Massawa at 2:35 p.m. on 2 July 1934

Source

Fantoli (1966, appendix L, event 150).

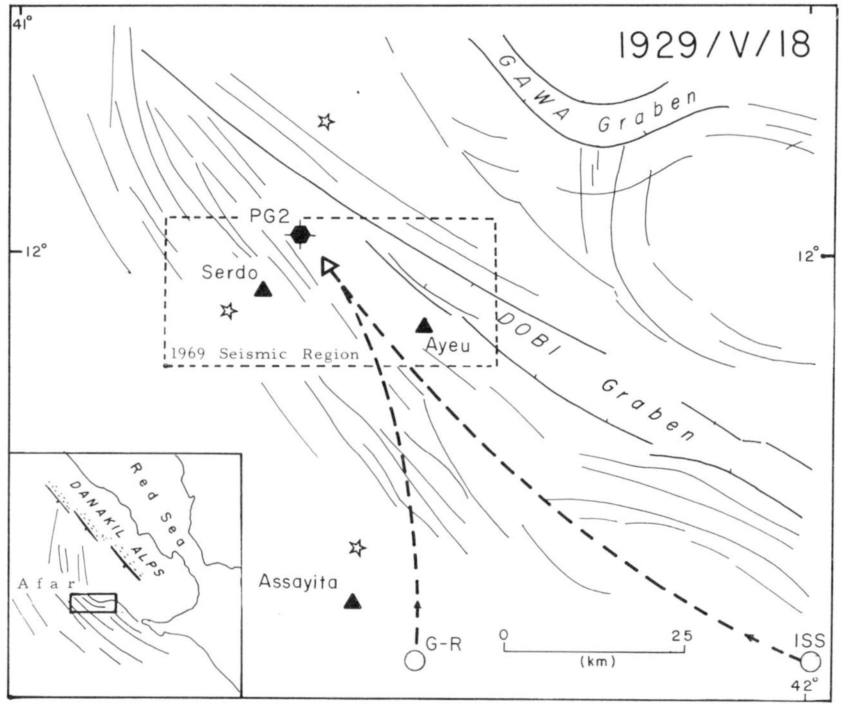

Fig. 79. Relocation of the original instrumental epicentre of 18 May 1929. The new site is within the 1969 seismic region near Serdo in central Afar.

1938/V/12

An earthquake of magnitude ≃ 5¾ (class d) occurred along the coast of Sudan at the foot of the northern prolongation of the Eritrean escarpment.

Sources

USCGS; Gutenberg and Richter (1949, p. 205); ISS (1953).

Comments

Three original solutions were available for the epicentre of 12 May 1938: Gutenberg and Richter, ISS, and the USCGS. Using the SPEEDY computer program, a relocation was attempted in 1974 based on the 49 station reports contained in the ISS data file. After three iterations, 31 sets of data were retained for the final solution. The parameters of the four epicentre determinations were:

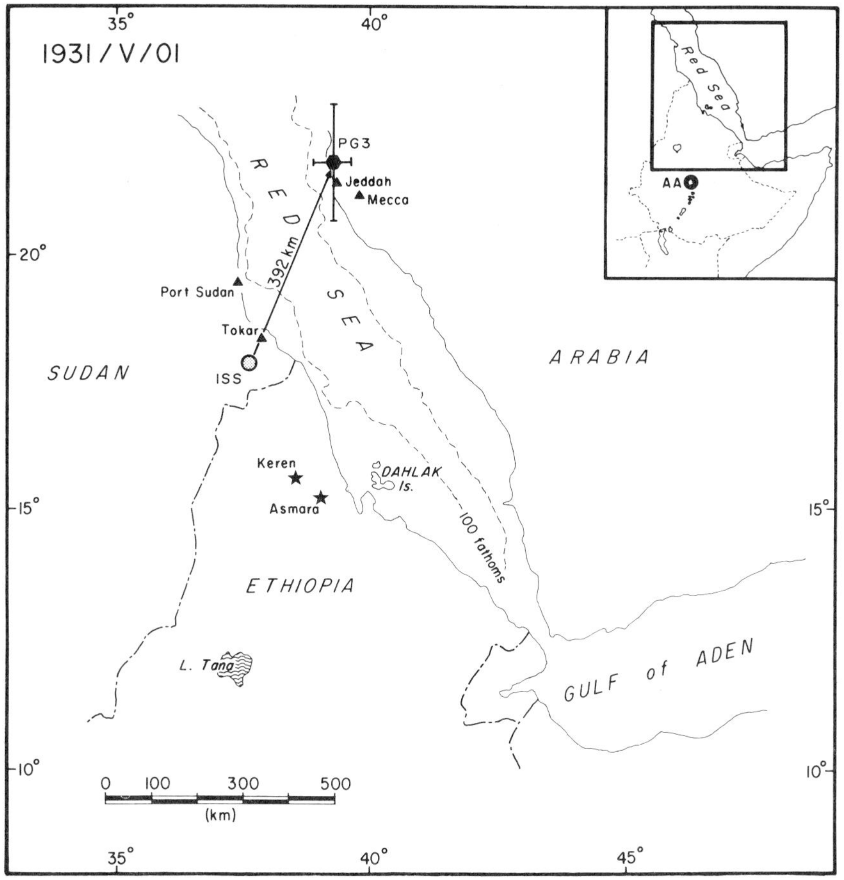

Fig. 80. Relocation of the original ISS epicentre for the Sudanese earthquake of 1 May 1931.

	H	Coordinates	
G & R	U.T. 21:31:35	N 18.5°	E 37.5°
ISS	U.T. 21:31:35	N 18°	E 37.5°
USCGS	U.T. 21:31:30	N 17°	E 38°
Relocation	U.T. 21:31:44.1	N 18.58 ± 0.38°	E 37.44 ± 0.25°

The recomputed values show almost no difference with the solution of Gutenberg and Richter, a total shift of 64 km almost due north (354°) of the original ISS epicentre, and a 100 km (341°) shift of the USCGS location.

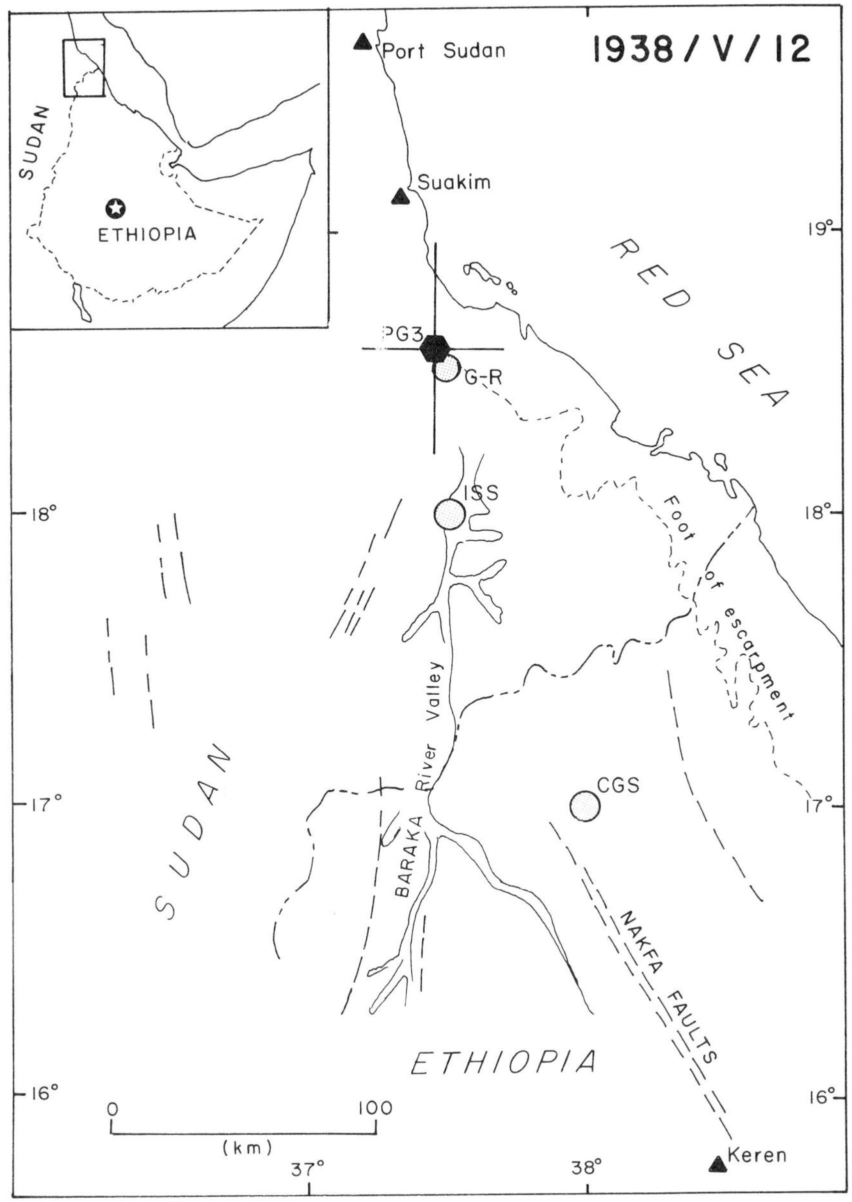

Fig. 81. Original and recomputed epicentre locations for the earthquake of 12 May 1938.

Figure 81 locates the recomputed epicentre at the foot of the prolongation of the Eritrean escarpment into Sudan. For information on the Red Sea marginal scarp in Sudan, see Whiteman (1968, p. 231–246).

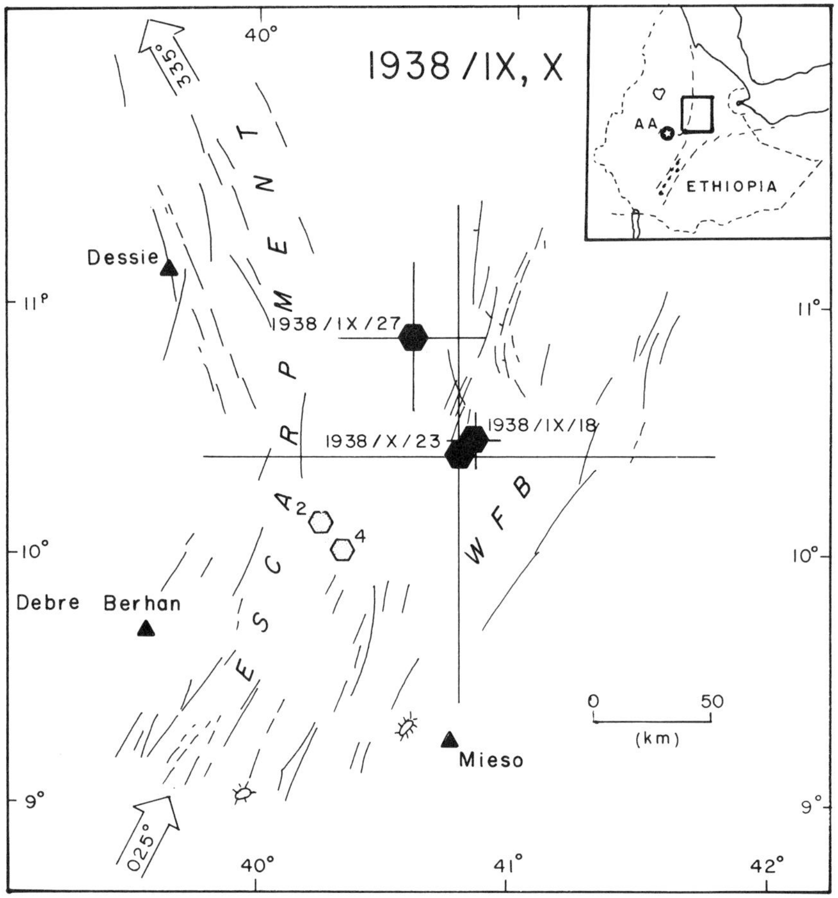

Fig. 82. Location map of the September-October, 1938 epicentres. The three events that are thought to be connected with the Wonji Fault Belt are indicated by full hexagons (⬢) with their estimated error bars; the two others connected with the Rift escarpment are indicated by shaded symbols (⬡). To avoid overloading the map, the original epicentral locations are not indicated.

1938/IX,X

In September and October 1938, five earthquakes of magnitudes ranging between 4.5 and 6.0 occurred between latitudes N 10° and N 11° and longitudes E 40° and E 41°. Those of 18 September and 23 October with epicentres at about N 10.4°, E 40.8° are directly related to the escarpment and therefore belong to Region A; the other three occurred in Region C and are presented here.

Sources

Gutenberg and Richter (1954); ISS (1953).

Comments

1. The five 1938 epicentres given by Gutenberg and Richter and by ISS were recomputed using the available data from 10–12 stations per event. The results shows possible errors as high as ± 1.2° in latitude and ± 0.3° in longitude. Although, at first sight, the range of possible errors in latitude may seem perturbing, from a seismicity point of view the meridional trend of the tectonic structures along which these earthquakes occurred minimizes the importance of the possible errors because a substantial shift in latitude may still keep the epicentres on the same tectonic structure.

Figure 82 is a regional tectonic sketch map on which the five relocated epicentres have been plotted. The distribution plot shows that the seismically active region of 1938 straddles two adjacent tectonic zones: the Plateau-Afar margin and the Wonji fault belt. The fact that the Addis Ababa contemporary newspaper *Corriere dell'Impero* did not mention these events even during war time is interpreted as a sign that they took place in a sparsely populated area. Their distribution according to tectonic regions, and their computation details are given later.

2. *Listing of Epicentral Locations*

For the sake of immediate information, the different solutions for the epicentres of September and October 1938 are classified by tectonic regions and the parameters compared.

Region: Wonji fault belt

(1) 18 September at U.T. 0:39: magnitude about 4½

ISS	N 09.5°	E 40.3°
Recomputed	N 10.37 ± 0.10°	E 40.81 ± 0.10° (PG2)
	N 10.37 ± 0.63°	E 40.83 ± 0.31° (PG3)

(3) 23 October at U.T. 02:25; magnitude class d (5.3–5.9)

G & R	N 10°	E 39.5°
ISS	N 09.5°	E 40.3°
Recomputed	N 10.39 ± 1.0°	E 40.75 ± 1.1° (PG3)

Region: Plateau-Afar escarpment

(2) 20 October at U.T. 13:15; magnitude class d (5.3–5.9)

G & R	N 10°	E 39.5°	
ISS	N 09.5°	E 40.3°	
Recomputed	N 10.1 ± 1.2°	E 40.2 ± 0.8°	(PG3)

(4) 23 October at U.T. 02:29; magnitude about 4½

ISS	N 09.5°	E 40.3°	
Recomputation	N 10.0 ± 1.0°	E 40.3 ± 0.3°	(PG3)

Region: Between escarpment and Wonji fault belt

(5) 27 September at U.T. 02:32; magnitude 6

G & R	N 11°	E 41°	
ISS	N 09.5°	E 40.3°	
Recomputed	N 11.1 ± 0.3°	E 40.7 ± 0.3°	(PG2)
	N 10.5 ± 0.7°	E 40.5 ± 0.3°	(PG3)

1944/IX/06

A magnitude 6 earthquake occurred in southern Ethiopia at U.T. 13:28:03, 6 September 1944. The location is by no means accurate.

Source

Gutenberg and Richter (1949, p. 205).

Comments

The original epicentre location by Gutenberg and Richter was N 06°, E 38°, a site of strong faulting at the south end of the main Ethiopian Rift. A recomputation (courtesy of Rothé 1965) based on seven station reports yielded quite a different pair of coordinates: G & R N 06°, E 38°; recomputation N 07.24 ± 1.08°, E 39.29 ± 0.52°. The two solutions show a discrepancy of 200 km in a SW-NE direction. The G & R site was located on the floor of the Rift; the recomputed site is on the SE Plateau. The locations are plotted on Fig. 83.

The war was in progress in 1944; no written information has been found that reports intensities that could help narrow down the location of the seismically active region. The instrumental locations are 200 km apart in two different tectonic provinces. The temptation is strong to neglect such an event but on the other hand one cannot simply brush aside a shock of magnitude 6, which occurred in a region that is now densely populated. A provisional "educated guess" concerning the epicentre can be based on circumstantial evidence. On one hand the region of the SE Plateau centred about N 07°, E 39° at present appears to be aseismic; on the other hand, this survey shows three shocks of magnitude ⩾ 6 on the Rift floor between

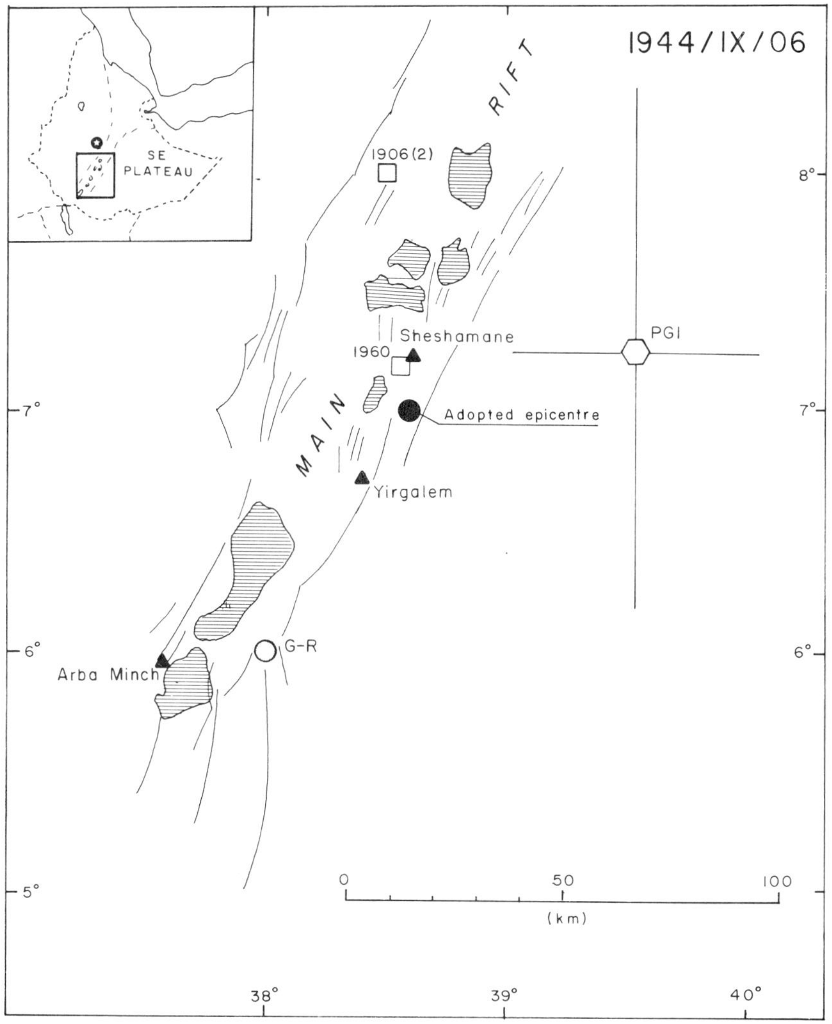

Fig. 83. Magnitude 6 earthquake of 8 September 1944. The full circle is the adopted epicentre location; the open squares indicate other epicentres that occurred at other dates in the same area.

N 07° and N 08°: two in August 1906 at N 08.0°, E 38.5° and one in July 1960 at N 07.2°, E 38.5°. As the Rift floor has been seismically active during the last decades between latitudes N 07° and N 08°, a provisional but probable location at N 07°, E 38½° has been adopted. It lies at the foot of the eastern escarpment, somewhere near the centre of the area of confidence defined by the error bars of the two teleseismic solutions. If more information becomes available, these coordinates will have to be revised.

This event is also indexed in Region B.

1950/IX/18

At U.T. 00:39:30, 18 September 1950, the BCIS located an epicentre at N 13.7°, E 42.2°, in the southern part of the Red Sea, barely 5 km from the Ethiopian coast. No other agency reported it; no magnitude is given.

Source

BCIS.

Comments

The location is off the coast of the Haitehnef Peninsula, 25 km away from two known seismotectonic zones: the Red Sea Axial Trench and the Zukur-Hanish fault line (see Fig. 91).

1951/XII/11

There was an earthquake of magnitude 4.7 in Central Afar.

Source

Rothé (1969)

Comments

Rothé's original epicentral solution was N 12.0°, E 40.5° with an estimated ± 0.5° possible error. In the absence of any other local control except surface tectonic features, the adopted location is N 12.0°, E 40.7° along the NW-SE trending Dagar ridge. (Refer to the entry 1969/III–V (Serdo–Central Afar) for comments on the seismicity of Central Afar).

1952/IX/10

There was an earthquake of magnitude 5.2 (USCGS) in the Danakil depression.

Sources

BCIS; ISS; Rothé (1969)

Comments

Three teleseismic solutions are available for this epicentre: ISS and Rothé at N 14.5°, E 39.5°, and BCIS at N 12.8°, E 39.6°. Recomputation in 1974 based on the 34 station reports in the ISS data file yielded, after 3 iterations of the SPEEDY program, the following parameters for an assumed focal depth of 33 km: origin time U.T. 09:06:17.5 ± 1.6 s; latitude N 14.524 ± 0.202°, longitude E 40.204 ± 0.185°. With respect to the locations

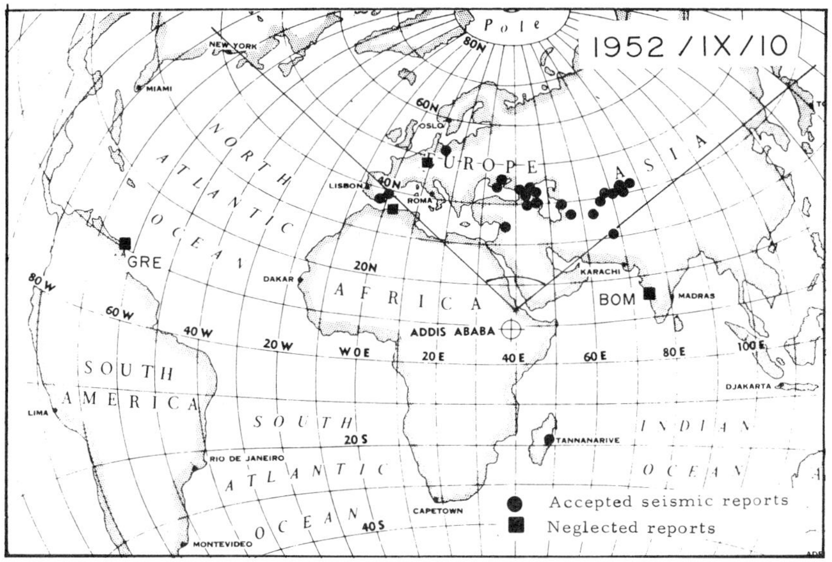

Fig. 84. Spatial distribution of the seismic stations that reported the earthquake of 10 September 1952.

obtained by ISS and Rothé, the SPEEDY results represent a total shift of 76 km almost due east (088°). The new site corresponds to the eastern margin of the depression, 10–15 km east of the Erta'Ale volcanic axis.

A typical distribution map of the seismic stations, which in the early 1950s would have reported to ISS earthquakes of magnitude 5 originating in East Africa, is introduced here (Fig. 84). Considering that the two stations that could give the best E-W control for an epicentre located in Ethiopia (Grenada in South America and Bombay in India) had to be discarded because of excessive arrival time residuals (322 and 43 respectively), only one station, Tananarive in Madagascar, is left out of the 100° sector to the north of Ethiopia where all the other stations are clustered. One can easily understand that with such a station distribution, if ISS gave equal weight to all station reports, a -322 s anomaly at Grenada would bias the epicentre westward. The present recomputation is certainly also biased by the nonhomogeneous spatial distribution of the reporting stations (Fig. 85).

1955/III/03

On 03 March 1955, at U.T. 00:43, an earthquake of magnitude 5 (BCIS) occurred in the central trough of the Red Sea, east of the Dahlak Island. Recomputations locate the epicentre along the eastern scarp of the

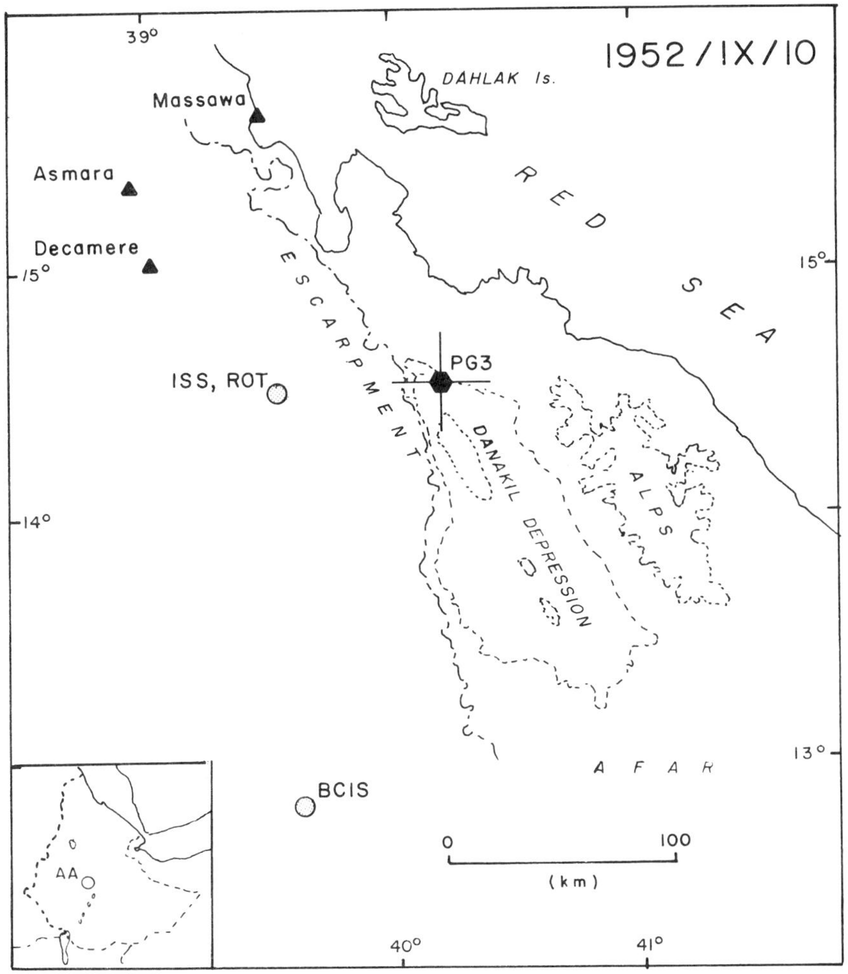

Fig. 85. *Relocation map for the epicentre of 10 September 1952 in the Danakil Depression.*

trough. The event is introduced in the data bank to avoid fringe effects in the seismicity mapping process.

Sources

BCIS; Fairhead and Girdler (1970); Sykes and Landisman (1964).

Comments

The locations given by the above mentioned agencies are: BCIS N 16.0°, E 41.5°; JED N 16.54°, E 41.25°; S & L N 16.46°, E 41.29°, s.e. ± 0.43 s (Fig. 86).

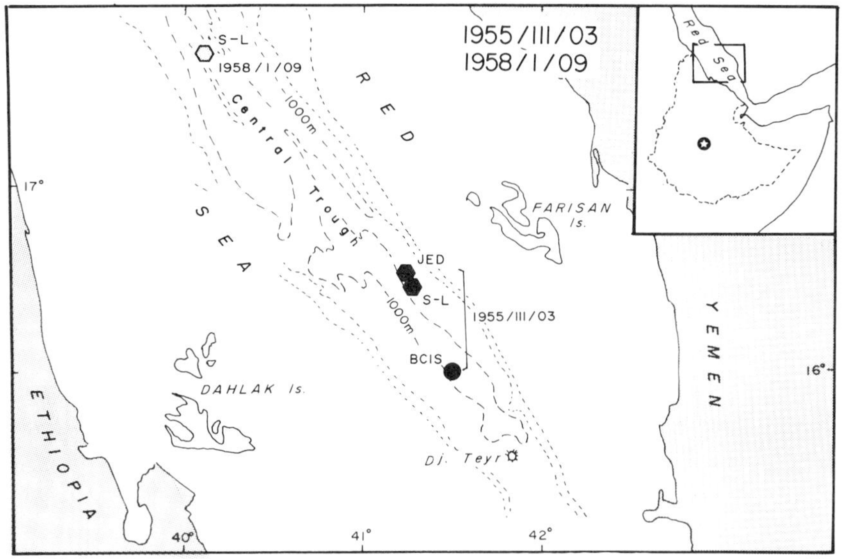

Fig. 86. Location of three teleseismic solutions for the epicentre of 3 March 1955 at U.T. 00:43.7. Added to the map is the epicentre of 9 January 1958. The isobaths are after Bäcker et al. 1975.

1957/III/14

At U.T. 00:11.5 on 14 March 1957, an earthquake of magnitude 5.2 (BCIS) was reported from the north end of the Danakil Alps in Afar. The location adopted from recomputations by three different agencies is N 14.89°, E 40.20°.

Sources

BCIS. (Fairhead and Girdler 1970; Rothé 1969; Sykes and Landisman 1964).

Comments

The original location indicated by the BCIS was N 15.0°, E 40.0°. The data were recomputed by Sykes and Landisman (1964), Rothé (1967), and Fairhead and Girdler (JED 1970) and the three recomputed locations lie within 10 km of a centre of gravity at N 14.885°, E 40.203°. This centre of gravity has been adopted as the most probable location for the earthquake of 14 March 1957 at U.T. 00:11.5 (Fig. 87).

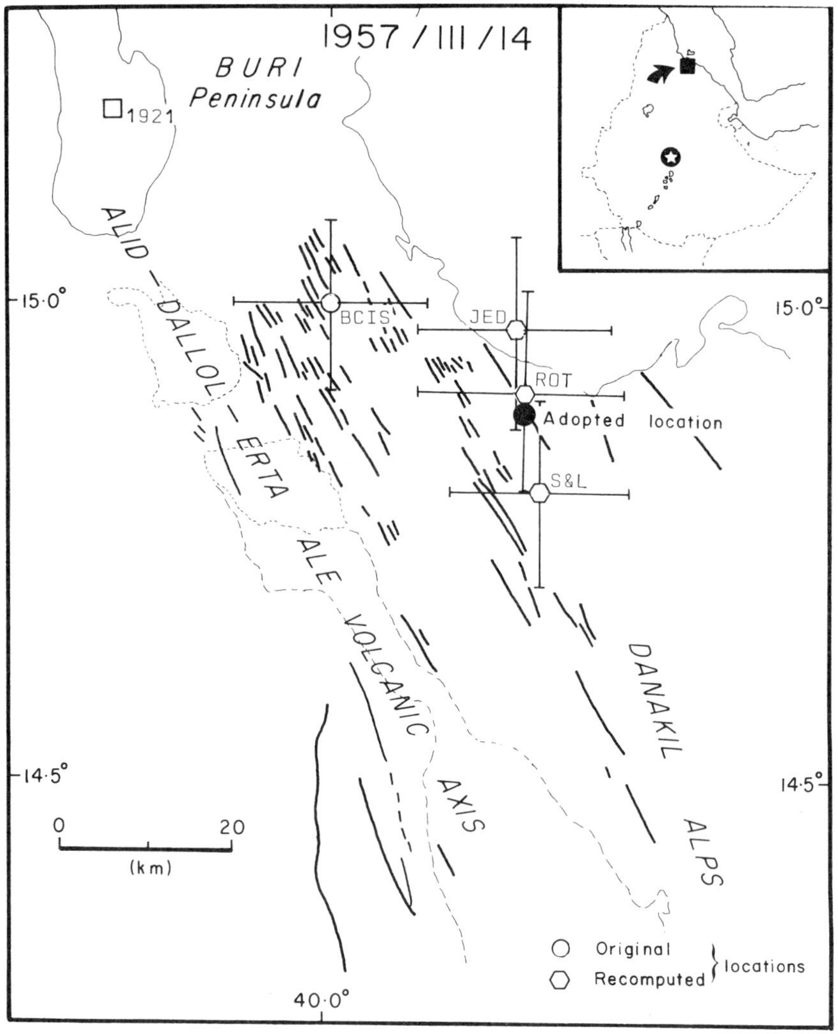

Fig. 87. Location of the instrumental solutions and adopted location for the earthquake of 14 March 1957, on the south shore of Hawachil Bay and along the fault that marks the southwest limits of the bay.

1958/I/09

At U.T. 07:56 on 9 January 1958, Sykes and Landisman (1964) reported an epicentre in the central trough of the Red Sea, at N 17.71°, E 40.12°. No magnitude is given. This epicentre has been included in the computations to avoid fringe effects in contouring. Its epicentre is plotted on Fig. 86.

1958/II/13

On 13 February 1958, a magnitude 5.2 earthquake occurred in the Red Sea at approximately N 14.23°, E 41.96°. The location is on or near the seaward prolongation of the Bidu-Dubbi volcanic line.

Sources

BCIS; ISS; Fairhead and Girdler (JED) (1970); Sykes and Landisman (1964).

Comments

1. Five solutions are available for this epicentre: two original ones (BCIS and ISS) and three recomputations based on different computing techniques and different data files. The coordinates obtained by the different agencies are: *original coordinates* — BCIS (N 13.75°, E 41.25°); ISS (N 14.35°, E 42.01°) (mean location N 14.05°, E 41.63°); *recomputed coordinates* — JED (N 14.26°, E 41.96°); S & L (N 14.34°, E 42.00°); PG3 (N 14.31 ± 0.18, E 42.00 ± 0.17) (mean location N 14.23°, E 41.9°).

The last recomputation was based on the 21 reports contained in the ISS data file, two of which were discarded after two iterations for excessive arrival time residuals (-20.8 and +5.2 s, respectively). A third station was ignored because its location could not be traced. The computing program was SPEEDY, which is currently used by ISC. Figure 88 shows the spatial distribution of the reporting seismic stations; they are satisfactorily distributed in longitude east and west of the epicentre but there is only one station to the south (Lwiro) for latitude control.

2. Figure 91 gives the epicentre location with respect to the NNE-trending alignment of five volcanoes: the Oyma at the southwestern end of the line; the Adaga, the Malahle, and the Nabro — currently known as the Bidu group; and the Dubbi complex at the northeastern end. To these five volcanoes on the mainland could be added the Kod'Ali islands, two spatter cones located a few kilometres off the coast. The Oyma and the Bidu volcanoes are silicic units and older in origin; the Dubbi complex and the Kod'Ali cones are basaltic and much younger.

The ERTS-1 imagery (Mohr 1974a, sheet 2) shows, in the vicinity of the Dubbi volcanic complex, lineaments transverse to the direction of the Danakil range but with the same north-northeastward trend as the volcanic alignment described above. No modern geological or tectonic map (Brinckmann and Kürsten 1970; Geological Survey of Ethiopia 1973; CNR-CNRS 1973; UNDP-Ethiopia Geothermal Survey, 1973) refers to these lineaments as actual faults, although it is recognized (CNR-CNRS

1973) that the NNE-trending feature under the Dubbi and Bidu volcanic groups *is an opened fracture occupied by the volcanoes mentioned above.*

The Bidu-Dubbi zone is active not only volcanically but also seismically (see entries 1400/Dubbi and 1861/V-IX). The present earthquake of m_b(CGS)5.2 occurred 45 km off the Ethiopian coast, slightly north (25–30 km) of the prolongation of the N26E Bidu-Dubbi fracture toward the centre of the Red Sea. A detailed plot of the seismic activity in the area is given in entry 1960/XII/16. It should be noted that the present epicentre apparently falls on a secondary tectonic line (CC′ in Fig. 91) some 20–25 km north of the volcanic axis. Let us keep in mind, however, that (1) the epicentre latitude control is inadequate, being controlled by only one station south of it; and (2) that a possible error of 25–30 km (¼°) would not be excessive in the case of the 1958 events. The location of the epicentre of February 1958 is plotted on Fig. 91.

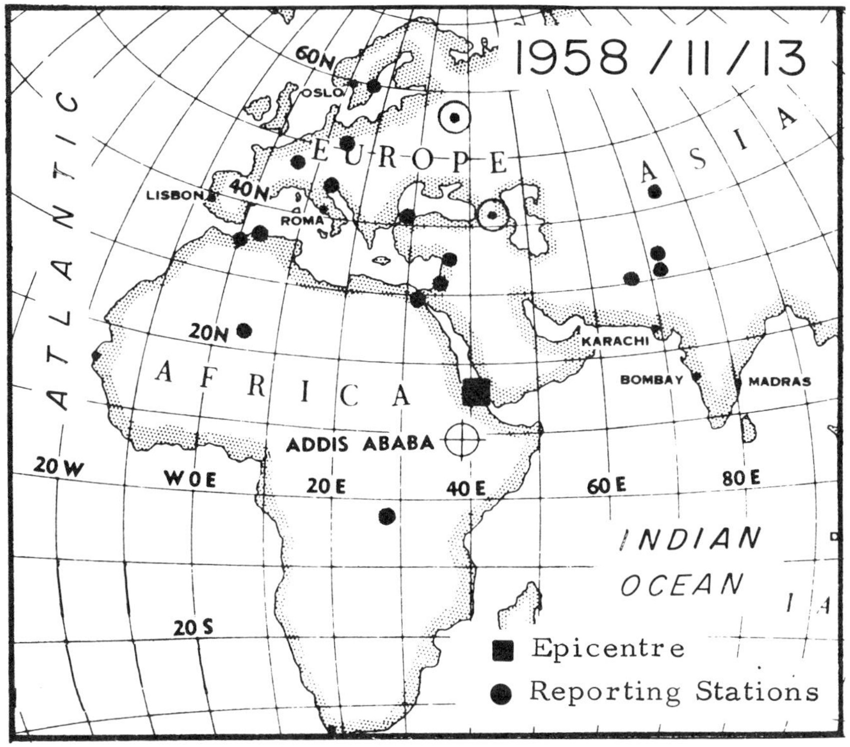

Fig. 88. Distribution of the seismic reporting stations from ISS data file.

1960/VII/14

At U.T. 18:39.5 on 14 July 1960, an earthquake of magnitude 6.3 (LWI) occurred on the floor of the main Ethiopian Rift Valley near the Chabbi volcano. Tremors were felt over a radius of 200 km. No surface fracturing nor structural damage was observed.

Aftershocks were recorded by the seismic station at Addis Ababa until the end of July.

Sources

AAE Data File; BCIS; ISS; LWI; MOS; USCGS. (Fairhead and Girdler 1970; Rothé 1964; Sykes and Landisman 1964).

Comments

1. Epicentral Location of the Main Shock

Teleseismic Solutions — Seven teleseismic solutions are available; their parameters are:

Agency	H	Coordinates		From AAE Az°	From AAE △ (km)
CGS*	18:39:34	N 07.00°	E 38.50°	188	227
ISS*	18:39:36	N 07.25°	E 38.45°	190	201
JED	18:39:49.1	N 07.18°	E 38.82°	182	206
MOS	18:39:35	N 07.00°	E 37.50°	212	265
ROT*	18:39:35.8	N 07.20°	E 38.50°	189	205
STR*	18:39:34	N 07.00°	E 38.50°	187	206
S & L*	18:39:35.8	N 07.17°	E 38.46°	189	209
Average of 7		N 07.11°	E 38.39°		217
Average of 5*		N 07.12°	E 38.48°		210 ± 10

Six instrumental locations are plotted in Fig. 89 (including inset C); the MOS epicentre is out of the map's frame and had to be neglected. Of the seven solutions, five (*) perfectly agree for a mean epicentre at N 07.12 ± 0.11°, E 38.48 ± 0.02°; JED differs from the others by 8° in azimuth with respect to Addis Ababa and MOS differs both in azimuth and epicentral distance. (It has often been observed in this survey that the USSR computed epicentres in Ethiopia are usually biased by about 1° westward). The magnitude M_s 6.3 was determined by Lwiro (LWI, Zaire) and accepted as such by the USCGS. No depth determination was attempted but the fact that a shock of magnitude 6.3 did not leave any scar on the earth surface might indicate a focal depth deeper than expected under a rift floor.

Adopted Epicentral Location — Although the intensity reports were far from being spatially well distributed, it appears that the area of observed maximum amplitude was located near Mt Chabbi (N 07.2°, E 38.4°) from where a tourist guide reported that *the tremors were so strong that he had to use his rifle as a crutch not to fall on the ground* (personal communication, Sheshamane 1960).

On the strength of five (*) out of seven teleseismic solutions and of the somewhat incomplete intensity distribution reports, an epicentral location at N 07.2°, E 38.5° was adopted. The location of epicentres in the central region of the Rift is socially important because of the vicinity of the provincial capital Awasa (N 07°, E 38.5°) to an area of potential volcanic and seismic activity. Seismically, the area is active; the present survey reveals it: see among others, entry 1906/Shoa. Moreover, it is a fact that Awasa . . . which did not exist in 1960; Yrgalem was then the capital of Arssi . . . is built on unconsolidated alluvial formations. It would not be exaggerated to compare its geological site to that of Managua, the ill-fated capital of Nicaragua, many times destroyed by earthquakes. The vicinity of Awasa to the Chabbi volcano is also to be considered. There are legends of recent volcanic activity in the Chabbi caldera, as recent as the 1930s; in which case, volcanic ash would have fallen over Sheshamane. Although intensive surveys among the older people in Sheshamane did not substantiate the authenticity of the phenomenon, it remains that Chabbi is a dormant volcano and that fumarolic activity still exists on its flanks and in its caldera. Dr Gibson is of the opinion from petrological studies (personal communication) that Chabbi is at a *dying state*. The fact that a shock of magnitude 6.3 followed by 2 weeks of aftershocks apparently did not modify in any way the volcano's regime of activity confirms Dr Gibson's opinion that Chabbi is dying. A dying volcano is not a dead volcano! For more details on Chabbi, see Mohr 1966b, di Paola 1972a, and Lloyd-UNDP 1973, p. 73–79.

2. *Aftershock Activity*

Twenty-eight aftershocks were recorded at Addis Ababa from 14 to 30 July inclusive; their epicentral distances ranged from 180 to 215 km.

3. *Local Travel Times (A Check)*

The only useful information supplied by the AAE seismograms is iP(Z) at U.T. 18:40:05; all other phases were wiped out. Using H_i computed by the different agencies and the equation $T_p = (4.6 \pm 0.5) + (\Delta / (7.95 \pm 0.03))$s an apparent distance of 203 km from Addis Ababa was obtained. This Δ compares fairly well with the above results ($\Delta^* = 210 \pm 10$ km).

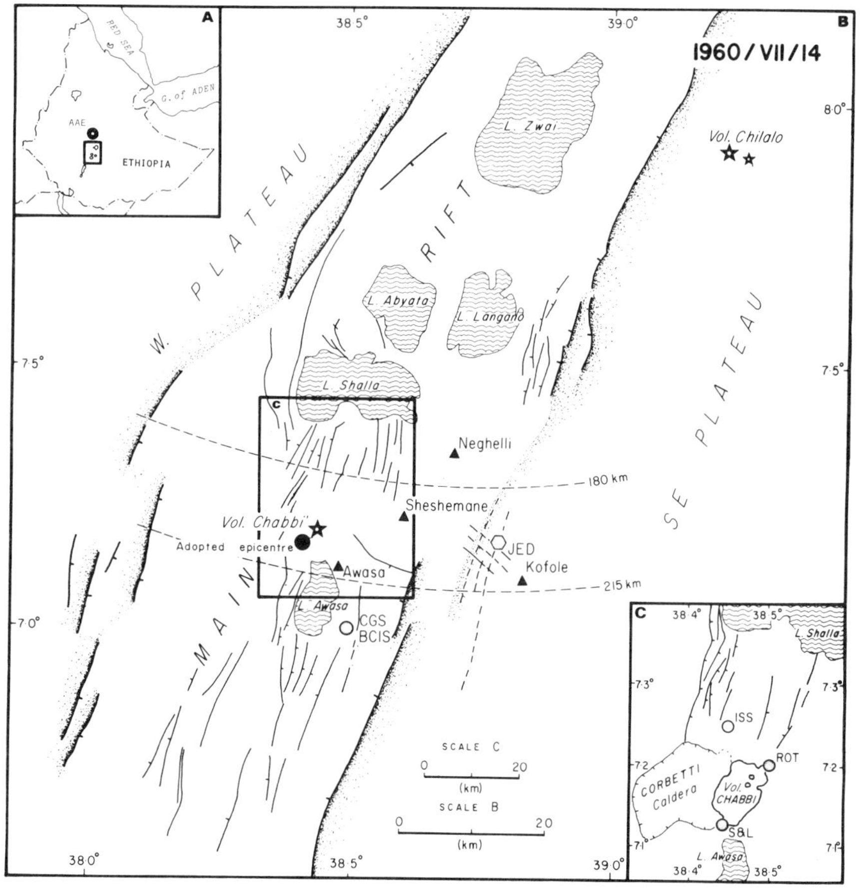

Fig. 89. Location map of Volcano Chabbi on the floor of the main Ethiopian Rift Valley and of the epicentre of 14 July 1960. The dashed line distance-arcs indicate the range of apparent epicentral distances for the aftershocks of 14–30 July 1960. (The fault pattern is taken from Mohr's unpublished 1:500 000 tectonic map of Ethiopia; the details of inset B are from di Paola 1972a.)

1960/VIII/13

Within 2 h on 13 August 1960, two earthquakes of magnitude M_L(AAE) 4.4 and 4.9 occurred near the Dahlak Islands in the Red Sea. Only the second one at U.T. 22:28.3 was reported by the International Agencies.

Sources

AAE Data file; BCIS; USCGS. (Sykes and Landisman 1964; Fairhead and Girdler 1970).

Comments

1. Solutions Based on Teleseismic Data

Three solutions based on 12 teleseismic records are available for the main shock. They are:

	H	Coordinates		△ from AAE
CGS	22:28:24.6	N 15.8°	E 40.2°	768 km
JED	22:28:19.4	N 15.11°	E 40.15°	692 km
S & L	22:28:13.6 ± 1.59	N 14.70°	E 40.16°	648 km

The longitudes obtained from these three independent solutions agree within ± 0.025°, that is within less than 3 km, but the latitudes differ by ± 0.55° or ± 60 km.

2. Solution Based on Local Travel Times

From the main event, Addis Ababa recorded an iPn(Z) at U.T. 22:29:55. As the observatory at the time was still using a radio-controlled Riefler pendulum as its time standard and the recording drums were driven by the city mains, a possible time error of ± 0.5 s is probable. Accepting H (CGS) as the origin time, a time interval (iP-H) of 95.8 ± 0.5 s is obtained. Comparing this Pn travel time to those of two well-determined epicentres that occurred in the same region, namely:

	Coordinates		△ (km) from AA	(iP-H) s
18 Sept 1967	N 15.69°	E 39.03°	741	97.7
07 Nov 1976	N 15.820°	E 41.423°	808	111.3

proportionally, for a P-H of 95.8 ± 0.5 s, one obtains △ = 696 ± 4 km from Addis Ababa, the same △ as that obtained in the JED solution. The adopted epicentre is therefore N 15.2°, E 40.2° near the western shore of Hawachil Bay. Because the foreshock registered exactly the same S-P time (110 s) as the main shock, both are presumed to have originated at the same site (Fig. 90).

The epicentre is very near that of 14 March 1957.

1960/X/23

At U.T. 19:21 on 23 October 1960, an earthquake of estimated magnitude ⩾ 4.5 occurred along the western scarp of the Red Sea central trough. Its epicentre almost coincided with that of 9 January 1958.

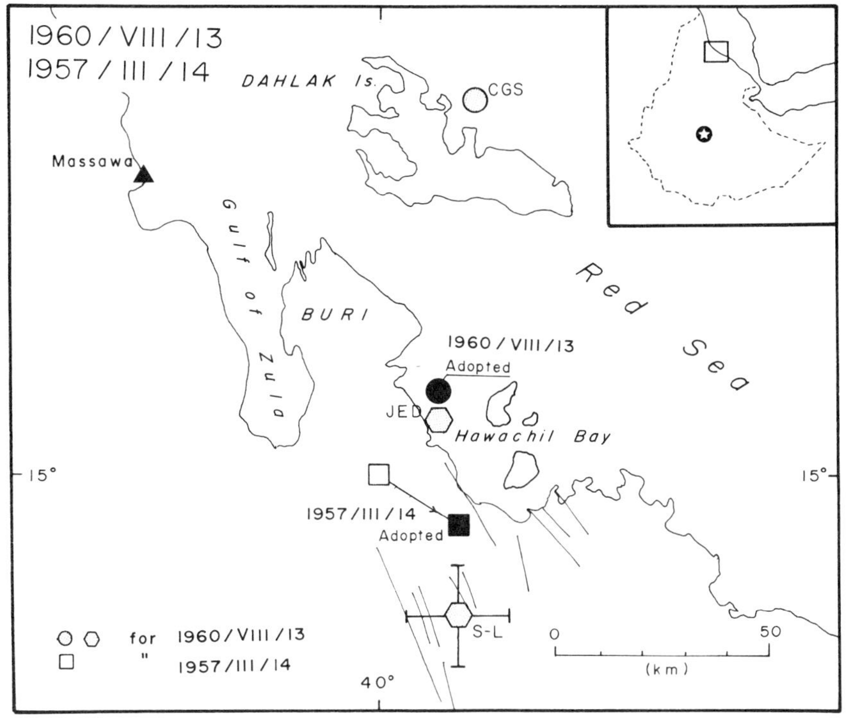

Fig. 90. Epicentre solutions for the earthquakes of 13 August 1960 and the adopted epicentre (■) for the earthquake of 14 March 1957.

Sources and Parameters of the Event

	H	Coordinates		h
USCGS	19:21:15.7	N 17.90°	E 40.90°	25 km
S & L	19:21:07.7	N 17.50°	E 40.07°	

Comments

Included in the computations to avoid fringe effects in seismicity map contouring.

1960/XII/16

At U.T. 16:29:25, a magnitude 5.1 earthquake occurred in the Red Sea at the southeastern corner of its axial valley. The original BCIS location was N 14.7°, E 42.9°. Three agencies recomputed the original data:

	H	Coordinates		h
JED	16:49:20.5	N 14.81±≥0.03°	E 42.42±≥0.04°	
ROT	16:49:15.0	N 14.70°	E 42.70°	29
SYK	16:49:15.0±1.82	N 14.65°	E 42.57°	29

No local seismograms are available due to a 5-day power failure in Addis Ababa.

Sources

BCIS; Rothé 1964; USCGS. (Sykes and Landisman 1964; Fairhead and Girdler (JED) 1970, 1971).

Comments

The earthquake of 16 December 1960 took place outside Ethiopian territory; it is included in the present catalogue for two reasons: (1) its vicinity to the Ethiopian shores and its magnitude ≥ 5 make it necessary to incorporate it in the seismicity computations to avoid fringe effects in the contouring process; and (2) its location suggests a possible extension of the inland Bidu-Dubbi volcanic axis NNE-ward across the Red Sea floor.

The three recomputed locations cluster within a circle of 20-km radius at the southeastern corner of the mid-Red Sea valley defined by the 200-fathom isobath. The centre of this epicentre circle lies on the prolongation of the N26E Bidu-Dubbi volcanic axis, which originates inland at N 13°, E 40.5° and under the sea (AA′ in Fig. 91) encloses the 200-fathom southern boundary of the Red Sea valley. In May 1861, the sequence of earthquakes that accompanied the eruption of volcano Dubbi (entry 1861/V–IX) marked the southwestern end of the AA′ zone.

This could be coincidence that the 1861 and 1960 epicentres line up along a N26E zone, which is a prolongation of an alignment of inland volcanic centres and marks the steep boundary of an underwater valley. It could also be coincidence that the 1958 epicentres and 1975 epicentres are located on a similar parallel tectonic feature. It might not be coincidence that all the isobaths that terminate the 500-, 200-, and 100-fathom contour lines in the southern sector of the Red Sea are all oriented N26E, a direction subparallel to that of the transform faults in the Gulf of Aden, to the Alula-Fartak trench, and to the Owen fracture zone in the north Indian Ocean.

It is also along such tectonic lines (C-C′, D-D′) that the central axis of the Red Sea trough is, at latitude N 10.5°, displaced dextrally 30 km westward and forms a > 100-fathom channel passing west of Hanish Island in the direction of the straits of Bab el Mandeb. Such a tectonic line (F-F′) has been identified by Gass (1970a, p. 370) near the Zukur and Hanish islands.

The nature of these N26E postulated faults cannot be ascertained without data on the mechanism of the earthquakes associated with them.

1962/VIII/25

On 25 August 1962 at U.T. 00:54:08, an earthquake of magnitude 4.8 (CGS) occurred in the Red Sea, north of the Dahlak Islands. It was followed on 11 November at U.T. 15:15:28 by another shock of magnitude 5.6. Twelve stations reported the first event; 95 reported the second. These two earthquakes, although they occurred at sea, have to be included in this

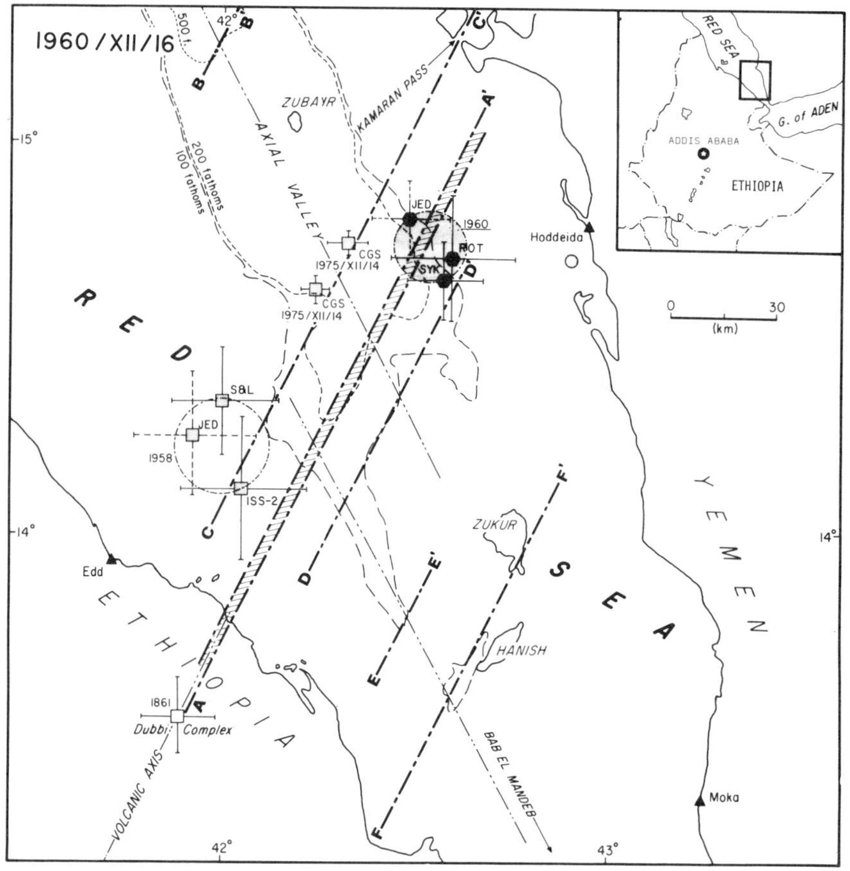

Fig. 91. Epicentre locations and postulated N26E tectonic lines in the southern Red Sea.

survey because of their proximity to the Ethiopian coast and to the Dahlak Islands.

The main reason for introducing a location map — assuming all source conditions besides magnitude are equal — is to illustrate how in 1962 the overall accuracy in epicentre determination was dependent on the number of reporting stations (Fig. 92). The epicentre parameters are given in Part II; they are not repeated here. The location map suffices to illustrate the point.

For computation pùrposes, in both cases, Fairhead and Girdler's JED epicentres have been adopted as the most probable solutions:

	H	Coordinates		m_b	N
25 Aug	00:54:17.5	N 17.12°	E 40.14°	4.8	11
11 Nov	15:15:33.9	N 17.22°	E 40.58°	5.6	95

1962/XI/11

At U.T. 15:15:34 there was an earthquake of magnitude m_b 5.8 at the southern end of the central trough of the Red Sea (if the central trough is defined by the 1000-m isobath).

Comments are presented in entry 1962/VIII/25.

1964/VII/11

In the early morning of 11 July 1964 (3:24 a.m.) a tremor of intensity III was felt in Nazreth (N 08.5°, E 39.3°), on the floor of the Rift Valley.

Source

AAE Data File

Comments

The shock was recorded by AAE at U.T. 00:24:52.0, had an apparent epicentral distance of about 100 km in a southeasterly direction from Addis Ababa, and a magnitude M_L=2¼. Nazreth is located on the Wonji Fault Belt, the axial zone of the Ethiopian Rift (see Fig. 93).

1964/VIII/03

At U.T. 10:45, a shock was reported from northeast Ethiopia. Basing its computation on 25 station reports, the BCIS located the epicentre on the western shelf of the Red Sea near the Ethiopian coast. Moscow, on the other hand, located it 250 km to the south of the BCIS location, inland, near Decamere. There were no reports from Decamere or its surroundings although the estimated magnitude was about 4½.

No epicentre determination was attempted by ISS nor USCGS.

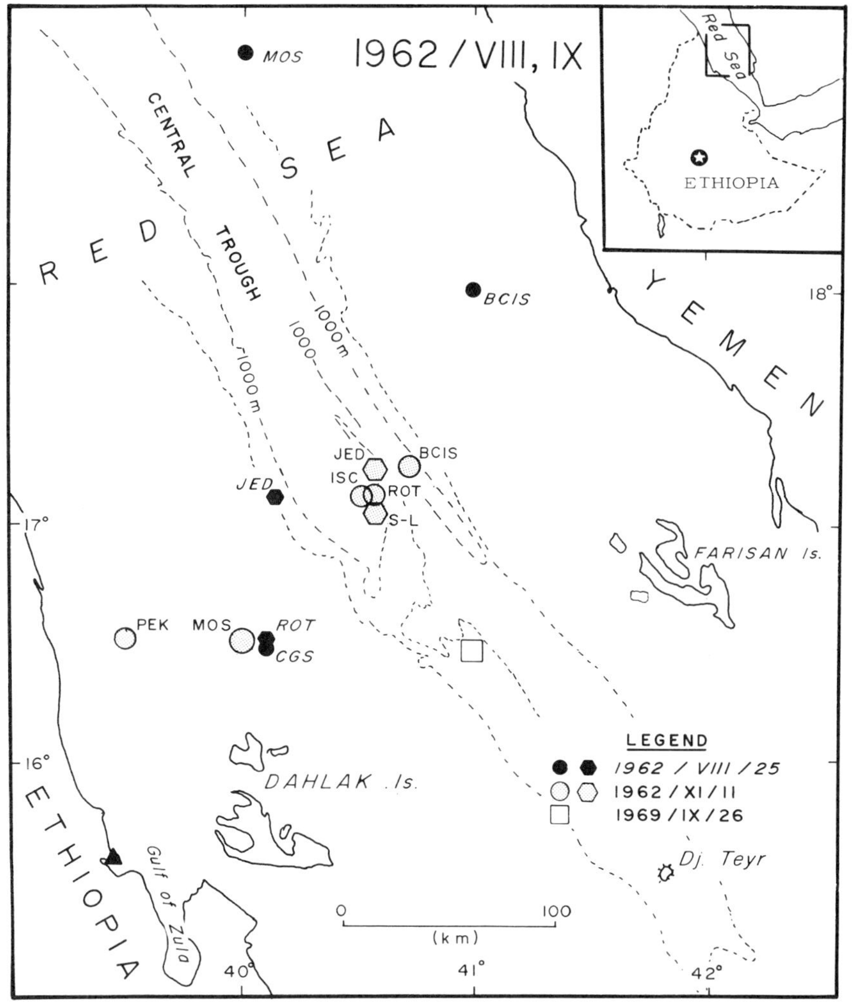

Fig. 92. Location of the instrumental epicentres calculated by different agencies for the earthquakes of 25 August and 11 September 1962. Notice the dispersion in the solutions for the first event (VIII/25, closed symbols) for which only 11 station reports were available, and the crowding of five of seven solutions for the second (shaded symbols) at the south end of the central trough. The bathymetric contour outlines are after Bäcker et al. 1975.

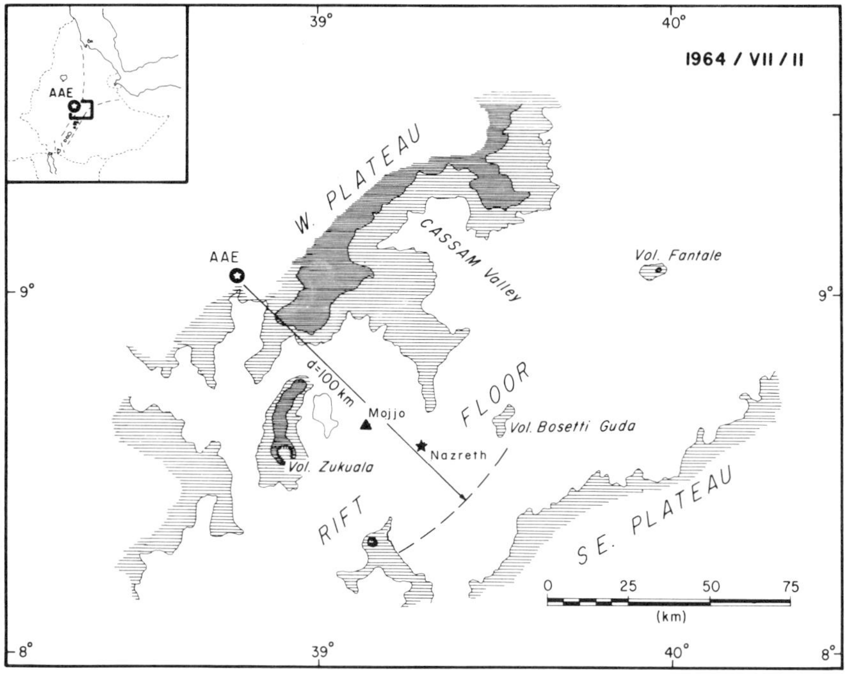

Fig. 93. The Nazreth region in the rift valley.

Sources

BCIS (1964, p. 3372); ISC (1964(2), p. 201).

Comments

The parameters of the solutions are:

	H_o	Coordinates		Stations	Distance from AAE
BCIS	10:45:09	N 17.5°	E 39.1°	25	942 km
MOS	10:45:14	N 15.2°	E 39.0°	4	686 km

The BCIS solution has been adopted for the following reasons: (1) the BCIS solution is based on 25 station reports; MOS used only 4; and (2) there were no tremors felt by the population at or near the site where MOS located the epicentre.

If the location of the epicentre as determined by BCIS is correct, it might indicate a prolongation of the Zula-Massawa Channel seismic zone northward along the Eritrean coast (see Fig. 78).

1965/VI/07

There was seismic activity in Central Afar at the southern end of the Tendaho graben. The main shock occurred at U.T. 13:43:58 ± 1, was of magnitude m_b 5.0, and was followed during the next 7 h by four aftershocks with M_L(AAE) = 4.0, 3.3, 4.0, and 3.1, respectively. Tremors of intensities III–IV were felt at Assayita (N 11.5°, E 41.4°), Serdo (N 12.0°, E 41.3°), and Dubbi (N 11.7°, E 41.1°).

Sources

AAE Data File and Information Cards; ARPA; BCIS; ISC (1965(1), p. 235); USCGS (PDE 52-65). (Fairhead and Girdler JED 1970 and 1971; Dakin 1975).

Comments

The following table gives the epicentre parameters obtained from teleseismic data; they all agree with ± 12 km. The aftershocks were also restricted to a very limited region; S-P time values ranged between 46.0 and 48.5. For computation purposes, the activity was considered as one shock of magnitude 5.1 at the centre of the region (N 11.51°, E 41.47°) (see Fig. 94).

	Coordinates		h (km)	m_b	Stations
ARPA	N 11.4°	E 41.5°	40		
BCIS	N 11.6°	E 41.5°			
ISC	N 11.48 ± 0.084°	E 41.51 ± 0.064°	42 ± 18	4.9	41
JED	N 11.63°	E 41.38°			
USCGS	N 11.43 ± 0.12°	E 41.48 ± 0.12°	40 ± 29	5.1	8
Mean coordinates:	N 11.51°	E 41.47°			

The aftershock sequence has been commented upon the Dakin (1975, sequence 9). For information on the historical seismotectonic activity of the area, see entries 1631/II/10 and 1969/III–V (Serdo–Central Afar).

1966/IV/09

At U.T. 19:11:11 on 9 April 1966, an earthquake of magnitude m_b 4¾ occurred at the north end of the Danakil Alps on the northeastern coast of Ethiopia. The region is very sparsely populated; no felt reports are available.

Sources

AAE Data File; ISC (1966(1), p. 213); USCGS, (PDE 23-66, p. 14). (Fairhead and Girdler JED 1970).

Comments

1. Three instrumental locations based on teleseismic records are available:

	Coordinates		From AAE Δ°	From AAE Az°	h (km)
ISC	N 14.50 ± 0.07°	E 40.90 ± 0.13°	5.8	20.7	27
JED	N 14.46°	E 40.74°	5.8	19.4	
USCGS	N 14.37 ± 0.08°	E 40.79 ± 0.15°	5.7	20.2	33

These solutions agree within a standard deviation range of ± 0.07° in latitude and ± 0.08° in longitude; they also agree with the epicentral distance of 5.7° (635 km) from Addis Ababa obtained by the empirical equation applicable to AAE seismograms for earthquakes originating in northeastern Ethiopia: $\Delta° = (T_{Sn-Pn} - 3.636) / 0.1823$. The adopted location is N 14.45°, E 40.75° at an azimuth of about N20E with respect to Addis Ababa (Fig. 95).

2. The main shock of 9 April was preceded 4 days earlier by a foreshock of magnitude M_L(AAE) 3.7 at an epicentre distance of 625 ± 10 km from Addis Ababa.

3. Both the main shock and the foreshock registered Sn phases at AAE. If the theory is right that the Sn phase is sensitive to or highly attenuated by discontinuities in the lithosphere, it has to be concluded that no important lithospheric break exists along the great circle path heading 020° from Addis Ababa to the Danakil Alps, more specifically along the path passing between the volcanoes Dallafilla and Erta'Ale. Such a discontinuity is proved to exist at the north end of the Depression, see entries 1967.5–1968.5 and 1968/V/23. In conclusion, if the lithospheric anomaly north of the Salt Plain is wide enough to absorb all or almost all the energy of the Sn phase, then at latitude N 13.7°, between the volcanoes Dallafilla and Erta'Ale, the same discontinuity that is expected to propagate southeastward along the central axis of the Depression is either absent or so restricted in width that it hardly interferes with the passage of Sn. It also means that the magma chambers of both volcanoes are restricted in extent.

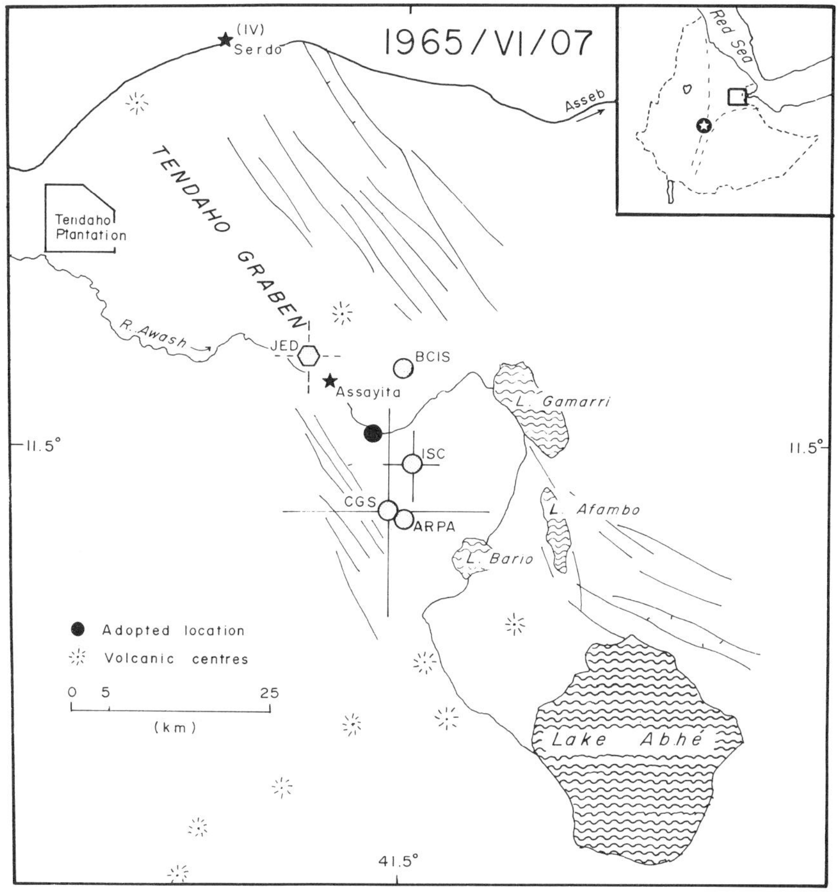

Fig. 94. Instrumental locations and adopted epicentre for the earthquake of 7 June 1965 in southeastern Afar at the southern end of the Tendaho-Assayita graben.

1967/IX/15

Around one o'clock in the morning of 15 September 1967, an earth tremor was felt in Asmara (N 15.5°, E 39.0°) and Decamere (N 15.1°, E 39.0°). No damage was observed. As the epicentre was located along the Plateau escarpment, the entry is to be found in Region A.

1967/IX/21

At U.T. 18:36.5 on 21 September 1967, an earthquake of magnitude m_b(CGS) 4.4, M_L(AAE) 4.6, had its epicentre in the central trough of the Red Sea.

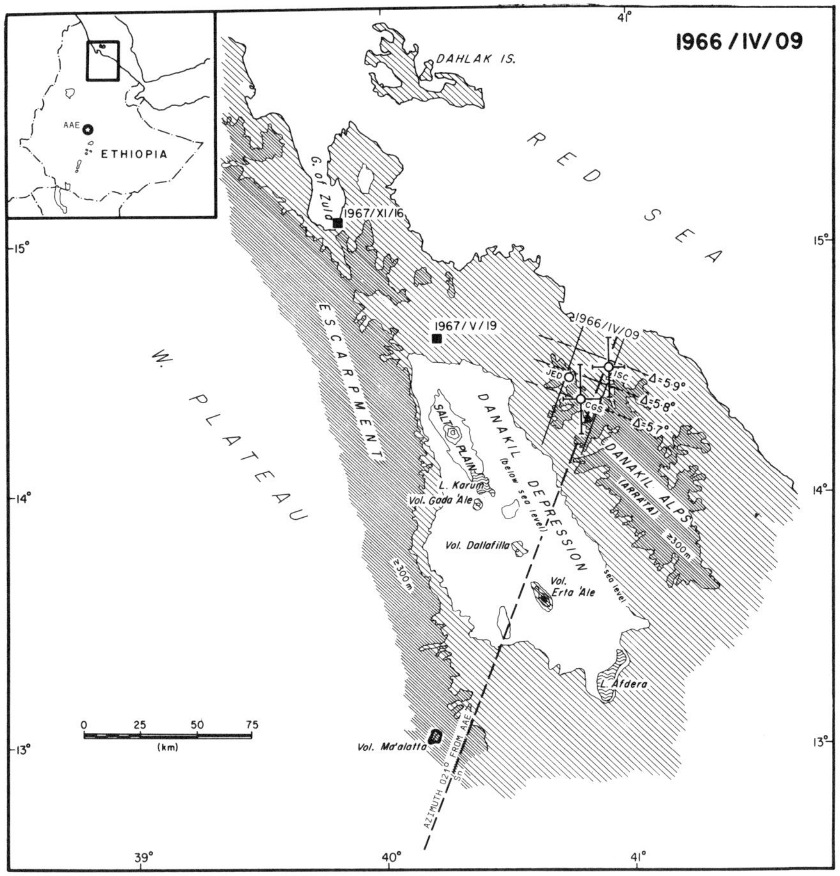

Fig. 95. Location of the epicentre on 9 April 1966, and its seismic ray-path to AAE admist the volcanoes of northern Afar.

Sources

AAE Data File; ISC (1967(2), p. 222); USCGS (EDR 67-67, p. 5). (Fairhead and Girdler JED 1970).

Comments

This earthquake is included here for two reasons: (1) because of possible fringe effects in the seismicity contouring process; and (2) as a check on local travel times and magnitudes.

The parameters of the present epicentre are very well determined:

	H	Coordinates		h (km)
CGS	18:36:26.1 ± 1.1	N 17.94 ± 0.13°	E 39.97 ± 0.16°	16 ± 34
ISC	18:36:32.0 ± 2.1	N 17.84 ± 0.08°	E 40.30 ± 0.15°	80 ± 25
JED	18:36:27.0	N 18.05	E 40.13°	
AAE	18:36:35.3	N 18.1	E 40.1°	
Mean coordinates		N 17.94°	E 40.13°	

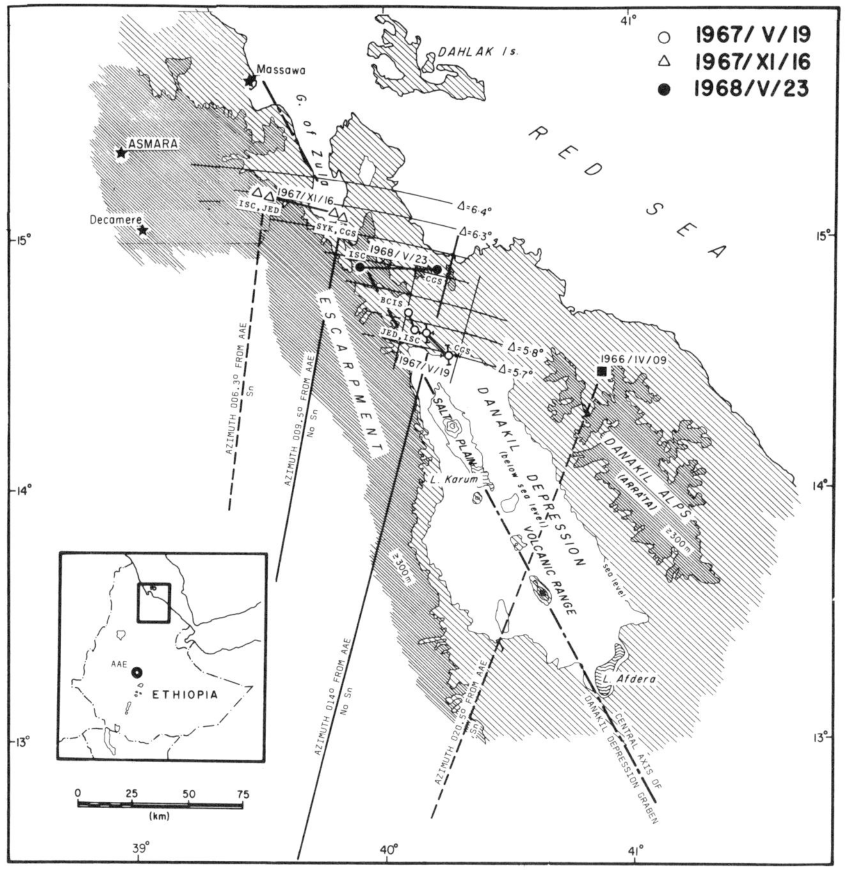

Fig. 96. Location map of the three outbursts of seismic activity that occurred in northern Afar during 1967-78. Such features of the Depression as the sea level contour line, the salt plain, and the axial active volcano range are outlined; seismic ray-paths from the three centres of activity to the station of Addis Ababa are also plotted.

In the seismicity computations, the mean value of all solutions was adopted as the epicentre location.

1967/XI/16

At U.T. 02:22:04, an earthquake of magnitude m_b 5.1, followed within 36 h by seven aftershocks of magnitude M_L(AAE) 3.5, occurred at the south end of the Gulf of Zula. The instrumental epicentres of the main shock by four agencies are:

	H	Coordinates	
ISC	02:22:05.2	N 15.19 ± 0.06°	E 39.49 ± 0.09°
JED	02:22:06.2	N 15.17°	E 39.53°
SYKES	02:22:03.1	N 15.10°	E 39.80°
USCGS	02:22:03.1	N 15.09 ± 0.04°	E 39.82 ± 0.06°

Details and discussions are to be found in entry 1967.5–1968.5.

1967.5–1968.5
Seismic Activity at North End of Danakil Depression, Afar

From mid-May 1967 to the end of May 1968, three outbursts of seismic activity were reported from northern Afar. Because the epicentres occurred within a restricted tectonic province, and carry similar implications concerning the crustal structure under the Depression, they are analyzed here, chronologically.

Sources

AAE Data File and Information Cards; BCIS; Fairhead and Girdler (JED 1970); ISC; USCGS.

1. Earthquake Activity Centred on 19 May 1967

Time and Location of the Main Shock — On 19 May 1967, USCGS reported an earthquake of magnitude m_b(CGS) 5.1 in the Kaynie-Da sector of the Danakil Depression, northeast of the Salt Plain. The shock occurred at U.T. 15:52:36 ± 0.3. Epicentral solutions based on teleseismic data yielded the following parameters:

	Coordinates		From AAE		h
			Δ°	Az°	(km)
BCIS	N 14.70°	E 40.10°	5.8	12.8	
ISC	N 14.62 ± 0.03°	E 40.17 ± 0.04°	5.8	13.7	43 ± 11
JED	N 14.87°	E 40.14°	6.0	12.8	
USCGS	N 14.53 ± 0.03°	E 40.26 ± 0.04°	5.7	15.2	13

These four solutions are plotted on Fig. 96; they agree within ± 15 km in longitude, ± 8 km in latitude, and ± 13 km in epicentral distance from Addis Ababa. The mean location (N 14.68 ± 0.14°, E 40.17 ± 0.07°) is considered very well determined.

Local Records — The AAE seismograms of the main shock show an iPn onset at U.T. 15:54:00.4, *no* Sn trace, and an Sg/Lg onset at 15:55:24.4 on the SP-NS component and at 15:55:26.0 on the SP-EW. The epicentre-to-station ray-path across the Afar Depression is also plotted on Fig. 96 at the azimuth 014°.

Earthquake Sequence — Dakin (1975) considered the earthquake of 19 May 1967 as the main shock of a sequence that initiated at the beginning of May, lasted up to June, and consisted of 79 shocks of magnitude $3.1 < m_b < 5.1$: namely four foreshocks, one main shock on 19 May, and 74 aftershocks.

The only local records available in 1967 were the AAE seismograms on which Pn and Sn phases for events originating from the northeastern sector of the Danakil Depression are highly attenuated, if not completely absent, which prevents any reliable azimuth determination. In such a case, it must be assumed that events that were recorded during a restricted period preceding or following a main shock, that produced similar trace patterns, and were located at equivalent epicentral distances from Addis Ababa, originated from the same seismic region as the main shock. Undoubtedly, such working assumptions carry uncertainties; there is no way of avoiding such assumptions when records from only one local station are available.

Using the same assumptions, I question the fact that the four events listed by Dakin as foreshocks were directly related to the shock of 19 May: the epicentral distances are roughly equal but the trace patterns are different. Sn phases appear on the "foreshock" seismograms; they are absent on the others. For this reason, the four events, as well as three aftershocks, have been eliminated from the sequence, the latter because the epicentral distances differed by 1° or more.

Earthquake Activity Centred on 16 November 1967

Time and Location of the Main Shock — On 16 November, USCGS reported an earthquake of magnitude m_b 5.1 at the south end of the Gulf of Zula, the northwestern continental end of the Afar Depression. Intensities

IV were reported from Asmara (N 15.5°, E 39.0°) and Decamere (N 15.1°, E 39.1°). The origin time was U.T. 02:22:04 ± 01; the focal depth, normal.

Four epicentral solutions based on teleseismic data are available for this earthquake; they are listed here and plotted on Fig. 96.

	Coordinates		From AAE		
			Δ°	Δ km	Az°
USCGS	N 15.09 ± 0.04°	E 39.82 ± 0.06°	6.1	683	9.5
SYKES	N 15.10°	E 39.80°	6.2	684	9.4
ISC	N 15.19 ± 0.06°	E 39.49 ± 0.09°	6.2	689	6.3
JED	N 15.17±≥0.03°	E 39.53±≥0.04°	6.2	688	6.9

The AAE seismograms *do not* exhibit any Sn trace for this event.

Notice that all the teleseismic solutions agree extremely well on epicentral distances from AAE (Δ = 6.2°) and on the latitude component; their mean value is N 15.09 ± 0.05°. The agreement is not as good on longitude, and two groups of solutions are clearly defined: USCGS and Sykes centred on E 39.81° on the eastern shores of the Gulf; and ISC and JED (Fairhead and Girdler) centred on 39.51° on the western side of the Gulf. A difference in longitude of the order of 0.3° resulting from solutions based on teleseismic data is of common occurrence and would not need to be discussed here were it not for the fact that, in this particular case, a difference of 30–35 km in longitude means locating the epicentre either east or west of the Gulf of Zula. This difference in location is of great importance in the controversy over whether or not the Gulf of Zula is the continuation of the Danakil Depression graben, NNWward into the Massawa Channel and the Red Sea.

As indicated above, the AAE seismograms for this earthquake show *no* Sn trace. If the epicentre was located on the western shore of the Gulf, the absence of Sn would be difficult to explain because the seismic epicentre-to-AAE ray-path no doubt intersects crustal anomalies such as the Plateau-Afar escarpment but, as far as we know, does not cross any discontinuity in the lithosphere. From the eastern shore of the Gulf, the epicentre-to-station path crosses the prolongation of the Depression at present covered by the lava flows of Volcano Alid; the absence of Sn on AAE seismograms could be explained along that path by the presence of a lithospheric discontinuity. Such an observation strengthens the arguments for a NNW-continuation of the Depression lithospheric discontinuity at least as far north as the Gulf of Zula.

From these speculations, it follows that the USCGS/Sykes epicentral locations are more realistic than the two others. Sykes' epicentre location has been adopted.

Extent of the Seismic Activity — The main shock occurred on 16 November at U.T. 02:22:03. In the following 36 h, it was followed by seven aftershocks in the magnitude range of 3.1 < m_b <5.1 (Dakin 1975).

2. Earthquake Activity Centred About May 1968

Time and Location of the Main Shock — On 23 May 1968, another shock of magnitude m_b(CGS) 4.8 occurred in the northeast sector of the Danakil Depression. The origin time was at about U.T. 23:16. Light tremors were felt in Decamere (N 15.05°, E 39.05°).

Among the epicentral solutions based on teleseismic records, ISC and USCGS give the following parameters (plotted on Fig. 96):

	Coordinates		From AAE		
			Δ°	Δ km	Az°
ISC	N 14:860±0.038°	E 39.900±0.058°	5.9	660	11
USCGS	N 14:747±0.049°	E 40.217±0.075°	5.9	655	14

It is apparent, at first sight, that a situation similar to that of 16 November 1967 developed: there is good agreement in the teleseismic solutions on the latitude component and the epicentral distance from AAE, but there is a discrepancy of about 35 km in longitude. A discrepancy in longitude carries the same implication here as it did in the two previous cases in relation to the location of the epicentre either east or west of a presumed lithospheric discontinuity, in this case the Depression central graben.

The answer to that question, based on local data, is the same as for the two other epicentres: the AAE seismograms show an attenuated but clear Pn impulse at 23:37:33.7, *no* Sn, and a Sg/Lg at 23:39:03.4. It follows, therefore, that if the argument concerning the absorption of Sn phases by discontinuities in the lithosphere is valid, the epicentre had to be located east of E 40.1°.

Seismic Activity During May 1968 — Given a situation identical to that described above, the same assumptions were made concerning the identification of events of lower magnitude; namely, that when azimuth determination is practically impossible, shocks with comparable trace patterns and equivalent epicentral distances from AAE occurring during a restricted period of time about the main shock, probably have their origin in the same seismic region.

Following these assumptions, Dakin (1975) determined that the event of 23 May at U.T. 23:36 was the main shock of a sequence of 80 earthquakes (18 foreshocks and 61 aftershocks) generated between 8 and 31 May 1967. Figure 97 shows the frequency-magnitudes relationship for the 80 earthquakes considered to belong to the May seismic sequence; the b-value is 0.96, a typical value for rift zones.

Keeping in mind the uncertainties inherent in the assumptions described above, and therefore proportional uncertainties in the identification of each event, it was observed that during this sequence some 85% of the seismic traces at AAE did not reveal any Sn phases; some 15% did. This observation suggests, if the identifications are right, that both sets of faults forming the eastern and western limits of the Depression graben could have been seismically active during that period.

1967/XII/17

On 17 December 1977, an earthquake of magnitude M_L(AAE) 4.7 occurred at U.T. 18:03:43 (local time: 21 hours) at the north end of the Gemu-Gofa rift, in the Bala Plains. As the location is on the western side of the Gemu Highlands, the entry is classified under Region E. It is also indexed here because the tremors were felt across the main Rift and on the SE Plateau, and because the north end of the Gemu-Gofa rift is terminated by strong NNE and NE curvilinear faulting that cuts through the Gemu Highlands and joins the main Ethiopian Rift escarpment west of Lake Abaya (see Fig. 152).

1968/V/23

An earthquake of magnitude m_b 4.8 occurred at U.T. 23:16, in the north sector of the Danakil Depression. Light tremors were reported from Decamere (N 15°, E 39°). Two epicentral locations have been determined by international agencies:

	H_0	Coordinates	
ISC	23:36:08.4 ± 0.3	N 14.86 ± 0.04°	E 39.90 ± 0.06°
CGS	23:36:06.4	N 14.75 ± 0.05°	E 40.22 ± 0.08°

Details and discussions are to be found in entry 1967.5 –1968.5.

1969/III/16

A tremor of intensity III was felt in Batie (N 11.2°, E 40.0°) at about seven o'clock in the morning. Batie is located halfway up the Afar-Plateau escarpment on the Combolcia-Asseb road.

Source

Christian Mission in Batie (personal communication); AAE Data File.

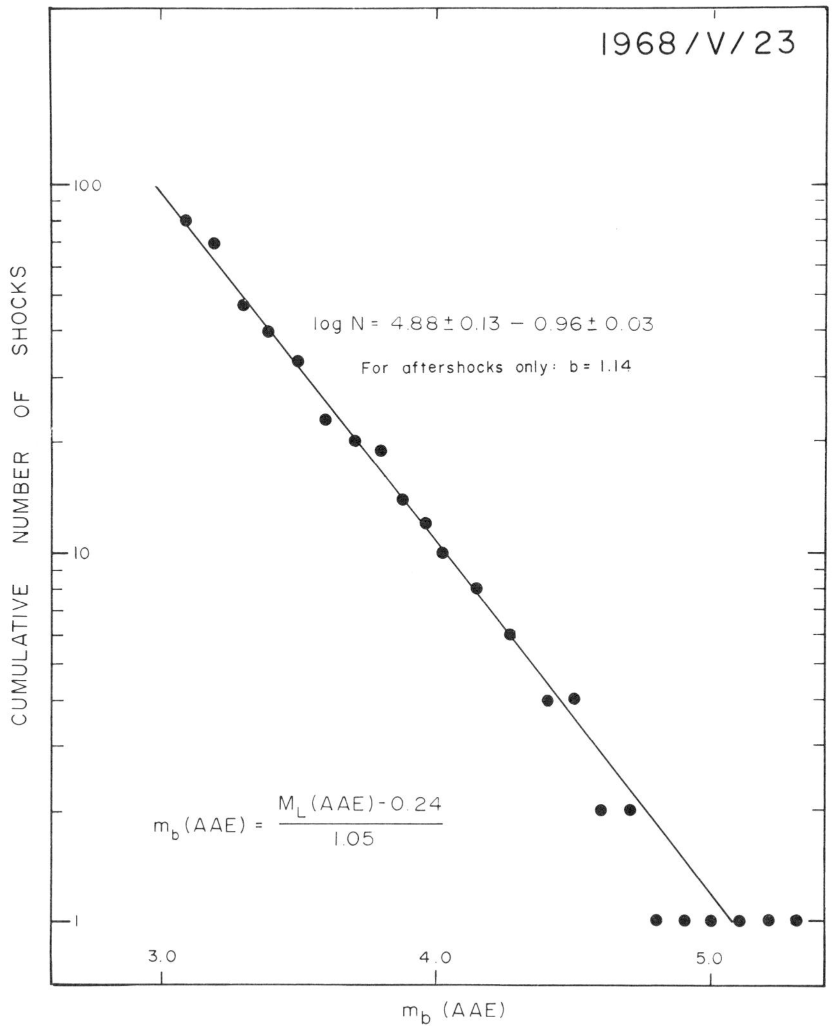

Fig. 97. Frequency-magnitude relation graph for the seismic activity in the Afar Depression in May 1968.

Comments

The event was recorded at U.T. 04:02:18.1 at AAE. The apparent epicentre was located at 2.6–2.7°, NNE-NE of Addis Ababa. The Pn onset was of low amplitude and emergent; no accurate azimuth determination was possible.

The magnitude M_L(AAE) was about 4.

1969/III–V (Serdo–Central Afar)

The town of Serdo (N 11°57′, E 41°19′) in the Danakil Desert was completely destroyed by an earthquake of magnitude m_b 5.9. Out of a population of 420 people, 24 lost their lives and 167 were injured within minutes of the main shock at 2 p.m. (U.T. 11:05), 29 March; 15 others died later of injuries suffered during the collapse of their houses. Cracks, fissures, and subsidences appeared mainly in the plain southwest of the ridge on which Serdo was built. A second shock of magnitude m_b 6.2 on 5 April at U.T. 02:18 (5 a.m.) opened three important faults about 8 km east of Serdo; one of these showed a vertical displacement of 75 cm downthrown to the northeast and a left lateral shear displacement of 65 cm. These faults crossed the highway. The same shock triggered rockslides along the highway and over quite a large area.

Over 250 aftershocks were recorded at the WWSS station in Addis Ababa; two reached magnitude m_b 5.2.

Within 2 h of the disastrous shock of 29 March, help arrived from the nearby plantation in Tendaho and from the Christian Mission in Batie. The dead were buried immediately after identification by their families; the survivors were injected against typhoid and cholera and evacuated to Logghia, 40 km west of Serdo, where a refugee camp was organized on the compound of the plantation's stores.

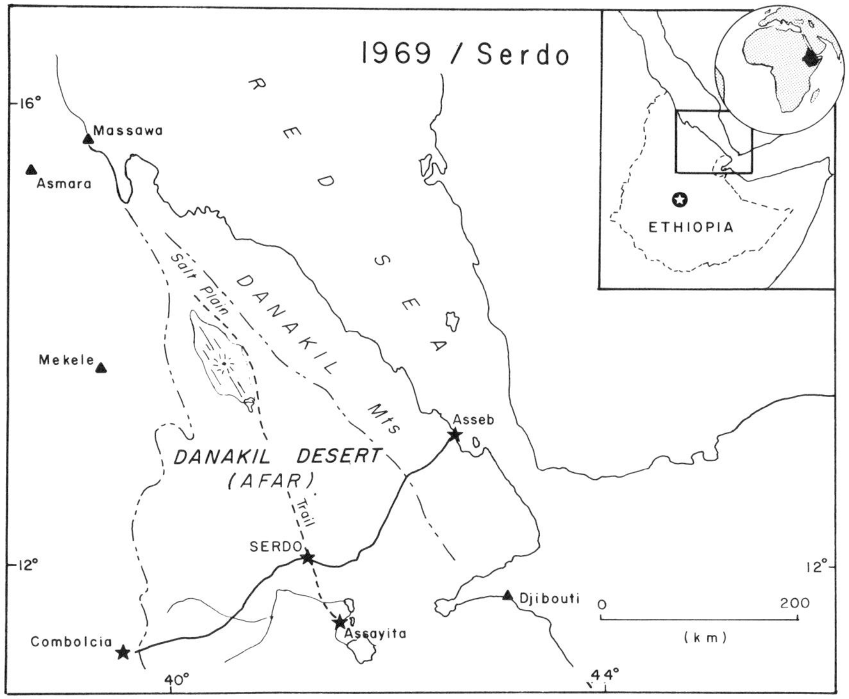

Fig. 98. Location map of Serdo in the Danakil Desert of northeastern Ethiopia.

Sources

Descriptive

Gouin (1969, 7 and 17 April) (Official Reports on the situation in Serdo); Dakin et al. (1971); Searle and Gouin (1971a).

Instrumental

AAE Data File; BCIS; ISC; Fairhead and Girdler (1970); Gouin (1975); Sykes (1970); USCGS.

1. Geographic Location of Serdo

The village of Serdo is situated near the geometrical centre of the "Afar triangle," midway between the port of Asseb on the Red Sea and Combolcia on the Ethiopian Plateau (Fig. 98). Its geographic coordinates are N 11.97°, E 41.31°; the elevation 400 m above sea level. Historically, Serdo grew up at the junction of the north-south caravan route from the salt plain in the Danakil Depression to the provincial capital Assayita, and the important east-west commercial trail from the Red Sea coast to the Ethiopian highlands.

In a way, Serdo was a unique town; its profile over the eastern horizon, outlined by a concrete water tower and other masonry buildings dominating the plain of Kurub from the top of an elevated ridge, was, to say the least, unexpected in the middle of the desert. Serdo owed its older masonry structures to the road construction company that built and maintained the Asseb-Combolcia highway in the late 1930s. Newer masonry buildings of dubious structural design have been added recently. The majority of the houses in the village were light wooden structures that were top-heavy and had 20–30 cm of earth and cinders for thermal insulation. Figure 99 identifies by name the different sectors (fedo, in the Afar language) prior to the 1969 earthquakes.

2. Regional Tectonics and Geology

Central Afar is covered by volcanics (Fig. 100). Basalts form ridges, cinder cones, and extensive flows of both central and fissural origin, and rhyolite lavas and tuffs are associated with larger scattered centres and appear to overlie all but the youngest basalts. These volcanic flows are highly dissected by fresh normal faults into steeply sided blocks; the faults are variable in throw direction and the blocks generally tilt toward the SW or SSW. Between the fault-blocks lie planar areas formed by fluvial and lacustrine sedimentation.

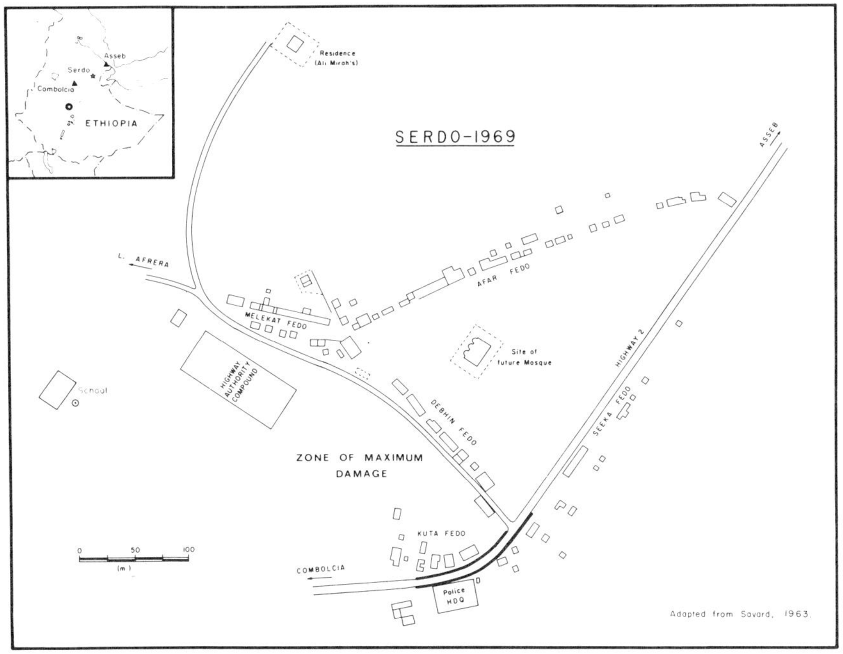

Fig. 99. The desert town of Serdo before the earthquakes. The basic layout was taken from the field notes of Georges C. Savard, head of the Department of Sociology, University of Addis Ababa.

Serdo was built on and around one of these fault-blocks trending N40W and tilted toward the southwest. The best masonry and concrete structures, such as the 30 ft (10 m) water tower, the police headquarters, and the highway authority compound, were erected on the ridge, a rhyolite block of almost vertical flow-banded structure (Fig. 101); the new school, a masonry structure too, was located in the plain northwest of the ridge (Fig. 103); the rest of the village was mainly concentrated in the plain to the east, enclosed between two ridges.

3. Regional Seismic and Volcanic Activity

Historical Survey

What is presently known of the seismic and volcanic activity in Aussa prior to 1969 is summarized in this paragraph. Details are to be found in the entries corresponding to the dates indicated; some of the epicentres are plotted in Fig. 100.

1608, 23 December (15 Remadam A.H. 1017): volcanic activity in the form of smoke coming out of the ground, east of Mt Waraba, near a lake, in Aussa.

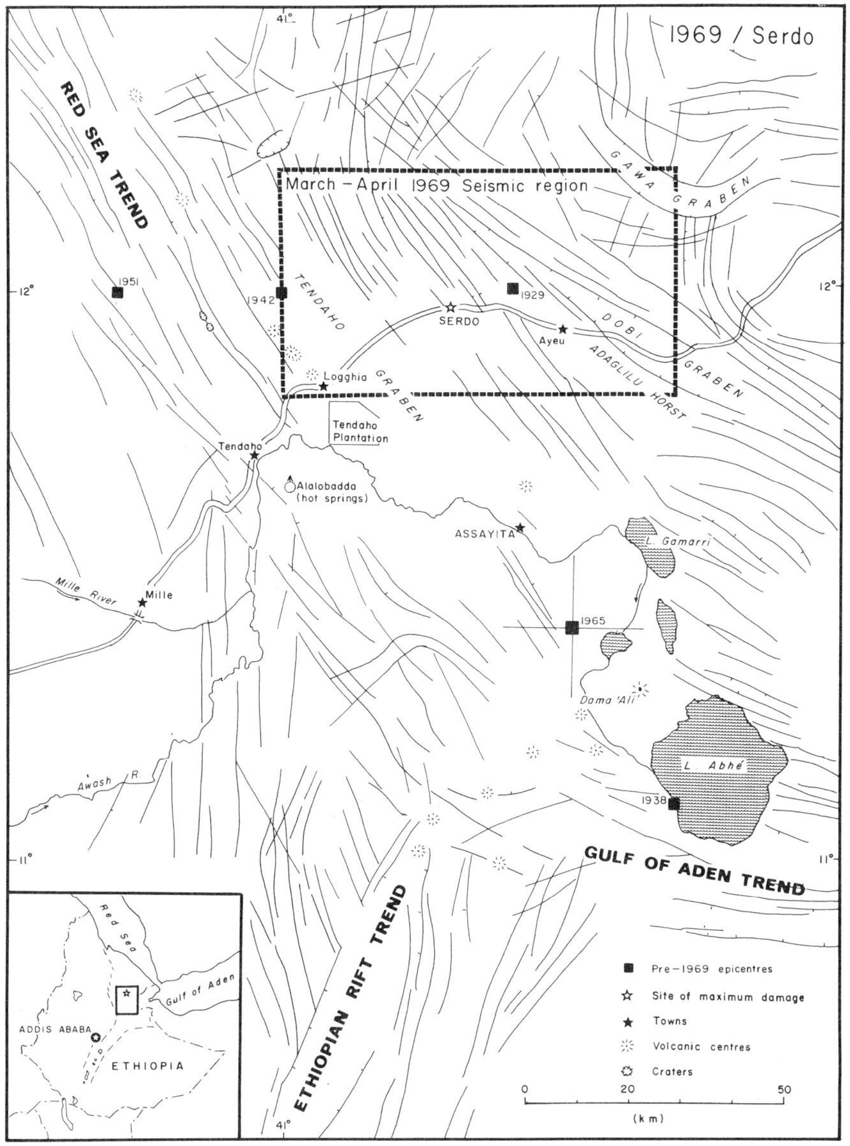

Fig. 100. Tectonic outline of Central Afar after Dakin et al. (1971). Plotted on this map are the epicentres known to have occurred before 1969. Standard error bars are indicated only for the epicentre of 1965; the others could not practically be determined.

1627 From 18 March to midsummer, seismovolcanic activity in Aussa destroyed all houses in Waraba; 50 people are presumed to have lost their lives.

1883 Strong earth tremors were felt near Mt Janghudi (N 10.5°, E 41.1°) and in Asseb (N 12.9°, E 42.7°). It is probable that the epicentral region was located in Central Afar.

1929, 18 May U.T. 01:02:15. Epicentre at N 12.0°, E 41.4°; M = 6.

1938, 27 September U.T. 02:31:51. Epicentre at N 11.1°, E 41.7°; M = 6.

1942, 18 November U.T. 12:01:20. Epicentre at N 12°, E 41°; M =5½.

1951, 11 December U.T. 02:34:36. Epicentre at N 12°, E 40.7°.

1965, 7 June U.T. 13:43:57. Epicentre at N 11.43° $\pm$ 0.12°, E 41.48 $\pm$ 0.12°; h = 40 $\pm$ 29 km.

Seismic Activity Near Serdo in 1969

Instrumental Data — Of the 250 foreshocks and aftershocks of magnitudes $\geqslant$ 3 recorded during the period of seismic activity around Serdo, 12 have been listed by various agencies. Selected parameters are given in Table 4. Because the USCGS epicentre locations have been observed to best fit the AAE station recordings and were used in the determination of regional travel times (Searle and Gouin 1971), the residuals for P arrival times as calculated by USCGS are also indicated. The epicentral distances (Δ) given in Table 4 are reckoned from Addis Ababa.

The instrumental epicentres given in Table 4 are plotted in Fig. 104.

Field Observations — On 4 April 1969, a low-altitude air survey of the earthquake striken area was made by USAID personnel. At the time, no apparently fresh geological surface disturbances were observed other than the cracks south of Serdo. On 5 April, the first of the three investigation teams headed by myself arrived at Serdo; the others followed in April–May 1969 and in 1970. During the field investigations, local residents and caravan teams were questioned, damage to structures examined, and all accessible tracks covered in the search for ground effects. Because the ruggedness of the terrain precluded access by ground vehicles to regions outside the main trails, the surveys are somewhat incomplete.

Casualties. The absolute number of casualties, some 40 dead and 160 wounded, is not excessively high, but if considered as a percentage of a total population of 420 people, it means that half of the population was either killed or wounded. Had the destructive shock of 29 March occurred before noon during school hours, the toll would have easily risen by another 20–30 children killed or injured. The total collapse of the school building leaves no doubt about it (see Fig. 101).

Damage to Man-Made Structures. All the masonry structures located on the Serdo ridge in the Kuta, Debhin, and Melekot *fedos*, including the highway authority compound and the police headquarters buildings, were completely destroyed on 29 March (Fig. 103). The school located in the plain northwest of the town was likewise destroyed (Fig. 103). The reinforced-concrete water tower stayed up, but its brick facing fell off and the whole structure is now inclined by about 5° from the vertical in a SW direction. Houses with flexible wooden structures, unless heavily topped with earth, suffered much less. Only two structures stayed virtually intact; they were entirely wooden: the house of H.E. Bitweded Ali Mirah, and the Telecommunications Office (Fig. 102).

In the region of Serdo, as well as 8 km to the east in the vicinity of faults F_1 and F_2 described below, bridges and culverts were damaged (main structures cracked, head walls broken, supporting walls fissured, etc.).

The damage was concentrated in a rather restricted area. No damage was reported from Logghia, 40 km southwest of Serdo, one small crack was observed at Ayeu, 25 km to the northeast; on the Tendaho Cotton Plantation, at about the same distance from Serdo as Logghia, a few adobe houses collapsed (Fig. 103), some walls were fissured, and cracks ran across the pavement of the club house, the fish pool, and the swimming pool. They are discussed below under "cracks." The damage at the plantation occurred on 5 April.

Geologic Surface Effects

Cracks. In this context, "cracks" are defined as small tensional fissures with no significant vertical or lateral displacement. Several were observed in the tarmac of the highway. They were up to 20 metres long, 30 cm wide and about 20 cm deep. All but two were parallel to the road. In Serdo, one was perpendicular to the road but parallel to a nearby culvert; it was certainly caused by the dislocation of the concrete culvert structure from the road pavement. The other was located about 0.5 km west of Serdo; its strike (N60W) made an angle of about 10° with the axis of the road. All these cracks, with the possible exception of the last one, are believed to have been entirely controlled by slumping of the highway understructure; they are not considered to have any directional tectonic significance.

A crack, 5–10 cm wide, striking north-south, opened at the northwest end of the Serdo ridge, running through the ground and the masonry foundation of the main building on the highway authority compound. This crack was due to slumping of the foundation structure.

Cracks not on the surface of the road, and therefore of probable significance as regards directional tectonic trends, were observed at two locations: (1) about 500 m south of the town; and (2) about 8 km to the east of it. At both sites, they occurred in loose sediments. At location (1) in the sediment covered plain south of Serdo, cracks (C_1) developed in a zone about 400-m long by 10-m wide (Fig. 105B). Individual cracks were 0.5 – 3.0 m long, 15–20 cm deep, and 1–20 cm wide. The centre of the zone subsided by 20–25 cm. The general trend of the zone was N40W; it changed to roughly east-west at its northwestern end. The direction of individual cracks

Table 4. Parameters for the 12 largest events during the Serdo sequence.

		H_O	Coordinates		h(km)	St.	m_b	M_s	Δ(km)	O–C
(1) 29 Mar	CGS	09:15:54.1 ± 0.17	11.97 ± 0.03°	41.18 ± 0.03°	33	72	5.8	6.3	420	−1.6
	ISC	09:15:54 ± 1.0	11.91 ± 0.03°	41.21 ± 0.03°	35 ± 10	204	5.9		417	
	JED		12.02°	41.18°					426	
	PG		11.93°	41.27°					423	
(2) 29 Mar	CGS	11:04:47.9 ± 3.7	11.96 ± 0.04°	41.29 ± 0.055°	4 ± 22	54	5.6		427	−0.5
	ISC	11:04:52 ± 1.1	11.92 ± 0.03°	41.36 ± 0.032°	35 ± 11	166	5.5		429	
	JED		12.03°	41.24°					429	
	PG		11.94°	41.30°					426	
(3) 29 Mar	CGS	11:07:30.0 ± 0.3	11.99 ± 0.03°	41.21 ± 0.85°	33	21	5.3	5.8	424	
	ISC	11:07:45 ± 6.0	12.01 ± 0.07°	41.1 ± 0.14°	164 ± 56	55	4.9		418	
	JED		12.06°	41.48°					449	
	PG		11.91°	41.57°					443	
(4) 29 Mar	CGS	13:08:11.4 ± 3.9	11.94 ± 0.03°	41.52 ± 0.09°	4 ± 23	23	5.1		495	−0.7
	ISC	13:08:17 ± 1.2	11.94 ± 0.04°	41.31 ± 0.05°	43 ± 12	88	5.1		427	
	JED		12.03°	41.41°					441	
	PG		11.85°	41.49°					433	
(5) 29 Mar	CGS	18:30:42.2 ± 0.4	12.00 ± 0.08°	41.38 ± 0.12°	33	8	4.6		437	
	ISC	18:30:49 ± 5	11.87 ± 0.09°	41.40 ± 0.13°	95 ± 52	19			427	
	JED		12.01°	41.34°					435	
	PG		11.89°	41.43°					431	
(6) 05 Apr	CGS	02:18:29.9 ± 1.9	12.15 ± 0.03°	41.20 ± 0.04°	17 ± 14	46	6.2	6.1	437	−2.3
	ISC	02:18:30 ± 2.8	12.00 ± 0.04°	41.35 ± 0.04°	19 ± 20	175	5.8		435	
	JED		12.19°	41.13°					448	
	PG		12.12°	41.26°					446	
(7) 05 Apr	CGS	20:06:23.8 ± 0.4	11.91 ± 0.08°	41.14 ± 0.11°	33	6	4.3		412	+0.3
	ISC	20:06:24 ± 2.8	12.0 ± 0.15°	41.2 ± 0.20°	32 ± 33	13			424	
	JED		12.02°	41.10°					420	
	PG		11.95°	41.19°					419	
(8) 05 Apr	CGS	20:14:35.9 ± 0.4	12.03 ± 0.04°	41.47 ± 0.07°	33	19	4.9	5.2	447	+1.2
	ISC	20:14:41 ± 1.3	12.02 ± 0.05°	41.28 ± 0.07°	70 ± 14	35	4.8		431	
	JED		12.11°	41.49°					454	
	PG		11.96°	41.59°					449	
(9) 06 Apr	CGS	16:51:45.5 ± 2.0	12.03 ± 0.04°	41.12 ± 0.06°	20 ± 15	39	5.2	5.4	421	−0.1
	ISC	16:51:47 ± 1.1	11.99 ± 0.03°	41.40 ± 0.039°	41 ± 11	104	5.1		437	
	JED		12.07°	41.11°					424	
	PG		12.00°	41.21°					425	
(10) 07 Apr	CGS	06:23:53.4 ± 1.2	11.98 ± 0.10°	41.28 ± 0.08°	33	14	4.6		428	−0.3
	ISC	06:23:55 ± 1.8	11.92 ± 0.10°	41.40 ± 0.09°	58 ± 20	24			432	
	JED		12.04°	41.30°					434	
	PG		11.93°	41.39°					432	
(11) 08 Apr	CGS	02:13:58.7 ± 1.3	11.93 ± 0.06°	41.37 ± 0.07°	34 ± 13	18	4.8		430	−0.4
	ISC	02:14:01 ± 1.1	11.88 ± 0.05°	41.42 ± 0.05°	56 ± 12	52	4.8		430	
	JED		12.02°	41.37°					438	
	PG		11.90°	41.62°					446	
(12) 05 May	CGS	02:45:38.9 ± 1.3	11.94 ± 0.08°	41.26 ± 0.06°	35 ± 15	20	5.2	5.0	423	
	ISC	02:45:40 ± 1.5	12.07 ± 0.04°	41.34 ± 0.06°	38 ± 15	83	4.9		440	
	JED		12.01°	41.28°					431	
	PG		11.90°	41.37°					428	

Fig. 101. Ruins of Serdo after the main shock of 29 March 1969. Nothing was left intact except the wooden shack of the telecommunications board (upper left, plate 1) and a wooden house north of the town. In Afar Fedo, the light structures stood the shocks. On the upper right of plate 2, the silhouette of the section of the village erected on the Serdo ridge can be seen. The letter A locates the highway authority compound, B the reinforced concrete water tower leaning about 5° SW, and C the village school that had housed 30–35 students 2 h earlier. Plates 3 and 4 show the water tower, deprived of its brick facing, but still standing among the ruins of the post office and police headquarters.

Fig. 102. Fault F_1, some 8 km east of Serdo. It developed on 5 May (striking N40W subparallel to the ridge axis) and had a vertical displacment measured on the highway of about 75 cm and a left-lateral displacement of about 65 cm. Northwest of the highway F_1 could be traced for about 1 km before it became indistinct among the hillwash (plates 5 and 9). Plate 6. Fault F_2 crossing the road obliquely. Fracturing plus slumping reached a vertical displacement of some 95 cm. Plate 7. A broken boulder along the fault. Plate 10. A rock apparently freshly broken by impact on a harder specimen after having been projected up by a vertical seismic motion. The point of impact was clearly observed (also see plate 19). Plate II. A close-up of the north wall of the road-cut through Serdo ridge showing the almost vertical and highly contorted flow-banded rhyolite that forms the ridge.

Fig. 103. Plate 12. Fault F_1 along the hillside; direction N40W. Plate 13. Vertical differential displacement along the highway. Plate 14. Ruins of the highway authority compound. Notice the steel flagpole, 7.5 cm at its base, inclined to the SW. The cement base was cracked, and the local people ascertained that the bending of the flagpole was a result of the earthquake. Plate 15. Cracks in the tarmac; they tend N30W. Plate 16. Whole field of freshly broken boulders about 20 km north of Serdo. Another site where broken boulders, were found was about 10 km north of Serdo, on the prolongation of F_1. Plate 17. Upheaved riprap highway slope near F_3. The bridge sill and headwall had cracked, the capped section of the retaining wall had fallen, and the wall itself had fissured. Plate 18. An example of apparent rotational motion. Plate 19. Freshly broken rock showing the point of impact of a second boulder projected by sudden vertical seismic motion.

in this C_1 zone varied within ± 30° of the general trend. Local people reported that these cracks had developed on 29 March. They occurred in loosely packed alluvium overlying gravel beds, thus they were probably entirely due to compaction caused by the tremors and merely reflected the surface and underground regional topography.

At location (2) about 8 km east of Serdo and close to the main road (Fig. 105C) a zone of cracks (C_2) occurred in loose hillwash at the foot of a steep escarpment. These cracks appeared to be associated with fresh normal faulting that resulted from the shock of U.T. 02:18:29.9 on 5 April. This faulting is described below under "faults." The cracks were similar in length, width, and depth to those found at location (1). They did not, however, lie in a continuous band but occurred as individual, isolated units or in short discontinuous bands. Their general trend was N40W, but variations of ± 60° were observed.

A crack was observed running through the hard soil floor and the baking oven of the Tehame Tekle Hotel in Ayeu, 25 km east of Serdo. Its direction was N60–70W.

Other cracks were observed at the Tendaho Cotton Plantation: one ran through the east porch of the club house in a N80E direction (this crack was reported to be existing before 5 April and to have merely widened on 5 April); a second one cut through the structure of the fish pool in a N40W direction; a third ran along the joints of the walls in the swimming pool. All occurred on 5 April.

Faults. "Faults" in this context are fractures that show appreciable vertical and/or lateral displacement. Fresh faulting was observed at three positions about 8 km east of Serdo along the main road. Positions F_1 and F_2 are indicated on Fig. 104 and 105. The fault F_3, about 1 km east of F_2, is beyond the area of the maps.

Fault F_1, striking N40W, was almost perpendicular to the road and parallel to a major N40W fault scarp (Fig. 105C). Its vertical displacement, measured at the road, was about 75 cm and it showed a left lateral displacement of about 65 cm (50 cm was given in the preliminary report to Smithsonian Center for Short-Lived Phenomena, card 493/1969). This fresh fault was traced for about 100 m to the southeast of the road where it deteriorated into the series of cracks described above for location (2). To the northwest of the road, the fault continued for at least 1 km, becoming less distinct with increasing distance from the road, until it was eventually obscured by hillwash. The second fault (F_2), striking N50W, cut obliquely across the main road about 200 m east of F_1. F_2 had a vertical displacement of 90–95 cm, showed no detectable lateral displacement, and could not be traced on either side of the road (Fig. 105). About 500 m east of F_2 one small fault (F_3) showing a left lateral displacement of 10 cm was observed in the tarmac of the road. The strike of this fault was N50W, almost perpendicular to the road.

All the faults described above developed in the morning of 5 April.

Rockslides. One rockslide occurred 4 km east of Serdo; two more, on the southeast side of the Dobi graben (see Fig. 105) at 39 and 40 km east of Serdo by road. They were triggered during the afternoon of 5 April. Other rockslides and landslides occurred at different places as signaled by clouds of dust rising above the hills; their location could not be identified because the sites were inaccessible by land vehicles.

Ground motions (horizontal, rotational, and vertical)

(1) Horizontal motion: It was reported by the villagers of Serdo that on 29 March, people, animals, and walls were projected in an eastward or northeastward direction. The permanent inclination toward the southwest of both the 3-inch (7.5-cm) steel flagpole of the highway authority compound and the reinforced concrete tower (Fig. 101; Fig. 102) shows that their bases were suddenly jerked northeastward.

(2) Rotational motion was observed on the highway authority compound: one end of a 2-m long wall was rotated 10° counterclockwise. Such an angle of rotation may partly be due to unequal dislocation of the masonry.

(3) Vertical displacements: Fig. 103 shows two small freshly broken rocks, photographed in situ, near the main fault F_1. The fractures in the rocks were apparently caused by sudden upwards motion. Number 16 shows a rock sample that must have been projected upward and broke when it fell on a harder rock; number 19 shows the point of impact where a projected rock hit a small bolder and cracked it.

On either side of fault F_1, which was located 8 km east of Serdo, were first-order vertical control benchmarks: M35 at km 5.41 and L35 at km 9.92. In 1960, the USCGS survey team measured a difference in elevation of 450.3 cm between the two sites (USCGS 1957–1961, Ethiopian Geodetic Survey, Washington); in 1970, our releveling showed a difference of 510 cm, representing an increment of 60 cm over a 10-year period. This increment is presumed to have occurred rather suddenly at the beginning of May 1969. The regional extent of the subsidence with respect to the mean sea level reference in Asseb is unfortunately not known; there was no possibility of remeasuring a few hundred kilometres of first-order elevation profile.

Other phenomena observed during the 1969 period of seismic activity.

Noises. On 29 March, noises like explosions were reported in both Serdo and Dubti, near the Tendaho Plantation; people were under the impression that they were coming from the Alalobadda geothermal region (Fig. 100). Investigation of the Alalobadda site revealed no sign of explosion such as freshly broken rocks or a change in the level of the springs.

Smoke. "Smoke" was reported as having been observed over the hills surrounding Serdo, both on 29 March and 5 April. At the same time there

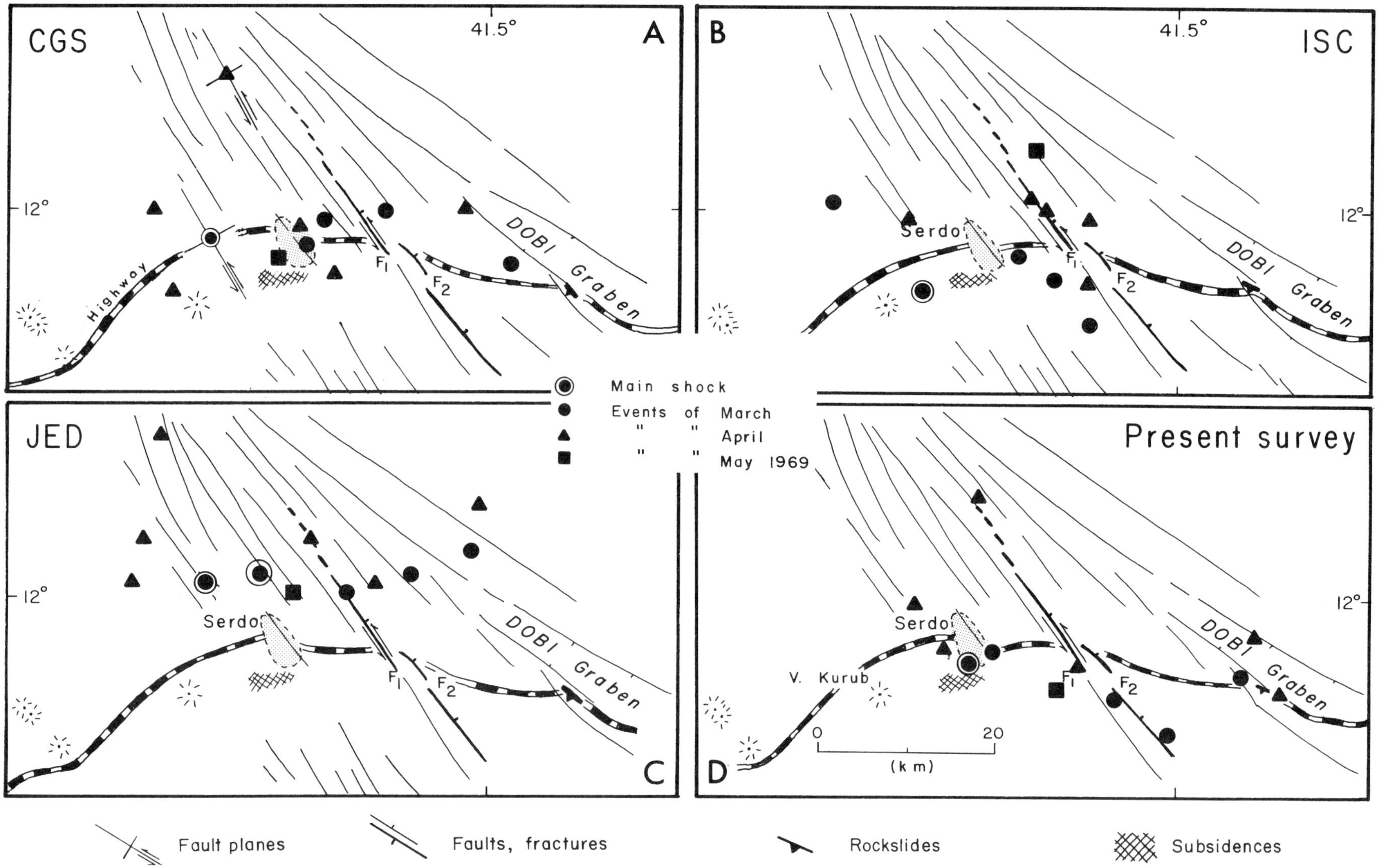

Fig. 104. Distribution plots of the Serdo epicentres as determined by: (A) the USCGS; (B) the International Seismological Center; (C) Fairhead and Girdler; and (D) Gouin.

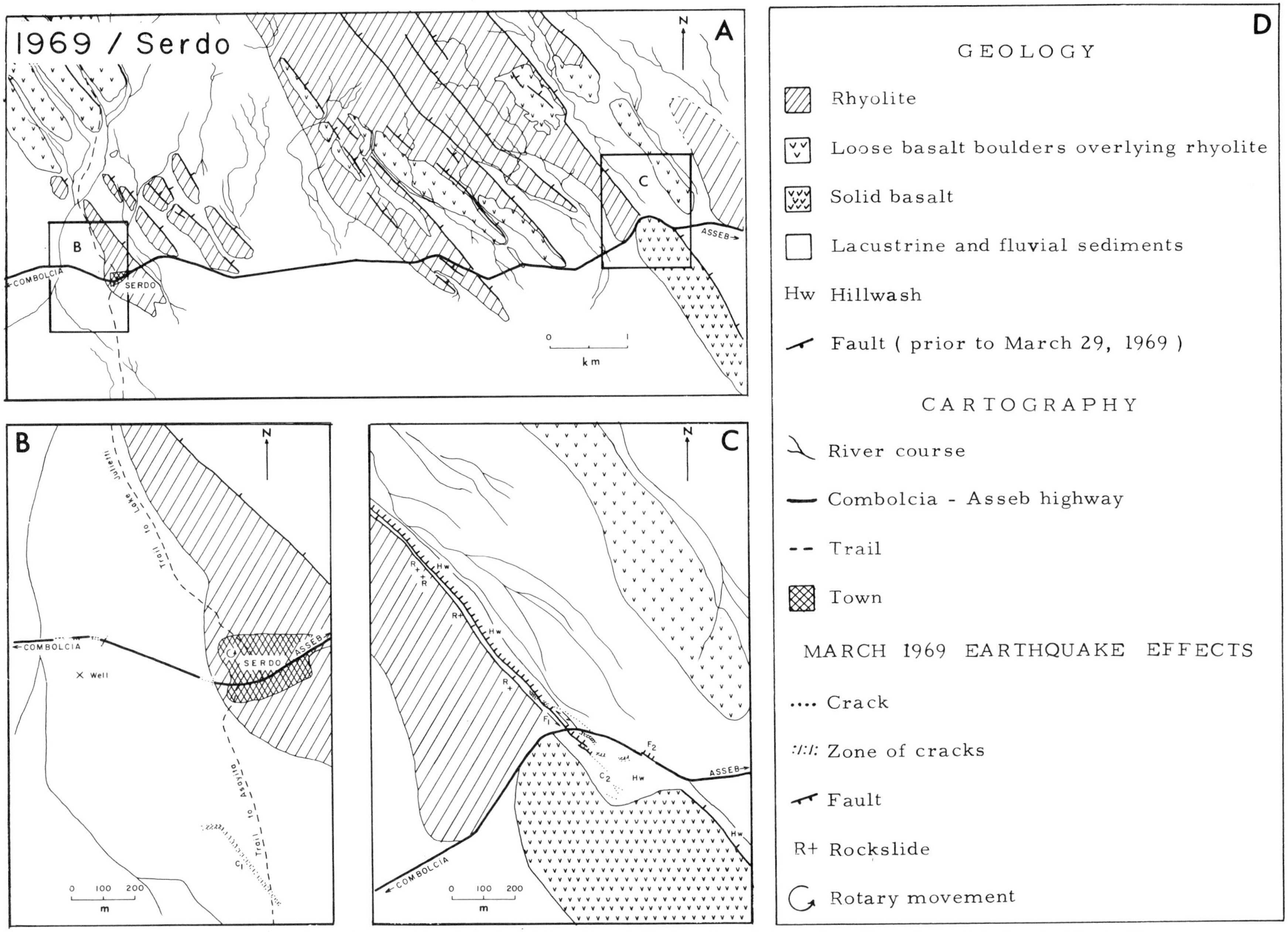

Fig. 105(A). Geological map of Serdo and vicinity. (B and C). Location of the surface effects observed during the March–May earthquake sequence of 1969 (after Dakin et al. 1971).

were rumors of renewed volcanic activity; the rumors were unfounded and it is believed that the "smoke" was dust thrown up by rockslides.

Smell. On 29 March, sulfurous smells were reported from Serdo. No satisfactory explanation was found.

Seismic Activity Around Serdo After 1969

The quiescent level of seismic activity in Central Afar did not in any way appear altered by the 1969 sudden outburst. Shocks at epicentral distances between 420 and 450 km from a northeasterly direction — the distance of Serdo from Addis Ababa is 432 km at an azimuth of 041° — were, and still are, observed at about the same rate as before on AAE seismograms. Their epicentral locations can be more accurately determined since the installation of the seismic network in Djibouti.

A microseismic survey of the Serdo-Tendaho region was conducted by the University of Durham during the period of February–September 1974. A network of four stations recording on magnetic tape was installed at Mille (N 11.420°, E 40.752°), Det Bahri (N 11.561°, E 41.208°), Tendaho (N 11.690°, E 40.958°), and Serdo (N 11.957°, E 41.359°). The location and geometry of the network are given in Fig. 106.

Due to technical and climatological problems, the operation of the network was intermittent; the records are good for 124 days. During that

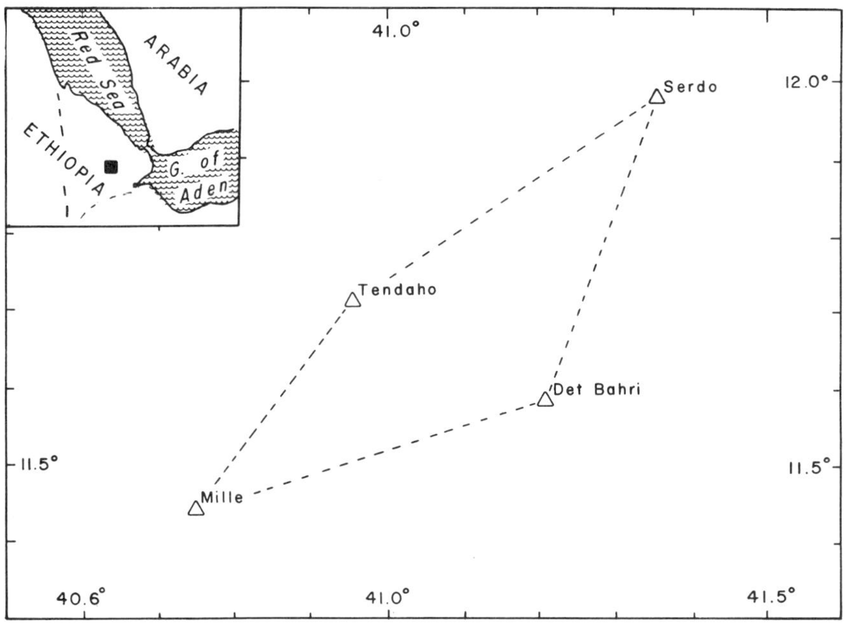

Fig. 106. Geographical location of the Durham seismic network in Central Afar during the microearthquake survey of 1974.

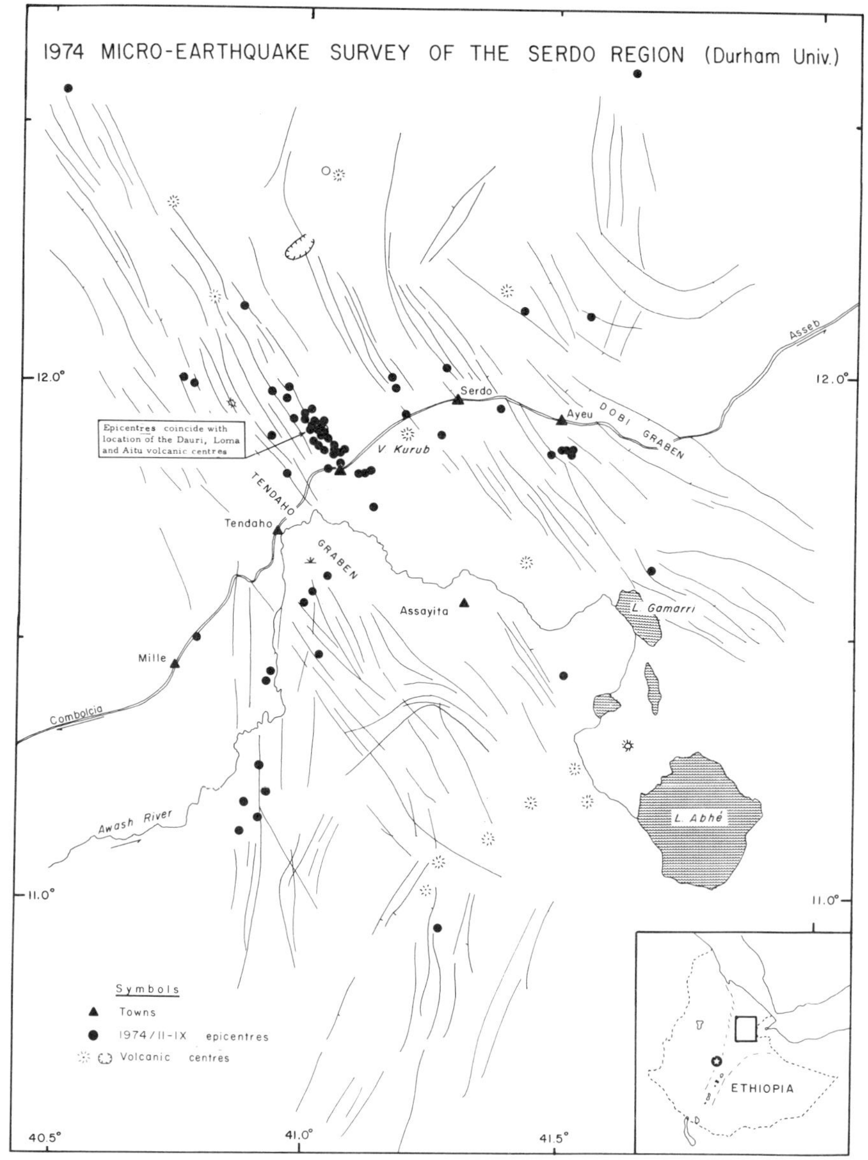

Fig. 107. Low magnitude epicentres in the Serdo region located from February to September 1974 by the Tendaho local network. No calculated focal depth exceeded 10 km. This 1974 epicentre plot is introduced here as a complement to the 1969 Serdo seismicity map.

period, over 400 local shocks were recorded. Epicentres could be determined for about 200 events and accurate depths for 25. More accurate details are intentionally not given here because the survey was part of a PhD research project by William G. Rigden and his thesis has not been defended yet.

Figure 107 is the location plot of the low magnitude epicentres determined by the Tendaho local network in the Serdo region.

4. Notes on Some Earthquake Source Parameters During the 1969 Serdo Sequence

Focal Depths — Although focal depths computed from teleseismic records are always subject to rather high uncertainties mainly resulting from "unknowns" neglected in the seismic earth models used, nevertheless, the values published by reliable agencies give a clue about the depth range of hypocentres in determined seismic regions. In the case of the 1969 Serdo earthquakes, the published values ranged between 4 ± 22 km and Normal for the USCGS, 32 ± 33 and 164 ± 56 km for ISC (Table 4). When one looks at the local crustal structure deduced from deep seismic soundings (Berckhemer et al. 1975, profile IV, p. 99 — see following table and Fig. 108) hypocentres deeper than 50 km become suspicious. The Durham University survey revealed that out of 25 focal depths that could be determined in the Serdo area from February to September 1974, none was larger than 10 km.

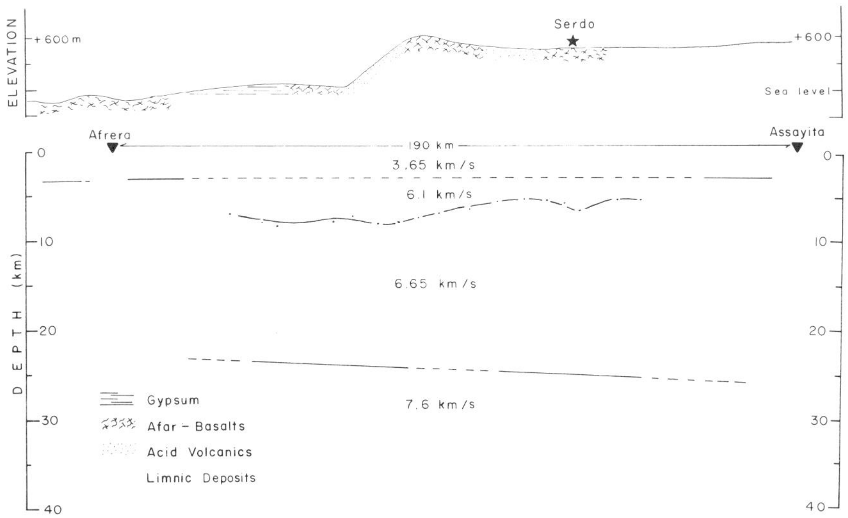

Fig. 108. Seismic depth-velocity meridional profile from Lake Afrera in the Depression to Assayita via Serdo obtained by Berckhemer et al. (1975) from deep soundings conducted in 1972 (adapted from Berckhemer 1975, Fig. 3 and 4, p. 99 and 101).

Layers	Depth (km)	Seismic velocity (km/s)
1	0–3	4.5
2	3–12	6.3
3	12–23	6.8
4	23–possibly 50*	7.4

*The extension in depth of the low-velocity layer is deduced from the adjustment of observed-to-computed gravity data (Makris et al. 1975, p. 138).

Fault Plane Solutions— McKenzie et al. (1970) computed fault plane solutions for the two largest Serdo earthquakes, those of 29 March at U.T. 11:04.8 and 5 April at U.T. 18:30.7. The two epicentres were about 20 km apart and their solutions identical, namely a first nodal plane heading N63E with a 70° dip to the SE and a second plane striking N27W with a 90° dip (McKenzie et al. 1970, p. 243–248) (see Table 5, Fig. 109).

McKenzie et al. chose the NNW-SSE (333°) direction as the primary nodal plane because the left lateral strike-slip motion it indicated was consistent with the field reports by Gouin and Dakin (1969) on the direction of the fresh faults that the earthquakes had triggered and with the general trend of local surface faulting. Gouin (1975) adopted the same interpretation.

Fairhead and Girdler (1971), Girdler (1977, personal communication) selected the alternate 063° direction for the primary nodal plane. They based their argument on: (1) the eastnortheasterly alignment of the larger

Table 5. Fault plane solutions available in 1971 for northeast African epicentres.

	Coordinates		m_b	Plane striking ENE		Plane striking NNW	
				Strike	Dip	Strike	Dip
1967/III/13	N 19.79°	E 38.72°	5.8	053°	82° SE	141°	83° NE
1962/XI/11	N 17.22°	E 40.58°	5.6	049°	78° SE	141°	81° SW
1969/III/29	N 12.02°	E 41.18°	5.8	063°	70° SE	153°	90°
1969/IV/05	N 12.19°	E 41.13°	6.2	063°	70° SE	153°	90°
1959/XII/21	N 13.98°	E 41.18°	6.7	030°	70° SE	123°	75° NE

Sources:
1967 and 1962 — Fairhead 1968; 1969 — McKenzie et al. 1970; and 1959 — Sykes (personal communication).

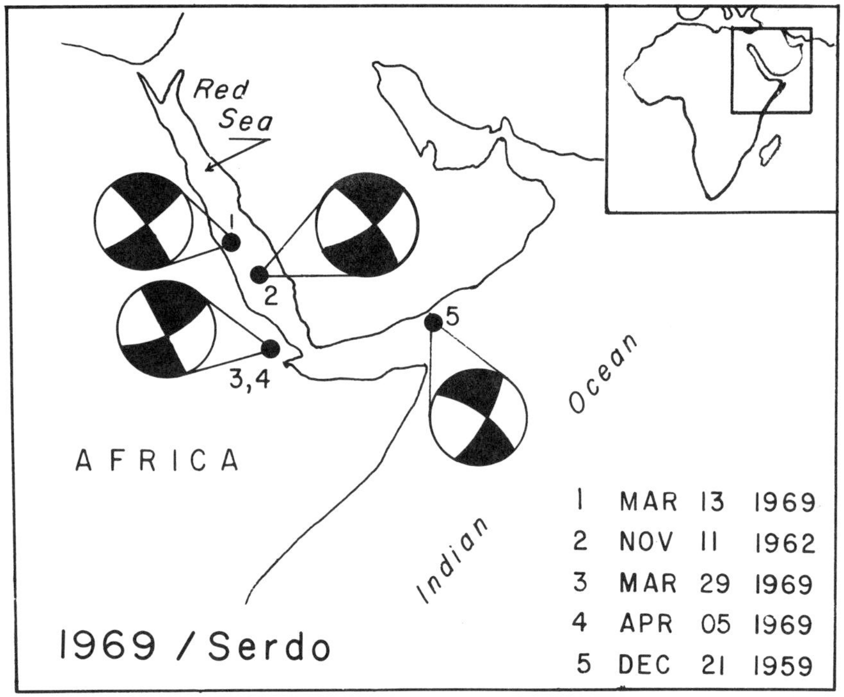

Fig. 109. Fault-plane solutions for five epicentres in the magnitude range $5.6 < m_b < 6.7$ available for the region of northeastern Africa. The plots are equal area projections of the lower hemisphere of the focal sphere.

Serdo earthquakes, which is very apparent on the USCGS and JED epicentre distribution pattern (Fig. 104), but not so evident on the ISC distribution pattern; and (2) the postulated existence of an ENE-trending transform fault under the Bidu-Dubbi volcanic line that would have caused the separation of the Danakil horst from the Nubian tectonic plate and allowed the observed 15° counterclockwise rotation of the Danakil horst with respect to the rest of Africa. They argued that a SW-NE right lateral motion at depth, as revealed by the larger Serdo shocks, could have produced northwesterly left-lateral faults on the surface similar to those observed 8 km east of Serdo. Such an explanation certainly agrees with Anderson's rhombic fault pattern theory and is backed by laboratory experiments (Courtillot et al. 1974). However, the general surface faulting pattern in the Serdo region north of N 11.9° and west of E 41.5° does not suggest any such northeasterly trend. Anywhere else to the north or east of the 1969 seismic region boxed by dotted lines on Fig. 100, the assumption of such a northeasterly fault at depth, that is below the thin crust of Afar, would be strengthened by the presence of curvilinear and ENE trending geologic structures. Even an idealized two-block laboratory model such as the one used by Courtillot et al., when progressively dislocated, leaves, on the surface of the clay layer that covers it, fault marks subparallel to the directions of the three sectors of the transform fault model (see Courtillot et al. 1974). Only the accurate plotting of aftershocks of a sequence in the same region by a local seismic network could reveal the identity of these fault planes. The same reasoning applies for the five almost identical solutions illustrated in Fig. 109.

Frequency-Magnitude Distribution and Strain Release — Fairhead and Girdler (1970) computed the frequency-magnitude (m_b) relation for the 12 epicentres determined from teleseismic data; Dakin and Gouin (1975) computed the same relation for 251 events whose original M_L (AAE) magnitudes were converted to m_b by the empirical equation: M_L (AAE) = $0.24 + 1.05\ m_b$ (CGS). The results were practically the same: $\log N = 4.71 - 0.76\ m_b$ (Fairhead and Girdler); $\log N = 4.62 - 0.73\ m_b$ (Dakin and Gouin) (Fig. 110).

The mean value of the b-coefficient (about 0.75) found for the 1969 Serdo sequence corresponds to values obtained over continental rift zones (Karnik 1969; Hattori 1974) and over Eastern Africa in particular, even if the authors used different limits for the East African regions considered, different periods of sampling, and different magnitude ranges. The published values for this region are:

b	Sampling period (years)	Regions	Authors
0.65		Western Rift	Wohlenberg 1968
0.79		East African Rift	Wohlenberg 1968
0.77		East Africa	Kaila and Narain 1971
0.84	60	Ethiopia and Horn of Africa	Gouin 1975
0.62	100	Ethiopia and Horn of Africa	Gouin 1975
0.71	343	Ethiopia and Horn of Africa	Gouin 1975

Other Source Parameters — The seismic moment, source dimensions, average dislocation, and stress drop were determined by Maasha and Molnar (1972) for the Serdo earthquake of 5 April 1969 using the theory developed by Brune (1970). A shear modulus μ = 3.3 × 10 cgs units and a density ρ = 2.71 g cm^{-3} were assumed. Maasha and Molnar's results were:

Moment:	P =	1.9(6)	× 10^{25} dynes cm^{-1}	Average = 1.43
	S =	1.1(8)	× 10^{25} dynes cm^{-1}	
Corner frequency:	P =	1.95	× 10^{-1} hz	
	S =	1.1(8)	× 10^{-1} hz	
Radii:	P =	11.0	km	Average = 11.9
	S =	12.2	km	
Area:		450	km^2	
Dislocation:		10	cm	
Stress drop:		3.4	bars	

The limits of confidence for these values are discussed by Maasha and Molnar (1972, p. 5737–39). Their conclusions on the low amplitude of the stress drops they observed during the Serdo earthquakes and others in the African rift system are worth noting. Low stress drop is characteristic of earthquakes caused by tectonic rifting along plate boundaries; higher stress drops accompany earthquakes that occur within tectonic plates where the crust is thicker and the strength of the material greater. From the gradual increase in stress drop from the Red Sea to South Africa, Maasha and Molnar concluded that the "northern part of the rift system marks a narrow zone of weakness separating two stable, aseismic plates, whereas the southern part of the rift is not such a plate boundary yet or a zone weakness . . ." (p. 5739).

1969/IX/26

On 26 September at U.T. 04:54.5, an earthquake of magnitude m_b 5.1 occurred at the south end of the central trough of the Red Sea, northeast of the Dahlak Islands. Four teleseismic solutions are available; they agree within ± 0.1° in longitude and latitude. The adopted epicentre is the average value of the four solutions: N 16:46°, E 41.05°; it is plotted on Fig. 92.

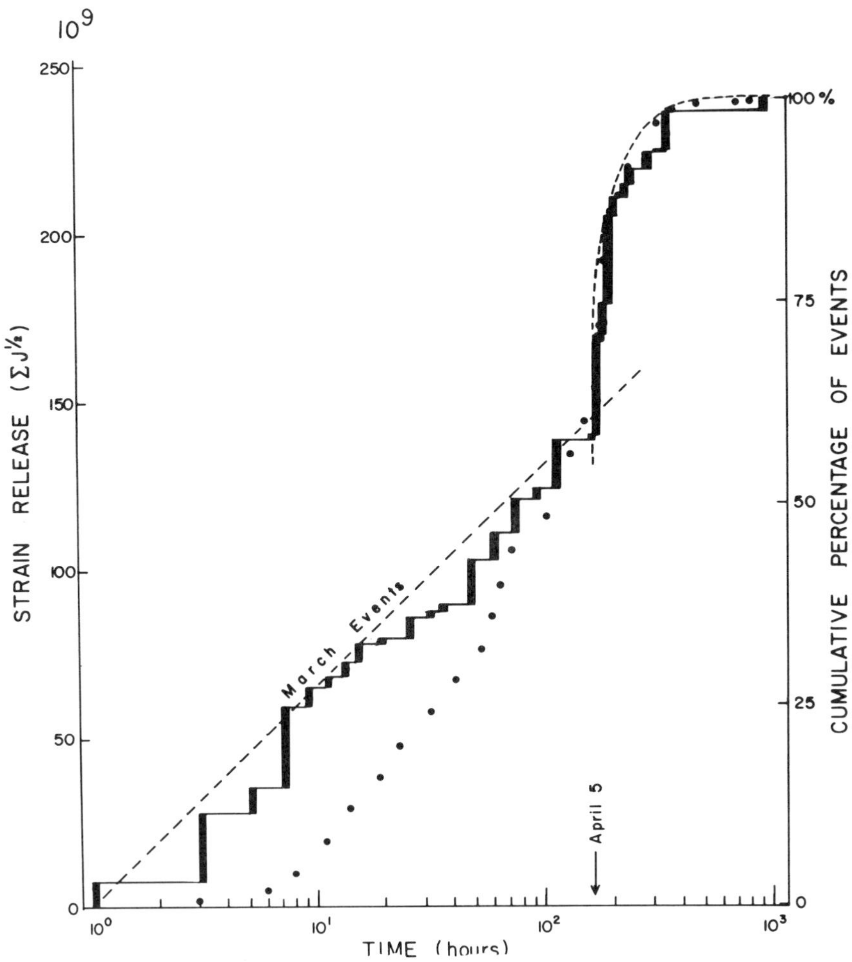

Fig. 110. Strain released during the 1969 March–April period of seismic activity (after Gouin 1971, p. 34).

Sources

BCIS; ISC (1969, p. 355–356); USCGS (EDR 63-69, p. 50).

Comments

As an evaluation of the earth models used by different agencies, it is to be noted that AAE recorded an iP(z) at 04:56:28.2 giving a P-H residual of only –0.6 s for the CGS model as compared to –1.5 s for the ISC model.

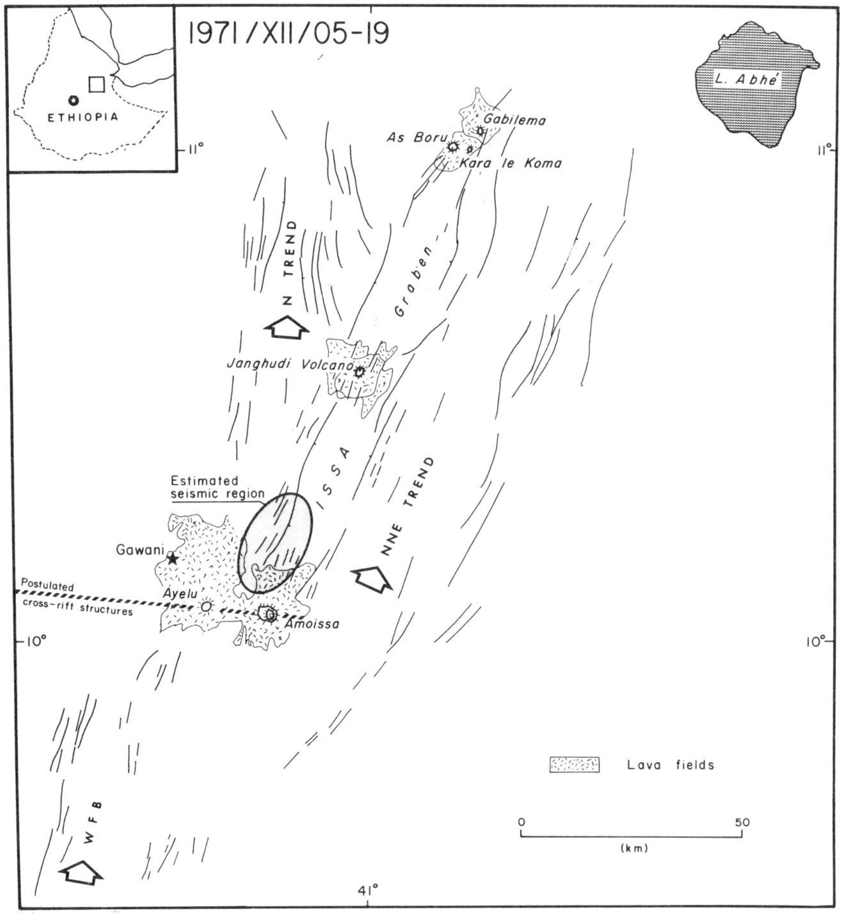

Fig. 111. Location of the 1971 seismic region in the vicinity of cross-rift structures that di Paolo (1970) postulated to explain the alignment of volcanoes Ayelu and Amoissa. These cross-rift structures would also coincide with a dextral displacement and a change in orientation of the Wonji Fault Belt.

1971/XII/05–19

There was seismic activity along the northern sector of the Wonji Fault Belt. The activity centred on N 10.2°, E 40.5°, in SW Afar near volcano Ayelu. Tremors were felt in Gawani village (N 10.2°, E 40.6°); no damage was reported (Fig. 111).

Source

AAE Data File.

Comments

1. The seismic activity started on 5 December at U.T. 20:13:58.5 with a shock of M_L (AAE) 3.9 (revised value; the AAE file indicates 4.3), at a distance of 230 km 055° from Addis Ababa, a site in the vicinity of Gawani. A minimum of 26 shocks of M_L (AAE) ranging from 1.9 to 3.9 left traces on the seismograms clear enough to be identified. The activity apparently ended on 19 December 1971.

Only two shocks traced iP's of amplitudes large enough to allow an azimuth determination (055 ± 005°). The bulk of the earthquakes originated from epicentral distances between 220 and 235 km from Addis Ababa. A few isolated events spreaded out between 190 and 250 km.

2. Earthquakes were previously recorded from the Gawani region: see entry 1938/IX, X and some of the epicentres in Wollo. The tremors felt near Mt Janghudi (N 10.5°, E 41.1°) in 1883 might very well have originated from the Gawani area rather than from Central Afar.

3. The 1971 seismic region, referred to as the Gawani seismic region, is located on the sector of the Wonji Fault Belt (W.F.B.), which has penetrated NNEward into SW Afar. The fault pattern in the area is dense and complex. Predominant faults are the W.F.B. proper, dextrally displaced at ≃ N 10° latitude and heading NNE toward the lakes Abhé and Gamarri. At N 10.2°, west of the W.F.B., a zone of faults and lineaments runs almost due north or slightly east of north between N 10° and N 11° on a course subparallel to the western escarpment of Afar. Coinciding with the zone of dextral displacement of the W.F.B., di Paolo (1970, p. 41) postulated WNW-ENE cross-rift structures, as yet undetected on the surface, to explain the Ayelu-Amoissa volcanic alignment; these features would join similar features observed about 30 km SE of volcano Amoissa.

Tectonic movements have occurred in recent times (di Paola 1970, p. 42) and the area offers geothermal potential.

1972/I/11

Tremors of intensity III were reported from Yirga Alem (N 06.8°, E 38.4°), on the floor the main Rift. Their origin was a shock of magnitude M_L (AAE) $\leqslant 4$ located along the western Plateau-Rift escarpment. The entry for this shock is classified in Section A.

1973/IV/03

Earth tremors were reported from Dubti (N 11.8°, E 41.1°). Logghia (N 11.7°, E 41.0°), Mille (N 11.4°, E 40.8°), and Serdo (N 11.6°, E 41.2°) on 2 and 3 April 1973. Only one shock was powerful enough to record an identifiable trace at Addis Ababa.

Sources
AAE Data File and Information Cards.

Comments
The main shock was recorded by AAE at U.T. 01:50:24.0 on 3 April; its apparent epicentral distance 457 ± 10 km at an azimuth of N40–44E. The M_L (AAE) was 4½. The adopted epicentral location is N 12.1°, E 41.2°.

1974/II–IX
Microseismic Survey of Central Afar

The survey was conducted from February to September 1974 by the University of Durham, England. A seismic network comprised of four stations was installed in the vicinity of Tendaho (N 11.69°, E 40.96°); details are given in entry 1969/III–V (Serdo–Central Afar).

Ninety-nine shocks were identified during the 124 days of intermittent operation; their epicentres are plotted on Fig. 112. An enlarged section of the Serdo area can be found on Fig. 107. Focal depth could be evaluated for 25 shocks; it ranged from 0.25 to 9.5 km. The depth average was about 4 km.

Source
The data were kindly supplied by W. Rigden, chief investigator of the survey (1978).

Comments
In the numerical listing, the Tendaho network is coded TLN (Tendaho Local Network). Not to be confused with the seismic station at Talang (TLN C), Sumatra.

1974/III–IV

Earth tremors were reported from southern Ethiopia during March and April 1974. The reports came from Bulki (Feleke Neway, N 06.2°, E 36.6°), Hagere Mariam (N 05.6°, E 38.3°), Warancha (N 06.3°, E 38.4°), and Arba Minch (N 06.2°, E 37.7°). No damage was observed.

Sources
AAE Data File and Information Cards.

Comments
The reported tremors refer only to a few earthquakes out of the 57 identified on the AAE seismograms as originating from the same region over a period of 5 weeks, from 9 March to 29 April 1974. The epicentral distances ranged from about 300–400 km in a general direction slightly west of

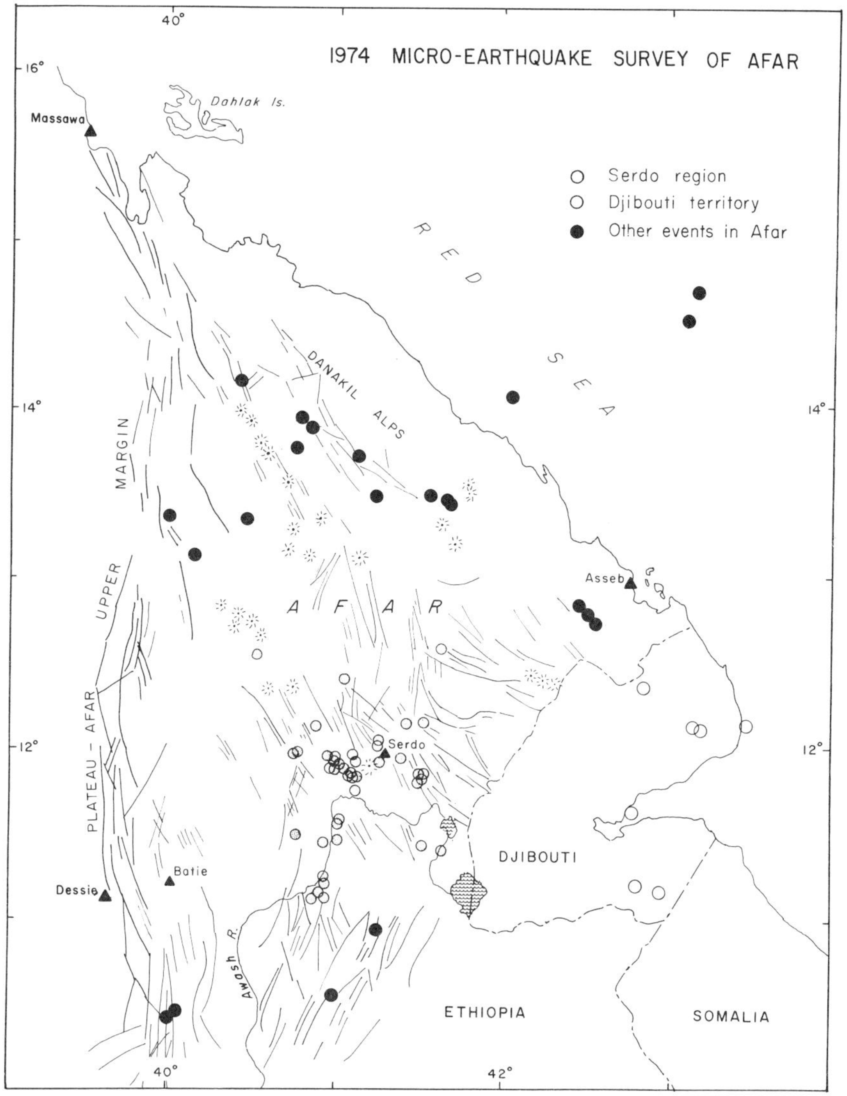

Fig. 112. Epicentres of low magnitude events determined by the Tendaho local seismic network (TLN) operated during 124 days from February to September 1974 by the University of Durham. A depth value could be estimated for 25 events; no focus was any deeper than 10 km.

south from Addis Ababa (Fig. 113). Their magnitudes M_L (AAE) were estimated between 2.0 and 4.5 (revised values). No apparent pattern of epicentre migration with time within the seismic region was observed.

The data obtained from the information cards are not accurate enough to relate the felt tremors to any particular shock of the swarm nor are they complete enough to draw isoseismal maps.

Combining the instrumental data and the macroseismic evidence, it can safely be inferred that the region of activity was along the western escarpment of the main Ethiopian Rift, between Chencha and Arba Minch.

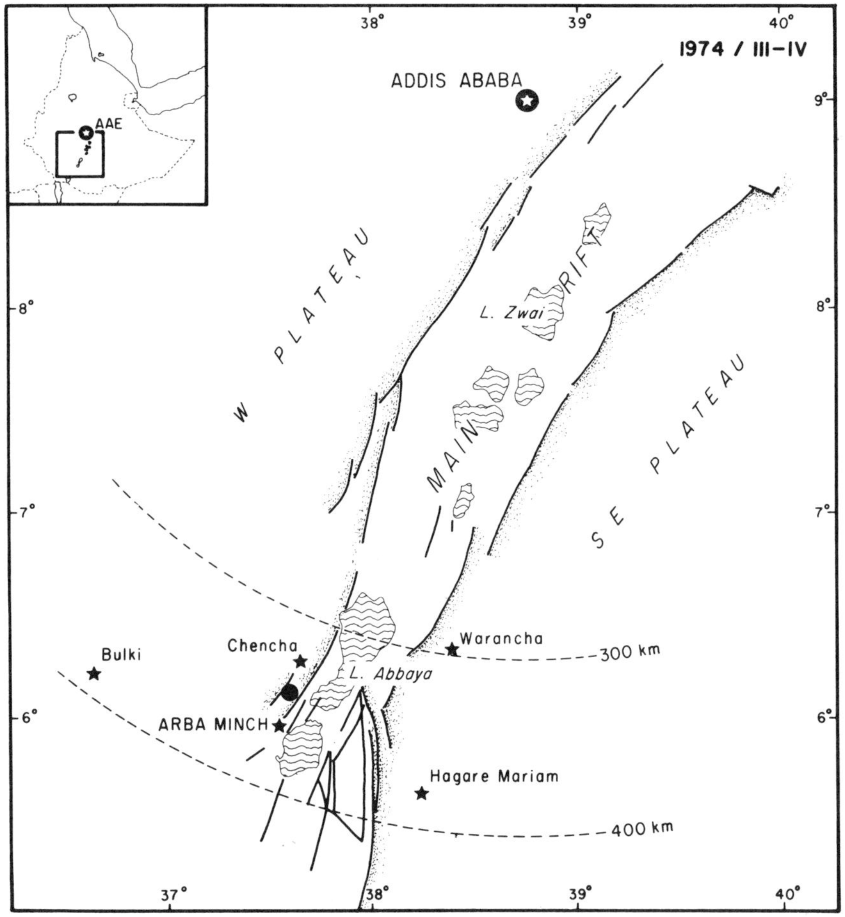

Fig. 113. Location of the seismic activity during March and April 1974. The star symbols identify the sites from where tremors were reported; the solid circle indicates the estimated centre of the seismic region.

In the computation of the regional seismicity (Gouin 1976), the whole swarm was considered as a unique shock of m_b 4.8 (sum of the energy released by 57 events) located at the centre of the seismic region, N 06.1°, E 37.6°. (Note: It is fully realized that one event of magnitude m_b 4.8, located at a definite site, does not produce the same macroseismic effects as 57 smaller events distributed over a large area. It is felt that such approximations are justifiable when the exact location of the other shocks cannot be determined.)

1974/IV/17,26

Two earthquakes of magnitude m_b 5.0 and 5.4 were located by ISC and USGS on the African side of the Red Sea axial trough, about 150 km off the coast of Ethiopia (Fig. 114). All the teleseismic solutions are listed in Part II of this catalogue; only those for which the standard errors are available are given here.

	U.T.	Coordinates	h (km)	m_b
17 April 1974				
ISC	18:27:34	N 17.30 ± 0.02° E 40.30 ± 0.03°	27	5.1
USGS	18:27:33.7	N 17.26 ± 0.03° E 40.37 ± 0.03°	33	5.0
26 April 1974				
ISC	18:08:18	N 17:10 ± 0.06° E 40.44 ± 0.06°	48	4.4
USGS	18:08:16.9	N 17:14 ± 0.04° E 40.38 ± 0.05°	33	4.8

1974/VI/30

At U.T. 13:26.5 on 30 June 1974, a magnitude m_b 4.5 earthquake occurred in the Red Sea near the Ethiopian coast, some 50 km northeast of Massawa. The geographic coordinates of its epicentre published by the International Seismological Centre (ISC) and by the National Epicentre Information Service (USGS) agree within 4.7 km in latitude and 2.3 km in longitude. They are: ISC N 16.013 ± 0.062°, E 39.361 ± 0.093° (for h = 33 km); and USGS N 15.970 ± 0.075°, E 39.610 ± 0.098° (for h = 33 km).

Sources

AAE Data File; ISC 1974 (1, p. 118); USGS (PDE 53-74).

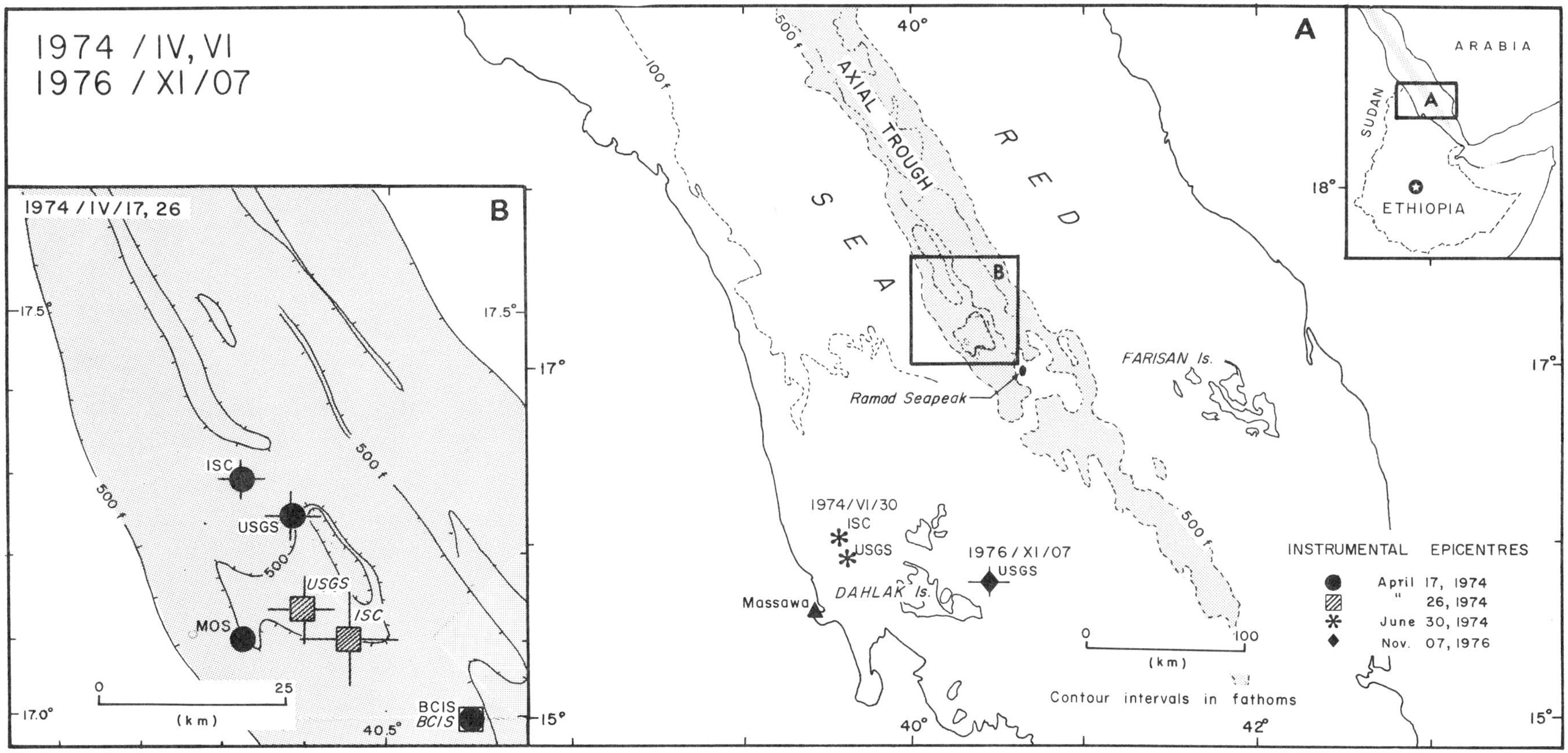

Fig. 114. Instrumental epicentre solutions for the earthquakes of 17 April, 26 April, and 30 June 1974 in the Red Sea axial trough. The outline of map A has been adapted from Laughton (1970); the contours in map B are adapted from Bäcker et al. (1975).

Comments

1. Epicentral Region

Figure 114 shows that the 30 June epicentre determined by ISC and USGS lies halfway between the Gulf of Zula graben and the southern tip of the 100-fathom isobath that from N 18°, SSEward, parallels the Eritrean coast and at about N 16.6° folds back on itself and forms a narrow channel along the same axis as that of the Zula and Danakil Depression grabens. In this catalogue, this channel has been identified as the Massawa Channel. The fact that such a well-determined epicentre occurred at that particular location strengthens the argument for the prolongation of the Danakil Depression tectonic feature north-northeastward via the Zula graben before being displaced dextrally toward the central trough of the Red Sea.

The 1974 epicentral region coincides with those of 1884 and 1921.

HFS2 (Sweden) located the June 1974 epicentre inland, at N 15.0°, E 39.0°, in a highly populated region of Eritrea. Had it occurred there, it would have been felt and reported to the Geophysical Observatory in Addis Ababa.

2. Local Seismic Records

A highly attenuated eP_n phase onset was recorded by AAE at U.T. 13:28:12.5, 107.7 s after the origin time published by the USGS. Sn was virtually absent.

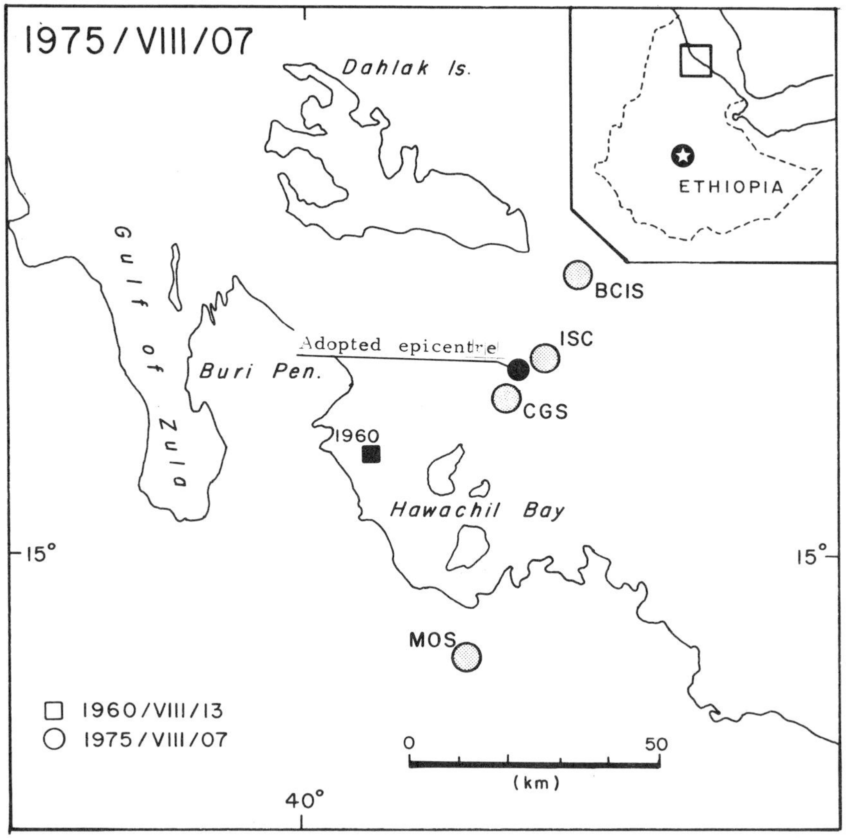

Fig. 115. Teleseismic solutions for the epicentre of 7 August 1975 in Hawachil Bay, northeastern Ethiopia.

1975/VIII/07

On 7 August 1975, an earthquake of magnitude m_b 4.6 was reported off the Ethiopian coast in the channel between the Dahlak Islands and Hawachil Bay. The epicentre parameters provided by BCIS, ISC, MOS, and USGS are listed in Part 2. The adopted coordinates are N 15.32°, E 40.42° (Fig. 115).

Sources

ISC; USGS (EDR 33-75).

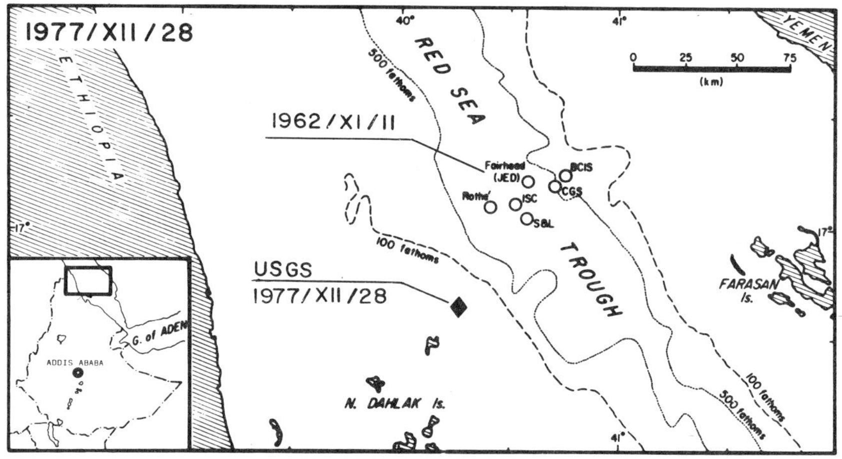

Fig. 116. NEIS preliminary determination of the epicentre for the earthquake of 28 December 1977 in the Red Sea.

1976/XI/07

An epicentre of magnitude m_b 4.8 was located by the USGS on the western shallow platform of the Red Sea (water depths less than 50 fathoms) east of the Dahlak Islands. Because this event should be included in the revision of seismic risks in Ethiopia, the USGS epicentral parameters are given here: *epicentral parameters* — H 05:53:07.6 ± 1.89 s, N 15.820 ± 0.091°, E 41.423 ± 0.116°, h 34 km, Δ°7.27 ± 0.13°, Az 20.6°; *phases recorded at AAE* — eP 05:55:59.0, S 05:56:26.3, (Lg) 05:58:03.1.

Source

USGS (EDR 17(3)-76, p. 261).

1977/XII/28

An earthquake of magnitude M_s 6.75 occurred in the Red Sea, northeast of the Dahlak Islands. The US National Earthquake Information Service located its epicentre at N 16.659 ± 0.022°, E 40.278 ± 0.022°, about halfway between the Dahlak Islands and the 1962 seismic region in the Red Sea trough (Fig. 116).

Source

USGS (EDR 19–77, p. 198).

Earthquake History of the Aden Western Sector (Region D)

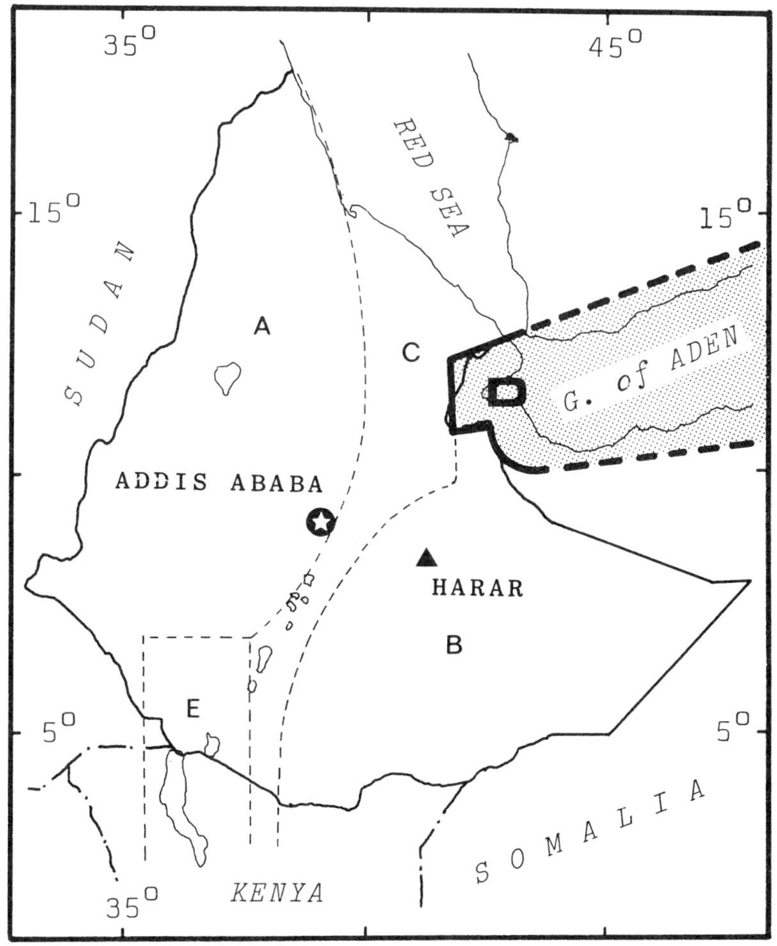

Changes in Regional Toponymy

At the time of the first entry for Region D, 1899, the territories surrounding the Gulf of Tadjoura, northwest of E43° 15′ (on the south shore of the gulf) were called La Côte Française des Somalis, abbreviated CFS and commonly known to English speaking geographers as French Somaliland. Later, the name was changed to Territoire Français des Afars et des Issas, abbreviated TFAI. Finally, as a newly independent country, it chose the name of Republic of Djibouti with Djibouti town as capital.

Southeast of E43° 15′ is the Republic of Somalia. It was formerly known as British and Italian Somalilands.

1899/II,XI

Earth tremors that Dreyfuss qualifies as "violent" were felt in Djibouti on 11 February and 21 and 23 November 1899. Walls in many houses were fissured during the February tremors but no damage was reported in November. The intensity at Djibouti in February was estimated as VI.

Sources

Heudebert (1901, p. 48); *Réveil de Djibouti*, quoted by Dreyfuss (1931, p. 342); Palazzo (1915, p. 323–324), quoting *Il Messaggiero di Roma* for 21 and 23 November 1899; Dainelli (1943, III, p. 706).

Comments

A note should be made here about the intensity evaluations for the town of Djibouti. The geological foundation under most of the town is unconsolidated limestone of coral origin. Such a geologic environment acts as a natural amplifier of seismic waves.

1906

Dreyfuss mentions that the people in Djibouti claim to have felt a *rather violent* earth tremor during 1906. It must be noted that Djibouti Government files are very well documented for that period and that there is no document in them supporting that statement. It is presumed that the reference is to the seismic activity of January 1907.

Sources

Dreyfuss (1931, p. 342). (Dainelli 1943, III, p. 706).

1907/I/18

The Office of the Governor of the *Côte Française des Somalis* (CFS; now called the *Republic of Djibouti*) reported that two earthquakes were felt in his territory on 18 January 1907: a strong one at 12:15 and another one of unspecified intensity at 6:35 p.m.

Sources

Annexe au Journal Officiel de la République Française, 1909, p. 167. (from Palazzo 1915, p. 333; Sieberg 1932, p. 887).

Comments

For what it is worth, the official report states that the felt ground motion was westward during the first shock and eastward during the second.

1907/X/28

In Djibouti about 10 tremors were felt on 28 October. No intensities were indicated.

Sources

Annexe au Journal Officiel de la République Française, 1909, p. 167 (from Palazzo 1915, p. 333; Sieberg 1932, p. 887).

1909/II/19

On 19 February 1909 at 6 a.m., an earth tremor was felt in Djibouti.

Source

Annexe au Journal Officiel de la République Française, 1910, p. 1431. (from Palazzo 1915, p. 335).

Comments

The direction indicated in the report is NE-SW, a likely direction for earthquakes happening at the west end of the Gulf of Aden.

Sieberg (1932, p. 887) mentions an earth tremor felt on 19 February 1908. The date may have been confused (1908 for 1909) as the Official Report of the Governor of the CFS does not mention any activity between October 1907 and February 1909.

1910

Three light tremors were felt in Djibouti on: 18 February at L.T. 08:15 (very light); 22 April at L.T. 13:15 (very light); and 6–7 August at L.T. midnight (strong enough for observers to estimate a northeastward direction for the incoming seismic wave).

Sources

Annex to *Journal Officiel de la République Française*, 1911, p. 1563, quoted by Palazzo (1915, p. 335).

1912/V–VIII

Seismic tremors were felt in Djibouti from 12 May to the end of August 1912. The centre of activity was located in the Gulf of Tadjura, at the west end of the Gulf of Aden. The following shocks were reported: 12 May at L.T. 18:45 (rather strong); 14 May at L.T. 01 (light); 28 May at L.T. 04; 22 June at L.T. 02:40 (light); and 27 August at L.T. 11:08 (strong).

Sources

Journal Officiel de la France d'Outre-Mer, p. 1207 (from Palazzo 1915, p. 336; Seiberg 1932, IV).

1924/IX–XI

From 26 September to 19 November 1924, eleven tremors of intensity II–III were felt in Djibouti; of these, nine were felt on 19 and 20 October. None were reported between 20 October and 19 November. On the other hand, 18 tremors were reported from Zeila (N 11.33°, E 43.47°; Somalia) between 24 and 28 October with a peak of 12 on the 27th. Zeila is located on the coast, 43 km east-southeast of Djibouti. For the location of Zeila, see Fig. 118.

Sources

Meteorological Office Report for the tremors in Djibouti. For those of Zeila: Taylor 1931, p. 34. (MacFayden 1933, p. 38).

Comments

The fact that the tremors of 24–28 October were reported only from Zeila and not from nearby Djibouti where a well-organized Meteorological Office kept a close inventory of all seismic activity in the Republic (then called *Côte Française des Somalis*) is somewhat suspicious. It was first thought that Taylor, writing 7 years after the events, could have mixed the years 1924 and 1926 because on 30 and 31 October 1926, two earthquakes of magnitude 5.3–5.9 and one of magnitude larger than 4.5 occurred off the coast of Somalia. This assumption becomes less probable if we also consider that the dates of the month quoted by Taylor — 24–28 October with a peak of activity on the 27th — neither correspond to those given by ISS and by Richter for October 1926 nor to those on which the tremors were felt in Djibouti in 1924. Personally, I am convinced that there is an error in Taylor's report. Djibouti is only 43 km from Zeila and Djibouti reported tremors on 24–28 October 1924 and on 30–31 October 1926. Therefore, Taylor must have been referring to one of these two periods of seismic activity.

1926/IV/28

The Meteorological Office in Djibouti reported an earth tremor of intensity II–III around midnight.

1926/X/30–31

On 30 October 1926, an earthquake of magnitude class "d" (5.3–5.9) occurred in the western sector of the Gulf of Aden; it was followed on the next day by two shocks of magnitude larger than 4.5.

From Djibouti, two tremors were reported 5 h before the main shock of 30 October at U.T. 01:38 (L.T. 4:38 a.m.), five during the main activity from 4:30 to 5:30 a.m., none at the time of the second shock, and five coinciding with the third earthquake. Eleven other tremors were felt sporadically until 3 December. All were reported to be of intensity III.

Sources

Bellamy (ISS, 1936); Gutenberg and Richter 1949; *Le Réveil de Djibouti*, 12 October 1963.

Comments

1. Location of the Epicentres

Three instrumental epicentres have been determined for 30–31 October 1926. Gutenberg and Richter (1949) located only one, the first, at N 11°, E 44°; ISS located all three at N 11.5°, E 43.5° (Bellamy 1936).

2. Reliability of the Epicentral Location

Tremors coinciding in time with these events were reported from Djibouti city. Considering that: (1) Djibouti is built on unconsolidated limestone that naturally amplifies seismic wave effects; (2) the tremors were of low intensity; and (3) the magnitude of the main shock was about 5½, it follows that the epicentre must have been at a minimum distance of

200–300 km from Djibouti as compared with 40 km for the ISS location and 115 km for that given by Gutenberg and Richter.

Attempts have been made to recompute epicentres of the same period; the results showed unreliable locations with standard deviations ranging between ± 1–10° (see entry 1930/X/24-31, Region B). The only control, therefore, on the location of the epicentre is macroseismic evidence, and that evidence points to a distance of 200–300 km from Djibouti. An educated guess would be that the seismic region was located at the eastern end of the Tadjoura Trench, in the vicinity of N 12°, E 45° (Fig. 117).

1927–28

The Meteorological Office reported from Djibouti nine light tremors during 1927–28. They were: 1927 — 6 January at night (5 tremors), 11 February at 6:40 (1 tremor), 12 March at 18:00 (1 tremor); 1928 — 8 April at 11:42 (1 tremor), 21 September at 23:45 (1 tremor).

1929/I–IX

On 22 January 1929 an earthquake of magnitude about 6 (G & R), located at the western end of the Gulf of Aden, rocked the town of Djibouti. During the following three days, 42 aftershocks were felt. The maximum intensity of damage in Djibouti was estimated at VII–VIII. It appears that the felt-tremors ended by 27 September.

Sources

Bellamy (1936, p. 3); Dreyfuss (1931, p. 342); Gutenberg and Richter (1949); *Le Réveil de Djibouti*, 12 October 1963; Reports from the Meterological Office in Djibouti; Report No. 211 of 24 January 1929, from the Ministry of Public Works in Djibouti.

Dreyfuss summarizes the events as follows:

It started on January 22 at L.M.T. 17:42 with a very strong shock which lasted six or seven seconds; cracks appeared in the walls of most houses and one house, the property of the C.F.E. (Chemin de Fer Franco-Ethiopien), was destroyed. Other less violent shocks followed at irregular intervals until January 24. During these three days, we counted 42 felt tremors.

Comments

1. Epicentral Location

Gutenberg and Richter (1949) and ISS located the epicentre at N 11.5° E 43.5°; tremors of intensity VII–VIII were reported from Djibouti (N 11.6°, E 43.15°); IV-V from Tadjoura (N 11.8°, E 42.9°). The epicentral

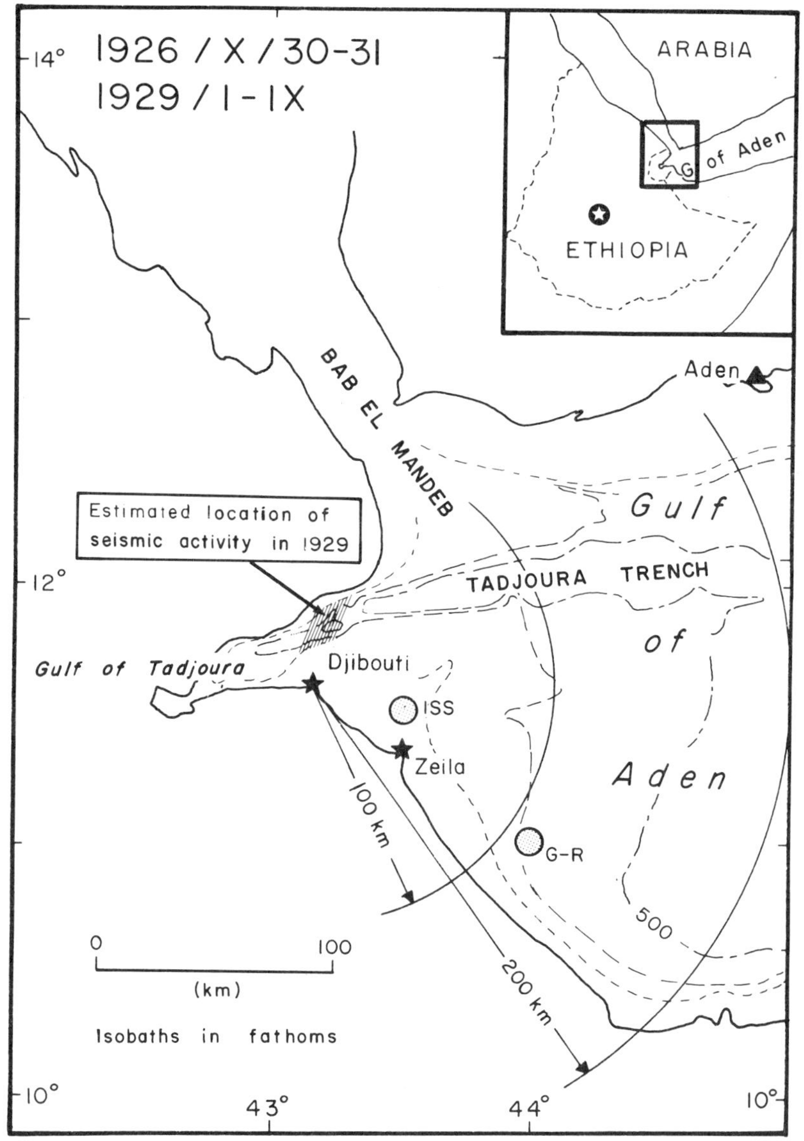

Fig. 117. Teleseismic epicentre solutions for the three larger shocks of 30–31 October 1926 in the Gulf of Aden, off the coast of Somalia. The proximity of the calculated epicentres to Djibouti and Zeila does not tally with the reported intensities of III. More realistic epicentre locations would be anywhere, offshore, along or beyond the 200 km arc centred on Djibouti.

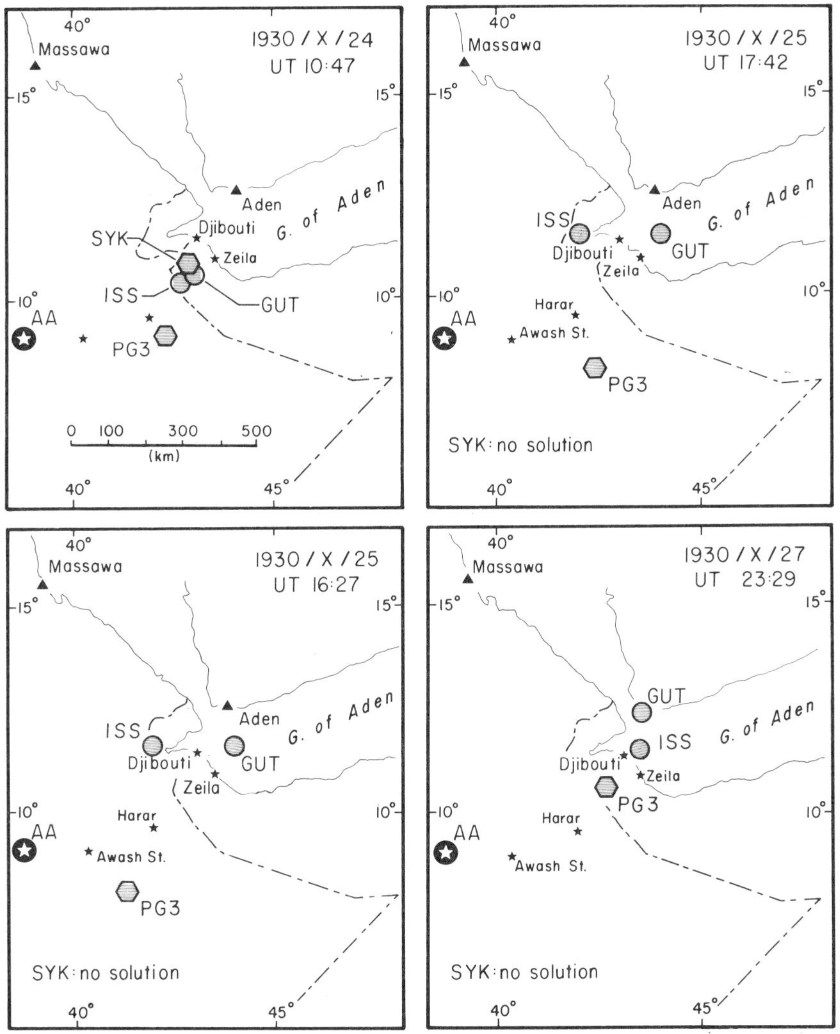

Fig. 118. Plots of the original and recomputed teleseismic locations for the epicentres of late October 1930 at the west end of the Gulf of Aden.

location at N 11.5°, E 43.5° of a shock of magnitude about 6 fits the intensities VII–VIII at about 40 km and IV–V at 70 km. However, from a geological viewpoint, no seismically active structure is known to exist in the vicinity of the calculated site and the bathymetric maps show no anomaly in their isobaths. In such a case, one feels justified in looking for a more congruent epicentre location within the limits of probable errors in the original solution. Slightly north of N 11.5° is the seismically active Tadjoura Trench. A more probable epicentral location would therefore be N 11.7°, E 43.2°, at the eastern border of the 1973 seismic region. Such a location also satisfies the relative distances linked with the reported intensities at Djibouti and Tadjoura (see Fig. 117).

2. Aftershocks Corresponding to Tremors Felt in Djibouti

These tremors are presumed to have originated from the same seismic region; they are listed by date, number of tremors, and maximum intensity: 23 Jan. (14) max. V; 24 Jan. (5) max. IV; 25 Jan. (3) max. IV; 26 Jan. (1) max. III; 28 Jan. (1) max. III^{+}; 29 Jan. (1) max. III^{+}; 30 Jan. (1) max. III^{+}; 2 Feb. (3) max. III; 7 Feb. (2) max. III; 8 Feb. (2) max. III; 21 Feb. (1) max. III; 19 Apr. (1) max. III; 25 Apr. (1) max. III; 5 Aug. (1) max. III^{+}; 27 Sept. (1) max. III^{+}. In total, 42 tremors were felt in Djibouti during 1929.

1930/V/30

An earth tremor of intensity III was reported from Djibouti at 1:15 p.m., 30 May 1930.

Source

Meterological Office, Djibouti.

1930/X–XI

From 17 October to 18 November 1930, numerous shocks were felt throughout the present Republic of Djibouti. Fifty five were registered by the Meterological Office, none with intensities higher than IV. Tremors were also reported from Zeila. In Central Ethiopia, they were apparently felt as far away as Awash Station (N 09°, E 40.2°).

Three out of four instrumental epicentres were originally located by Gutenberg and Richter and by the International Seismological Summary on Djibouti Territory:

25 Oct.	U.T. 16:29	N 11.5°	E 44°	(G&R)
		N 11.5°	E 42°	(ISS)
25 Oct.	U.T. 17:42	N 11.5°	E 44°	(G&R)
		N 11.5°	E 42°	(ISS)
27 Oct.	U.T. 23:29	N 11.5°	E 43.5°	(ISS)

These three instrumental epicentres were relocated, to the south, on the Aisha Block inside the Ethiopian border. The entire series is analyzed in entry 1930/X/24–31, Region B; plots of the different teleseismic solutions are given in Fig. 118.

Sources

Meteorological Office Report for the years 1924–1932; *Le Réveil de Djibouti*, 13 October 1963.

Comments

Referring to the seismic activity of October 1930, *Réveil de Djibouti* in its edition of 13 October 1963, mentioned damage corresponding to intensity VII in Tadjoura and Djibouti. The official report from the Meteorological Office did not grade any intensity higher than IV during that period.

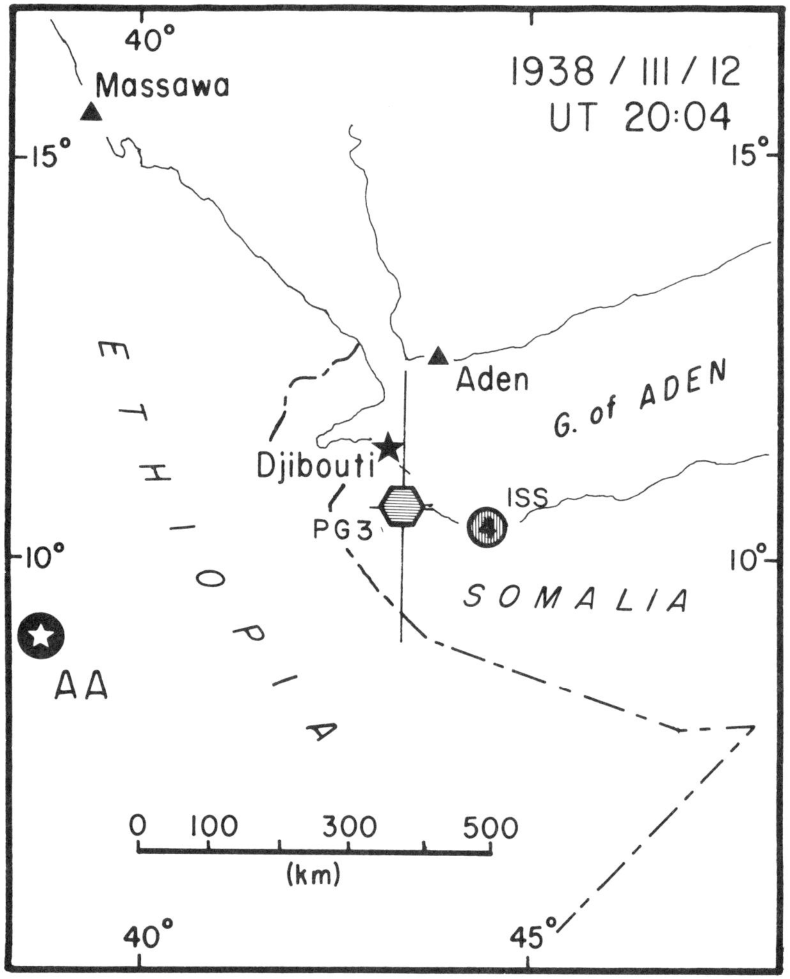

Fig. 119. Location of the only epicentre of March 1938 in northern Somalia and the Gulf of Aden for which a relatively reliable solution could be obtained. Numeral 4 inside the circle indicates that the four largest earthquakes of the sequence were located at the same site by ISS.

1931

Fourteen earth tremors were reported in Djibouti during 1931; the largest were of intensity III–IV.

Source

Meteorological Office Report from Djibouti.

1932

Thirty-two tremors were reported in Djibouti during 1932; none higher than intensity IV.

Source

Meteorological Office Report from Djibouti.

1934/X/6–7

During the night of 6–7 October 1934, three shocks of intensity II–III were reported from Djibouti.

Source

Meteorological Office Report from Djibouti.

1935–36

Nine tremors were felt in the Republic of Djibouti:

		Location	Time	N	Intensity
1935	16 Jan.	Djibouti	Midnight	2	III
	20 Jan.	Djibouti	05	2	III
	25 Apr.	Djibouti	03:48	2	V; noise
	25 Apr.	Tadjoura	about 04	2	IV
1936	24 June	Djibouti	07:30	3	III
	03 Nov.	Djibouti	02:10	1	V; noise
			03:05	1	III

Sources

Meteorological Office Report from Djibouti; Telegram No. 117 from Commanding Officer in Tadjoura, 25 April 1935.

1937

The Meteorological Office in Djibouti reported 13 very light tremors of intensity II–III from March to December 1937.

1938/III/11–14

From 11 March in the evening until noon, 14 March 1938, the Meteorological Office in Djibouti reported nine earth tremors of intensities II and III. They were caused by a main shock and three aftershocks located by ISS at N 10.5°, E 44.5°, an epicentral distance of 190 km from Djibouti. A recomputation of the ISS Data File using the SPEEDY computer program in 1974 showed that no solution could be obtained for the first three events of 11 March at U.T. 16:51:38, and 12 March at 12:37:34 and 13:04:40. The solution for the last one, which must have been of a magnitude equivalent to that of the first earthquake (12 stations reported both events), is: U.T. 20:04:23.7 ± 8.9 s, N 10.8 ± 1.1°, E 43.3 ± 0.3°. The location is on the coast of northern Somalia, about 90 km from Djibouti (Fig. 119).

Source

ISS for 1938, p. IV.

Comments

1. As tremors were reported only from Djibouti (N 11.6°, E 43.1°) and were of intensity II–III, an epicentre of magnitude about 5 in northern Somalia, a region that has recently shown no sign of being seismically active is, to say the least, suspicious.

2. Five other light tremors of intensity II–III were felt in Djibouti during the remainder of 1938.

1940/VIII/14

The International Seismological Summary located an earthquake of unspecified magnitude on or near the Somali coast. Twelve stations reported the event.

Source

ISS for 1940, p. VIII.

Comments

The original ISS epicentral location was N 11.0°, E 46.0° near the southern shore of the Gulf of Aden in Somalia. A recomputation of the data from the ISS files using the SPEEDY computer program, after three iterations, yielded the following parameters for h 33 km: U.T. 08:49:21.9 ± 14.5 s; N 09.9 ± 1.9°, E 46.2 ± 0.3° (Fig. 120).

The epicentral location, inland, is suspicious even if it lies on the southwesterly possible extension of the transform faults from the Gulf of Aden.

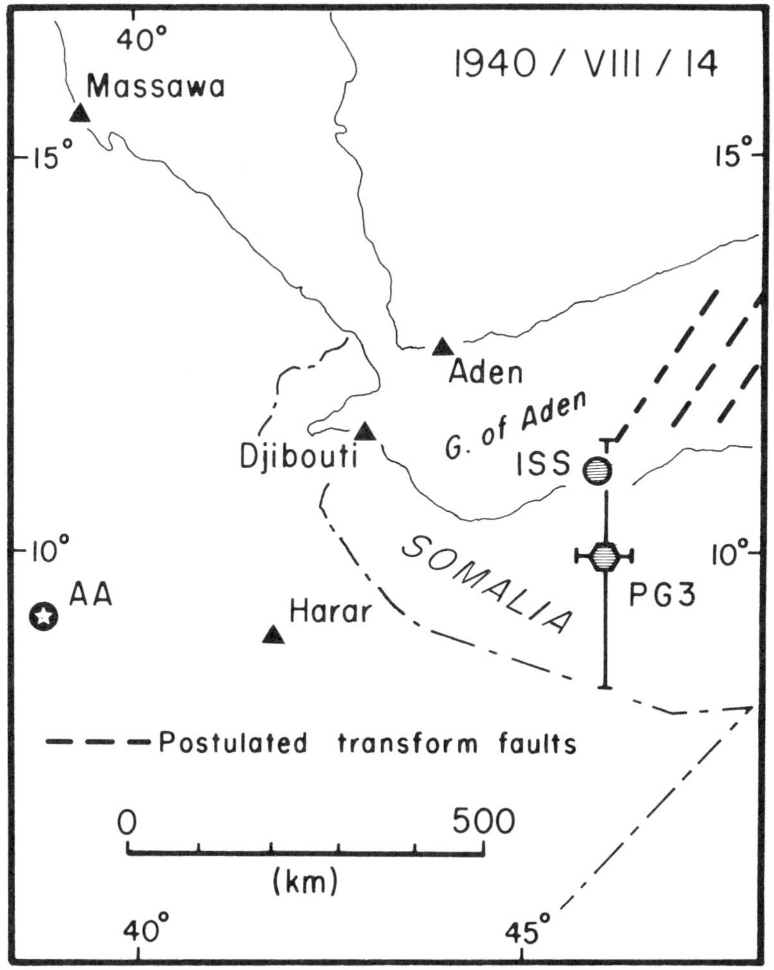

Fig. 120. Location of the instrumental epicentres obtained for the earthquake of 14 August 1940.

1941/III/19

On 19 March 1941, an earthquake of magnitude 5½ occurred in the western sector of the Gulf of Aden.

Sources

Gutenberg and Richter 1954; ISS.

Comments

Gutenberg and Richter located the epicentre at N 12°, E 43½°; ISS at N 10½°, E 44½°. A recomputation (PG3) based on the ISS data file yielded a new location at N 10.9 ± 1.0°, E 44.0 ± 0.4°. The northern latitude of the PG3 limits of confidence corresponds to Gutenberg and Richter's original location at a very likely site on the ridge of the Gulf of Aden median rift valley. The adopted coordinates are therefore those of Gutenberg and Richter (Fig. 121) (see comments in entry 1938/III/11–14).

1944/I,X

The Meteorological Office in Djibouti reported: (1) An isolated shock of intensity IV with a NS component at 7:30 a.m. on 1 January 1944; and (2) A swarm of 41 tremors from 27–31 October 1944; 30 of these occurred on 28 October. They were reported from Obock, Tadjoura, Balaoh, Dikhil, and Loyada, that is from practically every corner of the Republic. The fact that no international agency reported any epicentre in this region during the second half of October 1944 is an indication that the seismic activity consisted of an earthquake swarm with no shock of magnitude exceeding about 4.5. The centre of the seismic region must have been in the vicinity of the Ghubbet el Kharab in order to explain the fact that low magnitude earthquakes could be felt from Dikhil to Obock, a radius of 50–60 km.

1945/X/27–31

During the last 5 days of October 1945, 41 shocks were reported by the Meteorological Office in Djibouti. The most violent, of intensity VI–VIII (R.F. VII), happened at 3:15 a.m. on 28 October; it was reported from Obock, Tadjoura, Balaoh, and Dikhil on the western border and Loyada.

Source

Meteorological Office Report, Djibouti; Gutenberg and Richter (1954); ISS.

Comment

The origin of the tremors on 28 October was an earthquake of magnitude 5.6 (Pas) originally located by Gutenberg and Richter at N 10.0°, E 42.5° and by ISS at N 11.2°, E 42.7°. Two independent recomputations, the first was based on the USCGS file and the second on the ISS file, were

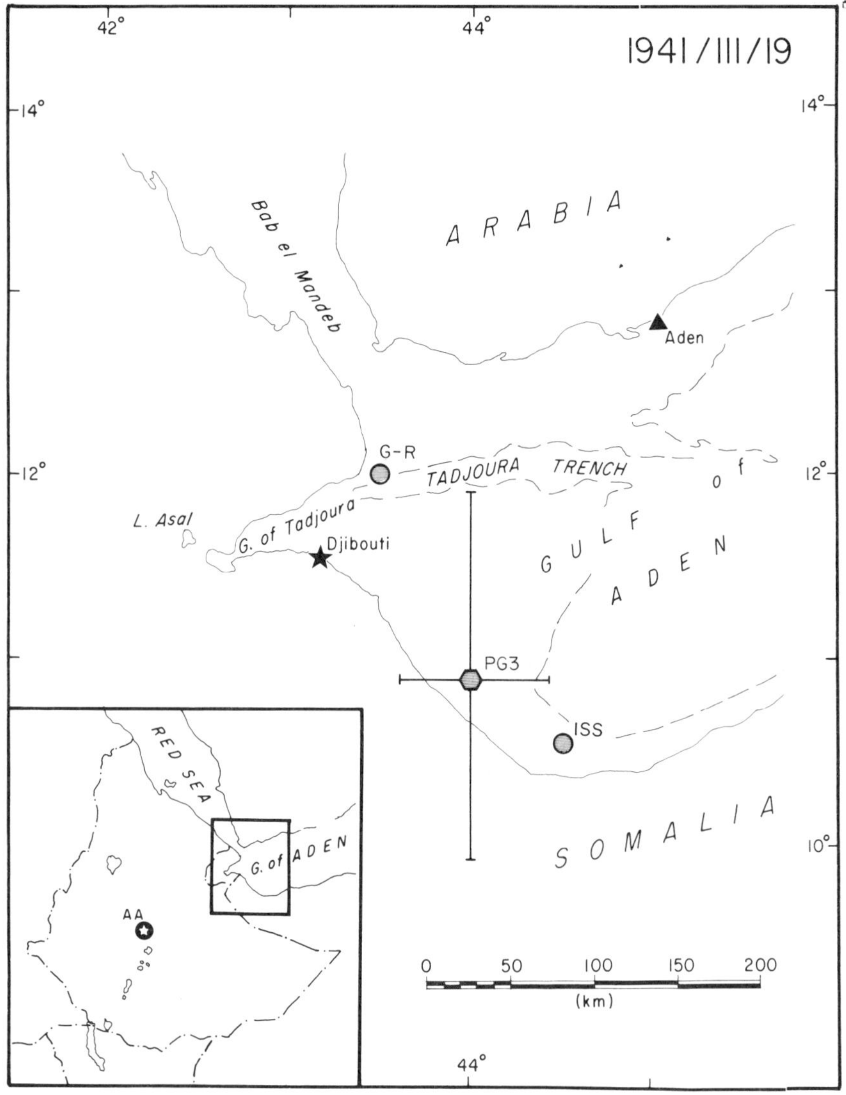

Fig. 121. Instrumental locations of the epicentre of 19 March 1941.

attempted. They yielded the following results: USCGS — U.T. 00:17:10.8, N 11.27 ± 0.3°, E 42.69 ± 0.3°; ISS — U.T. 00:17:15.4, 10.99 ± 1.7, 42.67 ± 0.7.

The results of the recomputations agree perfectly with the longitude component of the epicentre, but there is over 1° discrepancy in latitude; the reason is that all 16 stations used in the records recomputation cluster within an azimuth angle of 118° centred almost due north (012°) of the epicentral region. The meridional control is therefore very poor.

A clue to the most probable latitude value is the intensity VI–VII experienced in Djibouti. An intensity VI (MM) would normally be felt at about 40–50 km from the epicentre of a shock of magnitude 5.6. These distances were plotted around Djibouti; they indicate that the ISS original value and the recomputed one based on the USCGS file (courtesy of Dr Lynn Sykes) are the most likely location. The latter has been adopted. Note that later, in 1973, a magnitude 1.5 microearthquake was detected by the Djibouti network at almost the same site, N 11.18°, E 42.70°, indicating that the area is seismically active (Pontoise et al. 1975, Fig. 2, p. 844).

The location of these epicentres is somewhat unexpected (Fig. 122). In a personal letter, Ruegg and Lépine, the seismologists responsible for the Djibouti network, commented: *It is not evident that the 1945 earthquake is in anyway related to tectonic activity in the Gulf of Tadjoura. We think that it could as well be the result of an isostatic re-adjustment of the Aisha Horst. However, we do not reject the possibility that it could mark the prolongation, inland, of one of the transform faults presently active in the Gulf of Aden* (personal letter dated 8 February 1975).

Twenty eight tremors were felt in Djibouti during November and December, some with intensities V–VI. There is no way to ascertain if they originated from the same region as those of 27–31 October.

1945/XI–XII

The Meteorological Office in Djibouti reported 27 earth tremors in the town from 8 November to mid-December, 1945. The most violent ones, intensity VI, occurred on 8 and 14 November.

Source

Meteorological Report 168/B/MET/47, dated 17 March 1947.

1946–47

During February 1946, two tremors of unspecified intensity were felt in Djibouti. In 1947, two others of intensity IV and II during the night of 15–16 February and two very light ones on 17 March.

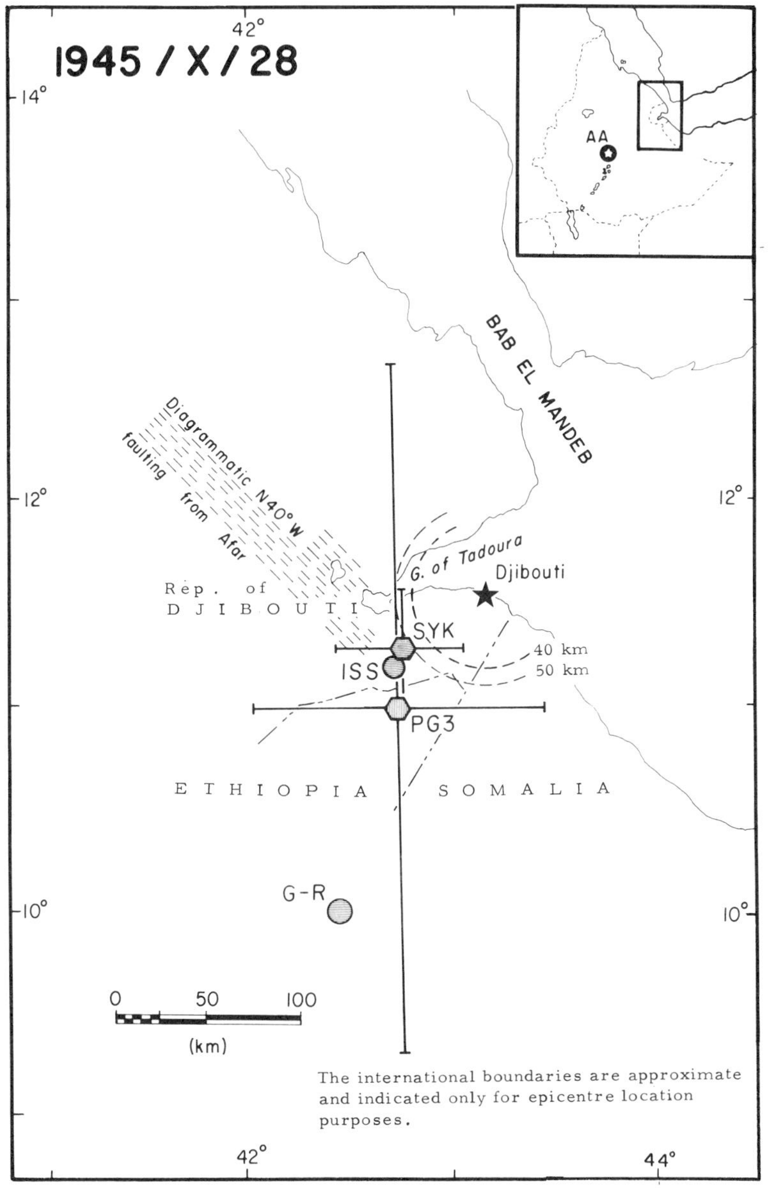

Fig. 122. Location of the four instrumental solutions for the epicentre of 28 October 1945. Also indicated are the intensity-VI distance-arcs centred on Djibouti.

Source

Meteorological Report 168/B/MET/47, dated 19 March 1947.

Comments

Djibouti often reports tremors of intensity as low as II. This is surprising, even on the Rossi-Forel scale, because intensity II is usually considered as a grade below the threshold of human detection.

1949/IV/22

During the night of 22 April 1949, a tremor of intensity V was reported from Djibouti and Obock, on either side of the Gulf of Tadjoura. The shock of L.T. 00:55 was the strongest; many others were felt in Obock from half past midnight to 2 a.m.

Source

Report No. 218 issued by the *Cabinet Civil* of Djibouti, 25 April 1949.

1949/VI/16–17

Earth tremors of maximum intensity V, accompanied by thunderlike noises, were felt on 16 and 17 June 1949. Trees were seen swaying, but no damage was reported.

Sources

Report 129/CL/MET/50 from the Meteorological Office in Djibouti; BCIS; ISS; Rothé (1954).

Comments

The source of the tremor was an earthquake of estimated magnitude about 5½ located by ISS at N 11.2°, E 42.7° and by Strasbourg at N 10.75°, E 42.5°. A recomputation of the ISS data file using the SPEEDY computer program, after three iterations, yielded the following parameters for an assumed depth of 33 km: U.T. 17:58:09.2 ± 4.5 s; N 11.57 ± 0.62°, E 42.57 ± 0.19°.

The new location is the southeastern end of the northeast escarpment of the Asal rift where it joins Ghoubet-al-Kharab (Fig. 123). It is a region of recent uplift, which separated Lake Asal from the Ghoubet-al-Kharab and is apparently still active. Normal fault systems and NW-trending open fissures cut through this recent uplift (Stieltjes 1973).

Recent microseismic activity appears to be concentrated along the centre of the Asal rift (Ruegg and Lépine, personal communication 1973). The range of standard errors in the computed coordinates of the present 1949 earthquake does not permit any better epicentre evaluation than a probability that it took place along the southeastern sector of the Maysattaka Boyna escarpment. (For more information on the geology, tectonics, and seismicity of the Asal Rift, see Stieltjes 1973; Lépine and Ruegg 1973; Ruegg and Lépine, 1973; Ruegg, 1975; Delibrias et al. 1975, and the references quoted in each of these papers.)

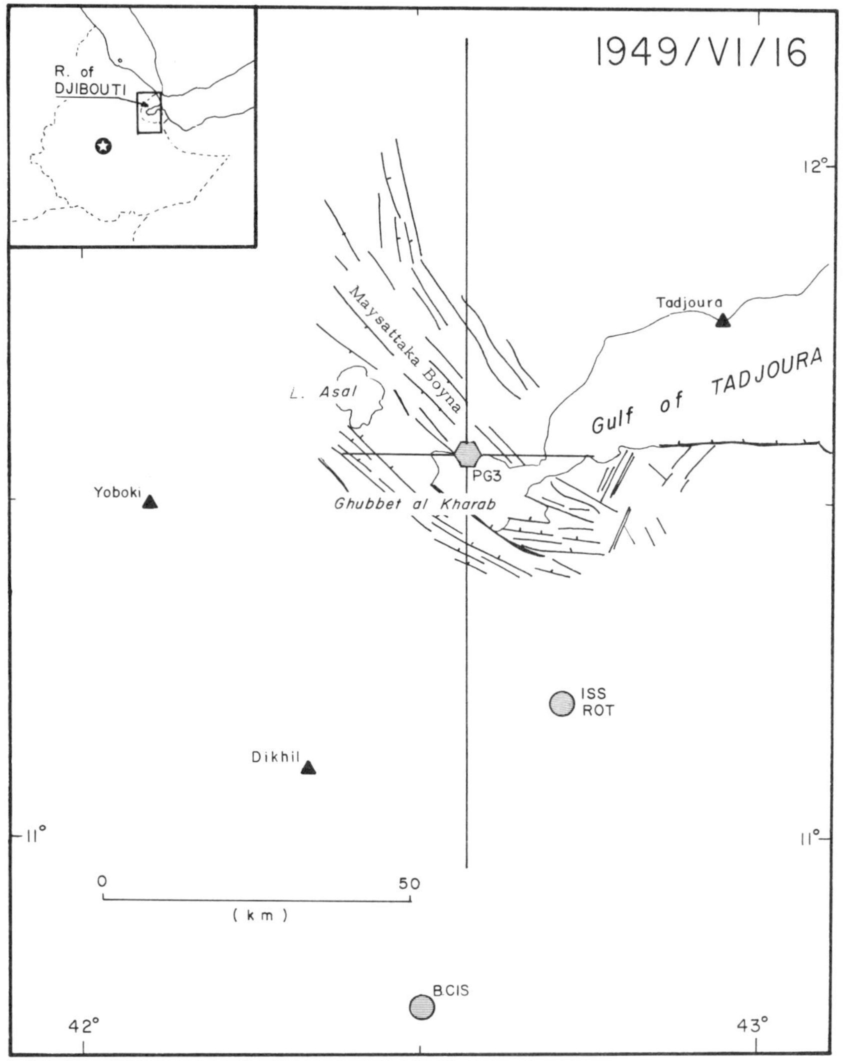

Fig. 123. Probable location of the 16 June 1949 earthquake along the southeastern sector of the Asal Rift escarpment.

Three aftershocks were felt in Djibouti on the next day, 17 June U.T. 11:59, 23:40, and 23:44; all were of intensity III. The timing is accurate because it was read from the weight-barograph trace at the Meteorological Station.

1950/II,VII

Four tremors were reported from Djibouti during 1950: three of intensity II–III on 1 February and one of intensity V at 10:30 p.m. on 8 July.

Source

Meteorological Office Report 389/ET/MET/50, dated 25 July 1950.

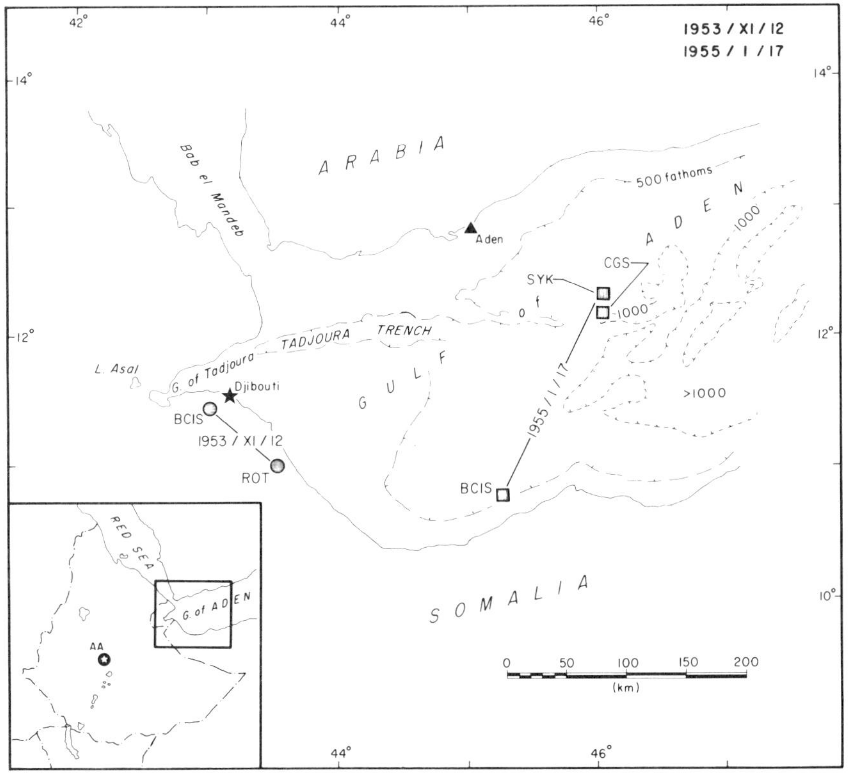

Fig. 124. Instrumental locations of the epicentres of 12 November 1953 and 17 January 1955.

1953/XI/12

During the night of 11–12 November 1953, a few shocks were reported from Djibouti (N 11.6°, E 43.15°); the sharpest one occurred at U.T. 01:17 (timed by a 25-mm impulse on the weight-barograph trace). The majority of the population was awakened, floors creaked, doors and windows opened and closed. Objects were tipped over. The intensity was estimated as R.F. V–VI, that is V on the Mercalli-Modified scale.

Sources

Meteorological Report MET/642/ET, dated 14 November 1953; BCIS; Rothé 1954; *Le Réveil de Djibouti*, 12 October 1963.

Comments

The source of seismic activity was located by Strasbourg at N 11.5°, E 43.0° and by Rothé at N 11.0°, E 43.5° (Fig. 124). For these solutions, no standard deviations have been published.

1954/XII–1955/V

From 27 December 1954 to 1 May 1955, ten tremors were felt in the Republic of Djibouti:

		Local time	Location	Intensity
1954	27 Dec.	18:35	Obock & Djibouti	III
		20:30	Obock	III
1955	20 Jan.	01:50	Yoboki (N 11.5°, E 42.1°)	III
	23 Feb.	22:55	Yoboki	V?
		23:10	Yoboki	III
	27 Apr.		Djibouti (4 shocks)	II,III
	28 Apr.		Djibouti (4 shocks)	II,III
	01 May	07:15	Djibouti	Weak

Sources

Meteorological Office Report from Djibouti, No. 2/CL/55 and 293/CL/55; Telegram 50016 from Dikhil dated 23 February 1955.

Comments

During this period, USCGS and BCIS located an earthquake of undetermined magnitude on 17 January 1955 at N 12.23°, E 46.02° in the Gulf of Aden, 320 km ENE of Djibouti. There is no reason to believe that the seismic activity experienced in the Republic was in any way connected with that earthquake. The origin of the tremors felt in Obock and Djibouti was most likely the Gulf of Tadjoura; for those reported from Yoboki, the region of the Lake Asal graben. For the location of Yoboki, see Fig. 123.

1957/IV/12–15

During 4 days, earth tremors were felt in Arta, Djibouti, and Tadjoura; they were of intensity VI in Arta and Djibouti on 12 April at U.T. 15:57. No damage was reported.

Sources

Meteorological Report 226/CL/57, dated 17 April 1957; BCIS; ISS; MOS; Sykes and Landisman (1964); Fairhead and Girdler (1970).

Comments

1. Reported Bursts of Seismic Activity

From the reported tremors, four bursts of activity can be distinguished:

	Day	Time (U.T.)	Arta	Djibouti	Tadjoura
(1)	12	15:59	VI	VI	
		16:05	IV	IV	Felt
		17:10	II		
(2)	13	About 01	Many		
(3)	14	23:40	IV; many in 10 min		
(4)	15	a.m.	II–III (4)		

2. Main Shock

The main shock of the series occurred at about U.T. 15:58:45 on 12 April, was of magnitude 5.0 (MOS) or 5.2 (BCIS), and caused tremors of intensity VI in Arta (N 11.53°, E 42.85°) and Tadjoura (N 11.79°, E 42.89°). Five agencies computed its epicentre; the five sets of coordinates ranged between N 11.47° and 11.54° and E 43.00° and 43.09°. The mean value of the four solutions has been adopted as the most probable location, namely N 11.50°, E 43.09° (Fig. 125).

The magnitude of this event was estimated at about 5.1, which is apparently too high for the intensity VI recorded at Arta and Tadjoura at epicentral distances of only 20 km. My estimate of magnitude would be about 4¾.

3. Previous Activity in 1957

Two reports on tremors felt in Djibouti on 2 January (intensity III) and 20 February (intensity IV) were sent to Strasbourg (Reports 7 and 112/CL/57 from the Meteorological Office).

1958/V/24–25

On 24–25 May, three earthquakes occurred at the south end of the Bab el Mandeb, or if seen from the northwest corner of the map, at the location

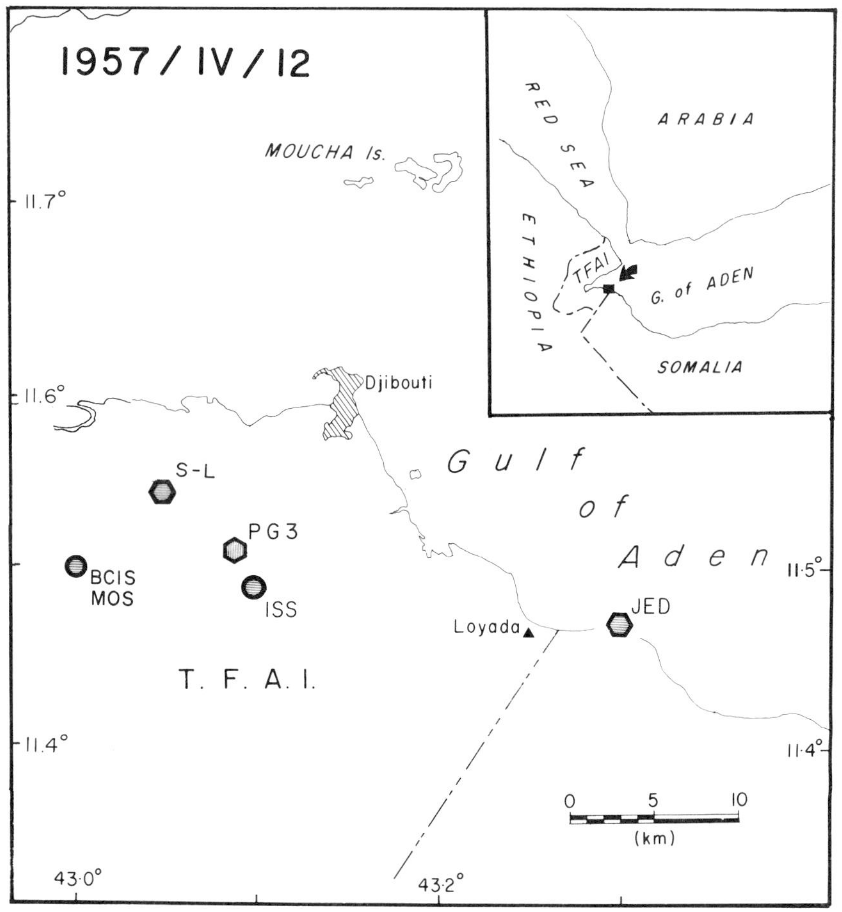

Fig. 125. Teleseismic original and recomputed solutions for the earthquake of 12 April 1957.

where an extrapolated Red Sea central axis would join the Tadjoura Trench in the Gulf of Aden. The magnitudes of these shocks were about 4.5, 5.5, and 5.0.

Sources

BCIS; ISS; Sykes and Landisman (1964); Fairhead and Girdler (1970, 1971).

Comments

Sykes and Landisman (1964) located the foreshock on 24 May U.T. 22:25:32.6 at N 12.17°, E 43.58° with a possible accuracy of ± 10 km.

Seven agencies located the main shock of magnitude 5.5 (MOS) at U.T. 23:53.7; three of these are recomputations using different computing techniques and data sets. The recomputed epicentre parameters are given below:

Agency	H (U.T.)	Coordinates		Stations
JED*	23:53:42.7	N 12.13 ± 0.04°	E 43.82 ± 0.03°	44
PG3	23:53:46.0	N 12.27 ± 0.27°	E 43.60 ± 0.18°	53
S-L	23:53:38.0	N 12.14 ± 0.10°	E 43.59 ± 0.10°	44

Parameters from the other agencies are given in Part II, p. 189. All available solutions are plotted on Fig. 126.

The aftershock on 25 May at U.T. 02:53.8 was located by six agencies. Again only the relocation parameters are given here:

Agency	H (U.T.)	Coordinates		Stations
JED*	02:53:53.1	N 12.16 ± 0.04°	E 43.88 ± 0.03°	37
PG3	02:53:55.3	N 12.14 ± 0.15°	E 43.73 ± 0.10°	47
S-L	02:53:48.4	N 12.13 ± 0.10°	E 43.69 ± 0.10°	37

As adopted locations, the mean of the recalculated coordinates were chosen: 24 May, foreshock at U.T. 22:25:32.6 (N 12.17°, E 43.58°); 24 May, main shock at U.T. 23:53.7 (N 12.18°, E 43.67°); and 25 May, aftershock at U.T. 02:53.8 (N 12.15°, E 43.77°).

*The standard deviation indicated for Fairhead and Girdler's locations are those of the master event used for the Joint Epicentre Determination.

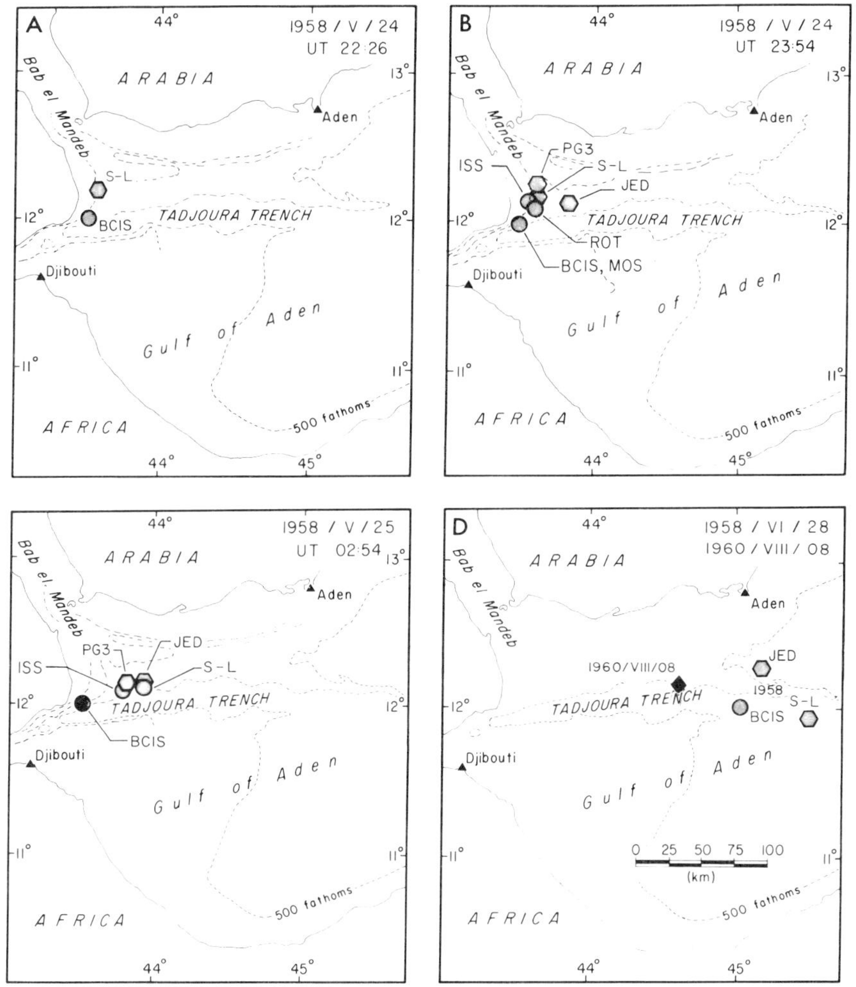

Fig. 126. Maps A, B, and C are epicentre plots of all teleseismic solutions for the events of 24–25 May. Map D locates the epicentre of 28 June of the same year.

1958/VI/28

On 28 June 1958, an earthquake of magnitude ⩾ 4.5 occurred in the same region as the epicentres of 24–25 May. Sykes and Landisman (1964) and Girdler and Fairhead (1970) recalculated the original data and gave the following locations: Sykes and Landisman (N 11.94°, E 45.44°); Fairhead and Girdler (N 12.26°, E 45.13°).

Sources

Sykes and Landisman (1964, p. 1929); Fairhead and Girdler (1970, p. 64).

Comments

The epicentre location is given on Fig. 126D.

1959/XII/31

On 31 December 1959, earth tremors were felt in Arta (N 11.5°, E 42.8°) and Djibouti (N 11.5°, E 43.0°). No damage was reported.

Sources

BCIS; Rothé (1964, p. 2538).

Comments

The source of the tremors was identified by BCIS as an epicentre located at N 11.6°, E 42.8° in the Gulf of Tadjoura. The BCIS considers this earthquake as a foreshock of activity that followed during January 1960.

1960/I/04-29

Seismic activity at the western end of the Gulf of Tadjoura started with three foreshocks on 31 December 1959, and one on 4 January 1960, at U.T. 06:08. Two of these were felt in Arta and Djibouti. The main shock occurred at U.T. 06:16.5 on 4 January; it was reported from Arta (N 11.5°, E 42.8°), Ali Sabieh (N 11.1°, E 42.7°), Djibouti (N 11.6°, E 43.1°), and Tadjoura (N 11.8°, E 42.9°). It caused damage to houses in Arta and triggered landslides on nearby slopes. The tremors were accompanied by loud noise. Aftershocks lasted until the end of January.

No magnitudes were determined because the AAE seismographs had yet to be calibrated. It is inferred, however, from the number of stations (56) that reported the event to ISS and from the macroseismic effects observed, that the main shock was of magnitude 4.5 or slightly higher.

Sources

AAE Data File; BCIS; Dakin (1975, p. 51–70); Fairhead and Girdler (1970); ISS (1960, p. 12); Rothé (1967); Sykes and Landisman (1964, p. 1927–40).

Comments

Three foreshocks were registered on 31 December 1959 by the Observatory in Addis Ababa; the first one (iP 15:12:05) was reported felt in Djibouti. On 4 January, Sykes and Landisman (1964) and Fairhead and Girdler (1970) located a foreshock at U.T. 06:08 practically on the same meridian (E 42.8°) but 30 km apart in latitude (N 11.50° versus N 11.83°). The main shock occurred at U.T. 06:16.5 on the same day. The 65 aftershocks that followed were located between 490 and 575 km from Addis Ababa; the last one was recorded on 29 January (Fig. 127).

1960/VIII/08

On 8 August 1960 at U.T. 12:28:11, an earthquake of magnitude 5.4 (PAS) occurred on the northern ridge of the Gulf of Aden rift. The epicentral mean value of seven teleseismic solutions based on 41 station reports is N 12.08 ± 0.03°, E 44.54 ± 0.06°. The event is included in this catalogue: (1) to avoid fringe effects in the amplitude contouring process; and (2) as a check on local travel times.

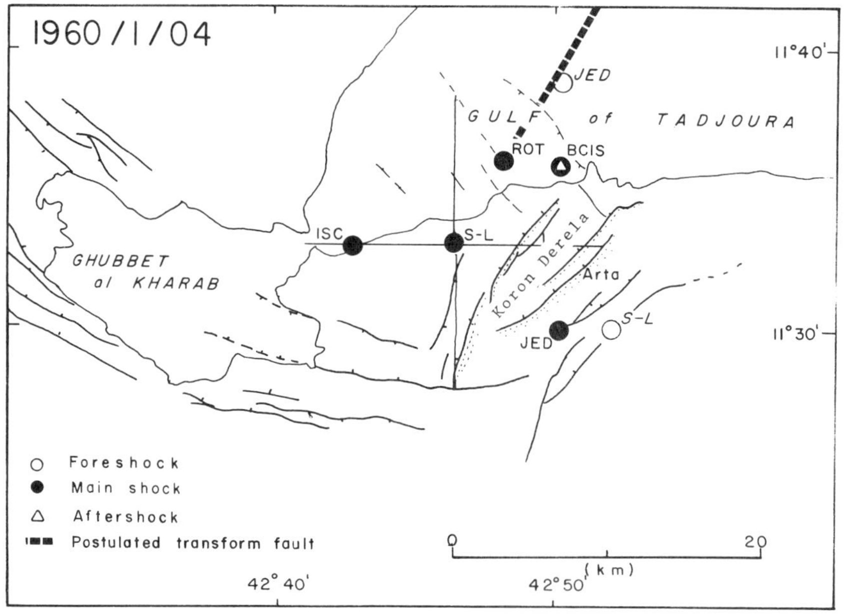

Fig. 127. Location of the instrumental epicentres as calculated from teleseismic data by different agencies. The tectonic background is adapted from Muller and Boucarut (1975).

Sources

AAE Data File; BCIS; CGS; ISS; MOS. (Fairhead and Girdler 1970; Rothé 1964; Sykes and Landisman 1964).

Comments

The iP-onset of this event was recorded at Addis Ababa at U.T. 12:29:46 and the Lg at 12:31:30. The epicentral distance of the adopted epicentre (N 12.08°, E 44.54°) is 6.4° or 715 km from AAE. For a Pn velocity of 7.95 and an observed time delay of +4.6 s for Pn, the apparent Lg velocity was 3.60 km/s along the path Gulf of Aden–Addis Ababa. The adopted location for this event is plotted on Fig. 126D.

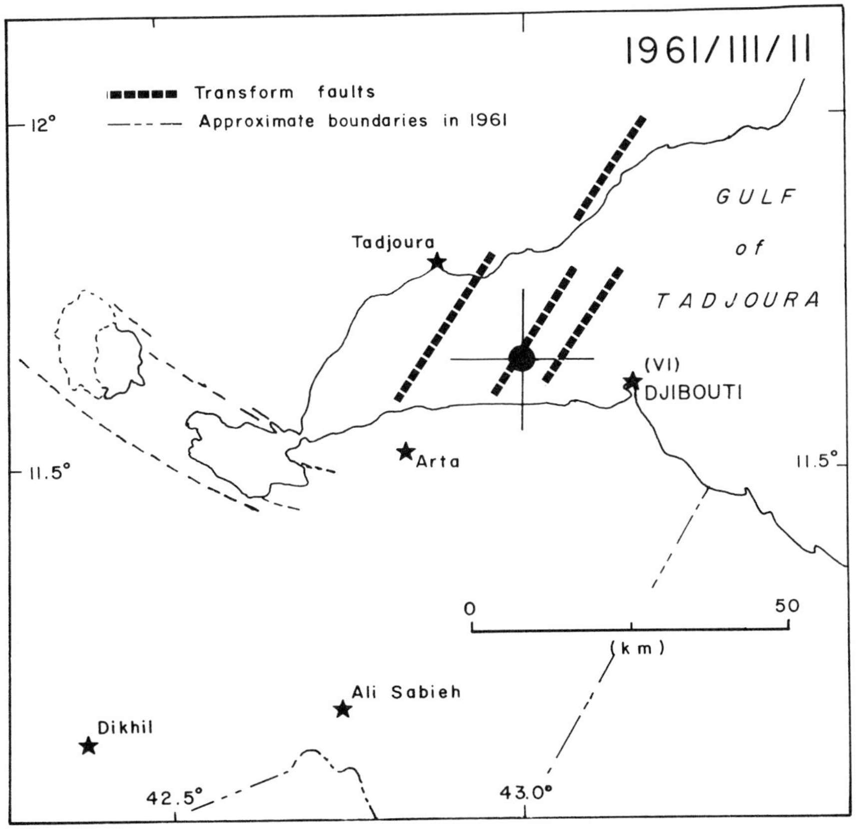

Fig. 128. Location of the adopted epicentre for the main shock of 11 March 1961. It is the average of eight sets of coordinates and coincides with the location recomputed by Sykes and Landisman (1964). The errors bars (± 10 km) are from the same authors.

1961/III/11–30

About noon on 11 March 1961, an earth tremor was felt throughout the Republic of Djibouti; the intensity was VI in the capital. This tremor marked the main shock of an earthquake sequence that featured almost no foreshock activity (only two) and 66 aftershocks, which occurred sporadically until the end of the month. Thirty-nine of the aftershocks were recorded by the weight-barograph at the Meteorological Station in Djibouti.

The magnitude of the main shock at U.T. 08:41 was estimated as 6¼ by Moscow, 6 by Quetta and Rothé, 5.9 by Sykes and Landisman; the approximate epicentre was N 11.65°, E 43.00° in the vicinity of a SW-NE transform fault in the Gulf of Tadjoura (Fig. 128).

Sources

AAE Data File; BCIS (1961, p. 361–62); Dakin (1975, p. 51–70); Fairhead and Girdler (1970, 1971); Meteorological Report 208/CL/61 from Djibouti; Sykes and Landisman (1964).

Comments

The epicentre of the main shock at U.T. 08:41 on 11 March was calculated by eight different agencies. All the solutions cluster around the mean position N 11.65°, E 43.00° in the vicinity of a postulated transform fault that cuts through the floor of the Gulf in a northeasterly direction. The March 1961 seismic region corresponds to that of January 1960. The parameters of each solution are listed in Part II.

1961/VI/20

An earthquake of magnitude m_b 6.1 occurred at U.T. 03:21.5 on 20 June 1961, along the northern scarp of the Tadjoura Trench in the Gulf of Aden. Six agencies reported it. Five solutions agree within ± 7 km in latitude and ± 9 km in longitude. The mean of the five sets of parameters has been adopted as the most probable location: N 12.22 ± 0.06°, E 44.29 ± 0.08°.

Because no tremors were reported from the continent, only the location map is given here (Fig. 129); details of each solution are given in PartII.

Sources

BCIS (1961, p. 979–80); Fairhead and Girdler (1970, p. 64); ISS (1961, p. 575–76); Sykes and Landisman (1964, p. 1930).

1963/II/21

An earthquake of magnitude m_b (AAE) about 4.8 occurred in the southern section of the Danakil Horst at U.T. 09:17:12 on 21 February 1963. The location given by the BCIS is N 12.0°, E 43.0°. It is surprising that tremor reports were not filed in from Tadjoura and Obock, at epicentral distances of 25 and 35 km, respectively!

Sources

AAE Data File; BCIS (1963, p. 325–26).

Comments

Only one teleseismic solution is available for this epicentre; it is based on eleven station reports. The epicentral distance from AAE is 568 km and corresponds to the apparent distance given by an S–P recorded by Addis Ababa: 565–575 km. A precise azimuth control is not possible. Despite the absence of felt reports from nearby Tadjoura and Obock, the coordinates given by the BCIS (N 12.0°, E 43.0°) are cautiously adopted. The location is plotted on Fig. 130.

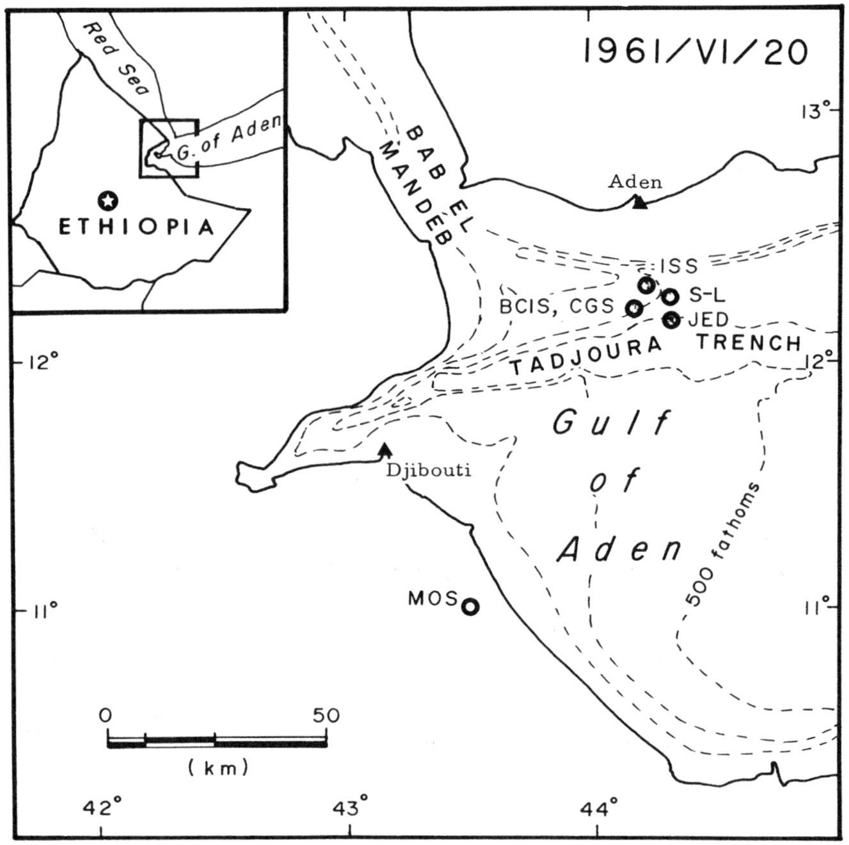

Fig. 129. Location of the instrumental epicentres available for the earthquake of 20 June 1961 in the Gulf of Aden.

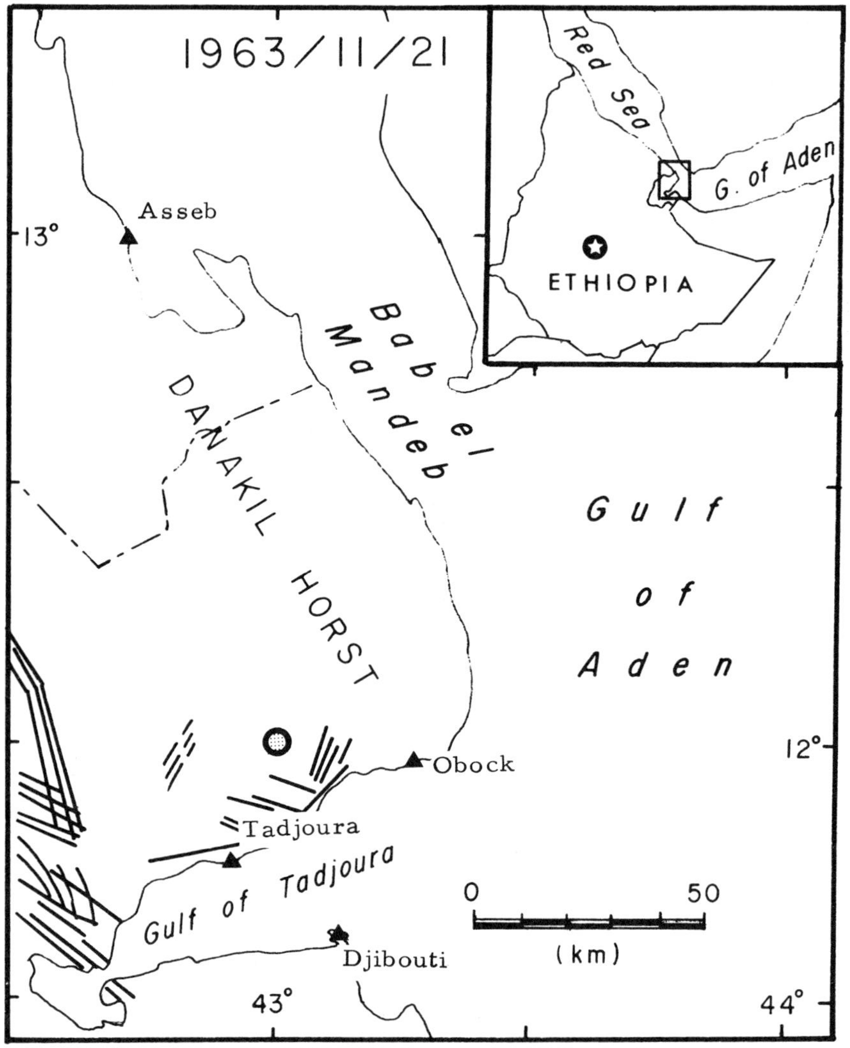

Fig. 130. Location by the BCIS of the instrumental epicentre for the earthquake of 21 February 1963. Only recent fault trends are indicated on the map.

Two aftershocks followed, one 10 min after the main shock and the other 8 h later; the S-P times for the three events were 60.2, 62.0, and 60.5 s, respectively. For a short description of the geology and structural evolution of the southern sector of the Danakil "Block", see Marinelli and Varet (1973, p. 1119–1122).

1963/X/05–23

At U.T. 14:57:47 on 5 October 1963, an earthquake of magnitude 5.3 (USCGS), 5.6 (UPP), occurred in the western sector of the Gulf of Tadjoura. It triggered landslides from the slopes near Arta on the south side of the Gulf to the Goda massif on the north shore. In the Gulf, the water was seen "boiling" and fishes were seen leaping into the air. The tremors were felt with intensity V in Arta (N 11.53°, E 42.85°) and IV in Djibouti (N 11.60°, E 43.15°). This event was the main shock of a period of seismic activity that initiated on 2 October and was monitored by the Observatory at Addis Ababa, 520 km away, until 23 October.

Sources

AAE Data File; BCIS (1953, p. 2333–36); ISS (1963, p. 902–04); *Le Réveil de Djibouti*, 12 October 1963; Dakin (1975, p. 51–70).

Comments

1. Location of the Seismic Region

There are five teleseismic solutions for the location of the main shock on 5 October. With the exception of the Moscow solution, located ½° west of the others, the instrumental epicentres cluster around N 11.57 ± 0.05°, E 42.82 ± 0.03° on the south shore of the Gulf of Tadjoura. The aftershock at U.T. 17:18:25 was located by the USCGS at N 11.7°, E 42.6° and by Fairhead and Girdler (1970) at N 11.70°, E 42.71° on the north side of the Gulf. The aftershock instrumental epicentre might lack the accuracy of the previous one — only seven station reports were used versus more than 120 for the main shock — but being located near the Goda massif from where land- and rockslides were observed gives confidence in the solution. The site is on the north side of the Gulf and, with respect to the main shock, on a SE-NW line subparallel to the axis of the Asal-Ghubbet al Kharab graben.

Figure 131 shows the regions within which the main shock and the largest aftershock of 5 October occurred. The first is located on the south shore of the Gulf of Tadjoura; the second on the north shore. Dust released by landslides and thrown up in the atmosphere was observed over each site by fishermen at sea near the pass of the Ghubbet. The locations of the various instrumental epicentres determined by international agencies are plotted on Fig. 134.

The epicentral distances of the aftershocks ranged from 500 to 535 km from Addis Ababa as indicated by the heavy dashed lines A and B on Fig. 131.

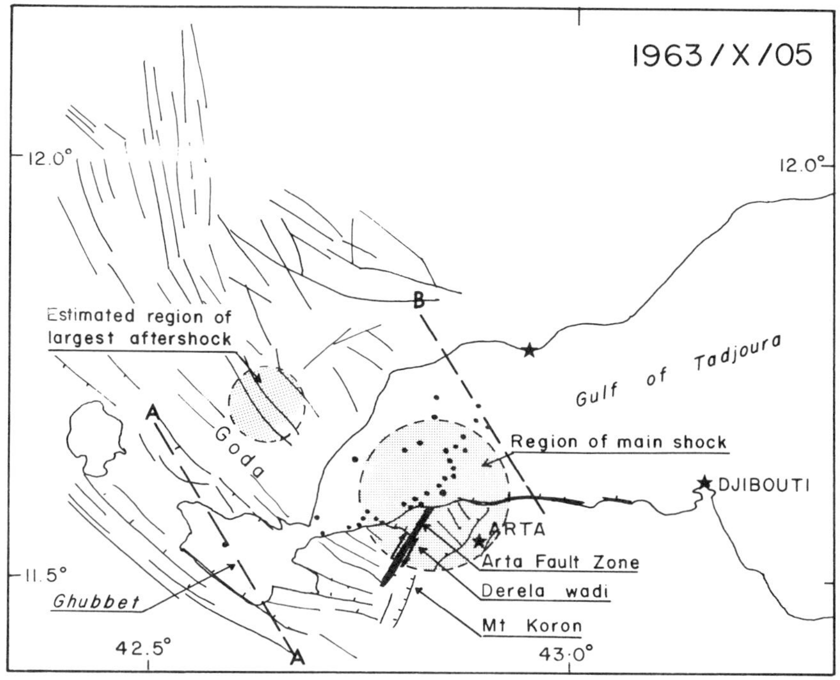

Fig. 131. Seismotectonic map of the October 1963 seismic activity at the west end of the Gulf of Tadjoura. Shaded circles are the estimated regions where the main shock and the largest aftershock of 5 October occurred; dashed lines A and B are the limits within which the other 267 aftershocks took place. The Arta Fault Zone is marked by a heavy line. The microearthquake epicentres marked by dots are taken from Lépine et al. (1976).

The main shock occurred at the northeast end of a fault zone running parallel to the Derela wadi. Recent microseismic activity as plotted on Fig. 131 reveals that the zone extends northeastward toward the central axis of the Gulf of Tadjoura. Lépine et al. (1976, p. 9–12) discovered that this fault zone was a transform zone with individual faults heading from 025° to 050°, and they postulated that this may express the displacement of the Asal-Ghubbet al Kharab graben, northeasterly, to the axial zone of the Gulf of Tadjoura.

Seismically, the southwestern sector of the Arta zone appears to have long been inactive; the microseismic activity of the last 5 years or so of the offshore sector is rather important and fault-plane solutions indicate a dextral motion with an azimuth of 050–055° (Ruegg, personal communication, March 1978).

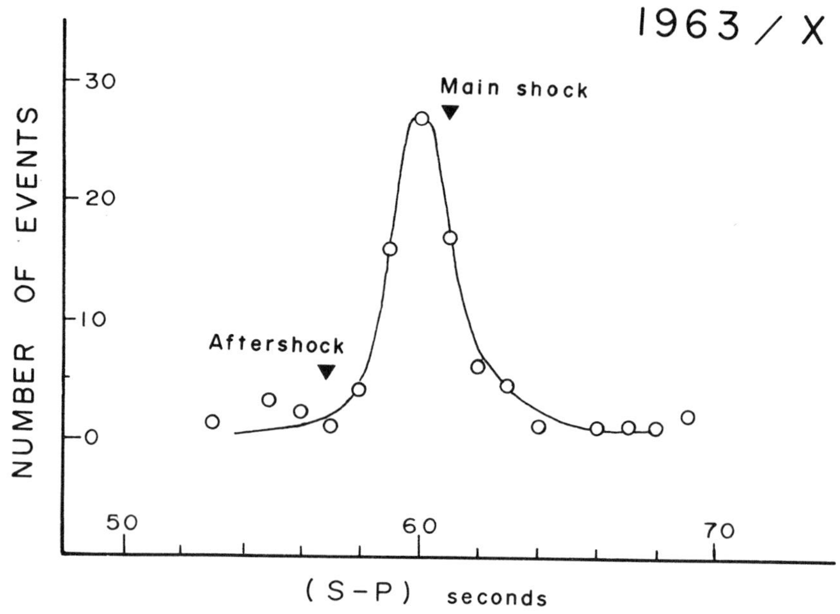

Fig. 132. S-P(AAE) time distribution during the October 1963 sequence indicating the relative distribution of epicentral distances from Addis Ababa and therefore the corresponding radial distance limits of the seismic region. No accurate azimuth determination can be made from a unique station.

2. The Earthquake Sequence: Some Statistical Data

Foreshocks: One on 2 October; two on 5 October.

Main Shock: 5 October 1963, at U.T. 14:57:57.4, (CGS); m_b 5.3; intensity V in Arta, IV in Djibouti; landslides near Arta.

Aftershocks: (1) reported by international agencies — one on 5 October at U.T. 17:18:25.0, m_b 5.3 (CGS), caused landslides along Guda massif; (2) reported felt in Djibouti — eight on 5 October, seven from 6–8 October; (3) recorded at AAE, Addis Ababa — 267 in the magnitude range $3.0 < m_b$ (AAE) < 5 with epicentral distances between 500 and 535 km from Addis Ababa (Fig. 132); and (4) frequency-magnitude relations — in the aftershock sequence, the b-value of the frequency-magnitude relation, where $N = \exp(a - bM)$, was 0.73 ± 0.04 for the cumulative frequency relation and 0.66 when calculated with Utsu's (1965) maximum likelihood method (see Fig. 133).

3. Friendly Controversy Concerning the Epicentral Location of the Main Shock on 5 October 1963

In a communication at Leeds University (England), a famed volcanologist provocatively questioned the reliability of instrumental epicentres

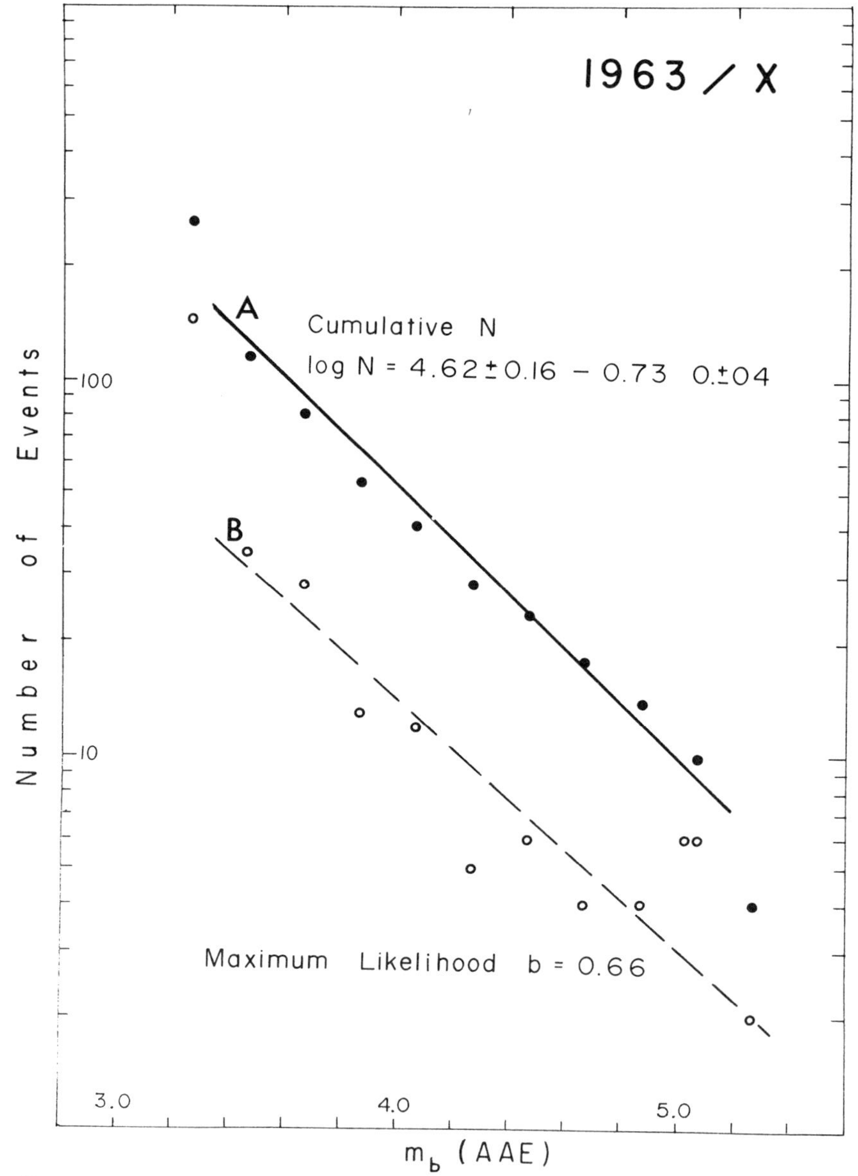

Fig. 133. Frequency-magnitude relation for the aftershocks during the October 1963 earthquake sequence. The solid regression line is the least-squares best fit to the cumulative frequency distribution; the dashed line is the maximum likelihood fit.

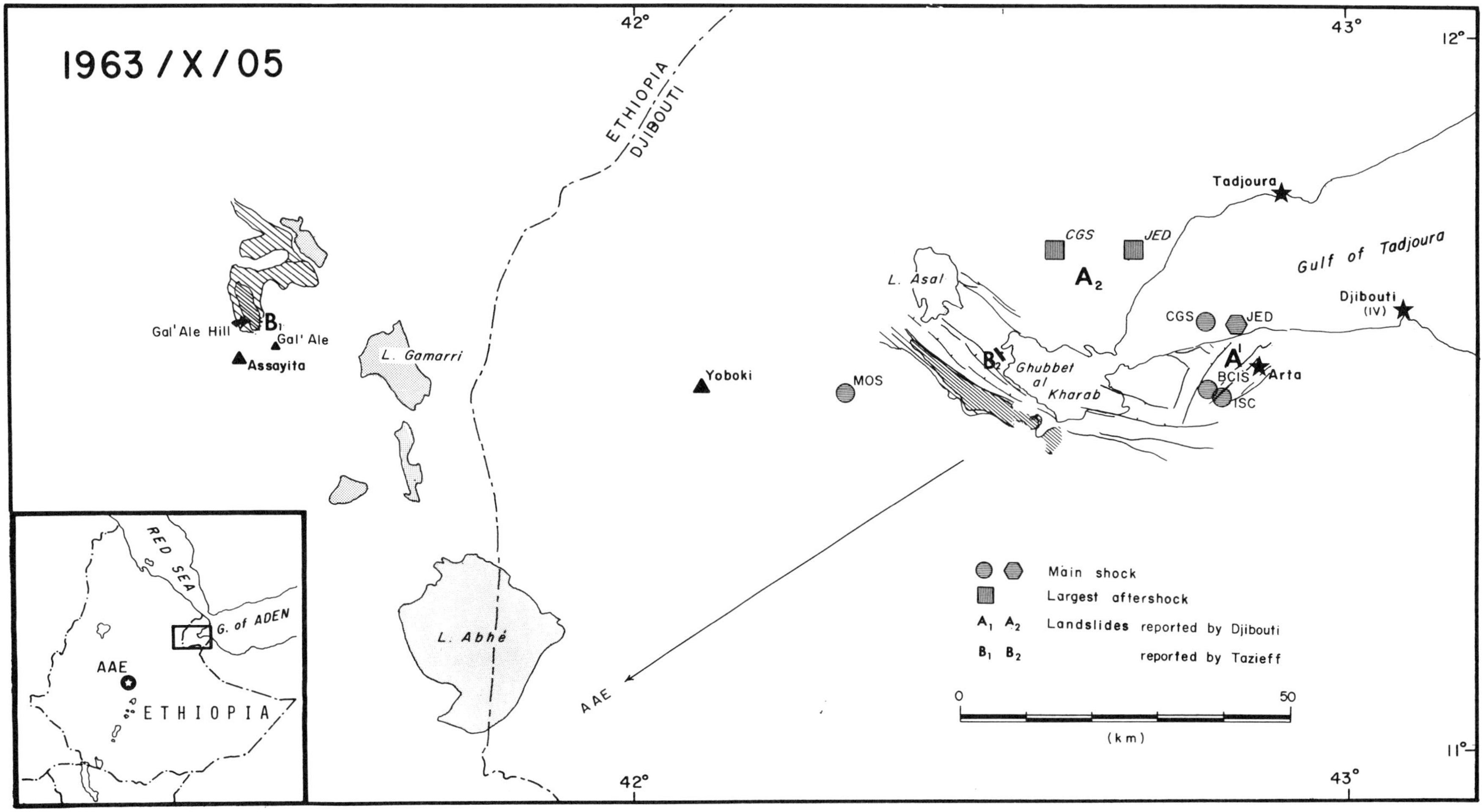

Fig. 134. Location map of the instrumental epicentres for the main shock and the largest aftershock of 5 October 1963. Also included are the locations mentioned in the text.

calculated from teleseismic records; among others that he mentioned is the epicentre of 5 October 1963 (Tazieff 1972, p. 171–182). I quote:

Before going ahead it is necessary to discuss the degree to which these epicenter locations can be safely used... In this particular case two main reasons urge caution... first, there is far too small a number of earthquake records concerned (half a dozen)[1]...

The first case is the magnitude 5.3 earthquake of October 5, 1963. Fairhead and Girdler's (1970) sketch map shows it plotted on the western part of the northern[2] *coast of the Gulf of Tadjoura... the authors give the coordinates N 11.60°, E 42.85°... but all eye witnesses are categorical; no tidal wave (tsunami)*[3] *was linked with this earthquake which on the other hand, was marked by rock-avalanches and appalling dust clouds towering above them.*[4] *These landslides happened behind the Ghubbet-al-Kharab,*[5] *in the Asal area. Impressive scars of this event are still perfectly observable, notably on the north-east slopes on Mount Galele, a tuff hill 250 meters high, which was denuded by the effects of the shocks*[6]... (p. 175).

Some of the statements put forward by Tazieff need to be slightly nuanced; a few have been referenced by numerals added to the original text quoted above:

(1) The *half a dozen* seismograms used in the solution of the main shock on 5 October at U.T. 14:57.8 turns out to be 125 for BCIS, 122 for ISS, and 21 for Fairhead and Girdler.

(2) Fairhead and Girdler's location at N 11.60°, E 42.85° happens to be near the south shore of the Gulf not on the north coast.

(3) The *Réveil de Djibouti* in its edition of 13 October, the first issue published immediately after the earthquakes, mentions that fishermen observed simultaneously: (a) the dust clouds rising above the landslides; and (b) the Gulf water literally boiling and frightened fishes leaping out of the water. Moreover, it is a little much to expect a tsunami from a magnitude 5.3 earthquake located near the shores of a shallow body of water!

(4,5) The eyewitnesses quoted by the local newspaper *Le Réveil* saw clouds of dust along the slopes of hills near Arta (N 11.5°, E 42.9°) on the southern horizon and in the direction of the Goda massif on the northern horizon. (Landslides in Goda might have been triggered by the aftershock.) They did not mention the hills behind the Ghubbet al Kharab.

(6) A certain number of years after the 1963 earthquakes, Tazieff personally observed fresh scars on the slopes of a certain Mount Galale (or Gal'Ale). There are two sites by the name of Galale: one on the northwest shores of the Ghubbet at N 11°34′, E 42°31′, and a second in Afar near Assayita. The first site fits a previous statement by Tazieff that "the landslides happened behind the Ghubbet"; however, the author explicitly specifies that he is referring to the second site and he gives its coordinates as N 11°36′, E 41°24′, some 125 km inland. Both sites are located in areas of high earth tremor incidence and both hills could have been denuded by seismic shocks. The problem is to fix a precise date, for an event that occurred a few years earlier, from nomads in whose philosopohy time does not count!

Figure 134 illustrates some of the points discussed above.

1963/XI/16

The Ionospheric Station in Arta (N 11.53°, E 42.85°) reported an isolated but strong tremor (IV) at U.T. 02.30 on 16 November 1963. This could easily be an aftershock of the October sequence.

1965/V/16

An earthquake of magnitude M_L (AAE) 4.6 occurred on 16 May 1965, off the coast of Somalia, north of the port of Berbera. Although the epicentral location is somewhat unexpected, the convergence of three teleseismic

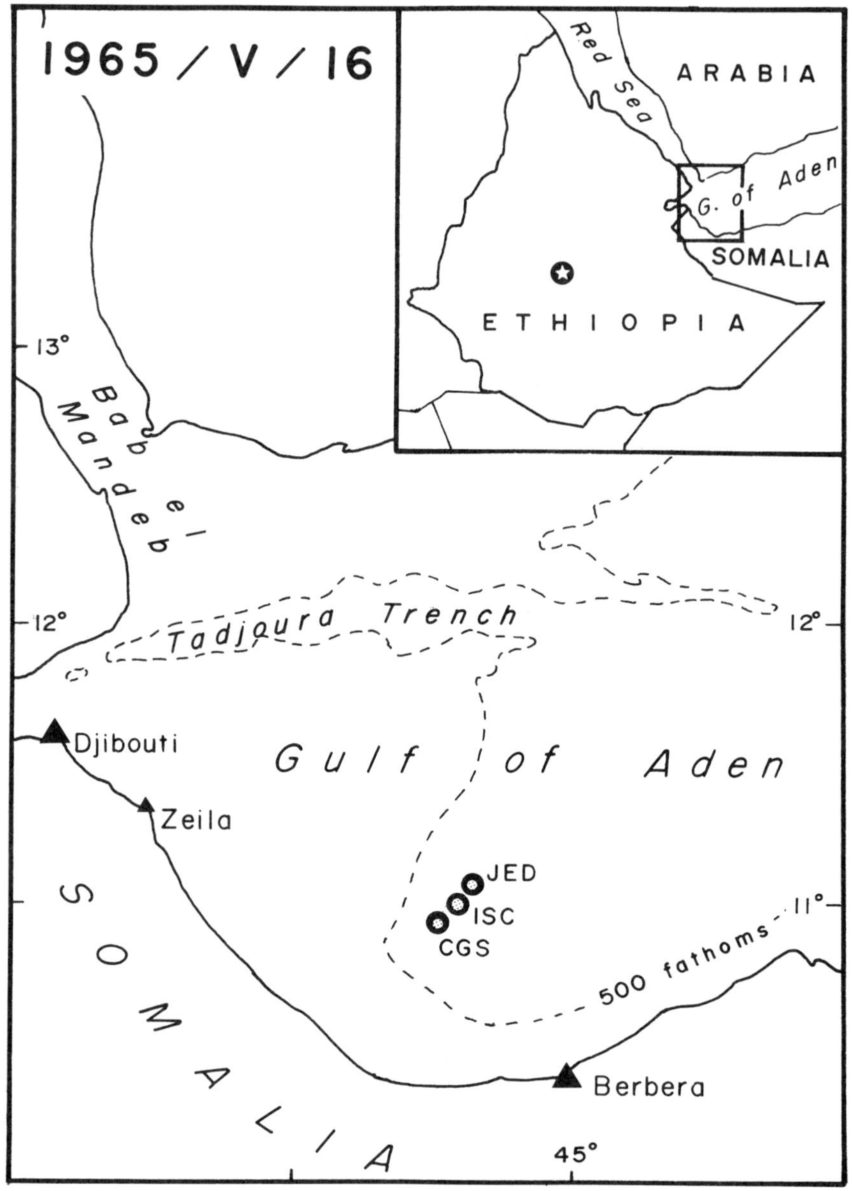

Fig. 135. Instrumental epicentres for the earthquake of 16 May 1965 off the north shore of Somalia.

solutions within a radius of less than 8 km confirms its reliability (Fig. 135). The parameters obtained from teleseismic data for a focal depth of 33 km are: CGS U.T. 00:45:56.8, (N 10.93± 0.14°, E 45.52 ±0.19°); ISC U.T. 00:45:57.1, (N 11.00 ± 0.14, N 45.60 ± 0.18°); and JED U.T. 00:45:58.0, (N 11.07, E 45.61).

Sources

CGS (EDR 50-65, p. 9); Fairhead and Girdler (1970, p. 64); ISC (1965 II(6), p. 128).

1965/VII/19

On 19 July 1965, at U.T. 15:49:36, an earthquake of magnitude m_b(AAE) 4.7 occurred on the eastern scarp of the Lake Asal graben at N 11.77°, E 42.52°.

Sources

AAE Data File; Fairhead and Girdler (1970); ISC; USCGS (PDE 65-66).

Comments

From seven station reports, CGS, ISC, and Fairhead and Girdler computed an instrumental epicentre at N 12.15 ±0.07°, E 42.58 ± 0.02°. Unfortunately, the published iP time arrival at AAE was erroneous, and in the USCGS computation the O-C residual for Addis Ababa was given as - 7.9 s.

Two recomputations were attempted based on a corrected AAE arrival time and on the data files from BCIS and ISC. The two recomputed locations were: N 11.700 ± 0.77°, E 42.483 ± 1.09 (ISC); and N 11.83 ± 0.08, E 42.55 ± 0.12 (BCIS). (The computations were made by courtesy of D. Fairhead and P. Peterschmidt.) The mean value of these two locations was accepted as the most probable epicentral location: N 11.77°, E 42.52° (Fig. 136).

For this event, some of the remarks made by Tazieff (1972, p. 176) proved to be correct; the originally published instrumental epicentre was off by some 60 km (see comments of entry 1963/X/05–23).

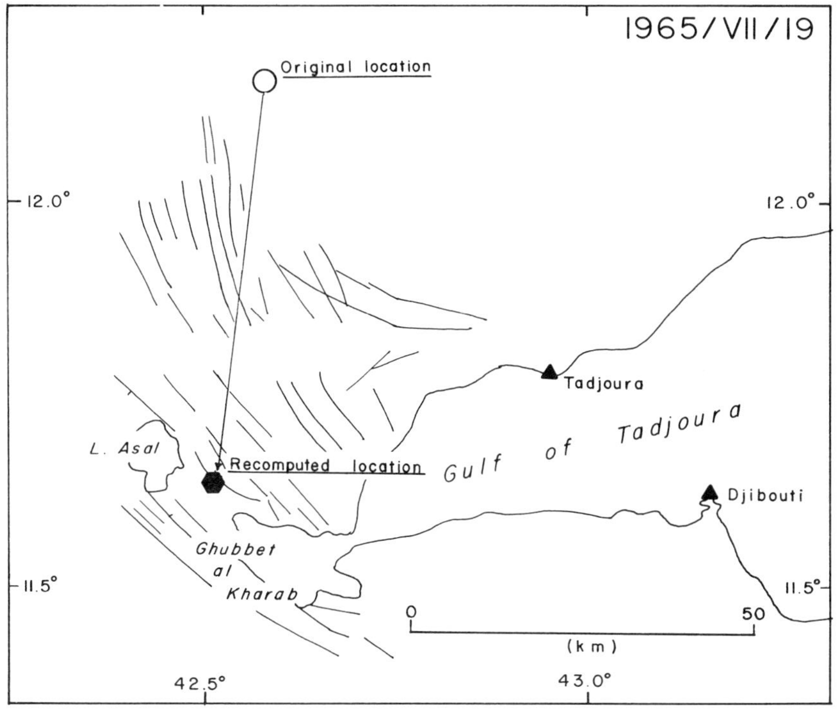

Fig. 136. Original and recomputed epicentral locations for the earthquake of 19 July 1965.

1966/I/15–31

There was seismic activity at the entrance of the straits of Bab el Mandeb. An isolated foreshock was registered on 15 January 1966. The swarm itself started on 20 January; it was characterized by three days of higher activity followed by more or less isolated shocks that lasted until 31 January. During that period the Observatory in Addis Ababa registered 157 earthquakes of magnitudes m_b (AAE) 3.1–5.0. The largest shocks were located at the eastern limit of the seismic zone that apparently extended from N 11.9°, E 43.2° to N 12.1°, E 43.8°. Twenty-two tremors were felt in Djibouti. The strongest were of intensity III–IV (Fig. 137).

Sources

AAE Data File; Meteorological Office at Djibouti (personal communication, dated 14 February 1966); Fairhead and Girdler (1970); ISC (1966, p. 213); USCGS (PDE 7-66); Dakin (1975, p. 51–70).

Comments

1. Seismic Region

Of the 157 events during this swarm, one epicentre was located by three agencies. The parameters are:

Agency	H (U.T.)	Coordinates		m_b	h(km)
CGS	12:39:43.1 ± 0.7	N 12:00 ± 0.09°	E 43:77 ± 0.10°	4.7	33
ISC	12:39:46 ± 3.5	N 11.7 ± 0.15	E 43.7 ± 0.17	4.7	74 ± 42
JED	12:39:44.9	N 12.11	E 43.64		

The seismic activity was located in the Tadjoura Trench, most probably along its northern escarpment between longitudes E 43.2° and 43.8° as can be estimated from the S-P time intervals (60–80 s) recorded at Addis Ababa.

2. Frequency-Magnitude Relationships

The cumulative number of events versus magnitude shows a steep change in b-value at m_b 4.2 (Fig. 138). For $m_b \leqslant 4.2$, the b-value is 0.92 ± 0.02; for the others, b is 1.55 ± 0.09. Using all the data, the overall non-cumulative frequency-magnitude relation computed by the maximum likelihood method (Utsu 1965) is 0.92.

1969/X/24

An earthquake of m_b 4.8 (CGS), 4.6 (ISC) had its epicentre in the western sector of the Gulf of Aden at the extreme southeast tip of the Tadjoura Trench on the 500-fathom contour line. The magnitude was too low to be felt on the African continent, but the event is included here because of possible fringe effects on the contouring process. The parameters are:

	U.T.	Coordinates		h(km)	m_b
CGS	10:12:41.8	N 11.87 ± 0.07°	E 44.86 ± 0.08°	25 ± 23	4.8
ISC	10:12:42.3	N 11.85 ± 0.07°	E 44.96 ± 0.10°	25	4.6

Moscow (MOS) located it at N 12.0°, E 45.4°.

1971/IV/25

An earthquake of magnitude m_b 3.9 (CGS), 4.3 (ISC), occurred at U.T. 11:42:15 on 25 April 1971 in the Tadjoura Trench at the south end of the Bab el Mandeb (Fig. 139). For computation purposes, the USCGS epicentre determination at N 12.02°, E 43.58° has been adopted because it is the solution for which the P travel-time to AAE shows the lowest residual (0.9 s).

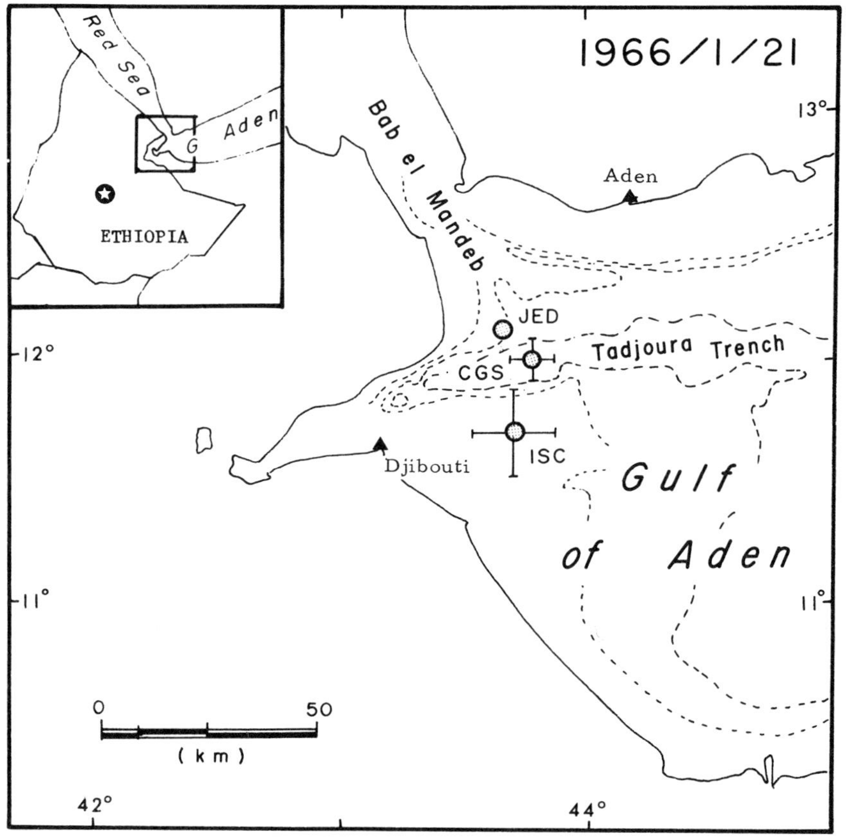

Fig. 137. Location of the January 1966 seismic region in the Tadjoura Trench.

Sources

AAE Data File; ISC (1971, p. 79); USCGS (EDR 33-71).

Comments

This epicentre falls in the vicinity of the projection of the Red Sea central axis, southeastward, into the Gulf of Aden. For other epicentres in the same vicinity, see entries 1958/V/24–25 and 1966/I/15–31.

1973/III–IV

From 27 March to 17 April 1973, a sequence of earthquakes occurred in the western half of the Gulf of Tadjoura. The number of shocks of magnitude M_L(IPG) $\geqslant$ 1.5 detected by the Djibouti seismic network was about

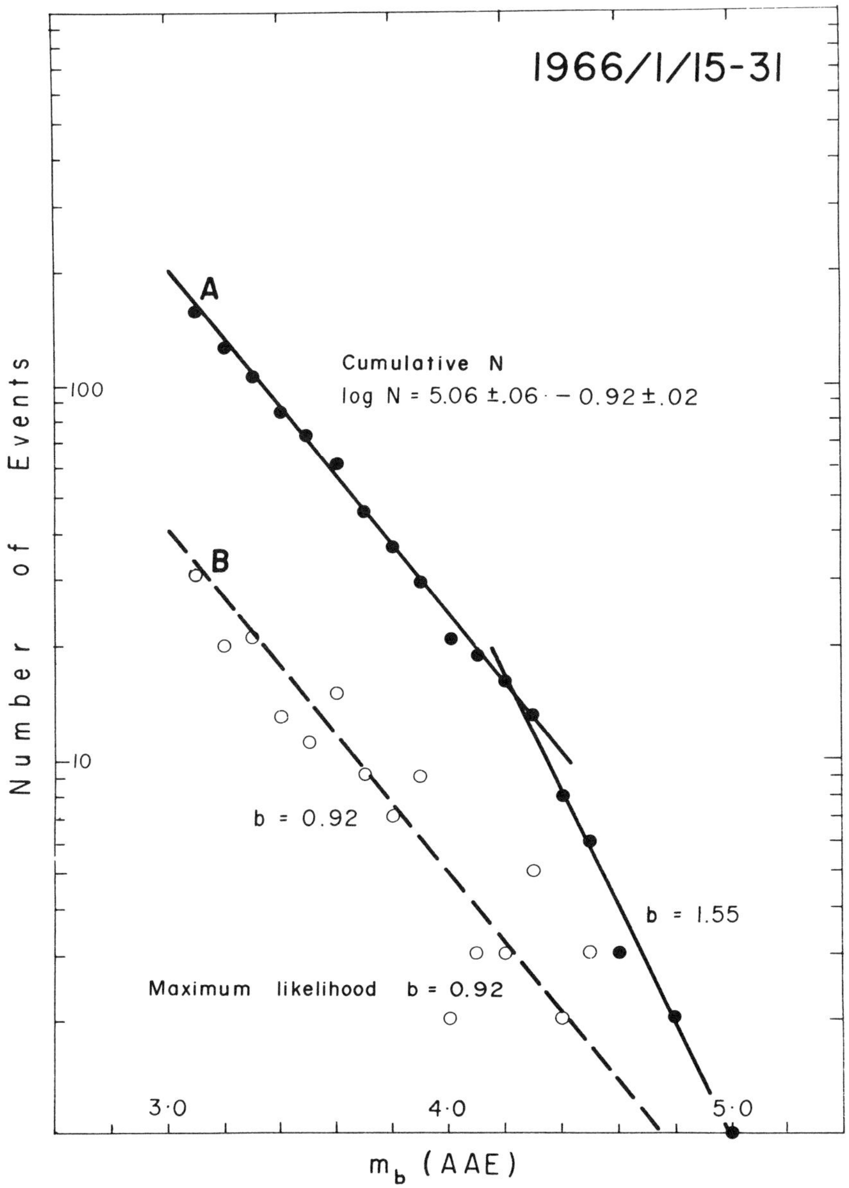

Fig. 138. Frequency-magnitude relationships during the swarm of January 1966 in the Gulf of Tadjoura. Curve A is the best least-squares fit for cumulative N versus m_b; curve B has been obtained by the maximum-likelihood method.

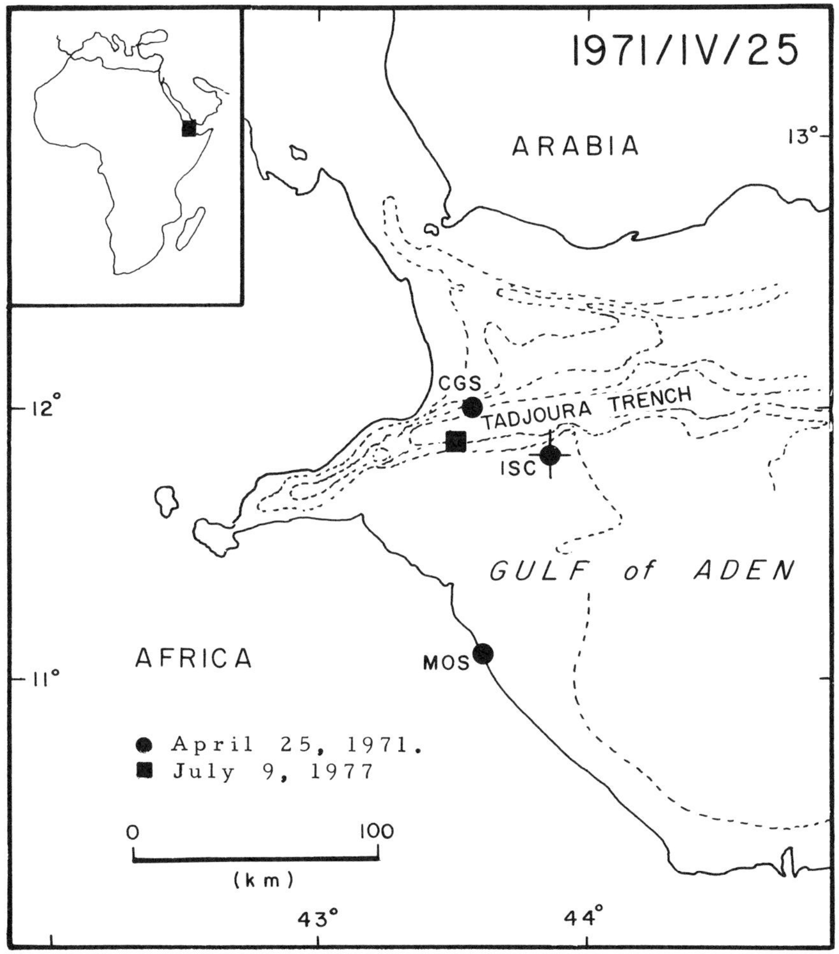

Fig. 139. Location of the three instrumental epicentres available for the earthquake of 25 April 1971.

13 000; 338 of m_b (AAE) $\geqslant$ 2.9 were recorded at Addis Ababa. The largest event was of magnitude m_b (CGS) 5.6.

Tremors were felt throughout the Republic of Djibouti and as far west inland as Dire Dawa (N 09.6°, E 41.9°) in Ethiopia. Djibouti suffered heavy damage.

Sources

AAE Data File; Lépine et al. (1973); Lépine and Ruegg (1973, p. 33–36, 1976, p. 841–845); Dakin (1975, p. 51–70); Ruegg (1975, p. 329–60);

ISS (1973); USCGS (EDR 21-31/73); *Le Réveil de Djibouti*, 31 March and 7 and 14 April 1973.

Comments

1. The Earthquake Sequence

The March-April 1973 earthquake sequence in the Gulf of Tadjoura was comprised of several episodes, each of which featured a larger earthquake that coincided with a general buildup of seismic activity (10–50 shocks per hour), and, curiously enough, was followed about 15 h later by a sudden increase of the activity, which reached 80–100 shocks per hour (Lépine et al. 1973). The outstanding episodes started on 28 March, 1 April, and 7 April.

In Fig. 140, the 6-hourly frequency of earthquakes in the magnitude range $2.9 < m_b$ (AAE) < 5.6 at intervals of 0.1 magnitude units is plotted. Curve A_1 represents the least squares fit to the cumulative number of events for the whole sequence from 25 March to 19 April. For that period, the b-value in Richter's (1958) equation, $\log N = a + bM$, is equal to -0.67 ± 0.01 for the events of magnitude lower than 5.5; the two largest events have been neglected. As the sequence was composed of several episodes whose aftershocks necessarily overlapped in time and increased the total number of smaller shocks which followed, for comparison sake an aftershock-uncontaminated sample was selected. Curve A_2 presents the least squares best fit to the data recorded during the first episode from 28 to 30 March; the regression coefficient is -0.63 ± 0.01, indicating, therefore, that the activity from the first episode had almost died out when the second initiated. Curve B is the noncumulative 6-hourly frequency-magnitude plot for the whole period of activity. The slope of its regression line obtained by Utsu's (1965) maximum likelihood method is -0.65.

All the b-values obtained for the 1973 seismic period by different methods are consistent among themselves (-0.62 to -0.65) and compare relatively well with the 0.73 ± 0.04 value obtained during the October 1963 swarm in the same area (this survey) and with the value of 0.75 obtained by Fairhead and Girdler (1971) for the 1969 Serdo sequence when the activity was associated with strike slip fault movement. They substantially differ from the 0.92 ± 0.02 value obtained for the aftershocks of January 1966 at the southern end of the straits of Bab el Mandeb.

2. The 1973 Seismic Region

Seismic Region Determined by the Local Seismic Network — Figure 141 shows the epicentral distribution determined by the local seismic network, which in 1973 was comprised of three stations: Tadjoura, Arta, and Atar. On this map, it is immediately apparent that most epicentres cluster within a narrow belt (10–15 km wide) highly correlated with the deepest trenches of the Gulf of Tadjoura. These trenches are bordered by normal faults of E-W direction; they are lined by submarine volcanoes and linked

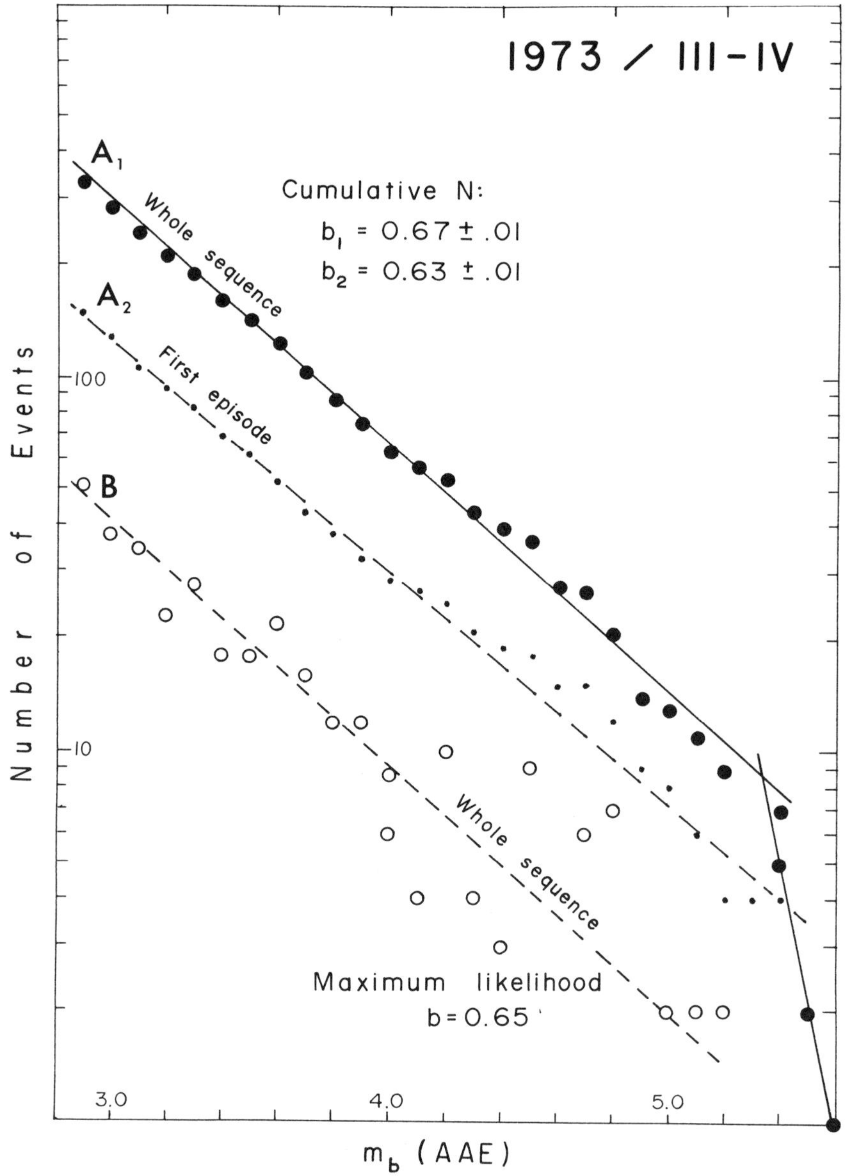

Fig. 140. Frequency-magnitude distributions of earthquakes during the March–April sequence of 1973 in the Gulf of Tadjoura.

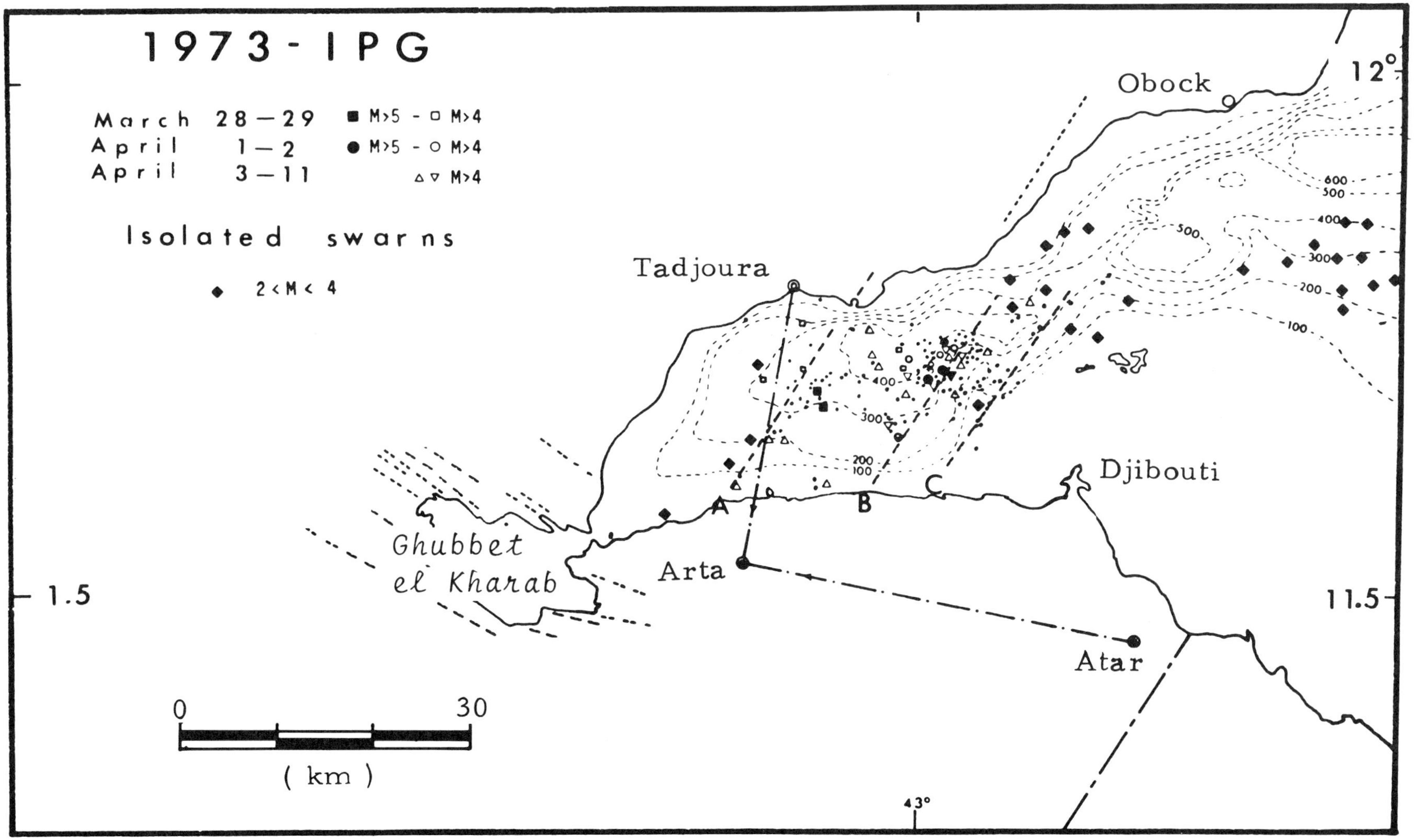

Fig. 141. Epicentre distribution plot during the earthquake sequence of March–April 1973 in the Gulf of Tadjoura as determined from local network records by Lépine and Ruegg (1976).

together by a network of NNE to NE faults that displace the axial trough of the Gulf of Aden, southward, before reaching the Ghubbet el Kharab. The tectonic structure in the Gulf of Tadjoura has been interpreted as a westward extension of the transform fault system and expansion zone that separates the African and Arabian plates (Lépine and Ruegg 1976, p. 841–45).

On Fig. 141 one also observes that most epicentres align either: (1) along the W-E scarps of the trenches as indicated by the bathymetric isolines; or (2) in the vicinity of the SSW-NNE to SW-NE transform faults. Lépine and Ruegg (1976, p. 843) add that some clusters of epicentres coincide with the location of submarine volcanoes that they themselves recently discovered by detailed bathymetric soundings.

Finally, we also observe that while the epicentres during the first episode of activity spread out over a distance of 80–90 km in longitude, the bulk of the other epicentres clustered around transform fault B.

Seismic Region Determined by Teleseismic Records — Fourteen earthquakes of the 1973 sequence were large enough to allow teleseismic determinations of their epicentres; of these, only six could be solved from the records of the local IPG seismograph network. The larger events saturated the network equipment because the sensitivity was intentionally set for shocks of lower magnitudes. For the sake of comparison, the determinations from the U.S. National Oceanographic and Atmospheric Agency, consistently coded CGS in this study, and from the International Seismological Centre at Edinburgh (ISC) have been selected and compared with the determinations of the local seismic network of the Institut de Physique du Globe (IPG). Table 6 lists the epicentre parameters given by these three agencies; the teleseismic data are plotted on Fig. 142.

Comparing the epicentre distributions of Fig. 141 and 142, it is observed that dispersion is much higher in the teleseismic solutions than in the local determinations. The 10–15 km narrow band of epicentres of the IPG pattern widens to 25 and 75 km, respectively, in the CGS and ISC patterns. The IPG distribution is restricted to the confines of the Gulf of Tadjoura; the others spread out over the continent. It is also observed that the CGS solutions are biased toward the northwest with respect to the IPG epicentre distribution; the ISC solutions are biased toward the southeast. This point is illustrated in the vector diagrams in Fig. 142 and 143.

3. Comparison of the Teleseismic Solutions with the Local Network Epicentre Determinations

The March–April 1973 earthquake sequence is the first instance in the history of seismic activity in the Horn of Africa when epicentres could be simultaneously determined from both local and teleseismic records. It is therefore appropriate to compare the epicentre determinations based on teleseismic information to those exclusively calculated from the local network. There is no doubt that the local seismograph network solutions, especially when the events took place within the configuration of the stations, are the most reliable. In the present case, we will compare the six epicentre determinations by the IPG network (Lépine et al. 1972) to those of the international agencies most commonly used in Africa, the USCGS and ISC.

Table 7 compares the teleseismic solutions to one another and then to

Table 6. Gulf of Tadjoura earthquake sequence of 1973/III–IV.

No.	Date	CGS			ISC			IPG		
		N	E	h(km)	N	E	h(km)	N	E	m_b
1	28 March	11.819°	42.698°	33	11.47°	42.85°	33			4.6
2		11.673°	42.830°	33	11.53°	43.07°	37	11.688°	42.903°	5.0
3		11.739°	42.746°	33	11.77°	42.89°	70			5.3
4					11.50°	42.64°	187			5.4
5		11.703°	42.927°	33	11.69°	42.93°	52			5.4
6		11.748°	42.870°	33	11.76°	42.78°	39			5.2
7	01 April	11.613°	42.927°	33	11.42°	43.01°	74	11.707°	43.003°	4.8
8		11.663°	43.035°	31	11.63°	43.60°	65	11.713°	43.022°	5.5
9		11.645°	42.931°	33	11.76°	43.03°	33	11.725°	43.022°	4.9
10		11.819°	42.794°	33	11.9°	43.3°	0			4.7
11	07 April	11.690°	43.021°	33	11.64°	42.95°	33	11.710°	43.033°	4.7
12		11.757°	42.786°	33	11.78°	42.84°	33			4.5
13	11 April	11.783°	42.847°	33	11.68°	43.00°	33	11.663°	42.968°	4.6
14	13 April	11.9°	43.8°	33	11.93°	43.83°	33			4.8

the determinations obtained by the local network. The differences are given in kilometres and directions from the reference points of origin. The numerical values given in Tables 6 and 7 are illustrated by the vector diagrams of Fig. 142 and 143.

Table 7. Comparison of the instrumental epicentres during the earthquake sequence of March-April 1973. (This comparison is expressed in terms of distances (km) and azimuths (0-360°) of the compared teleseismic solutions with respect to reference values. These references are in the first case the CGS solutions and in the second case the IPG solutions.)

		CGS vs ISC			IPG vs CGS & ISC					
No.	Date		km	Az°		km	Az°	km	Az°	St[a]
1	28 March	0	42	157						46
2		0	30	122	0	8	258	25	134	72
3		0	16	078						164
4										
5		0	1	166						151
6		0	10	278						126
7	01 April	0	23	157	0	13	218	32	178	56
8		0	62	093	0	6	166	64	098	199
9		0	17	040	0	13	228	4	013	52
10		0	56	081						12
11	07 April	0	10	234	0	3	211	12	229	83
12		0	6	067						7
13	11 April	0	20	125	0	19	315	4	062	15
14	13 April	0	5	044						75

[a]St = number of stations used in the computations.

Of the 13 teleseismic solutions that could be compared, three practically coincide, five lie within 10 km, and eight are within 20 km. The five others are from 30 to 62 km apart with an obvious southeastward bias of the ISC epicentres with respect to the CGS ones (Fig. 142). Curiously enough the differences do not decrease with larger numbers of reporting stations. Event 8 was reported by 199 stations and yet the ISC-CGS difference in location is the largest, 62 km; it is also the largest error when compared with the IPG local network solution.

Table 7, shows that five of six CGS epicentres agree within 13 km with the IPG local solutions; the outlier is within a range of 20 km. Three ISC epicentres agree within 12 km with the IPG solutions, but the others spread out as far away as 64 km. With respect to IPG, the CGS locations are biased westward; the ISC, southeastward (Fig. 143). It would be imprudent and unrealistic to determine a mean vector correction to be generally applied to the CGS or ISC epicentre determinations for the Horn of Africa. However, it is clear that: (1) the range of possible errors in distance is much higher for ISC determinations, almost 3:1 when compared with IPG-CGS differences; and (2) in probable errors larger than 15 km, the errors are biased to the northwest for CGS and the southeast for ISC.

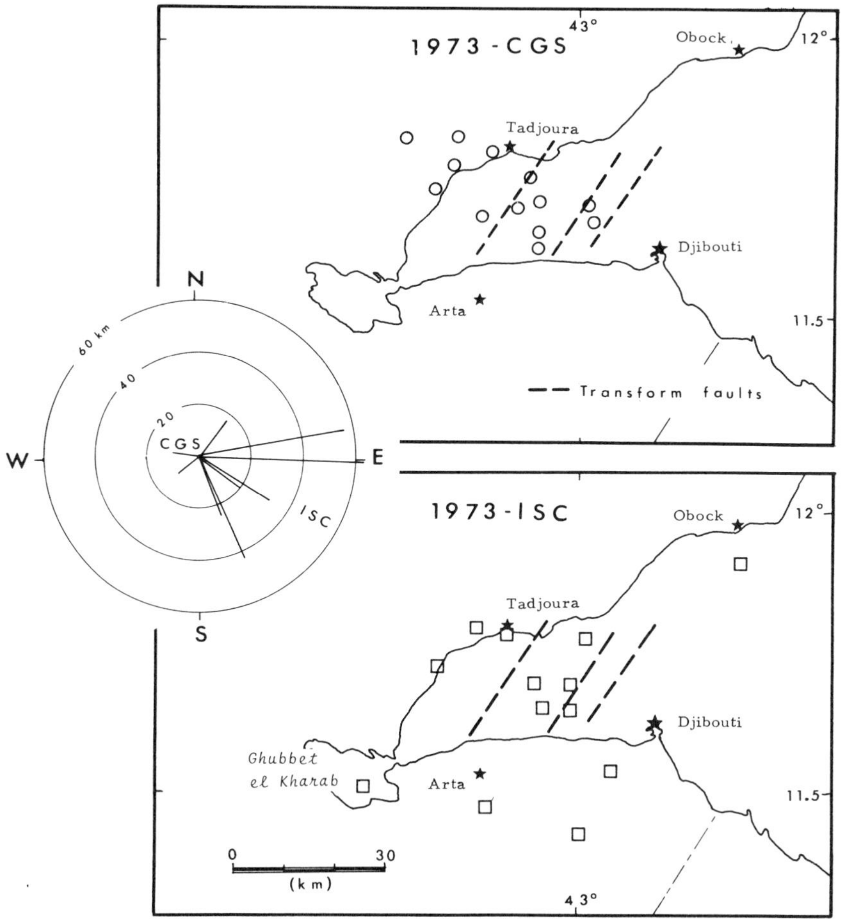

Fig. 142. Plots of the instrumental epicentres obtained by the USCGS and ISC from teleseismic data. The vector diagram given in the inset illustrates the relative locations of the epicentres calculated by ISC with respect to those of the CGS. The distance circles are 20 km apart.

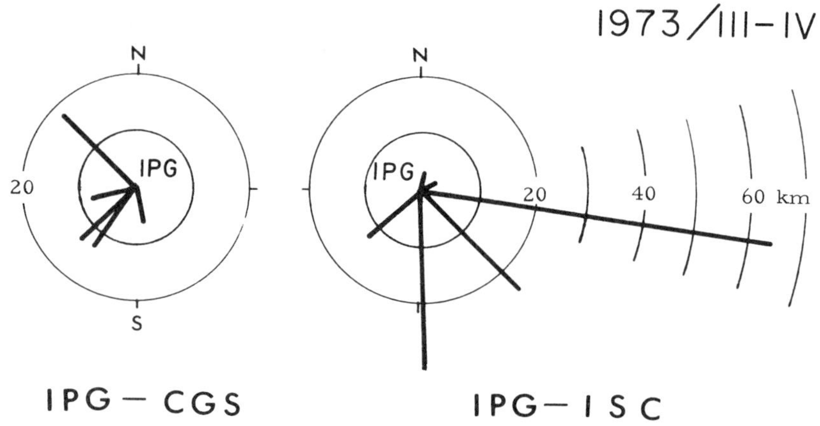

Fig. 143. Vector diagrams showing the differences between the CGS and ISC teleseismic epicentre determinations and the local IPG determinations. The distance between each circle is 10 km.

When no IPG solution is available, it appears perfectly justifiable to adopt, as instrumental epicentral location for an earthquake, the mean of the CGS and ISC parameters because both sets of data are usually biased in opposite directions.

4. Intensity of the Damage in Djibouti

The intensity of the damage to the harbour and houses in Djibouti was higher than normally expected. The explanation is that the structures damaged by the 1960, 1963, and 1965 earthquakes had been simply patched up rather than properly repaired. The government authorities were aware of the situation and of the resulting danger; they prudently ordered the evacuation of the public buildings, schools, and hospitals.

1974/VII/02

On 2 July 1974, the local IPG seismic network in Djibouti located an epicentre of magnitude M_L (IPG) 3.3 off the northeastern coast of Somalia. The location is N 10.89 ± 0.14°, E 43.89 ± 0.14° along the 100-fathom isobath, some 70–75 km SSE of the town of Zeila (Fig. 144).

Source

IPG report, Djibouti.

Comments

More than one epicentre has been located lately in the region of the Somali coast around N 11°, E 44°; among others, see entries 1938/III/11–14, 1940/VIII/14, 1941/III/19, and 1965/V/16. The presence of well-determined epicentres in that area takes up anew the question of the aseismicity of the Somali coast near Zeila commented upon earlier, especially in relation with Taylor's report, entries 1924/IX–XI and 1930/X–XI.

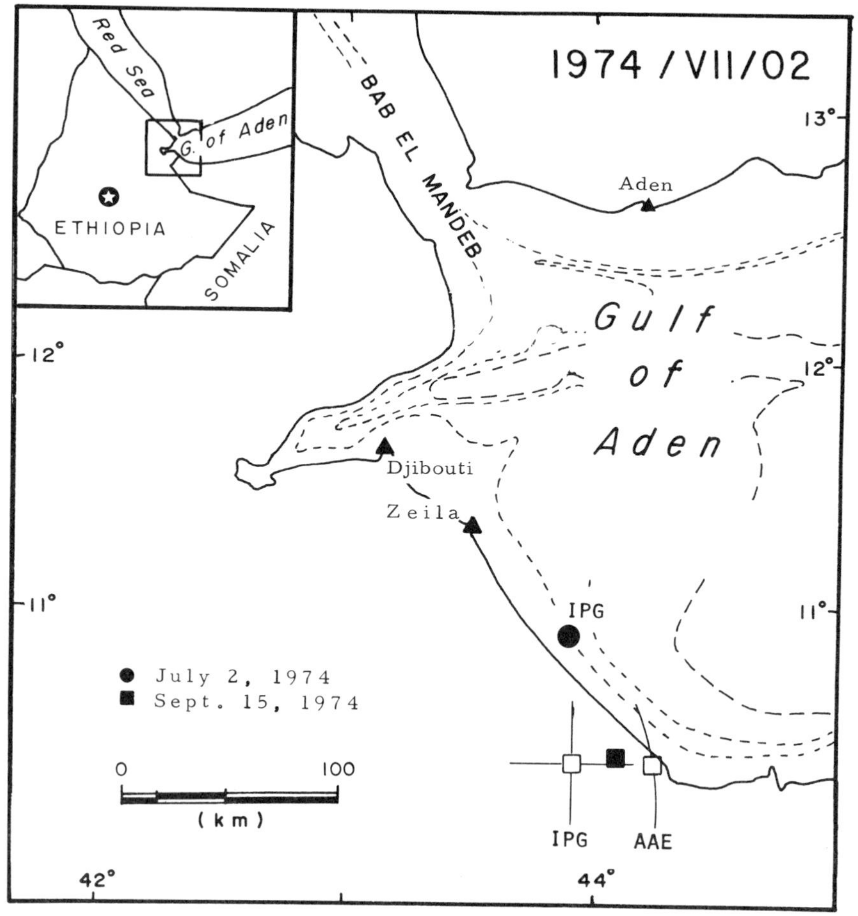

Fig. 144. Location of the 2 July 1974 epicentre off the northeastern shores of Somalia. The solutions for the epicentre of 15 September 1974 are also given.

1974/VII/09

On 15 September 1974, Arta reported an epicentre of magnitude M_L (IPG) 3.3 at N 10.35° ± 0.23°, E 43.95 ± 0.23° in Somalia. For that event, AAE gave a location at about 620 km N75E of Addis Ababa, further east than the IPG position. The mean of both solutions has been adopted: N 10.36°, E 44.15°. The location is plotted on Fig. 144.

Sources

AAE and Arta Data Files.

Comments

Remarks on the aseismicity of Somalia presented in entry 1974/VII/02 apply to this entry as well.

1977/VII/09

The local IPG seismic network in Djibouti reported an earthquake of magnitude M_L (IPG) 3.5 at N 11.867°, E 43.503°, in the Gulf of Tadjoura.

Sources

Lépine (personal communication by radio, same day), IPG-Arta, Djibouti.

Comments

Despite its low magnitude, the event is reported because of the accuracy of its location. Its epicentre is indicated by a full square on Fig. 139.

Earthquake History of the Gemu-Gofa and Turkana (Lake Rudolf) Rifts (Region E)

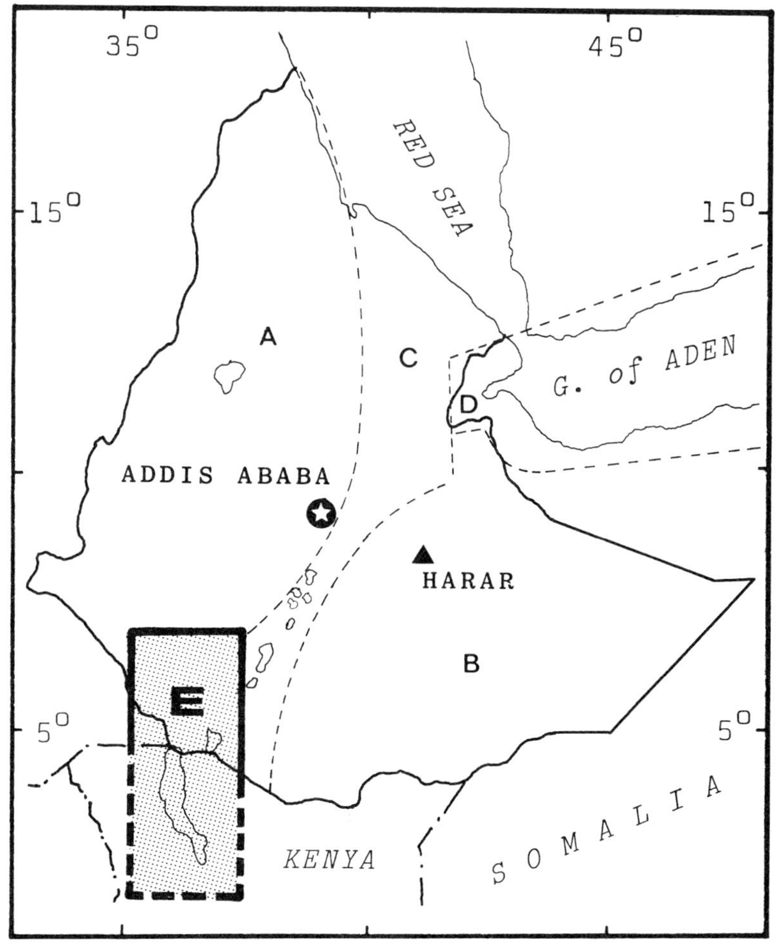

1913/IX/16

An earthquake of magnitude 6.2 was reported near the Ethiopia-Kenya border.

Sources

Gutenberg and Richter (1954); International Seismological Summary Data File (England) and BCIS File (France).

Comments

The instrumental epicentre locations for the 16 September 1913 earthquake calculated by seven different agencies (full circles on inset B, Fig. 145) scattered over 325 km in latitude and 650 km in longitude near the borders of Sudan, Ethiopia, and Kenya. The arithmetic mean value of the seven sets of parameters is N 04.3°, E 36.7°, a site at the south end of the Gemu-Gofa rift.

A recomputation based on reports from 10 unevenly distributed seismic stations was attempted (PG1, courtesy of Prof. Rothé, 1966) and yielded the following parameters: H (U.T.) 11:56:24.6 ± 11.5 s, N 03.56° ± 1.5°, E 36.28° ± 0.6°. Because the recomputation does not carry any more weight (± 1.5°) than the arithmetic mean of the seven original solutions, the latter has been adopted for computation purposes.

The region is seismically active; this is confirmed by at least two recent events: 08 March 1959, U.T. 22:37:16.9, N 03.78°, E 36.94° (Sykes); and 08 Sept. 1960, U.T. 12:05:13.5, N 04.25°, E 37.0° (LWI). The magnitude of the last earthquake was m_b (LWI) 4.5.

1919/VI/30

On 30 June 1919, ISS located two earthquakes at N 06°, E 37° in southern Ethiopia.

Source

Bellamy (ISS, 1936).

Comments

Despite the fact that 1919 epicentre determinations are subject to rather high uncertainties, the ISS locations are accepted at their face value (one full square symbol represents both events on Fig. 145). No alternative is available because the number of station reports is insufficient to justify a reliable recomputation.

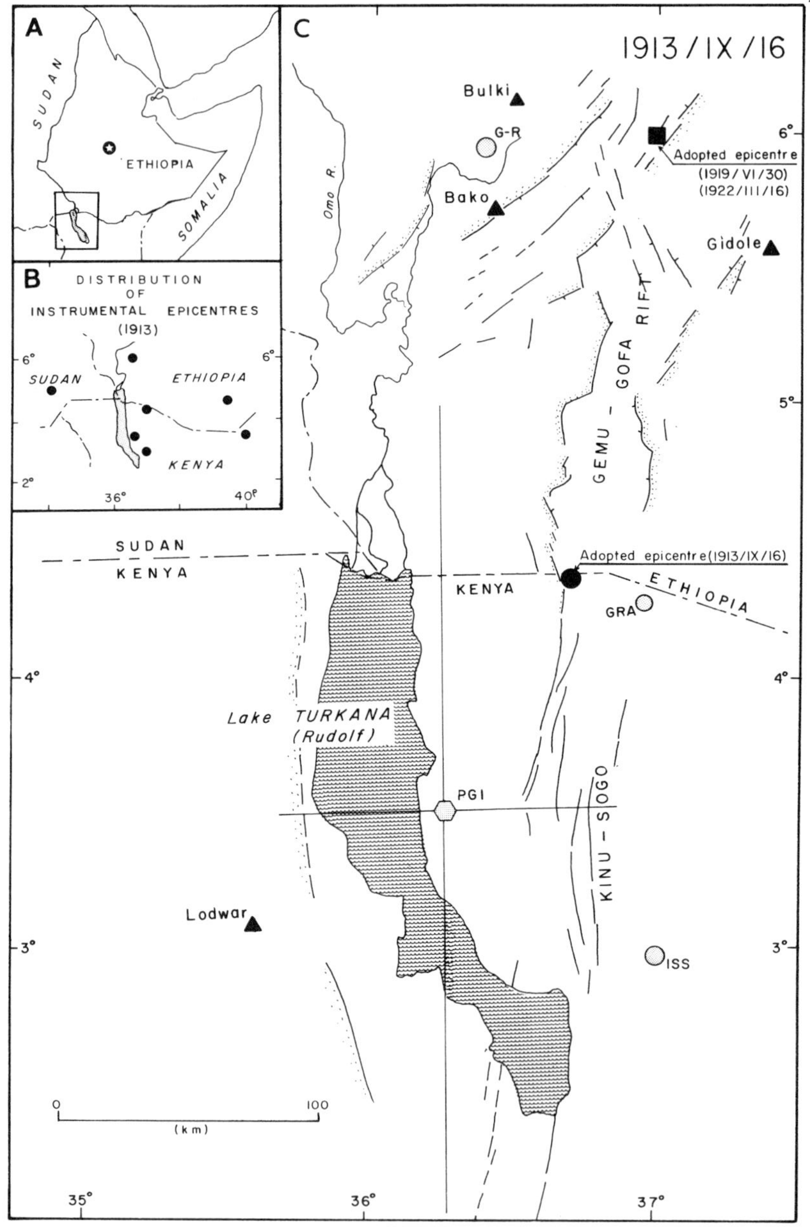

Fig. 145. Location map of the computed and adopted epicentre for the earthquake of 16 September 1913 on the southern Ethiopia border.

1922/III/16

On 16 March 1922, an earthquake was reported from southern Ethiopia.

Source

Bellamy (ISS, 1936).

Comments

The epicentral location given by ISS for this event is the same as for the earthquakes of 30 June 1919, namely N 06°, E 37°. It is accepted as such for the reasons given in entry 1919/VI/30.

1928/X/04

On 4 October 1928, an earthquake of magnitude 6 (G & R) was located by Gutenberg and Richter (1949) and by ISS in the southern part of the Ethiopia Plateau, west of Lake Abaya (Margharita).

Sources

Bellamy (1936); Gutenberg and Richter (1949, p. 205).

Comments

The epicentre of this 4 October 1928 earthquake was given by Gutenberg and Richter (1949) as N 07°, E 38° and by ISS as N 06.5°, E 37°. A recomputation at Strasbourg (PGI, courtesy of Prof Rothé, 1965) gave the new location as N 06.86 ± 0.20°, E 36.85 ± 0.19° for h equals 33 km (Fig. 146). The epicentral location is the Bako-Bulki region, a zone of recent horst-graben development, considered by Mohr (1974, p. 14) to be an expression of major crustal extension outside the Ethiopian rift.

For the computation of the seismicity of Ethiopia, a location at N 06.9°, E 36.9° has been adopted.

1937/XI/30

On 30 November 1937, 94 stations reported an earthquake of magnitude 6¼ at the southern end of the main Ethiopian Rift Valley.

Source

Gutenberg and Richter (1954); ISS 1953.

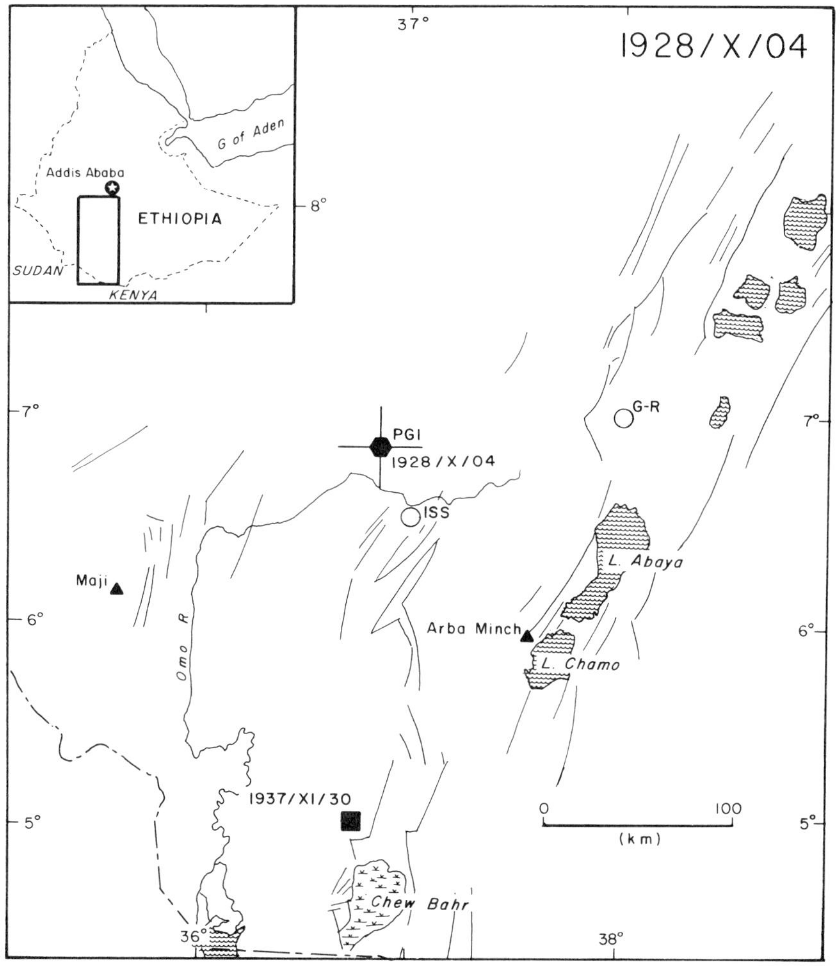

Fig. 146. Location of the three instrumental epicentre solutions for the earthquake of 4 October 1927 in southern Ethiopia. The fault pattern has been taken from the Geological Map of Ethiopia, Geological Survey, Addis Ababa, 1972.

Comments

Six epicentre solutions are available for the earthquake of 30 November 1937: the original solutions from CGS, GUT, and ISS, and three recomputed locations (courtesy Dr Lilwall and Prof Rothé). The maximum dispersion of the six locations is ±0.13° in latitude and ±0.4° in longi-

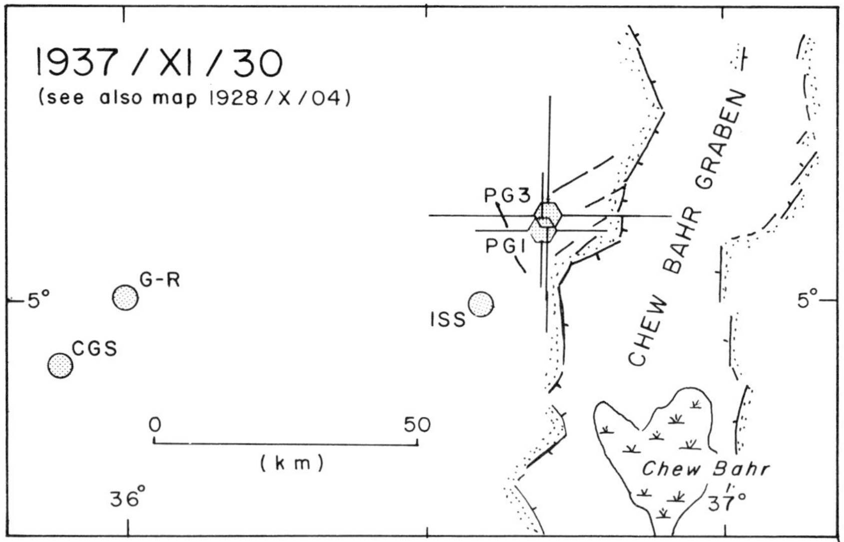

Fig. 147. Location of the original and recomputed instrumental epicentres for the earthquake of magnitude 6¼ on the western shoulders of the Chew Bahr (Stefanie) graben. The tectonic pattern is after Davidson et al. 1973.

tude; the recomputed solutions, surprisingly enough for a 1937 event, agree within better than ± 0.015° (Fig. 147).

The parameters of the recomputed locations are: PG3 (Lilwall in 1974) for h equals 33 km, N 05.15 ± 0.20°, E 36.71 ± 0.20°, PG1 (Rothé in 1965) for h equals 0.33, N 05.12 ± 0.10°, for h equals 0, E 36.71 ± 0.12°, and for h equals 33, E 36.69 ± 0.11°.

The epicentral location is indicated by a solid square on Fig. 146; the tectonic outline of the site on the western escarpment of the Chew Bahr graben and the locations of all solutions are given in Fig. 147. Note that Gutenberg and Richter as well as the USCGS located the event some 60 km away, on the western side of the Turkana rift, south of the Korath volcanic range.

1954/VI/30

An earthquake of magnitude between 5.6 (CGS) and 6.2 (BCIS) occurred in southern Ethiopia, in the Gemu-Gofa (L. Stefanie) rift. Damage of intensity V–VI was reported from Gidole (N 05.6°, E 37.5°). Strong tremors were also reported from Hagare Mariam (N 05.6°, E 38.2°) and Soddu (N 06.9°, E 37.7°). Aftershocks lasted until February 1955.

Sources

ISS (Kew Observatory 1961, p. 370–372); USCGS. Details on the intensity of damage and on the length of the seismicity period are taken from the reports of Miss Mary Frydenlund and Mrs M. Marlow addressed to the author during the survey. In 1954, Miss Frydenlund was stationed at the Norwegian Mission in Gidole and Mrs Marlow, at the S.I.M. Mission in Soddu. The description from Gidole is instructive:

It occurred in Gidole during the last days of June '54; I think it was the 30th, a Wednesday. I was teaching 15 women in my small house. Suddenly, I fell dizzy . . . and understood that there was an earthquake; I rushed the women outside. The time was about 16:30. As the floor of my house was not any higher than the ground, it did not sway much. A plate fell from a shelf in the kitchen. A larger house with concrete foundations oscillated ostensibly, cracked in a few places but was otherwise undamaged. Some masonry houses built in the 1930's were badly cracked and a complete wall from the Abouna's residence collapsed . . .

. . . At the mission station near Hagere Mariam, the same was happening . . . and word came from the town that tremors were also felt there . . .

. . . During the week that followed, there were other strong tremors but not as strong as the first; "murmurs in the ground" and tremors lasted until the next February, that is for 8 months. During the day, we did not notice it too much, but at night we could hear weak noises approaching, shaking the house slightly and passing away. I admit it was somewhat hard on the nerves . . . (Personal communication from Miss Frydenlund, dated Yrgalem, 18 June 1971. The original is in Norwegian; the translation was made by Mrs Hilde Porter).

Comments

1. Epicentral Location

Two original locations based on teleseismic records are available: ISS, BCIS (13:26:51), N 05.8°, E 37.3°; CGS (13:26:50), N 06°, E 37° (Note that NOAA'S earthquake data file indicates N 07° instead of N 06°).

Two recomputations were made, one based on the BCIS data file (courtesy of Prof Rothé in 1965) and one at Edinburgh (in 1974) based on the ISS data file. For an assumed focal depth of 33 km, the recomputed parameters are: PG1 (1965) (13:26:50.9 ± 0.5), N 05.88 ± 0.07°, E 37.22 ± 0.08°, 102 stations; and PG3 (1974) (13:26:56.1 ± 1.0), N 05.60 ± 0.15°, E 37.24 ± 0.12°, 95 stations.

It must be noted that the 1954 original data are remarkably close to the recomputed values. The first recomputation located the epicentre slightly north of the northwest apex of the Tsamahai Basin, a region where NNE-trending major faults close the NW side of the basin and impose a definite NNEasterly direction to the northern sector of the Gemu-Gofa rift. Davidson et al. (1973, p. 16) consider the Tsamahai Basin as an integral part of the

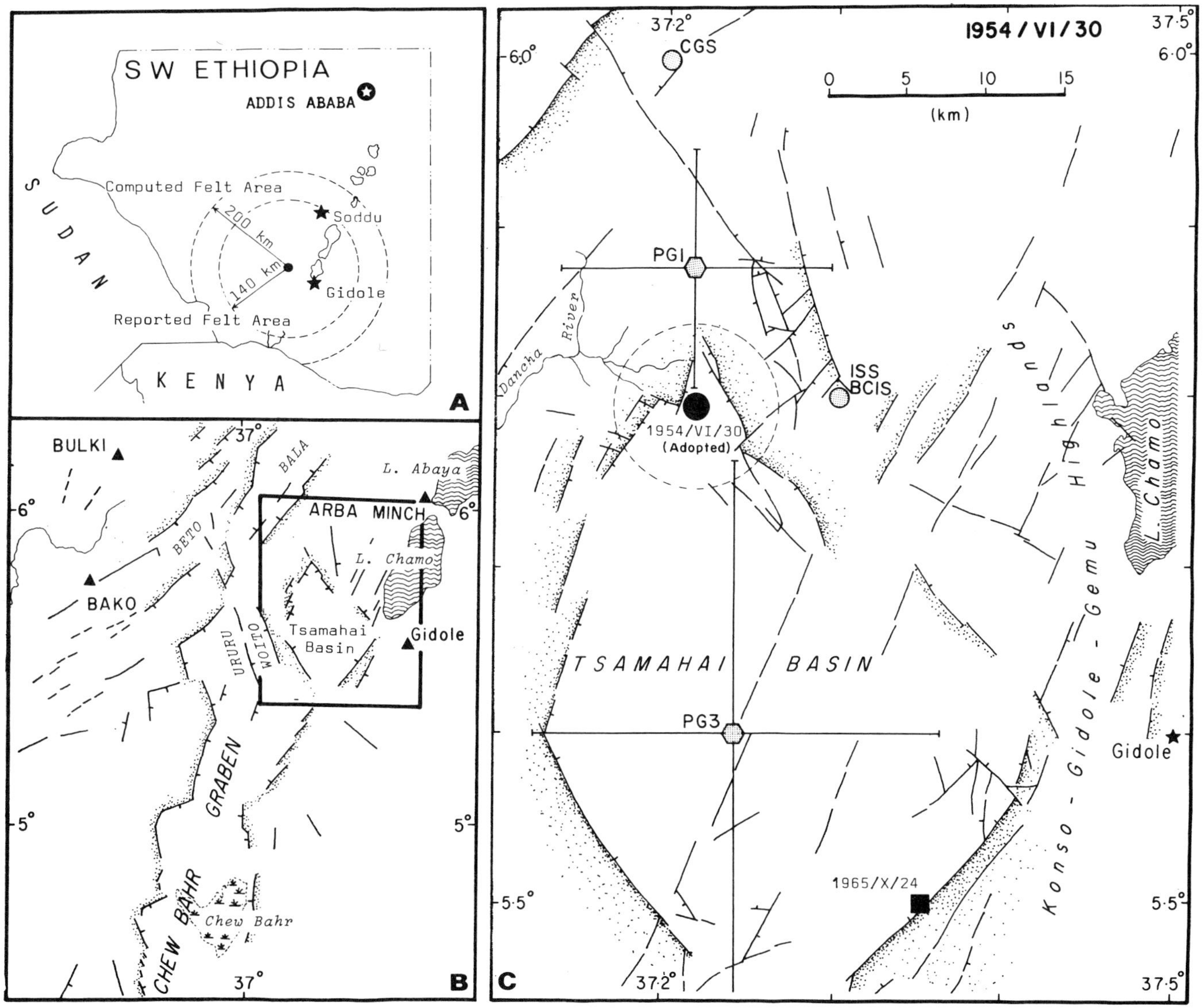

Fig. 148. Location map of the 1954 seismic region and of the epicentre of 30 June. The shaded symbols (●, ⬢)represent instrumental epicentre locations; the stars (★), the sites of reported intensities, the solid circle (●), the adopted location for the epicentre of 30 June 1954, and the solid square (■), the location of a subsequent earthquake in October 1965.

midsection of the Gemu-Gofa (L. Stefanie) rift valley. The second, using a different computing program (SPEEDY) and a different data file located the epicentre on the floor of the basin, 22 km south of PG1 site.

Giving proper weight to the fact that the three original solutions as well as the PG1 recomputed epicentre are located to the north of the Tsamahai Basin, as compared with the 1974 recomputed value near the centre of the basin, a compromise was attained by adopting as epicentral region the northern apex of the basin. Because no surface evidence could be found to decide whether only one or both escarpments of the triangle had been seismically active, a 5-km radius circle was drawn, encompassing the major faults in the area, and its centre, N 05.79°, E 37.22°, was adopted as the epicentral location for the earthquake of June 1954 (Fig. 148).

2. Magnitude and Energy Attenuation Rates

Two magnitudes have been calculated for the earthquake of 30 June 1954: 5.6 (CGS) and 6.2 (BCIS). The range of these magnitude values is indicated by solid bars at the extremities of the M6 curve on Fig. 149. The family of curves plotted on that figure represents the energy attenuation rates for magnitudes M4–M7 under rift crustal conditions as given by the empirical expression:

$$\text{Acceleration} = (0.69e^{1.64M})/(1.1e^{1.10M} + \Delta^2)$$

where: e = constant (2.71827), M = M_s magnitude, and Δ = the epicentral distance in kilometres (Milne and Davenport 1969, p. 737; Gouin 1976, p. 19–22). The conversion of accelerations into intensities is made using the following relation: Intensity = 3 ($\log_{10}$ Acc. + 1.5). Figure 149 also gives the two intensity reports from Gidole and Soddu at epicentral distances of 37 and 140 km, respectively. It is evident that the two points perfectly match the attenuation curve for magnitude 6.1–6.2.

In conclusion: (1) the western-America energy-attenuation curves fit the attenuation observed in the Gemu-Gofa rift region; and (2) the magnitude of the event, based on observed intensities, is 6.1–6.2, for an epicentral location at N 05.79°, E 37.22°. Under these conditions, the maximum felt area defined by intensity III would have been about 200 km in radius.

1955/II/04

On 4 February 1955, at U.T. 05:21, an earthquake of magnitude 5.2–5.5 occurred in southern Ethiopia. No felt reports are available.

Sources

BCIS. (Sykes and Landisman 1964; Fairhead's M.Sc Dissertation, University of Newcastle upon Tyne, 1968).

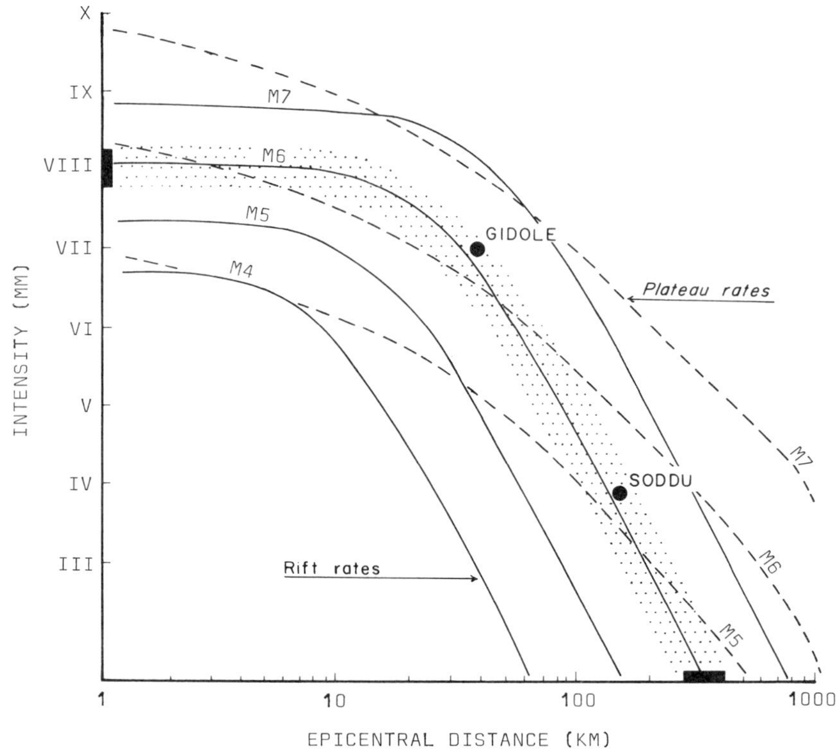

Fig. 149. Attenuation curves adopted for Ethiopia (Gouin 1976). The left sector of the curves corresponds to attenuation rates under rift crustal conditions; the right sector corresponds to attenuation rates under the plateaus. The solid bars and the corresponding stippled surface mark the range of magnitude calculated for the 30 June 1954 earthquake.

Comments

For this event, there are reports from 13 stations distributed over a solid angle of some 255° around the epicentral region, that is from 325° clockwise to 220°. There are no reports from the western sector. Recomputation of these data by Rothé (PG1), Sykes and Landisman (1964), and Fairhead (1968) yielded the following parameters and standard errors for a focal depth of 33 km: PG1 (U.T. 05:21:01.0 ± 1.8 s), N 05.74 ± 0.21°, E 36.43 ±0.35° (the solution for h = 0 changes both coordinates by – 0.04°); S & L (U.T. 05:21:00.9 ± 1.75 s), N 05.80 ± 0.11°, E 36.62 ± 0.11°; Fairhead (U.T. 05:21:04.8), N 05.59 ± 0.03°, E 37.00 ± 0.04°.

The standard errors assigned to Fairhead's JED coordinates are the standard errors of the master event used in his computation; to these

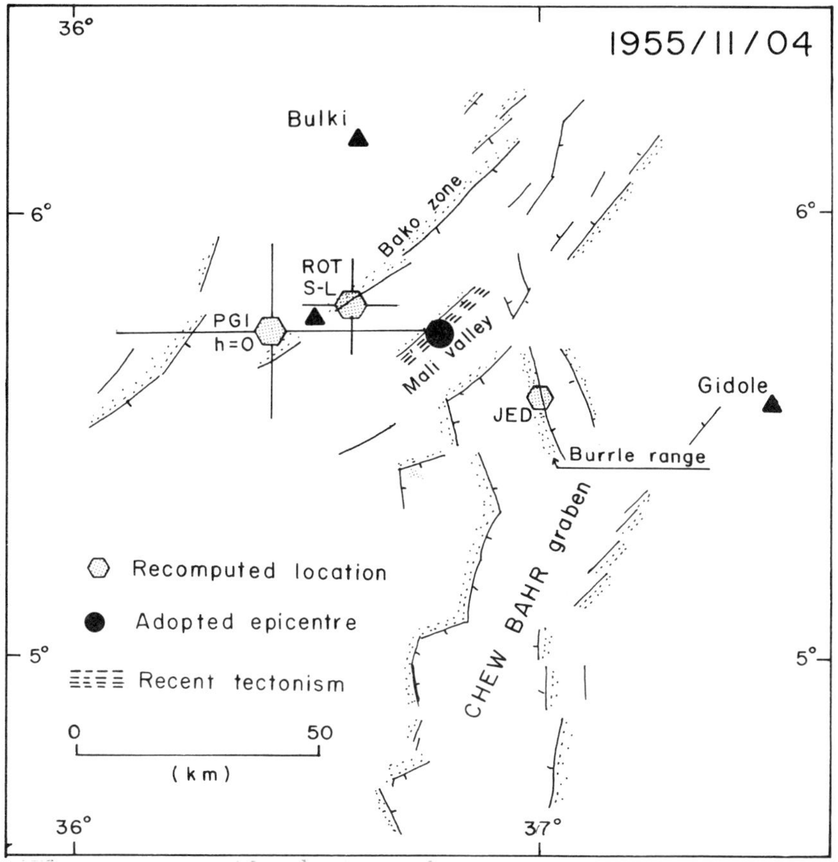

Fig. 150. Map showing the recalculated locations of the 4 February 1955 epicentre in relation to each other and to the major regional fault zones. The solid circle indicates the adopted location in the Mali valley.

should be added additional possible errors due to anomalies in seismic velocities between the location of the master event (S 03.88°, E 35.06°) and southern Ethiopia.

These locations are plotted in Fig. 150. The three locations fall within potentially seismic areas. The first two, near Bako at the southeastern end of strong curvilinear faulting that truncates the northern part of the Gemu-Gofa rift, are in a zone that Mohr (1974a) considers a zone of crustal extension. The third location is the steep escarpment of the Burrle range in the central portion of the same rift. The ambiguity in the choice of a location cannot, therefore, be solved by the instrumental data alone.

Surface evidence reveals dissection of recent clastic sediments at many places in and around the Gemu-Gofa rift; in particular, recent faulting of young red sediments along the steep scarp of the Mali valley. Until other evidence is obtained, the Mali escarpment has been adopted as the location of the 4 February 1955 earthquake at N 05.75°, E 36.80°.

1965/X/24

On 24 October 1965, an earth tremor of estimated intensity IV was reported from Gidole (N 05.7°, E 37.5°) and from the Konso area west of Gidole in southern Ethiopia. The tremors, which severely shook the houses, were accompanied by "roaring noises." No damage was reported.

Sources

AAE Information Cards and Data File.

Comments

The AAE seismograms indicate an epicentral distance of 415–425 km SSW of Addis Ababa (approximate azimuth 205°) and an M_L (AAE) of about 4.8. The calculated epicentral location is N 05.6°, E 37.2° and the origin time (H) U.T. 17:45:49.2.

The P-arrival time at NAI (S 01.27°, E 36.80°) is 17:47:27.3. If a $\overline{V}_P$ of 7.32 km/s along the axis of the Eastern Rift in Kenya is assumed, the epicentral distance from NAI would be 757 km and the point of intersection of the two distance-arcs from AAE and NAI would be at N 05.52°, E 37.37°, 37 km from Gidole where the tremors were felt with intensity IV. The attenuation curves of Fig. 149 indicate a corresponding magnitude m_b 4.3; the original M_L (AAE) was $\simeq 4\frac{3}{4}$.

A safe estimate for the epicentre of 24 October 1965 is the SE escarpment of the Tsamahai Basin, at N 05.50°, E 37.35° (solid square symbol on Fig. 148).

1967/V/07–08

Two earthquakes of M_L (AAE) 4.6 and 5.0 (M(BUL) 4.3) occurred in the vicinity of the western escarpment of the Turkana rift (Fig. 151). The epicentre was outside Ethiopia, 120 km south of the border. Because the intensity was V–VI along the border, the two events have been included in this catalogue.

Sources

AAE Data File; BCIS; Buluwayo Report; ISC (1967, 6/280).

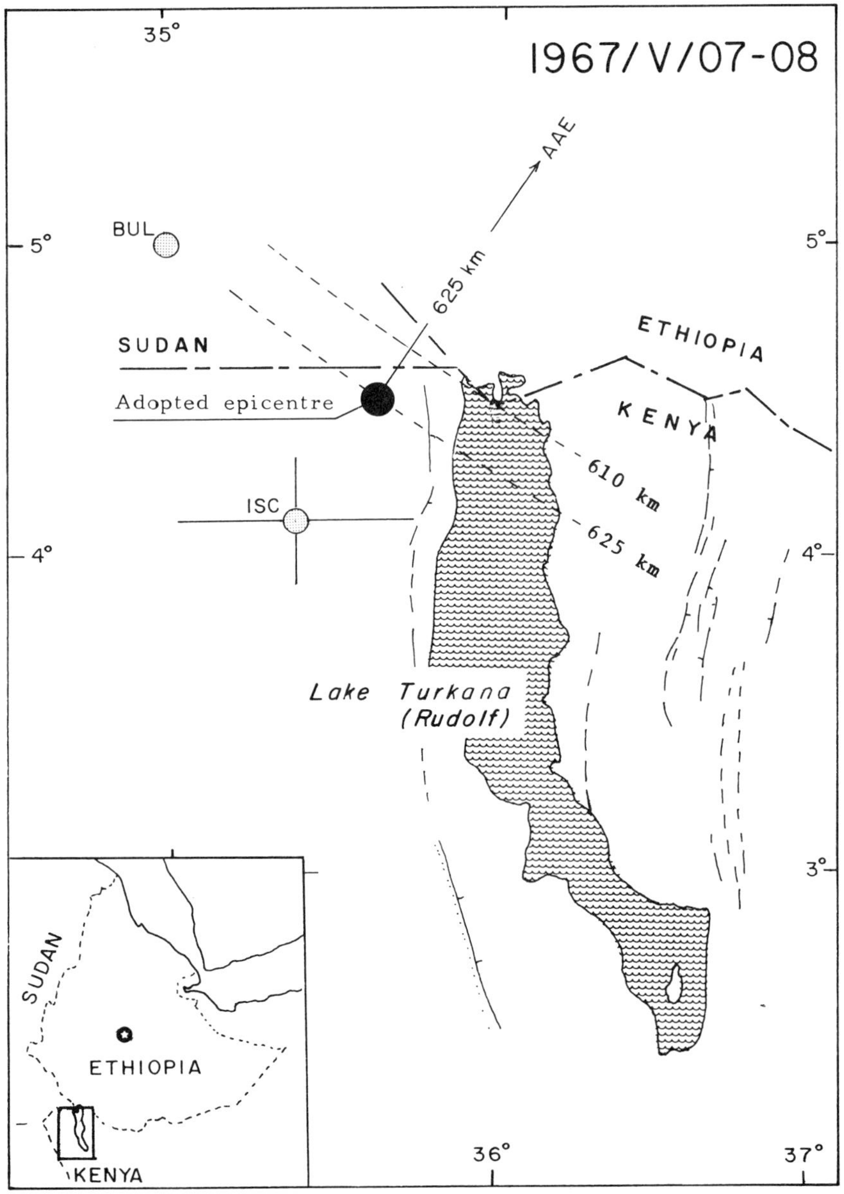

Fig. 151. Adopted location for the epicentres of 7 and 8 May 1967.

Comments

Two teleseismic solutions are available:

	H	Coordinates		M_s	Δ (from AAE)
BUL	13:58:29	N 05°	E 35°	4.3	611 km
ISC	13:58:27 ± 1.6	N 04.10 ± 0.21	E 35.40 ± 0.38		662 ± 40 km

AAE recorded the P-onset at U.T. 13:59:56.0. Assuming the origin times given by BUL and ISC and apparent $\bar{v}_P$ for h equals 0 of 7.32 km/s as observed for the 1965/X/24 event that originated from the same region, the apparent epicentral distance from AAE would be 610 km for H 13:58:29, and 625 ± 12 km for H 13:58:29 ± 1.6. For computation purposes, the coordinates N 04.3°, E 35.7° on the western escarpment of the Turkana (Rudolf) rift have been adopted as the most likely site. The epicentral distance is 625 km at an azimuth of 213° from Addis Ababa. As both events showed the same S-P time difference, they are considered to have the same epicentre.

1967/XII/17

During the evening of 17 December 1967, earth tremors were felt in Mizan Teferi (N 06.9°, E 35.5°), Arba Minch (N 06.1°, E 37.6°), and Kebre Mengist (N 05.8°, 3 39.0°). No serious damage was reported.

The source of the tremors was located some 59 km west of Arba Minch, in the Bala Plains. The epicentre was at N 05.86°, E 37.08°, a distance of 400 km from Addis Ababa.

Sources

AAE Data File and Information Cards; Buluwayo Seismic Report; Fairhead and Girdler (JED, 1970); ISC (1967, 13/533); USCGS (EDR 80-67).

Comments

1. The reported tremors were generated by a first shock that occurred at U.T. 18:03:42.6 ± 2.9 (CGS) on 17 December; a second shock occurred in the same region on the 19th at U.T. 02:30:04.5. No tremors were reported for the latter.

A mean epicentral location at N 05.90 ± 0.10°, E 37.06 ± 0.14° (Fig. 152) has been adopted based on the four teleseismic solutions listed below and the local AAE records:

	Coordinates		Δ°	Δ(km)	Az°	h(km)
BUL	N 06°	E 37°	3.5	389	210	
ISC	N 05.83 ± 0.09°	E 37.10 ± 0.12°	3.6	400	207	
JED	N 05.96	E 36.91	3.6	398	211	
USCGS	N 05.80 ± 0.09°	E 37.23 ± 0.10°	3.6	398	205	19 ± 24

2. The epicentral distance of this shock from southern Ethiopia has been so well determined by the International Seismological Center, Fairhead and Girdler, and the US Coast and Geodetic Survey (400, 398, 398 km, respectively) that the phases recorded at AAE should be used in the construction of regional travel-time tables. These phases recorded at AAE during the earthquake of 17 December 1967 were:

Body waves				Surface waves		
iPn 18:04:39.1		Sn 18:05:20.0		? 18:05:48		(EW)
P*	46	S*	32.0	LR	06	
Pg	52.5	Sg	37.2 (NS)	LQ	07	
			40 (EW)			

3. This earthquake is also indexed in Section C because its tremors were felt not only in southwest Ethiopia but also in the main Rift and as far east as Kebre Mengist on the SE Plateau. The location map in Region C emphasizes the correlation between this epicentre and others that occurred along the Plateau-Rift escarpment near Lakes Abaya and Chamo.

1971/IX/12

At 06:20 on the morning of 12 September 1971, an earth tremor was felt in three of the four *awaradjas* of Gemu-Gofa, southwest Ethiopia (*awradja:* administrative subdivision of a province). Intensities III–IV were reported from the Hamer-Bako region, about N 05.1°, E 36.5°.

Sources

AAE Data File and Information Cards: IRSAC/Lwiro seismic reports.

Comments

The shock was recorded at Addis Ababa (eP) at U.T. 03:18:18 suggesting an epicentral distance of 470 km at an azimuth of 221° and a magnitude M_L (AAE) 4.7. Lwiro (Zaire; S 02° 14.3′, E 28° 48.0′) recorded

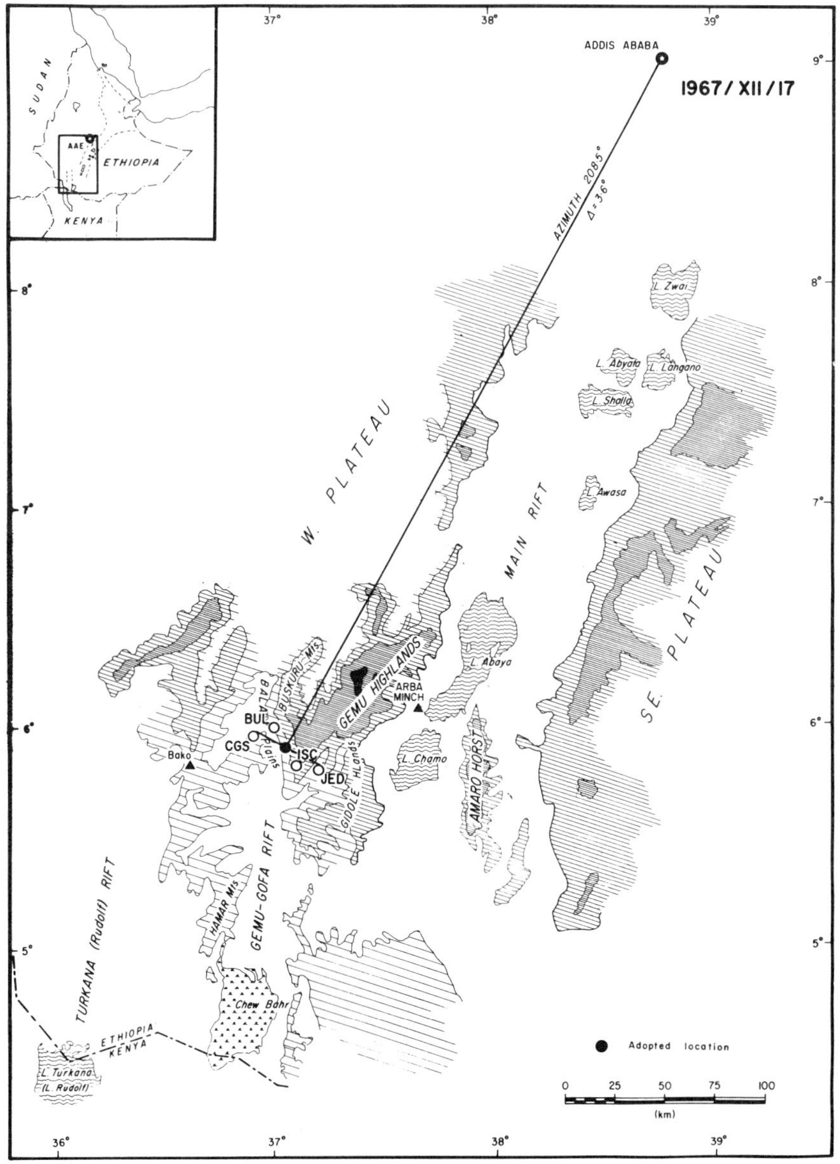

Fig. 152. Instrumental (○) and adopted (●) locations of the 17 December 1967 epicentre at the north end of the Gemu-Gofa rift and along the wall (most probably the western wall) of the Bala Plains.

the P-onset at U.T. 03:19:46.1. The LWI solution yielded an epicentral distance of about 1285 km and a magnitude of 4.8.

The arcs of the two solutions are indicated on Fig. 153; they overlap in southwest Ethiopia. The region in the centre of the confidence area generated by the two arcs is the Bako-Bulki zone at the north end of the northeastern branch of the Rudolf rift. Mohr describes the area as *a zone of extensive deformation of the Ethiopian Plateau ... where horst-graben development ... is witness to major crustal extension acting outside the (main Ethiopian) rift* (Mohr 1974, p. 40).

A realistic location of the epicentre of 12 September 1971 would be N 06.0°, E 36.3°.

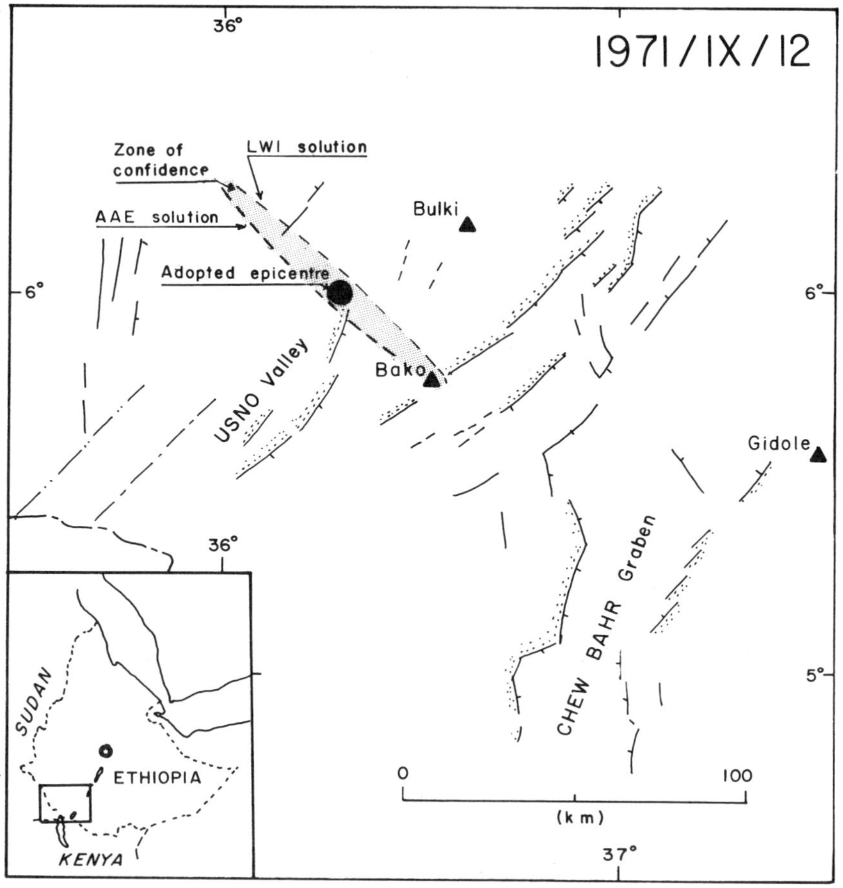

Fig. 153. Adopted epicentre for the earthquake of 12 September 1971. Its location is controlled by two distance-arcs centred on Addis Ababa and Lwiro, respectively.

1972/IV/13

An earth tremor was felt in Gidole (N 05.6°, E 37.5°), southern Ethiopia. No damage was reported.

Sources

AAE Data File; Regional Police Report.

Comments

The original shock was recorded at AAE at 01:52:04.5; the seismic wave phases are not recorded clearly enough to allow a reliable epicentral determination. For the location of Gidole, see Fig. 153.

1973/I/07

In midafternoon on 7 January 1973, earth tremors were reported from Saja (N 07.9°, E 37.5°), Bonga (N 07.3°, E 36.2°), Bulki (N 06.2°, E 36.6°), Hagare Mariam (N 05.6°, E 38.2°), and Yavello (N 04.8°, E 38.1°). Most of the population in Hagare Mariam felt the tremors and extensive rumbling was heard in Bulki (Fig. 154).

The USGS located the epicentre at N 05.266 ± 0.026°, E 36.849 ± 0.044°, a location on the upper margin of the Chew Bahr graben western escarpment (m_b (CGS) 4.9; focal depth 34 ± 8 km).

Sources

AAE Records and Data File; USGS (EDR 3-73).

Comments

The USGS epicentre is 469 km SSW of Addis Ababa (azimuth 207°).

The AAE solution, based on an apparent $\bar{v}_P$ of 7.04 km/s obtained from 24 October 1965, for an event originating in the same area yielded an epicentral distance of 464 km from Addis Ababa at an azimuth of 210°. This confirms the fact that the epicentre is well established within the USGS ellipse of confidence.

EDR indicates an m_b of 4.9 corresponding to M_L (AAE) 5.2 and local m_b 4.7. On the other hand, the apparent radius of the reported felt area is 325 km, a radius normally corresponding to a magnitude 6.6 in rift regions and 5.3 in plateau regions (see Fig. 149). The anomaly in the amplitude of the felt area is too large to be realistic for the earthquake of U.T. 12:17. AAE seismograms show a second event of lower magnitude 40 min later; its traces are too much distorted by the tail waves of the first earthquake to allow a good epicentral determination. Nevertheless, it is probable that the reports from Saja and Bonga refer to the second event, which must have been located at the northern end of the Gemu-Gofa rift valley.

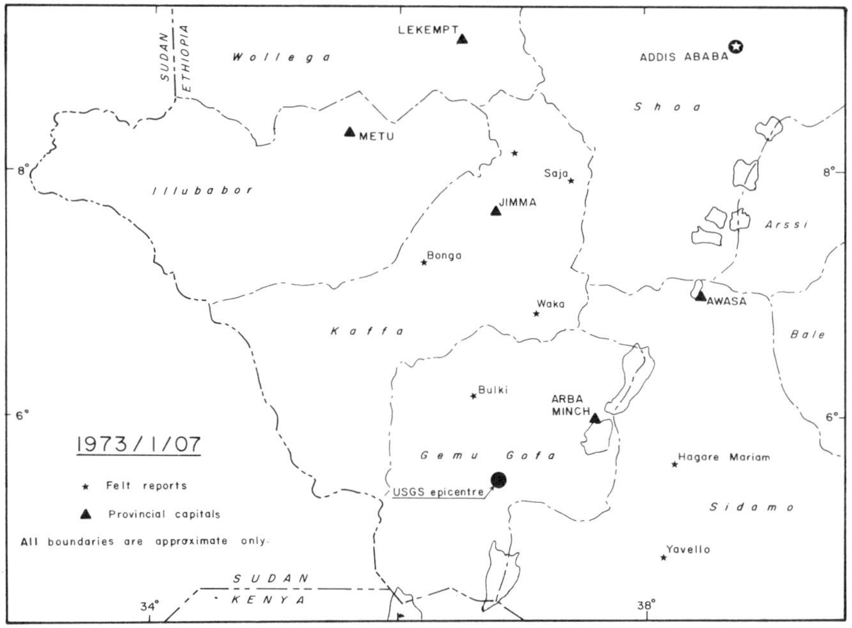

Fig. 154. Location of the instrumental epicentre calculated for the earthquake of 7 January 1973, in southwest Ethiopia, and of the sites from where tremors were reported to the Geophysical Observatory.

1974/III–IV

Earth tremors were reported from southern Ethiopia during March and April 1974. The reports came from Bulki (Feleke Neway, N 06.2°, E 36.6°), Hagare Mariam (N 05.6°, E 38.3°), Warancha (N 06.3°, E 38.4°), and Arba Minch (N 06.2°, E 37.7°). No damage was observed.

Sources

AAE Data File and Information Cards.

Comments

The tremors reported refer only to a few earthquakes out of 57 identified on the AAE seismograms as originating from the same region between 9 March and 29 April 1974. The epicentral distances ranged from about 300 to 400 km in a general direction slightly west of south from Addis Ababa. Their magnitudes M_L (AAE) were estimated between 2.5 and 5.0. No apparent pattern of migration of the epicentres within the seismic region along known fault systems was observed.

The data obtained from the information cards are not accurate enough either to relate the felt tremors to one particular shock of the swarm nor are they complete enough to draw isoseismal maps.

Combining the instrumental data and the macroseismic evidence, it can safely be inferred that the centre of activity was located along the western escarpment of the main Ethiopian Rift, somewhere between Chencha and Arba Minch.

These events are indexed under Region C, entry 1974/III–IV.

1974/VI/30

Earth tremors of intensity III were felt in Chencha (N 06.3°, E 37.7°) on 30 June 1974.

Source

AAE Information Cards.

Comments

The distance from Chencha to Addis Ababa is 340 km; this means that no earthquake with an epicentre near Chencha and a magnitude $M_L \leqslant 4$ would have registered a detectable trace at AAE. The location of Chencha is indicated on Fig. 113. (Chencha is the name of the older capital city; Arba Minch is the new capital.)

1975/XI–XII

There were major landslips at Bedero, Gemu-Gofa Province. A total of 86 out of some 750 families from the villages of Ela and Tohaparo (about N 06.2°, E 36.3°) had to abandon their homes as a result of major landslips that occurred on two consecutive years at the end of the rainy season. The area, named Bedero, is located about 20 km WSW of Feleke-Neway (Bulki), capital of the Gofa *awradja*.

Sources

Report on the investigation of landslips in the region of Bedero, Gemu-Gofa Province, submitted to the Relief and Rehabilitation Commission by Shaka G. Yohannee, Mola Bayeneh, F. M. Dakin, and M. Gilbert, December 1975, Addis Ababa; *Ethiopian Herald*, 26 December 1975, Addis Ababa.

Comments

The Bedero region is situated on Precambrian rock overlain by Tertiary volcanics generally dipping WSW and highly dissected into alternating steep-sided valleys and ridges by tributaries of the Omo river. Between Ela and Tohaparo, the ridge slope reaches 40°. Below the summit of the ridge, cracks opened; six of the largest ones were up to 30 cm in width, 250 cm in depth, and 500 m in length. Three thick mudflows poured down the slope and spread onto the Tohaparo terrace. The total area affected by cracks and mudflows measured about 2 km^2.

The landslips were caused by percolation of rainwater into the thick topsoil at Ela, washing out mud and weathered bedrock from beneath. This material poured out causing slope failures and mudflows. It should be emphasized that such destructive slope failures are common in Ethiopia.

Chronological Listing of Seismic and Volcanic Events

This section presents a computer printout of the information on seismic and volcanic activity within Ethiopia and the Horn of Africa for which some parameters could be quantitatively evaluated. Confidence limits are described at corresponding entries in Part I. The fields of information are the following:

Date of the Event(s)

Year, month, and day (Julian-Gregorian calendar).

Time of Day (H1)

*	For historical events or reported felt tremors: unless otherwise stated, the time indicated is the local mean time (U.T. + 3 h) or the apparent solar time. When the time indication is not definite in the original documents, the period of the day is indicated as follows: NT — nighttime; SR — sunrise; AM — morning; MD — midday; PM — afternoon; and SS — sunset.
U.T.	Indicates the origin time unless the numerals are followed by a P, in which case it indicates the P-phase recorded time at a particular station.

Geographical Coordinates

Greenwich system. Latitudes are North and longitudes East.

Focal Depth

H2	Given in kilometres.

Magnitude

MB	Body-wave magnitude either from teleseismic records or converted from local magnitude (see Explanatory Notes).
MS	Surface-wave magnitude from teleseismic records.
ML	Locally determined magnitude.
ME	Magnitude estimated from noninstrumental descriptions.

Intensity

I	Mercalli-Modified 1931 scale. Arabic numerals are used instead of Roman ones.

Location and Comments

Region

R	Refers to the five (A,B,C,D, and E) regional subdivisions used in Part I.

Reporting Agency (A)

BCIS	Bureau Central International de Sismologie, Strasbourg.
CGS	U.S. Coast and Geodetic Survey. To avoid confusion, this code is used throughout this file for all the NOAA agencies that published the PDE and EDR reports.
IPG	Djibouti network operated by the Institut de Physique du Globe, Paris.
ISC	International Seismological Centre, Edinburgh.
ISS	International Seismological Summary, England. The agency operated from 1913 to 1963.
JED	Acronym applied to agencies using the joint epicentre determination technique developed by Douglas, 1967. Mainly refers to Fairhead and Girdler.
LWI	Lwiro, Zaire. Formerly IRSAC: Institut pour la Recherche Scientifique en Afrique Centrale.
MOS	Institute of Physics of the Earth, USSR.
PG1	Recomputation. Courtesy of Prof J. P. Rothé using the BCIS computer program.
PG2	Recomputation. Courtesy of Dr Lynn Sykes of Lamont-Doherty Observatory, N.Y.
PG3	Recomputation. Courtesy of Dr R. Lilwall, Institute of Geological Sciences, Edinburgh.
SIE	A. Sieberg. Not to be confused with SIE: Siena station, Italy.
TLN	Tendaho local network in central Afar operated by the University of Durham. Not to be confused with TLN: Talang station, Sumatra.
ZZZ	Adopted epicentre parameters based on the information available to the author of this survey.

Note: The identification code for other seismic stations is that of the U.S. National Oceanic and Atmospheric Administration.

Y	M	D		H1	N	E	H2	MB	MS	ML	ME	I	LOCATION & COMMENTS	R	A
1400			*		13.50	41.87					6		DUBBI VOLCANO	C	ZZZ
1431			*										TREMORS FELT IN TIGRAY	A	
1433			*		14.1	38.7							N TREMORS FELT IN AXUM	A	
1518			*										N TREMORS FELT IN TIGRAY	A	
1553			*										N TREMORS FELT IN TIGRAY	A	
1593	09		*		13	39.5							TREMORS FELT IN GOJJAM	A	
1608	12	23											SECONDARY VOLCANIC ACTI-	C	
1608	12	23											VITY AT DOMA'ALI	C	
1631	02	10	*		11.25	41.70							VOLCANIC ACTIVITY AT DOMA	C	
1631	02	10	*										ALI. EARTH TREMORS	C	
1631	02	13	*	AM	11.25	41.70							FOOT OF DOMA'ALI	C	
1631	02	14	*	NT	11.25	41.70							FOOT OF DOMA'ALI	C	
1631	02	25	*	SS	11.25	41.70					5.7		FOOT OF DOMA'ALI. CASUAL-	C	ZZZ
1631	02	25										9	TIES. CITY DESTROYED.	C	ZZZ
1632	06		*										GONDAR (BEGHEMDER)	A	
1667	10	22	*		12.5	37.5							FELT IN GONDAR	A	
1703	09	10			12.5	37.5							GONDAR DAMAGED BY TORNEDO		ZZZ
1703	09	10											CASUALTIES 30.	A	ZZZ
1733	11	30	*		15.7	39.0					6		ERITREAN SCARP (N)	A	ZZZ
1733	11	30	*		15.7	39.0							ERITREA. LANDSLIDES AND		
1733	11	29	*										CASUALTIES	A	
1799	08	18	*	NT	12.5	37.5							FELT IN DEMBIA	A	
1802	05	28	*		12.6	37.5							FELT IN GONDAR	A	
1809	02	26	*		12.5	37.5							FELT IN GONDAR	A	
1818	06	30	*		14.2	38.8						6	FELT IN ADUA	A	
1832	05	04	*		15.0	39.5						4.5	FELT IN HALAI	A	
1836			*		11	38							FELT IN DEMBIA AND BE	A	
1836			*										GHEMDER	A	
1836			*		11.5	39.75						3.5	FELT IN FOGARA (3)	A	

Y	M	D		H1	N	E	H2	MB	MS	ML	ME	I	LOCATION & COMMENTS	R	A
1838	02	23	*	23 18	15.6	39.5						4	FELT IN MASSAWA	C	PAL
1838	02	25	*	23 20	15.6	39.5						3	FELT IN MASSAWA	C	PAL
1838	02	27	*	22 25	15.6	39.5						5	FELT IN MASSAWA	C	PAL
1838	02	27	*	22 48	15.6	39.5						4	FELT IN MASSAWA	C	PAL
1838	02	27	*	23 02	15.6	39.5						3	FELT IN MASSAWA	C	PAL
1838	03	01	*	03 00	15.6	39.5						6	FELT IN MASSAWA	C	PAL
1838	03	04	*	03 46	15.6	39.5						6	FELT IN MASSAWA, MONCULLO	C	PAL
1838	03	05	*	07 24	15.6	39.5						3	FELT IN MASSAWA	C	
1838	03	22	*	13 38	15.6	39.5						3	FELT IN MASSAWA	C	PAL
1841	04	23	*		09.5	39.8						4	NEAR ANKOBER (2)	A	
1841	04	24	*	AM	09.5	39.8						4	NEAR ANKOBER	A	
1842	07	07	*		09.5	39.8						5	ANKOBER (LANDSLIDES)	A	
1842	12	08	*		09.5	39.8						9	ANKOBER DESTROYED	A	
1844	10	23	*		15.6	39.4						5	FELT IN MUNCULLO	C	PAL
1844	10	23			15.5	39.5							MASSAWA CHANNEL	C	ZZZ
1845	02	12	*	PM	12.25	39					6.5		WOLLO. FELT RADIUS =650-	A	
1845	02	12											700 KM	A	ZZZ
1845	12	07	*	PM	09.3	37.2						4	4 TREMORS IN RARE + NOISE	A	
1846	01	26	*	17 06	08.2	37.2						3	FELT IN SAQA. KAFFA	A	
1848	08				15.5	39.6					4.		MASSAWA CHANNEL	C	
1848	07		*										RED SEA COAST,2 MONTHS OF		
1848	07		*										ACTIVITY IN MASSAWA	C	
1848	07		*										CHANNEL	C	
1848	07	13	*	MD	15.5	39.4							FELT ON W COAST GULF ZULA	C	
1848	08	01	*	NT	15.6	39.4						5	FELT IN MUNCULLO	C	
1848	08	02	*	02	15.6	39.4						4	FELT IN MUNCULLO	C	
1848	08	03	*	NT	15.6	39.4						3	FELT IN MUNCULLO	C	
1848	08	06	*	15	15.6	39.4						3	FELT IN MUNCULLO, MASSAWA	C	

Y M D		H1	N	E	H2	MB	MS	ML	ME	I	LOCATION & COMMENTS	R	A
1848 08 10	*	PM	15.6	39.4						4	FELT IN MUNCULLO	C	
1853 10	*		13.7	39.6						5	FELT IN AGULA, TAMBIEN	A	
1854	*										EARTHQUAKES IN TAMBIEN	A	
1854 02 02	*	MD	13.6	39.4						3	FELT IN BETALIHEM	A	PAL
1854 02 04	*	SR	13.6	39.4						3	FELT IN BETALIHEM	A	PAL
1854 02 05	*	SR	13.6	39.4						3	FELT IN BETALIHEM	A	PAL
1854 02 15	*	03 00	12.8	39.0					6.3		NEAR GUF GUF GRABEN	A	ZZZ
1854 02 15	*	MN	13.6	39.4						3	FELT IN BETALIHEM	A	PAL
1854 02 15	*	NT	13.7	39.6						6	FELT IN AGULA	A	
1854 02 15	*	NT	13.0	39.4						8	AMBA AFAJI (CRACKS)	A	
1854 02 16	*	NT	13.6	39.4							FELT IN BETALIHEM	A	PAL
1854 02 21	*		13.6	39.4							4 FELT IN BETALIHEM	A	PAL
1857 04	*										FREQUENT TREMORS IN	AC	PAL
1857 04	*										NORTHERN ETHIOPIA	AC	PAL
1857 04	*									4	FELT IN MASSAWA	C	PAL
1857 04	*										TOWN REPORTED DESTROYED	A	PAL
1857 04	*										IN TIGRAY	A	PAL
1861 05 08			13.74	41.55			5.5				VOLCANO DUBBI	C	GUT
1861 05 08	*		13.5	41.8							ERUPTION OF DUBBI VOLCANO	C	
											OVER 100 CASUALTIES	C	
1861 05 08											ACTIVITY LASTED MONTHS	C	
1861 05 08	*		13.9	41.6						5	EDD + 50 CM ASH + LAPILLI	C	
1861 05 08	*		14.8	42.9						3	HODDEIDA: TREMORS + ASH	C	
1861 05 08	*		15.6	39.5						3	MASSAWA: TREMORS + NOISE	C	
1861 05 08			13.3	43.3						3	MUKKA: TREMORS, ASH,NOISE	C	
1861 05 08			12.6	43.4							PERIM IS: NOISES	C	
1864 03 05	*	20 00	15.6	39.5						3	FELT IN MASSAWA	C	
1864 09 14	*	11 15	15.6	39.5						3	MASSAWA: FELT ON LAND AND	C	
1864 09 14	*										IN HARBOUR ON SHIP	C	PAL
1864 09 15	*	13 45	15.6	39.5						3	MASSAWA: FELT ON LAND AND	C	
1864 09 15	*										IN HARBOUR ON SHIP	C	
1864 10 21	*	08 45	15.6	39.5						5	FELT IN MASSAWA	C	
1875 11 02	*		16	38.5					6		ANSEBA FAULTS. AFTER-	A	
1875 11 02	*										SHOCKS TILL MARCH 1876	A	
1875 11 02	*										LANDSLIDES, CASUALTIES		
1875 11 02	*	11 00	14.2	38.9						6	ADUA (ROCKSLIDE)	A	
1875 11 02	*		15.8	38.5						8	KEREN (DAMAGE)	A	

Y	M	D		H1	N	E	H2	MB	MS	ML	ME	I	LOCATION & COMMENTS	R	A
1878	11		*										TIGRAY 6 DAYS OF TREMORS	A	
1880	06	30	*		14.1	38.7						6	TIGRAY. FELT IN AXUM	A	
1883	03				12	42					5.5		AUSSA	C	ZZZ
1883	03		*		13.0	34.5							REPORTED FELT IN ASSEB	C	
1883	03		*		10.5	41.1							FELT AT MT JANGHUDI	C	PAL
1884	01	02	*	NT	09.3	42.2						3	FELT IN HARAR	B	
1884	07		*		15.6	39.5							MASSAWA: 3 MONTHS OF	C	
1884	07		*										TREMORS. DAMAGE.	C	
1884	07	12	*	AM	15.6	39.5						3	MASSAWA	C	PAL
1884	07	12	*	NT	15.6	39.5						3	MASSAWA	C	PAL
1884	07	20	*	08 30	15.6	39.5						3	MASSAWA	C	PAL
1884	07	20		06 45	16	41			5.9					C	GUT
1884	07	20		06 45	15.7	39.6	15	5.9					MASSAWA CHANNEL	C	ZZZ
1884	07	20	*	09 30	15.6	39.5						8	MASSAWA.HEAVY DAMAGE.	C	
1884	08	18	*	09 00	15.6	39.5						5	FELT IN MASSAWA	C	PAL
1884	10	03	*		15.6	39.5						4	FELT IN MASSAWA	C	PAL
1884	10	14	*	04	13.5	39.5							TREMORS + SUBSIDENCE NEAR	A	PAL
1884	10	14	*									6	MEKELE	A	PAL
1886	05	08	*	PM	15.6	39.5						4	MASSAWA:4 TREMORS	C	PAL
1886	06	15	*	09 45	15.6	39.5							FELT IN MASSAWA	C	PAL
1886	08	08	*	SS	15.6	39.5						4	MASSAWA E MUNCULLO (2)	C	PAL
1886	12	07	*	06 25	15.6	39.5						4	FELT IN MASSAWA	C	PAL
1886	12	23	*	18 10	15.6	39.5						3	FELT IN MASSAWA	C	PAL
1887	01	30	*	23 44	15.6	39.5						4	FELT IN MASSAWA	C	PAL
1887	02	01	*	02 00	15.6	39.5						3	FELT IN MASSAWA (2)	C	PAL
1887	03	14	*	10	15.6	39.5							FELT IN MASSAWA	C	PAL
1887	12	28	*	02 45	15.6	39.5						4	FELT IN MASSAWA	C	PAL

Y	M	D		H1	N	E	H2	MB	MS	ML	ME	I	LOCATION & COMMENTS	R	A
1889	10	04	*	NT	15.6	38.5						3	FELT IN KEREN	A	
1891	02	12	*	MD	15.6	39.5						4	2 FELT IN MASSAWA	C	PAL
1891	04	27	*	04 40	15.6	39.5						3	FELT IN MASSAWA	C	PAL
1892	11	23	*		15.6	39.5						3	N FELT IN MASSAWA	C	PAL
1894	05	13	*	06 45	15.6	38.5							KEREN: TREMORS + NOISE	A	PAL
1894	07	04	*	NT	15.6	38.5							KEREN (NOISE)	A	PAL
1894	09	12	*	12 50	15.6	39.5						4	FELT IN MASSAWA	C	PAL
1894	11	03	*	23	15.0	39.4						5	FELT IN HALAI	A	PAL
1894	11	08	*	SR	15.6	38.5						3	FELT IN KEREN	A	PAL
1896	12	11	*	02 20	15.6	39.5						4	FELT IN MASSAWA	C	MED
1897	09	30	*	21 14	15.6	39.5						4	FELT IN MASSAWA	C	PAL
1899	02	11	*		11.5	43.0						6	FELT IN DJIBOUTI	D	
1891	11	21	*		11.5	43.0						6	FELT IN DJIBOUTI	D	
1899	11	23	*		11.5	43.0						4	FELT IN DJIBOUTI	D	
1900			*		08.3	39.4							PITCHSTONE FISURAL FLOWS SE OF KOKA RESERVOIR	C	AAE
1900	04	02	*	22	15.6	39.5						5	FELT IN MASSAWA	C	PAL
1900	04	02	*	22	14.9	38.8						3	FELT IN ADI UGRI	A	
1901	05		*	SR	15.8	38.8						6	FELT IN AMAZI	A	PAL
1901	11	11	*	05 10	14.8	40.2						5	3 FELT AT BUIA WELL	C	
1902	01	24	*	01 15	15.6	39.5						4	N FELT IN MASSAWA	C	PAL
1902	02	24	*	19	14.8	39.9						3	N FELT IN BABALA MADETO	C	PAL
1902	03	04	*	20 30	14.8	39.9						4	2 FELT IN UETEN	C	PAL
1902	03	05	*	02 20	14.8	39.9						3	FELT IN UETEN	C	PAL
1902	05	03	*	02	15.6	39.5						4	FELT IN MASSAWA	C	PAL
1902	05	26	*	09 25	15.6	39.5						4	FELT IN MASSAWA	C	PAL
1902	09	29	*	22 30	15.6	39.5						4	FELT IN MASSAWA	C	PAL
1902	01				15.1	39.7					5		GULF OF ZULA	C	ZZZ

Y	M	D		H1	N	E	H2	MB	MS	ML	ME	I	LOCATION & COMMENTS	R	A
1903	09	26	*	05 42	15.5	39.0							FELT IN ASMARA	A	
1904	05		*	PM	09	39						4	FELT IN ADDIS ABABA (2)	A	PAL
1904	06	30	*	13 55	09	39						3	FELT AT ADDIS ABABA	A	PAL
1905	03	09	*	PM	14.9	38.8						3	FELT IN ADI UGRI		
1905	03	09	*	PM	15.5	39.0							FELT IN ASMARA		
1905	12	23	*	MD	14	39						3	FELT IN ADI MALGUNDY	A	MED
1905	12	23	*	MD	15.5	39.0							FELT IN ASMARA (N)	A	
1906	03		*	NT	12.7	40.6					5.5		VOLCANIC E SEISMIC ACTIVI	C	ZZZ
1906	03										5.5		TY IN ALAYTA RANGE.SEVERE	C	ZZZ
1906	03												TREMORS THROUGHOUT AFAR	C	
1906	05	26	*	NT	11.2	39.7							FELT IN BORUMIEDA	A	PAL
1906	08	25		11 54 48	09	39			6.6					A	GUT
1906	08	25		11 55	09.1	38.7								A	STR
1906	08	25		11 54 48	08.0	38.5					6.5		ETHIOPIAN RIFT	A	ZZZ
1906	08	25	*	11 55	09	39						7	FELT IN ADDIS ABABA	A	DEM
1906	08	25		13 47 36	09	39			6.8					A	GUT
1906	08	25		13 50	09.1	38.7								A	STR
1906	08	25		13 49	08.0	38.5			6.8				ETHIOPIAN RIFT	A	ZZZ
1906	08	25	*	PM	09	39						8	FELT IN ADDIS ABABA	A	
1906	08	25	*	PM	07.85	38.70						7	ADMITULU (LANDSLIDES)	A	
1906	08	25	*		07.5	38.6						9	LANGANO (GEYZER)	A	
1906	08	25	*	PM	08.0	38.3							FELT IN SILTE	A	
1906	08	25	*	16 43	09	39						5	FELT IN ADDIS ABABA	A	D-M
1906	08	25	*	17 30	09	39						6	FELT IN ADDIS ABABA	A	D-M
1906	08	25	*	19 27	09	39						6	FELT IN ADDIS ABABA	A	D-M
1906	08	25	*	23 00	09	39						4	FELT IN ADDIS ABABA	A	D-M
1906	08	26	*	11 05	09	39						6	FELT IN ADDIS ABABA	A	D-M
1906	08	26		PM									CONTINUOUS TREMORS,SHOA	A	
1906	08	27	*	20 22	09	39						4	FELT IN ADDIS ABABA	A	D-M
1906	10	28	*	15 40	09	39						5	FELT IN ADDIS ABABA	A	D-M
1906	10	28	*	16 07	09	39						7		A	PAL
1906	11	08	*	02 50	09	39							2 TREMORS IN ADDIS ABEBA	A	PAL
1907	01	18	*	12 15	11.5	43.0						6	FELT IN DJIBOUTI REPUBLIC	D	PAL

Y M D		H1	N	E	H2	MB	MS	ML	ME	I	LOCATION & COMMENTS	R	A
1907 01 18	*	18 35	11.5	43.0							FELT IN DJIBOUTI	D	PAL
1907 04 12	*		09	39						3	FELT IN ADDIS ABABA	A	PAL
1907 04 17	*	MD	09	39						4	2 FELT IN ADDIS ABABA	A	PAL
1907 05	*										AFAR: TREMORS CAUSED BY		
1907 05	*										VOLCANIC ACTIVITY AT		
1907 05	*										ALAYTA	C	PAL
1907 10 28	*	PM	11.5	43.0						4	10 FELT IN DJIBOUTI	D	PAL
1908 12 28	*	19 30	15.7	38.5						4	FELT IN ELA BERED	A	PAL
1909 02 19	*	06	11.5	43.0						4	FELT IN DJIBOUTI	D	PAL
1910 02 04	*	NT	09	39						3	FELT IN ADDIS ABABA	A	
1910 02 18	*	08 15	11.5	43.0						3	FELT IN DJIBOUTI	D	PAL
1910 04 22	*	13 35	11.5	43.0						3	FELT IN DJIBOUTI	D	PAL
1910 08 06	*	NT	11.5	43.0						4	FELT IN DJIBOUTI	D	PAL
1911 03 02	*	08 20	09.6	41.9						3	FELT IN DIRE DAWA	B	
1912 03	*		09	39							FELT IN ADDIS ABABA	A	
1912 04 15	*	06 30	14.85	39.4							3 FELT IN ADI CAIEH	A	PAL
1912 04 20	*	12	15.7	38.6						4	FELT IN ELA BERED	A	PAL
1912 05	*										ACTIVITY W GULF ADEN FROM		
1912 05											MAY TO AUGUST	D	
1912 05 12	*	18 45	11.5	43.0						5	FELT IN DJIBOUTI	D	PAL
1912 05 14	*	01	11.5	43.0						3	FELT IN DJIBOUTI	D	PAL
1912 05 28	*	04	11.5	43.0						3	FELT IN DJIBOUTI	D	PAL
1912 06 22	*	02 40	11.5	43.0						3	FELT IN DJIBOUTI	D	PAL
1912 08 27	*	11 08	11.5	43.0						5	FELT IN DJIBOUTI	D	PAL
1912 12 28	*	23	15.5	39.0						5	FELT IN ASMARA	A	PAL
1913 01	*		15.5	39.0							MANY TREMORS IN ASMARA	A	
1913 02	*										FROM FEB TO MAY 457 TRE-	A	
											MORS FELT IN ASMARA ,	A	

Y M D		H1	N	E	H2	MB	MS	ML	ME	I	LOCATION & COMMENTS	R	A
											KEREN, ADI CAIEH, MASSAWA	A	
											MAX INTENSITY IN ASMARA=8	A	PAL
1913 02 04	*		15.5	39.0						6	FELT IN ASMARA	A	FAN
1913 02 10	*		15.5	39.0						7	FELT IN ASMARA	A	FAN
1913 02 18	*		15.5	39.0						7	FELT IN ASMARA	A	FAN
1913 02 27		16 22 54	17.5	39.0			5.8					A	GUT
1913 02 27		16 22 34	14.	39.								A	ISS
1913 02 27		16 22 34	16.	39.								A	PUL
1913 02 27		16 22 54	17.2	38.8	33		5.8		5.8		ERITREAN ESCARPMENT	A	ZZZ
1913 02 27	*	16 23	15.5	39.0						8	FELT IN ASMARA	A	FAN
1913 03	*		09	39						3	FELT IN ADDIS ABABA	A	
1913 03 04			15.5	39.0						6	FELT IN ASMARA	A	FAN
1913 03 05	*		15.5	39.0						6	FELT IN ASMARA	A	FAN
1913 03 23	*		15.5	39.0						7	ASMARA (PANIC)	A	FAN
1913 03 27		03 13	16.5	39.0			5.8					A	GUT
1913 03 27		03 12 45	15.9	39.0								A	ISS
1913 03 27		03 11 45	15.9	39.5								A	TIF
1913 03 27		03 13	16	39			5.5				ERITREAN ESCARPMENT	A	ZZZ
1913 03 27	*	03 15	15.5	39.0						8	FELT IN ASMARA	A	FAN
1913 05 13	*		15.5	39.0						7	6 TREMORS FELT IN ASMARA	A	FAN
1913 05 22	*	NT	15.5	39.0						4	3 TREMORS FELT IN ASMARA	A	FAN
1913 05 23	*	SR	15.5	39.0						3	2 TREMORS FELT IN ASMARA	A	FAN
1913 06											NEW SEISMIC STATION IN		
1913 06											ASMARA RECORDED 141 EVENTS		
1913 06											FROM JUNE 6 TO JULY 16	A	PAL
1913 09 16		11 56 60	05.0	34.0								E	ESK
1913 09 16		11 56 60	04.4	37.0								E	GRA
1913 09 16		11 56 42	06.0	36.5			6.2					E	GUT
1913 09 16		11 56 09	03.0	37.0								E	ISS
1913 09 16		11 56 25	03.56	36.28			6.2					E	PG1
1913 09 16		11 56 60	04.5	39.5								E	PUL
1913 09 16		11 56 60	03.5	40.0								E	TAS

Y M D		H1	N	E	H2	MB	MS	ML	ME	I	LOCATION & COMMENTS	R	A
1913 09 16		11 56 25	04.3	36.7			6.2				KINU-SOGO BELT (KENYA)	E	ZZZ
1915 05 21		04 18 06	06	31			6.6				SE SUDAN	A	GUT
1915 09 23		08 14 48	16	39			6.8					A	GUT
1915 09 23		08 14 48	16.0	38.5								A	ISS
1915 09 23		08 15	15.7	34.8								A	PUL
1915 09 23		08 14 48	16.	39.			6.8				NAKFA GRABEN	A	ZZZ
1915 09 23	*		15.5	39.0						6	DAMAGE IN ASMARA	A	AAE
1919 06 30		07 26 30	06	37								E	ISS
1919 06 30		07 26 30	06	37					5.8		S ETHIOPIA	E	ZZZ
1919 06 30		07 33 40	06	37								E	ISS
1919 06 30		07 33 40	06	37					5.0		S ETHIOPIA	E	ZZZ
1921 01 28	*	15 30	15	39						5	FELT AT ADI UGRI	A	CAV
1921 01 28	*	15 45	15	39						3	FELT AT ADI UGRI	A	CAV
1921 01 28	*	18 34	15	39						4	FELT AT ADI UGRI	A	CAV
1921 01 28	*	18 36	15	39						3	FELT AT ADI UGRI	A	CAV
1921 01 31	*	21 23	15	39						4	FELT AT ADI UGRI	A	CAV
1921 02 01	*	14 58	15	39						4	FELT AT ADI UGRI	A	CAV
1921 02 01	*	19 09	15	39						2	FELT AT ADI UGRI	A	CAV
1921 02 24	*	01 20	15	39						5	FELT AT ADI UGRI	A	CAV
1921 02 24		01 25	15	39						5	FELT AT ADI UGRI	A	CAV
1921 04 12		21	15	39						3	FELT AT ADI UGRI	A	CAV
1921 08 14		13 15 28	15.5	40.5			5.8					C	GUT
1921 08 14		13 15	15.5	39.0								C	ISS
1921 08 14		13 15 28	15.6	39.6			5.8				MASSAWA CHANNEL. VOLCANIC	C	ZZZ
1921 08 14											ACTIVITY ON SEA FLOOR	C	ZZZ
1921 08 14	*	13 15 20	15.6	39.5						9	FELT IN MASSWA	C	CAV
1921 08 14	*	18	15.6	39.5						5	FELT IN MASSAWA	C	CAV
1921 08 14	*	18	15.6	39.5						5	FELT IN MASSAWA	C	CAV
1921 08 14	*	20	15.6	39.5						3	FELT IN MASSAWA	C	CAV
1921 08 14	*	18 05	15.6	39.5						3	FELT IN MASSAWA	C	CAV
1921 08 15	*	00 25	15.6	39.5						3	FELT IN MASSAWA	C	CAV
1921 08 15	*	09	15.6	39.5						2	FELT IN MASSAWA	C	CAV
1921 08 17	*	23	15.6	39.5						4.5	FELT IN MASSAWA	C	CAV
1921 08 18	*	23 25	15.6	39.5						3	FELT IN MASSAWA	C	CAV
1921 09 03	*	PM	15	39						4	3 FELT IN ADI UGRI	A	CAV
1921 09 03	*	23 21	15	39						3	FELT IN ADI UGRI	A	CAV
1921 09 04	*	01	15	39						5	FELT IN ADI UGRI	A	CAV
1921 09 04	*	14	15	39						3	FELT IN ADI UGRI	C	CAV

Y	M	D		H1	N	E	H2	MB	MS	ML	ME	I	LOCATION & COMMENTS	R	A
1921	09	08	*	13 32	15	39						4	FELT IN ADI UGRI	A	CAV
1921	09	08	*	14 47	15	39						3	FELT IN ADI UGRI	A	CAV
1921	09	21		11 01 31	14	39			5.7					C	GUT
1921	09	21		11 01 31	15.5	39.0								C	ISS
1921	09	21		11 01 31	15.6	39.7							MASSAWA CHANNEL	C	ZZZ
1921	09	21	*	11 01 30	15.6	39.5						9	FELT IN MASSAWA	C	CAV
1921	09	21	*	14 07	15.5	39.0						5	FELT IN ASMARA	A	CAV
1921	09	21	*	16 07	15.5	39.0						4	FELT IN ASMARA	A	CAV
1921	09	21	*	16 35	15.5	39.0						4	FELT IN ASMARA	A	CAV
1921	09	21	*	20 15	15.5	39.0						4	FELT IN ASMARA	A	CAV
1921	09	22	*	00 06	15.5	39.0						4	FELT IN ASMARA	A	CAV
1921	09	22	*	03 45	15.5	39.0						4	FELT IN ASMARA	A	CAV
1921	09	22	*	08 47	15.5	39.0						4	FELT IN ASMARA	A	CAV
1921	09	22	*	14 51	15.5	39.0						5	FELT IN ASMARA	A	CAV
1921	09	22	*	15 34	15.5	39.0						5	FELT IN ASMARA	A	CAV
1921	09	22	*	21	15.5	39.0						4	FELT IN ASMARA	A	CAV
1921	09	23	*	00 33	15.5	39.0						4	FELT IN ASMARA	A	CAV
1921	09	23	*	03 17	15.5	39.0						4	FELT IN ASMARA	A	CAV
1921	09	24	*	02 30	15.5	39.0						4	FELT IN ASMARA	A	CAV
1921	10	02	*	02 20	15.5	39.0						3	FELT IN ASMARA	A	CAV
1921	10	02	*	14 14	15.5	39.0						3	FELT IN ASMARA	A	CAV
1921	10	03	*	23 07	15.5	39.0						3	FELT IN ASMARA		CAV
1921	10	04	*	01 16	15.5	39.0						3	FELT IN ASMARA	A	CAV
1921	10	06	*	NT	15.5	39.0						4	FELT IN ASMARA	A	CAV
1921	10	06	*	03	15.5	39.0						3	FELT IN ASMARA	A	CAV
1921	10	06	*	06 23	15.5	39.0						3	FELT IN ASMARA	A	CAV
1921	10	06	*	07 30	15.5	39.0						3	FELT IN ASMARA	A	CAV
1921	10	07	*	02 20	15.5	39.0						2	FELT IN ASMARA	A	CAV
1921	10	07	*	09 40	15.5	39.0						4	FELT IN ASMARA	A	CAV
1921	10	07	*	11 32	15.5	39.0						2	FELT IN ASMARA	A	CAV
1921	10	11	*	07 22	15.5	39.0						2	FELT IN ASMARA	A	CAV
1921	10	15	*	19 20	15.5	39.0						2	FELT IN ASMARA	A	CAV
1921	10	17	*	23 15	15.5	39.0						3	FELT IN ASMARA	A	CAV
1921	10	18	*	02 20	15.5	39.0						5	FELT IN ASMARA	A	CAV
1921	10	22	*	13 13	15.5	39.0						3	FELT IN ASMARA	A	CAV
1921	11	01	*	21 51	15.5	39.0						4	FELT IN ASMARA	A	CAV
1921	11	02	*	04 45	15.5	39.0						4	FELT IN ASMARA	A	CAV
1922	01	23	*		11.8	39.6							FELT IN WEJJA (WOLLO)	A	AAE
1922	03	16		14 56 20	06	37								E	ISS
1922	03	16		14 56 20	06	37					5		S.ETHIOPIA	E	ZZZ
1924	03	13		12 47 40	16.0	38.5								A	ISS

Y	M	D		H1	N	E	H2	MB	MS	ML	ME	I	LOCATION & COMMENTS	R	A
1924	03	13			16.0	38.8							ERITREAN ESCARPMENT	A	ZZZ
1924	09		*										W GULF ADEN ACTIVITY FROM	D	
1924	09		*										SEPT TO NOV. FELT IN DJI-	D	
1924	09												BOUTI AND ZEILA	D	
1926	04	28	*	MN	11.5	43.0						3	FELT IN DJIBOUTI	D	
1926	10	30		01 38 10	11.	44.			5.6					D	GUT
1926	10	30		01 38 10	11.5	43.5								D	ISS
1926	10	30		01 38 10	11.8	43.5			5.6				TADJOURA TRENCH	D	ZZZ
1926	10	30		NT	11.5	43.0						3	FELT IN DJIBOUTI	D	
1926	10	30		NT	11.0	43.5							FELT IN ZEILA (N)	D	
1926	10	31		11 43 15	11.5	43.5							2 SHOCKS	D	ISS
1926	10	31		11 43 15	11.7	43.5					5.0		TADJOURA TRENCH	D	ZZZ
1926	10	31	*		11.0	43.5							N FELT IN ZEILA	D	
1926	10	31		17 22 38	11.5	43.5								D	ISS
1926	10	31		17 12 38	11.7	43.5					5.0		TADJOURA TRENCH	D	ZZZ
1926	10	31	*		11.0	43.5							N FELT IN ZEILA	D	
1927	01	06	*	NT	11.5	43.0							5 TREMORS FELT DJIBOUTI	D	
1927	02	11	*	06 40	11.5	43.0							FELT IN DJIBOUTI	D	
1927	03	12	*	18 00	11.5	43.0							FELT IN DJIBOUTI	C	
1928	04	08	*	11 42	11.5	43.0							FELT IN DJIBOUTI	C	
1928	09	21	*	23 45	11.5	43.0							FELT IN DJIBOUTI	C	
1928	10	04		18 22 58	07.	38.			6.0					E	GUT
1928	10	04		18 23	06.5	37.								E	ISS
1928	10	04		18 22 58.1	06.87	36.85	0							E	PG1
1928	10	04		18 23 03.1	06.86	36.85	33							E	PG1
1928	10	04		18 23 03	06.9	36.9			6.0				BAKO-BULKI REGION	E	ZZZ
1929	01	22		14 43 05	11.5	43.5			6.0					D	GUT
1929	01	22		14 43	11.5	43.5								D	ISS
1929	01	22		14 43 05	11.7	43.1			6.0				TADJOURA TRENCH. 42 AFTER		ZZZ
1929	01	22											SHOCKS REPORTED FROM DJI-		
1929	01	22											BOUTI	D	
1929	01	22	*	14 42	11.5	43.0						8	DAMAGE IN DJIBOUTI	D	
1929	01	22	*	14 42	11.8	42.9						5	DAMAGE IN TADJOURA	D	

Y	M	D		H1	N	E	H2	MB	MS	ML	ME	I	LOCATION & COMMENTS	R	A
1929	01	23	*		11.5	43.0							N AFTERSHOCKS IN DJIBOUTI	D	
1929	05	18		01 02 12	11.5	41.5		6.0						C	GUT
1929	05	18		01 02	11.5	42.0								C	ISS
1929	05	18		01 02 15.2	12.03	41.36	0		6.0					C	PG2
1929	05	18		01 02 15	12.03	41.36			6.0				AFAR (SERDO REGION)		ZZZ
1929	08	05	*	PM	11.5	43.0						3	N FELT IN DJIBOUTI	D	
1929	09	27	*		11.5	43.0						3	FELT IN DJIBOUTI	D	
1930	05	30	*	13 15	11.5	43.0						3	FELT IN DJIBOUTI	D	
1930	10		*										FROM OCT 17 TO NOV 18,55		
1930	10		*										TREMORS IN DJIBOUTI	D	
1930	10	24		10 47 21	10.5	43.0			5.6					BD	GUT
1930	10	24		10 47 16	10.3	42.7								BD	ISS
1930	10	24		10 47 12.3	10.66	42.97	0							BD	PG2
1930	10	24		10 47 06.2	09.047	42.327	33							BD	PG3
1930	10	24		10 47 06	10.4	42.8			5.8				AISHA HORST	B	ZZZ
1930	10	24	*	10 47	11.1	42.7							FELT IN ALI SABIEH	B	
1930	10	24	*	10 47	09.0	40.2						3	FELT AT A WAS H STATION	B	
1930	10	24	*	10 47	09.5	42.0						4	FELT IN DIRE DAWA	B	
1930	10	24	*	10 47	11.5	43.0						5	FELT IN DJIBOUTI	E	
1930	10	24	*	10 47	11.4	43.0						5	FELT IN TADJOURA	E	
1930	10	24	*	10 48	11.0	43.5							10 REPORTED FROM ZEILA	E	
1930	10	25		16 28 46	11.5	44.0			5.6					E	GUT
1930	10	25		16 28 46	11.5	42.0								E	ISS
1930	10	25		16 28 46									NO RELIABLE SOLUTION		PG2
1930	10	25		16 28 18.7	08.021	41.364	33							B	PG3
1930	10	25		16 28 46	10.3	42.7			5.6				AISHA HORST	B	ZZZ
1930	10	25	*	16 28	11.5	43.0							7 FELT IN DJIBOUTI	E	
1930	10	25	*	16 29	11.0	43.5							7 REPORTED FROM ZEILA	E	
1930	10	25		17 41 55	11.5	44.0			5.6					E	GUT
1930	10	25		17 41 55	11.5	42.0								E	ISS
1930	10	25		17 41 31.4	08.360	42.349	33							B	PG3
1930	01	25		17 42	10.4	42.7							AISHA HORST	B	ZZZ
1930	10	27		23 28 41	12.5	43.5			5.6					E	GUT
1930	10	27		23 28 41	11.5	43.5								E	ISS
1930	10	27		23 28 36.5	10.649	42.592	33							B	PG3
1930	10	27		23 28 41	10.6	42.6			5.6				AISHA HORST	B	ZZZ
1931	05	01		09 47 55	18.0	37.5								C	ISS

Y M D		H1	N	E	H2	MB	MS	ML	ME	I	LOCATION & COMMENTS	R	A
1931 05 01		09 48 29.5	21.189	39.158	33							C	PG3
1931 05 01		09 48 29	21.2	39.2							WESTERN COAST OF ARABIA	C	ZZZ
1931	*		11.5	43.0							14 TREMORS FELT IN DJIBOU-		
										4	TI DURING 1931	D	
1932 02 01		07 38 24	10.3	42.7								B	ISS
1932 02 01		07 38 24									NO RELIABLE SOLUTION	B	PG3
1932 02 01		07 38 24	10.5	42.6					5.0		AISHA HORST	B	ZZZ
1932	*		11.5	43.0							32 TREMORS DJIBOUTI	D	
1932	*									4	DURING 1932	D	
1934 02 01	*	23 58	15.5	39.0						4	FELT IN ASMARA	A	FAN
1934 07 02	*	14 35	15.6	39.5						4	FELT IN MASSAWA	C	FAN
1934 10 06	*	NT								3	3 TREMORS IN DJIBOUTI	D	
1935 01 16	*	MN	11.5	43.0						3	2 TREMORS IN DJIBOUTI	D	
1935 01 20	*	05	11.5	43.0						3	2 TREMORS IN DJIBOUTI	D	
1935 04 25	*	03 48	11.5	43.0						5	2 IN DJIBOUTI + NOISE	D	
1935 04 25	*	03 48	11.8	42.9						4	2 IN TADJOURA	D	
1935 06 22	*		11.8	39.6							FELT IN WEJJA (WOLLO)	A	AAE
1936 06 24	*	07 30	11.5	43.0						3	3 TREMORS IN DJIBOUTI	D	
1936 11 03	*	02 10	11.5	43.0						5	TREMOR + NOISE DJIBOUTI	D	
1936 11 03	*	03 05	11.5	43.0						3	FELT IN DJIBOUTI	D	
1937 11 30		12 57 48	04.9	35.9								E	CGS
1937 11 30		12 57 46	05	36			6.3					E	GUT
1937 11 30		12 57 48	05.0	36.6								E	ISS
1937 11 30		12 57 48.1	05.12	36.71	0							E	PG1
1937 11 30		12 57 53.1	05.12	36.69	33							E	PG1
1937 11 30		12 57 54.6	05.15	36.71	33							E	PG3
1937 11 30		12 57 55	05.1	36.9			6.3				CHEW BAHR GRABEN	E	ZZZ
1937	*		11.5	43.0							13 TREMORS REGISTERED IN	D	
1937	*										DJIBOUTI DURING 1937	D	ZZZ
1938 03 11		16 51 33	10.5	44.5								D	ISS

Y	M	D	H1	N	E	H2	MB	MS	ML	ME	I	LOCATION & COMMENTS	R	A
1938	03	11	16 52 08.7	13.553	44.798	33						NOT RELIABLE	D	PG3
1938	03	11	16 52	10.8	43.3					4.5		S OF DJIBOUTI CITY	D	ZZZ
1938	03	12	12 37 34	10.5	44.5								D	ISS
1938	03	12	12 37 34									NO SOLUTION	D	PG3
1938	03	12	12 37 34	11.8	43.3					4.5		TADJOURA TRENCH	D	ZZZ
1938	03	12	13 04 40	10.5	44.5								D	ISS
1938	03	12	13 04 40									NO SOLUTION	D	PG3
1938	03	12	13 04 40	11.8	43.3					4.5		TADJOURA TRENCH	D	ZZZ
1938	03	12	20 04 13	10.5	44.5								D	ISS
1938	03	12	20 04 23.7	10.761	43.281	33							D	PG3
1938	03	12	20 04 24	11.8	43.3					4.5		TADJOURA TRENCH	D	ZZZ
1938	05	12	21 31 30	17	38								C	CGS
1938	05	12	21 31 35	18.5	37.5			5.7					C	GUT
1938	05	12	21 31 35	18.0	37.5								C	ISS
1938	05	12	21 31 44.1	18.580	37.438	33							C	PG3
1938	05	12	21 31 44	18.5	37.5	33		5.8				SUDAN RED SEA COAST	C	ZZZ
1938	09	18	00 38 35	09.5	40.3								C	ISS
1938	09	18	00 38 40.8	10.37	40.81	0							C	PG2
1938	09	18	00 38 47.8	10.368	40.827	33							C	PG3
1938	09	18	00 38	10.37	40.82					4.5		WONJI FAULT BELT	C	ZZZ
1938	09	27	02 31 49	11.	41.			6.0					C	GUT
1938	09	27	02 31 41	09.5	40.3								C	ISS
1938	09	27	02 31 51.2	11.13	40.74	0							C	PG2
1938	09	27	02 31 53.1	10.520	40.459	33							C	PG3
1938	09	27	02 31 53	10.83	40.60	33	6					FOOT OF AFAR W. SCARP	C	ZZZ
1938	10	20	13 14 58	10.0	39.5			5.6					A	GUT
1938	10	20	13 14 55	09.5	40.3								A	ISS
1938	10	20	13 15 05.0	10.109	40.226	33							A	PG3
1938	10	20	13 15 05	10.11	40.23			5.6				W PLATEAU MARGIN	A	ZZZ
1938	10	23	02 25 14	10.0	39.5			5.6					C	GUT
1938	10	23	02 25 12	09.5	40.3								C	ISS
1938	10	23	02 25 25.9	10.391	40.748	33							C	PG3
1938	10	23	02 25 14	10.39	40.75			5.6				WONJI FAULT BELT	C	ZZZ
1938	10	23	02 28 47	09.5	40.3								C	ISS
1938	10	23	02 25 56.7	10.011	40.333	33							A	PG3
1938	10	23	02 28 57	10.01	40.33	33				4.5		W PLATEAU MARGIN	A	ZZZ

Y	M	D		H1	N	E	H2	MB	MS	ML	ME	I	LOCATION & COMMENTS	R	A
1940	08	14		08 49 24	11.0	46.0								D	ISS
1940	08	14		08 49 21.9	09.912	46.158	33						(UNRELIABLE)	D	PG3
1940	08	14		08 49 22	11.0	46.0					4.5		GULF OF ADEN	D	ZZZ
1941	03	19		01 31 52	12.0	43.5			5.5					D	GUT
1941	03	19		01 31 40	10.5	44.5								D	ISS
1941	03	19		01 31 49.1	10.930	43.961	33						(UNRELIABLE)	D	PG3
1941	03	19		01 31 49	12.0	43.5	33		5.3				W GULF OF ADEN	D	ZZZ
1942	11	18		12 01 21	12.0	40.5			5.6					A	GUT
1942	11	18		12 01 07	10.0	38.0								A	ISS
1942	11	18		12 01 20.1	11.345	39.156	33							A	PG3
1942	11	18		12 01 20	11.35	39.16	33				5.6		W. PLATEAU MARGIN	A	ZZZ
1942	11	18	*		11.8	39.6							WALDIA (WOLLO)	A	AAE
1944	01	01	*	07 30	11.5	43.0						4	FELT IN DJIBOUTI	D	
1944	01	27	*										SWARM OF 41 SHOCKS IN REP		
1944	01	27	*										OF DJIBOUTI OCT 27 AND 28		
1944	01	27	*										FELT OBOCK TADJOURA,BALA-		
1944	01	27	*										OH,DIKHIL,LOYADA. MAX IN-		
1944	01	27	*									4	TENSITY = 4 IN DJIBOUTI	D	
1944	09	06		13 27 55	06	38			6					C	GUT
1944	09	06		13 28 02.7	07.24	39.29	0							B	PG1
1944	09	06		13 28 03	07.2	38.5	33		6				RIFT VALLEY	BC	ZZZ
1945	03	31		22 07 48	14.5	39.5								A	ISS
1945	03	31		22 07 56.9	14.870	39.541	33							A	PG3
1945	03	31		22 07 57	14.87	39.54	33				4.5		W. PLATEAU MARGIN	A	ZZZ
1945	10	28		00 17 10	11.0	42.5		5.6						D	GUT
1945	10	28		00 17 10	11.2	42.7								D	ISS
1945	10	28		00 17 10.8	11.27	42.69	0							D	PG2
1945	10	28		00 17 15.4	10.992	42.667	33							D	PG3
1945	10	28		22 11 02	11.13	42.68	33		5.5				REPUBLIC OF DJIBOUTI	D	ZZZ
1945	10	28	*		11.5	43.0						7	FELT IN DJIBOUTI	D	
1945	10	28	*										FELT THROUGHOUT REPUBLIC	D	
1945	10	28	*										41 AFTERSHOCKS	D	
1945	11		*		11.5	43.0							27 TREMORS IN DJIBOUTI		
1945	11												FROM NOV 8 TO MID DEC	D	
1945	11	08	*		11.5	43.0						6	FELT IN DJIBOUTI	D	
1946	02		*		11.5	43.0							2 TREMORS IN DJIBOUTI	D	

Y	M	D		H1	N	E	H2	MB	MS	ML	ME	I	LOCATION & COMMENTS	R	A
1947	02	15	*	NT	11.5	43.0						4	2 TREMORS IN DJIBOUTI	D	
1947	03	17	*		11.5	43.0						3	2 TREMORS IN DJIBOUTI	D	
1948	02	17		22 11 02	14.0	40.2								A	ISS
1948	02	17		22 11 06.2	13.711	39.851	33							A	PG3
1948	02	17		22 11 02	13.3	40.0			5.6					A	ROT
1948	02	17		22 11 06	13.7	40.0	33		5.6				MAGLALA-RENDA COMA	A	ZZZ
1949	04	22	*	00 55	11.5	43.0						5	FELT IN DJIBOUTI (N)	D	
1949	04	22	*	00 55	12.0	43.3						5	FELT IN OBOCK (N)	D	
1949	06	16		17 58 00	11.2	42.7								D	ISS
1949	06	16		17 58 09.2	11.570	42.569	33							D	PG3
1949	06	16		17 58 05	11.2	42.7		5.5						D	ROT
1949	06	16		17 58	10.75	42.50								D	STR
1949	06	16		17 58 09	11.57	42.57			5.5				NE SCARP ASAL RIFT	D	ZZZ
1949	06	16	*	NT	11.5	43.0						6	FELT IN DJIBOUTI	D	
1950	02	01	*		11.5	43.0						3	3 TREMORS IN DJIBOUTI	D	
1950	03	26		16 53 25	14.5	39.5								A	ISS
1950	03	26		16 53 29.8	14.318	39.316	33							A	PG3
1950	03	26		16 53 25	14.5	39.5		5.5						A	STR
1950	03	26		16 53 30	14.4	39.4	33		5.5				TIGRAY (ESCARPMENT)	A	ZZZ
1950	07	08	*	22 30	11.5	43.0						5	FELT IN DJIBOUTI	D	
1950	08	02		13 49 55	15.0	39.5								A	CGS
1950	08	02		13 49 55	14.5	39.5								A	ISS
1950	08	02		13 50 01.8	14.522	39.715	33							A	PG3
1950	08	02		13 49 45	15.0	38.0	150							A	P00
1950	08	02		13 49 43	12.0	43.0								A	PRA
1950	08	02		13 49 55	14.5	39.5		6.2						A	ROT
1950	08	02		13 49 58	14.5	40.0	500							A	STR
1950	08	02		13 50 02	14.52	39.72	33		6.2				TIGRAY (ESCARPMENT)	A	ZZZ
1950	09	18		00 39 30	13.7	42.2								C	STR
1950	09	18		00 39 30	13.8	42.5					4.5		S. RED SEA	C	ZZZ
1951	12	11		02 34 36	12.0	40.5			4.7					C	ROT
1951	12	11		02 34 36	12.0	40.7	33		4.7				CENTRAL AFAR	C	ZZZ
1952	09	10		09 06 13	14.5	39.5								C	ISS
1952	09	10		09 06 17.5	14.524	40.204	33							C	PG3
1952	09	10		09 05 58	14.5	39.5			5.2					C	ROT

Y M D		H1	N	E	H2	MB	MS	ML	ME	I	LOCATION & COMMENTS	R	A
1952 09 10		09 05 58	12.8	39.6								C	STR
1952 09 10		09 06 17	14.5	40.2		5.2					DEPRESSION E MARGIN	C	ZZZ
1953 05 28	*	01 24 46	09.2	41.9								B	CGS
1953 05 28		01 24 46	09.2	41.9			5.2					B	STR
1953 05 28		01 24 46	09.2	41.9			5.2				HARARGHE	B	ZZZ
1953 05 28	*	01 25	09.6	41.3						7	FELT IN DIRE DAWA	B	
1953 05 28	*	01 25	09.3	42.1						7.5	FELT IN HARAR	B	
1953 05 28	*		09.3	42.1							3 AFTERSHOCKS FELT IN	B	
										3.5	HARAR CITY	B	AAE
1953 11 12		01 20 32	11.0	43.5			4.7					D	ROT
1953 11 12		01 20 32	11.5	43.0								D	STR
1953 11 12		01 20 32	11.7	43.3			4.7				G. OF TADJOURA	D	ZZZ
1953 11 12	*	01 20 32	11.5	43.0						5.5	FELT IN DJIBOUTI	D	
1954 01 17		17 39 38	16.5	36.0								A	STR
1954 01 17		17 39 38	16.5	36.0					4.5		NE SUDAN	A	ZZZ
1954 06 30		13 26 50	06.0	37.0	33		5.6					E	CGS
1954 06 30		13 26 51	05.8	37.3								E	ISS
1954 06 30		13 26 50.9	05.86	37.19	0							E	PG1
1954 06 30		13 26 50.9	05.88	37.20	33							E	PG1
1954 06 30		13 26 56.1	05.597	37.235	33							E	PG3
1954 06 30		13 26 51	05.8	37.3			6.0					E	ROT
1954 06 30		13 26 55	06.0	37.2			6.2					E	STR
1954 06 30		13 26 56	05.78	37.23	33		5.9				S. ETHIOPIA	E	ZZZ
1954 06 30	*		05.6	37.5						5.5	DAMAGE IN GIDOLE	E	
1954 06 30	*		05.6	38.2						5	FELT IN HAGERE MARIAM	E	
1954 06 30	*		06.9	37.7						4	FELT IN SODDU	E	
1954 06 30	*										FELT AFTERSHOCKS UNITL		
1954 06 30	*										FEB 1955	E	
1954 12 27	*	NT	11.5	43.0						3	FELT IN DJIBOUTI	D	
1954 12 27	*	NT	12.0	43.3						3	2 FELT IN OBOCK	D	
1955 01 17		15 35 06	12.23	46.02								D	CGS
1955 01 17		15 35 13.2	12.32	46.02								D	S-L
1955 01 17		15 35 06	10.75	45.25								D	STR
1955 01 17		15 35 06	12.3	46.0	33				4.5		GULF OF ADEN	D	ZZZ
1955 01 20	*	01 50	11.5	42.1						3	FELT IN YOBOKI	D	
1955 02 04		05 21 04.8	05.59	37.00								E	JED
1955 02 04		05 21 01	05.78	36.47	0							E	PG1

Y M D		H1	N	E	H2	MB	MS	ML	ME	I	LOCATION & COMMENTS	R	A
1955 02 04		05 21 05.8	05.74	36.43	33							E	PG1
1955 02 04		05 21 00.9	05.80	36.60		5.5						E	ROT
1955 02 04		05 21 00.9	05.80	36.62								E	S-L
1955 02 04		05 20 42	06.0	37.5			5.2					E	STR
1955 02 04		05 21 06	05.75	36.80	33		5.3				S. ETHIOPIA	E	ZZZ
1955 02 23	*	NT	11.5	42.1						4	2 FELT IN YOBOKI	D	
1955 03 03		00 43 44.8	16.54	41.25								C	JED
1955 03 03		00 43 40.2	16.46	41.29								C	S-L
1955 03 03		00 43 40	16.0	41.5			5.0					C	STR
1955 03 03		00 43 45	16.54	41.23	33		5.0				RED SEA TROUGH	C	ZZZ
1955 04 27	*		11.5	43.0						3	4 FELT IN DJIBOUTI	D	
1955 04 28	*		11.5	43.0						3	4 FELT IN DJIBOUTI	D	
1957 03 14		00 11 38.5	14.97	40.19								C	JED
1957 03 14		00 11 33	14.8	40.2								C	ROT
1957 03 14		00 11 36	15.0	40.0			5.2					C	STR
1957 03 14		00 11 33.0	14.80	40.22								C	S-L
1957 03 14		00 11 39	14.89	40.20	16		5.2				BURI PENINSULA	C	ZZZ
1957 03 14		00 43 40.2	16.46	41.29								C	S-L
1957 03 14		00 43 40	16.46	41.29							RED SEA	C	ZZZ
1957 04 12		15 58 44	11.49	43.11								D	ISS
1957 04 12		15 58 47.3	11.47	43.30		5.0						D	JED
1957 04 12		15 58 35.0	11.50	43.00			5.0					D	MOS
1957 04 12		15 58 50.8	11.51	43.09	33							D	PG3
1957 04 12		15 58 43.5	11.54	43.05			5.0					D	S-L
1957 04 12		15 58 45	11.5	43.0			5.2					D	STR
1957 04 12		15 58 47	11.50	43.09	33			4.8			REPUBLIC OF DJIBOUTI	D	ZZZ
1957 04 12	*		11.5	42.8						6	FELT IN ARTA	D	
1957 04 12			11.5	43.0						6	FELT IN DJIBOUTI	D	
1957 04 12		16 06 00	11.5	43.0								D	STR
1957 04 12		16 06 00	11.7	43.0					4.5		G. OF TADJOURA	D	ZZZ
1957 10 01	*	PM	12.5	39.5						5	4 FELT IN MAI CHEW	A	AAE
1957 10 18	*		12.5	39.5						3	FELT IN MAI CHEW	A	AAE
1957 11 06	*		12.5	39.5						3	FELT IN MAI CHEW	A	
1958 01 09		07 56 27.2	17.71	40.12								C	S-L
1958 01 09		07 56 27	17.71	40.12					4.5		RED SEA TROUGH	C	ZZZ

Y M D		H1	N	E	H2	MB	MS	ML	ME	I	LOCATION & COMMENTS	R	A
1958 02 13		10 23 34	14.35	42.01								C	ISS
1958 02 13		10 23 36.5	14.26	41.92								C	JED
1958 02 13		10 23 38.9	14.112	42.005	33							C	PG3
1958 02 13		10 23 34	13.75	41.25			5.2					C	STR
1958 02 13		10 23 33.7	14.34	42.00								C	S-L
1958 02 13		10 23 36	14.24	41.98			5.2				RED SEA	C	ZZZ
1958 05 24		22 25 32.6	12.17	43.58								D	S-L
1958 05 24		22 25 40.0	12.00	43.50								D	STR
1958 05 24		22 25 33	12.17	43.58					4.5		BAB EL MANDEB FORESHOCK	D	ZZZ
1958 05 24		23 53 39	12.15	43.52	0							D	ISS
1958 05 24		23 53 42.7	12.13	43.82		5.5						D	JED
1958 05 24		23 53 27.0	12.00	43.50			5.5					D	MOS
1958 05 24		23 53 46.0	12.269	43.603	33							D	PG3
1958 05 24		23 53 38.0	12.10	43.60			5.5					D	ROT
1958 05 24		23 53 38.0	12.14	43.59			5.5					D	S-L
1958 05 24		23 53 38	12.0	43.5								D	STR
1958 05 24		23 53 38	12.18	43.67			5.5				BAB EL MANDEB	D	ZZZ
1958 05 25		02 53 48.5	12.10	43.73	0							D	ISS
1958 05 25		02 53 53.1	12.16	43.88		5.0						D	JED
1958 05 25		02 53 55.3	12.144	43.735	33							D	PG3
1958 05 25		02 53 40	12.0	43.5		5						D	MOS
1958 05 25		02 53 48.4	12.13	43.69			5.0					D	S-L
1958 05 25		02 53 48.0	12.0	43.5								D	STR
1958 05 25		02 53 48	12.15	43.77	33		5.0				BAB EL MANDEB	D	ZZZ
1958 06 28		17 05 24.5	12.26	45.13								D	JED
1958 06 28		17 05 16.2	11.94	45.44								D	S-L
1958 06 28		17 05 22.0	12.00	45.00								D	STR
1958 06 28		17 05 25	12.26	45.13					4.5		GULF OF ADEN	D	ZZZ
1959 12 31		15 10 52	11.6	42.8			4.5				FORESHOCK	D	STR
1959 12 31		11 10 52	11.8	42.8			4.5				GULF OF TADJOURA	D	ZZZ
1959 12 31	*	15 10 52	11.5	42.8						4	FELT IN ARTA	D	ROT
1959 12 31	*	15 10 52	11.5	43.0						4	FELT IN DJIBOUTI	D	ROT
1960 01 04		06 07 55.1	11.50	42.87							FORESHOCK	D	S-L
1960 01 04		06 08 00.7	11.83	42.84								D	JED
1960 01 04		06 08 00	11.7	42.8					4.3		REPUBLIC OF DJIBOUTI	D	ZZZ
1960 01 04	*	06 10	11.1	42.7							FELT IN ALI SABIEH	D	
1960 01 04	*	06 10	11.5	42.8							FELT IN ARTA	D	
1960 01 04	*	06 10	11.5	43.0						4	FELT + NOISE IN DJIBOUTI	D	
1960 01 04		06 16 30	11.55	42.77	33						MAIN SHOCK	D	CGS

Y	M	D		H1	N	E	H2	MB	MS	ML	ME	I	LOCATION & COMMENTS	R	A
1960	01	04		06 16 32.5	11.55	42.71	0							D	ISS
1960	01	04		06 16 35.3	11.51	42.84								D	JED
1960	01	04		06 16 30.9	11.60	42.80								D	ROT
1960	01	04		06 16 30.9	11.55	42.77								D	S-L
1960	01	04		06 16 35.0	11.60	42.80			5.5					D	STR
1960	01	04		06 16 35	11.51	42.84					4.7		VICINITY OF ARTA	D	ZZZ
1960	01	04	*	06 17	11.1	42.7							FELT IN ALI SABIEH	D	
1960	01	04	*	06 17	11.5	42.8						8	ARTA-LANDSLIDES,DAMAGE	D	AAE
1960	01	04	*	06 17	11.5	43.0							FELT IN DJIBOUTI	D	
1960	01	04	*	06 17	11.4	43.0							FELT IN TADJOURA	D	
1960	01	04		22 07 00	11.60	42.80							AFTERSHOCK	D	STR
1960	01	04		22 07 00	11.60	42.80					4.		REPUBLIC OF DJIBOUTI	D	ZZZ
1960	01												65 AFTERSHOCKS AT AAE	D	AAE
1960	07	14		18 39 34.0	07.00	38.50								C	CGS
1960	07	14		18 39 36.0	07.25	38.45								C	ISS
1960	07	14		18 39 49.1	07.18	38.82								C	JED
1960	07	14		18 39 36					6.3					C	LWI
1960	07	14		18 39 35	07.00	37.50			6.3					C	MOS
1960	07	14		18 39 35.8	07.20	38.50								C	ROT
1960	07	14		18 39 34.0	07.00	38.50			6.3					C	STR
1960	07	14		18 39 35.8	07.17	38.46			6.3					C	S-L
1960	07	14		18 39 36	07.2	38.5	40		6.3				RIFT FLOOR NEAR V. CHABBI	C	ZZZ
1960	08	08		12 28 08.5	12.10	44.60	15	5.4						D	CGS
1960	08	08		12 28 08.0	12.06	44.50								D	ISS
1960	08	08		12 28 12.4	12.06	44.49								D	JED
1960	08	08		12 27 40.0	12.10	44.60			5.4					D	MOS
1960	08	08		12 28 07.7	12.10	44.50	15	5.4						D	ROT
1960	08	08		12 28 07.7	12.06	44.49								D	S-L
1960	08	08		12 28 08	12.1	44.6			5.7					D	STR
1960	08	08		12 28 12	12.08	44.54	15		5.7				GULF OF ADEN	D	ZZZ
1960	08	13		20 04 25	15.2	40.1			4.9				HAWACHIL BAY		ZZZ
1960	08	13		22 28 19	15.27	40.00								C	AAE
1960	08	13		22 28 24.6	15.8	40.2	42							C	CGS
1960	08	13		22 28 19.4	15.11	40.15								C	JED
1960	08	13		22 28 13.6	14.70	40.16								C	S-L
1960	08	13		22 28 19	15.2	40.2	42				4.5		HAWACHIL BAY	C	ZZZ
1960	10	23		19 21 15.7	17.90	40.30	25							C	CGS
1960	10	23		19 21 07.7	17.50	40.07								C	S-L
1960	10	23		19 21	17.50	40.07	42				4.5		RED SEA TROUGH	C	ZZZ

Y M D		H1	N	E	H2	MB	MS	ML	ME	I	LOCATION & COMMENTS	R	A
1960 12 16		16 49 15.9	14.7	42.9	29		5.0					C	CGS
1960 12 16		16 49 20.5	14.81	42.48								C	JED
1960 12 16		16 49 15.0	14.70	42.60	29							C	ROT
1960 12 16		16 49 15.9	14.70	42.90	29		5.2					C	STR
1960 12 16		16 49 15.0	14.65	42.57								C	S-L
1960 12 16		16 49 16	14.81	42.48	29		5.2				S RED SEA	C	ZZZ
1961 03 11		08 41 00.4	11.2	43.3	18							D	CGS
1961 03 11		08 41 04.0	11.64	42.98	0							D	ISS
1961 03 11		08 41 07.9	11.71	43.25		5.9						D	JED
1961 03 11		08 41 08.0	11.5	43.5			6.2					D	MOS
1961 03 11		08 41 02	12.0	43.0			6.0					D	QUE
1961 03 11		08 41 03.6	11.70	43.00			6.0					D	ROT
1961 03 11		08 41 03.6	11.65	42.95		5.9						D	S-L
1961 03 11		08 41 06	11.8	43.0			6.1					D	STR
1961 03 11		08 41 04	11.65	43.00	18	6.0					GULF OF TADJOURA	D	ZZZ
1961 03 11	*	08 42	11.5	43.0						6	FELT IN DJIBOUTI	D	
1961 03 11	*	08 42									FELT THROUGHOUT REPUBLIC	D	
1961 05	*										FROM MAY-SEPT OVER 3500	A	AAE
1961 05											SHOCKS RECORDED AT AAE	A	AAE
1961 05											FROM WOLLO. KARA KORE HEA	A	AAE
1961 05	*										VILY DAMAGED, MAJETE DES-	A	AAE
1961 05	*										TROYED. NO KNOWN CASUALTY	A	AAE
1961 05 29		04 59 42.1	10.43	40.03		5.0						A	JED
1961 05 29		04 59 35.0	10.0	39.0			5.0					A	MOS
1961 05 29		04 59 38.9	10.50	39.74			5.0					A	S-L
1961 05 29		04 59 40	10.50	39.50								A	STR
1961 05 29		04 59 39	10.38	39.73			5.0				WOLLO	A	ZZZ
1961 05 29		10 52 01.2	10.4	39.4	25		5.0					A	CGS
1961 05 29		10 52 14.0	10.28	39.91	121							A	ISS
1961 05 29		10 51 49.4	10.43	40.16		5.5						A	JED
1961 05 29		10 51 53.0	09.00	39.50			5.5					A	MOS
1961 05 29		10 52 01.2	10.60	39.40	25		5.5					A	ROT
1961 05 29		10 51 59.5	10.39	39.81			5.5					A	S-L
1961 05 29		10 52 02	10.38	39.86	25		5.5				WOLLO	A	ZZZ
1961 05 29		11 39 56.2	10.62	40.06								A	JED
1961 05 29		11 39 47.5	10.17	40.12								A	S-L
1961 05 29		11 39 57	10.57	39.73	32				4.5		WOLLO	A	ZZZ
1961 05 29		11 59 09	10.50	39.50								A	STR

Y	M	D		H1	N	E	H2	MB	MS	ML	ME	I	LOCATION & COMMENTS	R	A
1961	05	29		11 59 09	10.50	39.50					4.5		WOLLO	A	ZZZ
1961	05	29		19 24 01.3	10.12	40.11	100							A	ISS
1961	05	29		19 23 53.9	10.46	40.18		5.0						A	JED
1961	05	29		19 24 01.3	10.50	39.84			5.0					A	S-L
1961	05	29		19 24 02	10.41	39.88			5.0				WOLLO	A	ZZZ
1961	05	29		19 26 05.5	10.4	40.0	52							A	CGS
1961	05	29		19 26 05.5	10.4	40.0			5.0					A	MOS
1961	05	29		19 26 05.5	10.50	39.84								A	S-L
1961	05	29		19 26 06	10.50	39.84	52		5.0				WOLLO	A	ZZZ
1961	05	29	*	NT	09	39						5	ADDIS ABABA: TREMORS, PA-	A	AAE
1961	05	29	*	19 30	09	39							NIC IN BOARDING SCHOOLS	A	
1961	05	29		19 29 04.8	10.60	39.90	33							A	CGS
1961	05	29		19 29 04.8	10.60	39.90	33				4.5		WOLLO	A	ZZZ
1961	05	29		19 40 29.0	10.46	40.16								A	JED
1961	05	29		19 40 19.0	10.50	39.50								A	MOS
1961	05	29		19 40 24.4	10.43	39.77								A	S-L
1961	05	29		19 40 25.0	10.50	39.50								A	STR
1961	05	24		19 40 28	10.41	39.86	33				4.5		WOLLO	A	ZZZ
1961	05	30		13 11 19.4	10.58	40.12								A	JED
1961	05	30		13 11 16.9	10.72	39.79								A	S-L
1961	05	30		13 11 18.0	10.50	39.50								A	STR
1961	05	30		13 11 17	10.53	39.82					4.5		WOLLO	A	ZZZ
1961	06	01		21 07 21.7	10.75	39.94								A	JED
1961	06	01		21 07 18.5	10.92	39.57								A	S-L
1961	06	01		27 07 20.0	10.30	39.90								A	STR
1961	06	01		21 07 21	10.70	39.64					4.5		WOLLO	A	ZZZ
1961	06	01		23 29 21.1	10.6	39.3	51	6.7					PDE	A	CGS
1961	06	01		23 29 21.2	10.40	39.90	51	6.4					REVISED EDR	A	CGS
1961	06	01		23 29 18.0	10.54	39.89	0							A	ISS
1961	06	01		23 29 21.7	10.53	40.17		6.4						A	JED
1961	06	01		23 29 15.0	10.00	39.50			6.7					A	MOS
1961	06	01		23 29 21.2	10.60	39.80			6.3					A	ROT
1961	06	01		23 29 05.0	09.00	39.00			6.5					7	SHL
1961	06	01		23 29 18.8	10.63	39.81			6.4					A	S-L
1961	06	01		23 29 21.0	10.30	39.90			6.5					A	STR
1961	06	01		23 29 20	10.48	39.87	51		6.5				WOLLO	A	ZZZ
1961	06	01	*	23 30	09	39						7	ADDIS ABABA	A	AAE
1961	06	01		23 56 48.6	10.1	39.6	60							A	CGS

Y	M	D	H1	N	E	H2	MB	MS	ML	ME	I	LOCATION & COMMENTS	R	A
1961	06	01	23 56 52.0	10.65	39.89								A	JED
1961	06	01	23 56 37.8	09.43	39.89								A	S-L
1961	06	01	23 56 52	10.60	39.59	60				4.5		WOLLO	A	ZZZ
1961	06	02	00 01 46.9	10.40	39.80	29							A	CGS
1961	06	02	00 01 47.3	10.49	39.96								A	JED
1961	06	02	00 01 42.8	10.46	39.63								A	S-L
1961	06	02	00 01 46	10.44	39.66	29				4.5		WOLLO	A	ZZZ
1961	06	02	00 08 57.9	10.4	40.0	64							A	CGS
1961	06	02	00 08 57.6	10.42	40.26								A	JED
1961	06	02	00 08 53.5	10.41	39.90								A	S-L
1961	06	02	00 08 57	10.37	39.96	64				4.5		WOLLO	A	ZZZ
1961	06	02	00 21 19.4	09.84	40.01								A	JED
1961	06	02	00 21 18.9	10.41	39.70								A	S-L
1961	06	02	00 21 24.0	10.50	40.00								A	STR
1961	06	02	00 21 19	09.79	39.71					4.5		WOLLO	A	ZZZ
1961	06	02	00 57 48.8	09.00	40.20	33							A	CGS
1961	06	02	00 57 55.3	09.89	40.19								A	JED
1961	06	02	00 57 53.2	10.03	39.20								A	S-L
1961	06	02	00 57 54.0	09.50	40.00								A	STR
1961	06	02	00 57 55	09.84	39.89					4.5		WOLLO	A	ZZZ
1961	06	02	01 16 07.1	09.40	40.20	33							A	CGS
1961	06	02	01 16 07	09.40	40.20	33				4.5		WOLLO	A	ZZZ
1961	06	02	01 16 10.6	09.7	39.7	60							A	CGS
1961	06	02	01 16 36.0	09.50	40.00								A	STR
1961	06	02	01 16 10.6	09.6	39.8	60				4.5		WOLLO	A	ZZZ
1961	06	02	02 35 24.4	09.00	40.00	33							A	CGS
1961	06	02	02 35 41.6	11.27	40.54								A	JED
1961	06	02	02 35 33.9	10.88	40.51								A	S-L
1961	06	02	02 35 42	11.22	40.24	24				4.5		WOLLO	A	ZZZ
1961	06	02	03 19 39.5	09.94	39.95								A	JED
1961	06	02	03 19 34.9	10.03	40.12								A	S-L
1961	06	02	03 19 30.0	09.50	40.00								A	STR
1961	06	02	03 19 39.5	09.89	39.65	33				4.5		WOLLO	A	ZZZ
1961	06	02	03 49 11.1	10.06	40.84								A	JED
1961	06	02	03 49 04.0	09.71	40.69								A	S-L
1961	06	02	03 49 18.0	09.50	40.00								A	STR
1961	06	02	03 49 11	10.01	40.54					4.5		WOLLO	A	ZZZ

Y	M	D		H1	N	E	H2	MB	MS	ML	ME	I	LOCATION & COMMENTS	R	A
1961	06	02		04 51 10.4	09.8	40.0	41	6.4					PDE	A	CGS
1961	06	02		04 51 14.8	10.30	39.80	41	6.4					EDR	A	CGS
1961	06	02		04 51 20.0	10.32	39.82	77							A	ISS
1961	06	02		04 51 04.8	10.31	40.13		6.2						A	JED
1961	06	02		04 51 00.0	09.00	39.00			6.3					A	MOS
1961	06	02		04 51 10.6	10.40	39.90			6.4					A	ROT
1961	06	02		04 50 56.0	09.00	38.00			6.3					A	SHL
1961	06	02		04 51 10.6	10.36	39.91		6.2						A	S-L
1961	06	02		04 51 14.0	10.30	39.90			6.3					A	STR
1961	06	02		04 51 04.8	10.26	39.83	41		6.4				WOLLO	A	ZZZ
1961	06	02	*	04 51									ROCSLIDES, SUBSIDENCES,	A	
1961	06	02	*	04 51									15 KM FISSURE ALONG HIGH-	A	AAE
1961	06	02	*	04 51									WAY NO. 1	A	AAE
1961	06	02		05 22 29.1	10.4	39.6	26							A	CGS
1961	06	02		05 22 42.0	10.21	39.89	99							A	ISS
1961	06	02		05 22 19.1	10.20	40.08								A	JED
1961	06	02		05 22 28.5	10.30	40.05								A	S-L
1961	06	02		05 22 29	10.15	39.78	26				4.5		WOLLO	A	ZZZ
1961	06	02		05 44 52.4	10.4	39.8	31							A	CGS
1961	06	02		05 45 10.0	10.58	40.14	156							A	ISS
1961	06	02		05 44 36.9	10.27	40.22		5.8						A	JED
1961	06	02		05 44 52.7	10.60	40.10	31		5.8					A	ROT
1961	06	02		05 44 52.7	10.59	40.06		5.8						A	S-L
1961	06	02		05 44 57.0	10.30	39.90			5.8					A	STR
1961	06	02		05 44 53	10.22	39.92	31		5.8				WOLLO	A	ZZZ
1961	06	02		06 17 13.3	10.50	39.70	36							A	CGS
1961	06	02		06 17 15.8	10.67	40.07								A	JED
1961	06	02		06 17 10.7	10.54	39.82								A	S-L
1961	06	02		06 17 16	10.62	39.77	36				4.5		WOLLO	A	ZZZ
1961	06	02		07 02 52.4	10.3	40.0	54						PDE	A	CGS
1961	06	02		07 02 49.9	10.00	40.00	54						EDR	A	CGS
1961	06	02		07 02 50.0	10.13	39.90	21							A	ISS
1961	06	02		07 02 47.1	10.20	40.19								A	JED
1961	06	02		07 02 49					5.5					A	MOS
1961	06	02		07 02 46.1	10.14	39.91			5.5					A	S-L
1961	06	02		07 02 50	10.0	40.0			5.5					A	STR
1961	06	02		07 02 52	10.15	39.89	54		5.5				WOLLO	A	ZZZ
1961	06	02		07 21 45.3	10.10	39.60	33							A	CGS
1961	06	02		07 21 49.8	10.73	40.04								A	JED
1961	06	02		07 21 43.7	10.50	39.74									S-L

Y	M	D	H1	N	E	H2	MB	MS	ML	ME	I	LOCATION & COMMENTS	R	A
1961	06	02	07 21 49.9	10.68	39.74	33				4.5		WOLLO	A	ZZZ
1961	06	02	22 19 34.5	10.43	40.28								A	JED
1961	06	02	22 19 30.0	10.31	39.89								A	S-L
1961	06	02	22 19 34	10.38	39.98					4.5		WOLLO	A	ZZZ
1961	06	02	23 32 32.1	09.50	39.80	33							A	CGS
1961	06	02	23 32 39.7	10.52	40.15								A	JED
1961	06	02	23 32 34.6	10.42	39.95								A	S-L
1961	06	02	23 32 36	10.49	39.85	33				4.5		WOLLO	A	ZZZ
1961	06	03	02 05 36.1	10.29	40.11								A	JED
1961	06	03	02 05 27.5	09.80	40.22								A	S-L
1961	06	03	02 05 34.0	10.30	39.90								A	STR
1961	06	03	02 05 34	10.24	39.81					4.5		WOLLO	A	ZZZ
1961	06	03	15 20 30.9	10.60	39.80	60							A	CGS
1961	06	03	15 20 28.7	10.43	40.16								A	JED
1961	06	03	15 20 22.4	10.17	39.97								A	S-L
1961	06	03	15 20 30.00	10.75	40.00								A	STR
1961	06	03	15 20 31	10.38	39.86	60				4.5		WOLLO	A	ZZZ
1961	06	03	15 23 15.8	09.80	39.60	50	5.7						A	CGS
1961	06	03	15 23 34.0	10.50	39.80	151							A	ISS
1961	06	03	15 23 03.6	10.62	40.23								A	JED
1961	06	03	15 23 16.4	10.50	39.90			5.8					4	ROT
1961	06	03	15 23 16.4	10.49	39.93		5.8						A	S-L
1961	06	03	15 23 18.0	10.75	40.00			5.6					A	STR
1961	06	03	15 20 30	10.57	39.93	60		5.8				WOLLO	A	ZZZ
1961	06	03	16 25 46.9	09.50	39.80	33							A	CGS
1961	06	03	16 25 56.0	10.59	40.13								A	JED
1961	06	03	16 25 51.8	10.61	40.01								A	S-L
1961	06	03	16 25 54.0	10.75	40.00								A	STR
1961	06	03	16 25 54	10.54	39.83					4.5		WOLLO	A	ZZZ
1961	06	04	00 41 43.1	10.34	40.02								A	JED
1961	06	04	00 41 39.2	10.40	39.82								A	S-L
1961	06	04	00 46 42.0	10.75	40.00								A	STR
1961	06	04	00 41 43	10.29	39.72					4.5		WOLLO	A	ZZZ
1961	06	06	17 46 43.4	10.88	39.86								A	JED
1961	06	06	17 46 42.8	11.08	39.19								A	S-L
1961	06	06	17 46 42.0	11.00	39.25								A	STR
1961	06	06	17 46 43	10.83	39.56					4.5		WOLLO	A	ZZZ

Y	M	D		H1	N	E	H2	MB	MS	ML	ME	I	LOCATION & COMMENTS	R	A
1961	06	07		15 01 16.5	10.54	40.17								A	JED
1961	06	07		15 01 12.2	10.59	40.04								A	S-L
1961	06	07		15 01 12.0	10.30	39.90								A	STR
1961	06	07		15 11 6	10.54	39.74					4.5		WOLLO	A	ZZZ
1961	06	14		20 32 24.0	10.8	40.1	56						PDE	A	CGS
1961	06	14		20 32 21.6	10.50	39.90	56						EDR	A	CGS
1961	06	14		20 32 29.0	10.58	39.73	92							A	ISS
1961	06	14		20 32 10.1	10.55	40.05								A	JED
1961	06	14		20 32 19.0	10.00	40.00			5.0					A	MOS
1961	06	14		20 32 21.6	10.60	39.80			5.7					A	ROT
1961	06	14		20 32 17.8	10.64	39.80								A	S-L
1961	06	14		20 32 18.0	10.75	40.00			5.7					A	STR
1961	06	14		20 32 23	10.50	39.75	56		5.7				WOLLO	A	ZZZ
1961	06	19		04 34 15.1	10.30	40.10	33							A	CGS
1961	06	19		04 34 14.3	10.44	40.21								A	JED
1961	06	19		04 34 11.6	10.46	39.96								A	S-L
1961	06	19		04	10.39	39.91					4.5		WOLLO	A	ZZZ
1961	06	20		03 21 34.3	12.20	44.20	33		6.1					D	CGS
1961	06	20		03 21 30.0	12.29	44.25								D	ISS
1961	06	20		03 21 28.8	12.17	44.37								D	JED
1961	06	20		03 21 21.0	11.00	43.50			6.0					D	MOS
1961	06	20		03 21 29.5	12.20	44.30			6.1					D	ROT
1961	06	20		03 21 29.5	12.23	44.34			6.1					D	S-L
1961	06	20		03 21 28.8	12.22	44.29	33		6.1				W GULF OF ADEN	D	ZZZ
1961	06	24		15 04 34.2	10.65	40.14								A	JED
1961	06	24		15 04 31.3	10.69	39.89								A	S-L
1961	06	24		15 04 34	10.65	40.14					4.5		WOLLO	A	ZZZ
1961	08	25	*	21 43 48.9									FELT IN ADDIS. I=IV,200KM	A	AAE
1962	08	25		00 54 08	16.49	40.12	33		4.8					C	CGS
1962	08	25		00 54 17.5	17.12	40.14		4.8						C	JED
1962	08	25		00 54 30	19.	40.			4.8					C	MOS
1962	08	25		00 54 08.0	16.50	40.10								C	ROT
1962	08	25		00 54 25	18.00	41.00								6	STR
1962	08	25		00 54 08.0	16.49	40.12			4.8					C	S-L
1962	08	25		00 54 17	17.12	40.14	33		4.8				RED SEA TROUGH	C	ZZZ
1962	09	09	*		15.33	37.53						3	FELT IN AGORDAT	A	AAE
1962	11	11		15 15 33.6	17.20	40.70	34	5.6						C	CGS
1962	11	11		15 15 34.0	17.12	40.52	43							C	ISS

Y	M	D		H1	N	E	H2	MB	MS	ML	ME	I	LOCATION & COMMENTS	R	A
1962	11	11		15 15 33.9	17.22	40.58								C	JED
1962	11	11		15 15 27.0	16.50	40.00			5.5					C	MOS
1962	11	11		15 15 20.0	16.50	39.50	15		5.2					C	PEK
1962	11	11		15 15 28.0	17.10	40.60	34		5.8					C	ROT
1962	11	11		15 15 30.0	17.25	40.75			5.8					C	STR
1962	11	11		15 15 28.0	17.05	40.58			5.6					C	S-L
1962	11	11		15 15 20.0	17.22	40.58	15		5.7				RED SEA	C	ZZZ
1963	02	21		09 17 12.0	12.00	43.00								D	STR
1963	02	21		09 17 12	12.00	43.00					4.5		S. DANAKIL HORST	D	ZZZ
1963	07	14		17 18 06	15.1	38.7			4.5					A	MOS
1963	07	14		17 18 10	15.6	39.0								A	STR
1963	07	14		17 18 10	15.6	39.1			4.5				ERITREAN ESCARPMENT	A	ZZZ
1963	07	14	*	17 20	15.5	39.0						4.5	2 FELT IN ASMARA	2	AAE
1963	10	02	*	14 57 47.4	11.6	42.8	33	5.3						D	CGS
1963	10	05		14 57 43.0	11.49	42.82								D	ISS
1963	10	05		14 57 47.8	11.60	42.85		5.3						D	JED
1963	10	05		14 57 48.0	11.64	42.83		5.3						D	JED
1963	10	05		11 57 49	11.5	42.3			6.					D	MOS
1963	10	05		14 57 47.4	11.60	42.80			5.8					D	ROT
1963	10	05		14 57 42	11.5	42.8								D	STR
1963	10	05		14 57 48	11.57	42.82	21	5.3					GULF OF TADJOURA	D	ZZZ
1963	10	05	*										WATER SEEN AS BOILING	D	
1963	10	05	*	14 58	11.5	42.8						5	ARTA: TREMORS,LANDSLIDES	D	AAE
1963	10	05	*	14 58	11.5	43.0						4	FEL IN DJIBOUTI	D	AAE
1963	10	05		17 18 25.0	11.7	42.6	33	5.3						D	CGS
1963	10	05		17 18 25.6	11.70	42.71								D	JED
1963	10	05		17 18 25.6	11.70	42.65	33		5.3				GODA MTS	D	ZZZ
1963	10	05	*		11.7	42.7						6	GODA MASSIF: LANDSLIDES	D	
1963	10	05	*		11.5	43.0							14 AFTERSHOCKS DJIBOUTI	D	AAE
1963	11	16	*	05 30	11.53	42.85						4	FELT IN ARTA	D	AAE
1964	07	03			11.0	39.4								A	AAE
1964	07	03		19 18 34.0	11.01	39.28	60	5.0						A	CGS
1964	07	03		19 18 32.7	11.05	39.70	54	4.9						A	ISC
1964	07	03		19 18 33.0	11.32	39.57		5.0						A	JED
1964	07	03		19 18 34.0	11.00	39.30	60							A	ROT
1964	07	03		19 18 33	11.0	39.4	60	4.2	5.0				WOLLO	A	ZZZ
1964	07	03	*	19 18	11.1	39.3						6	FELT IN DESSIE	A	AAE
1964	07	11		00 24 52 P				2.5					D(AAE) ABOUT 50 KM. WFB	C	AAE
1964	07	11	*									3	FELT IN NAZRETH.	C	AAE

Y M D		H1	N	E	H2	MB	MS	ML	ME	I	LOCATION & COMMENTS	R	A
1964 08 03		10 45 09	17.5	39.1								C	STR
1964 08 03		10 45 14	15.2	39.0								C	MOS
1964 08 03		10 45 09	17.1	39.0	33				4.5		FOOT OF ESCARPMENT	C	ZZZ
1964 10 17		17 41 46 P						3.0			D(AAE)=150 KM	A	AAE
1964 10 17	*		09	39						3	FELT IN ADDIS ABABA	A	AAE
1964 10 17		17 41 20	10.20	39.75				3			ESCARPMENT	A	ZZZ
1964 10 17		18 03 03	10.20	39.75				3			ESCARPMENT	A	ZZZ
1964 12 13		18 42 44	09.2	41.7	33	3.6		4.0			SE PLATEAU	B	ZZZ
1964 12 13	*	18 42 44	09.2	42.2						4	FELT IN BISIDIMO		AAE
1965 03 07		07 32 38.1	12.1	46.3	38	4.9						D	CGS
1965 03 07		07 32 36.2	12.16	46.27	13	5.2						D	ISC
1965 03 07		07 39 39.4	12.19	46.34		4.9						D	JED
1965 03 07		07 32 36	12.16	46.27							W GULF OF ADEN	D	ZZZ
1965 03 07		07 42 31.2	12.1	46.3	33	5.3						D	CGS
1965 03 07		07 42 32.4	12.15	46.34	39	5.4						D	ISC
1965 03 07		07 42 32.2	12.17	46.38		5.3						D	JED
1965 03 07		07 42 30	11.9	46.2		5.5						D	MOS
1965 03 07		07 42 32.2	12.15	46.36		5.3					GULF OF ADEN	B	ZZZ
1965 04 14		07 53 10.8P	11.3	39.6		3.9	4.3				W PLATEAU MARGIN	A	ZZZ
1965 04 14	*		11.1	39.3						3	FELT IN DESSIE	A	AAE
1965 05 16		00 45 56.8	10.93	45.52	33							D	CGS
1965 05 16		00 45 57.1	11.00	45.60	33							D	ISC
1965 05 16		00 45 58.0	11.07	45.61								D	JED
1965 05 16		00 45 56	11.00	45.57		4.2		4.6			OFF N COAST OF SOMALIA	D	ZZZ
1965 06 07		13 44 52.1P						5.5	4.8			C	AAE
1965 06 07		13 43 57.2	11.4	41.5	40							C	ARP
1965 06 07		13 43 57.2	11.43	41.48	40	5.1						C	CGS
1965 06 07		13 43 58.0	11.35	41.53	55		5.1				(RECOMPUTED)	C	CGS
1965 06 07		13 43 57.8	11.48	41.51	42		4.9					C	CGS
1965 06 07		13 43 59.0	11.63	41.38								C	ISC
1965 06 07		13 43 58.0	11.60	41.50								C	JED
1965 06 07		13 43 58	11.51	41.47	40		5.1	5.5			CENTRAL AFAR	C	STR
1965 06 07	*		11.6	41.4						3	ASSAYITA	C	ZZZ
1965 06 07	*	13 44	11.8	41.2						3.5	DUBTI-TENDAHO	C	AAE
1965 06 07	*	13 44	12.0	41.3						3	SERDO	C	AAE
1965 06 29	*	22 04 33.3P	16.3	38.9		3.8		4.2			D(AAE)=850 KM	A	AAE

Y M D		H1	N	E	H2	MB	MS	ML	ME	I	LOCATION & COMMENTS	R	A
1965 06 29		22 02 30	16.1	38.9		3.8					ERITREAN ESCARPMENT	A	ZZZ
1965 06 29	*	22 04 33	15.3	38.9						3	FELT IN ASMARA	A	AAE
1965 07 19		15 49 35.9	12.12	42.58	33							D	CGS
1965 07 19		15 49 35.9	12.1	42.6	33							D	ISC
1965 07 19		15 49 37.0	12.22	42.56								D	JED
1965 07 19		15 49 33.63	11.83	42.55							(BC1S)	D	PG
1965 07 19		15 49 36.5	11.700	42.483							(FAIRHEAD)	C	PG
1965 07 19		15 49 37	11.77	42.52	33				4.5		E MARGIN ASAL RIFT	D	ZZZ
1965 09 30	*	18 59 18	12.7	39.7				4.2			GUF GUF GRABEN. FELT	A	ZZZ
1965 09 30		21 23 55.6P	12.7	39.7	33			4.5				A	AAE
1965 09 30		21 22 55.6	12.7	39.7				4.5			GUF GUF GRABEN	A	ZZZ
1965 09 30	*	21 23 56	12.8	39.6						4	FELT IN KORBETTA	A	AAE
1965 09 30	*	21 23 56	12.7	39.55						4	FELT IN MAI CHEW	A	AAE
1965 09 30	*	21 23 56	13.5	39.5						3	FELT IN MEKELE	A	AAE
1965 09 30		22 17 45	12.7	39.7				4.0			GUF GUF GRABEN	A	ZZZ
1965 10 24		17 46 46 P	05.6	37.2		4.3					D(AAE)=420 KM	E	AAE
1965 10 24		17 45 50	05.5	37.4				5.1			DULEI GRABEN	E	ZZZ
1965 10 24	*		05.7	37.5						4	TREMORS + NOISE IN GIDOLE	E	AAE
1965 12 30		08 54 13.5	18.75	39.49	33		4.1					C	CGS
1965 12 30		08 54 14.5	18.87	39.71		4.1						C	JED
1965 12 30		08 54 14.5	18.87	39.71			4.1				RED SEA TROUGH	C	ZZZ
1966 01 21		12 39 43.1	12.00	43.77	33		4.7				LARGEST IN SWARM OF 157	D	CGS
1966 01 21											SHOCKS REGISTERED AT AAE	D	
1966 01 21		12 39 46.0	11.70	43.70	74		4.7					D	ISC
1966 01 21		12 39 44.9	12.11	43.64		4.7						D	JED
1966 01 21		12 39 45	12.11	43.77	33		4.7				N TADJOURA TRENCH	D	ZZZ
1966 01 21	*									4	22 TREMORS IN DJIBOUTI	D	
1966 02 05		04 20 10 P				3.8		4.2			D(AAE)=785 KM	A	AAE
1966 02 05	*	04 20	15.5	39.0						5	FELT IN ASMARA	A	AAE
1966 02 05	*	10 25	15.5	39.0						3	FELT IN ASMARA	A	AAE
1966 04 09		19 11 10.5	14.37	40.79	33	4.7						C	CGS
1966 04 09		19 11 11.0	14.50	40.90	27	4.8						C	ISC
1966 04 09		19 11 12.0	14.46	40.74		4.7						C	JED
1966 04 09		19 11	14.45	40.75	27	4.7					DANAKIL ALPS	C	ZZZ
1967 01 23		05 33 19.1P	08.4	36.5				4.1				A	AAE

Y	M	D		H1	N	E	H2	MB	MS	ML	ME	I	LOCATION & COMMENTS	R	A
1967	01	23		05 32 59.6	08.67	37.65				4.1			AMBO-GUDER REGION	A	ZZZ
1967	01	23	*	AM	09.0	37.9						5	FELT IN AMBO	A	AAE
1967	01	23	*	AM	08.4	36.3						4	FELT IN BEDELE	A	AAE
1967	01	23	*	AM	08.5	38.0							FELT IN GHION	A	AAE
1967	01	23	*	AM	09.1	36.5							FELT IN LEKEMT	A	AAE
1967	01	23	*	AM	08.2	37.8							FELT IN WOLKITE	A	AAE
1967	01	30	*	19 25	15.5	39.0					4.2	3	FELT IN ASMARA	A	AAE
1967	05	07		08 56 04 P				4.1		4.6			FORESHOCK	E	AAE
1967	05	07		08 55 08	04.3	35.7		4.1					W SCARP TURKANA RIFT	E	ZZZ
1967	05	08		13 58 29.0	05.0	35.0			4.3					E	BUL
1967	05	08		13 58 27.0	04.10	35.40	0							E	ISC
1967	05	08		13 59 00	04.3	35.7		4.5		5.0			W SCARP TURKANA RIFT	E	ZZZ
1967	05												4 FORESHOCKS	C	
1967	05	19		15 52 34.2	14.53	40.26	13	5.1						C	CGS
1967	05	19		15 52 39.0	14.62	40.17								C	ISC
1967	05	19		15 52 40.5	14.87	40.14		4.8						C	JED
1967	05	19		15 52 39	14.70	40.10								C	STR
1967	05	19		15 52 39	14.68	40.17	13	5.1					N END DANAKIL DEPRESSION	C	ZZZ
1967	05												74 AFTERSHOCKS	C	
1967	05	19		17 11 12.8	14.82	40.03				4.7			DEPRESSION E MARGIN	C	ZZZ
1967	05	19		21 21 34.2	14.67	40.12	13			4.2			DEPRESSION E MARGIN	C	ZZZ
1967	05	20		00 20 33.8	14.55	40.19	13			4.2			DEPRESSION E MARGIN	C	ZZZ
1967	05	20		03 38 36	14.74	40.08	13			4.3			DEPRESSION E MARGIN	C	ZZZ
1967	09	18		02 04 37.5P	15.79	38.97		4.9					D(AAE)=750 KM	A	AAE
1967	09	18		02 02 59.8	15.69	39.03	33	4.8						A	CGS
1967	09	18		02 03 04.0	15.78	38.94	58	4.9						A	ISC
1967	09	18		02 03 01.8	15.80	38.93		4.8						A	JED
1967	09	18		02 03 02	15.79	38.93	33	4.8					SABARGUMA GRABEN	A	ZZZ
1967	09	18	*	NT	15.5	39.						3	FELT IN ASMARA	A	AAE
1967	09	21		18 36 35	18.1	40.1				4.6				C	AAE
1967	09	21		18 36 26.1	17.94	39.97	16	4.4						C	CGS
1967	09	21		18 36 32.0	17.84	40.30	80							C	ISC
1967	09	21		18 36 27.0	18.05	40.13		4.4						C	JED
1967	09	21		18 36 27	17.94	40.13		4.4					RED SEA	C	ZZZ
1967	09	26		17 21 26.6	07.3	37.8	33			2.7			W ESCARPMENT OF RIFT	A	ZZZ
1967	09	26	*	17 21 25	07.5	37.8						3	FELT IN HOSANNA	A	AAE
1967	11	16		02 22 03.1	15.09	39.82	33	5.1						C	CGS
1967	11	16		02 22 05.2	15.19	39.49	33	5.1						C	ISC

Y	M	D		H1	N	E	H2	MB	MS	ML	ME	I	LOCATION & COMMENTS	R	A
1967	11	16		02 22 06.2	15.17	39.53		5.1						C	JED
1967	11	16		02 22 03.1	15.10	39.80	33	5.1						C	SYK
1967	11	16		02 22 03	15.09	39.81	33	5.1		5.9			GULF OF ZULA	C	ZZZ
1967	11	16	*	NT	15.5	39.0						4	FELT IN ASMARA	C	AAE
1967	11	16	*	NT	15.1	39.1						4	FELT IN DECAMERE	C	AAE
1967	12	17		18 04 39.1P				5.0						E	AAE
1967	12	17		18 03	06.	37.			4.7					E	BUL
1967	12	17		18 03 42.6	05.80	37.23	19							E	CGS
1967	12	17		18 03 43.8	05.83	37.10	19	4.9						E	ISC
1967	12	17		18 03 46.3	05.96	36.91								E	JED
1967	12	17	*	18 03 46	05.90	37.06	19	4.9					CHEW BAHR GRABEN	CE	ZZZ
1967	12	17	*	NT	06.1	37.6						4	FELT IN ARBA MINCH	E	AAE
1967	12	17	*	NT	05.8	39.0						3.5	FELT IN KIDANE MENGIST	E	AAE
1967	12	17	*	NT	06.9	35.5						3	FELT IN MIZAN TEFERI	E	AAE
1968	01	23		19 18 13.0	08.71	37.66	33		5.1					A	CGS
1968	01	23		19 18 14.4	08.69	37.41		4.9						A	ISC
1968	01	23		19 18 14.7	08.74	37.56		5.1						A	JED
1968	01	23		19 18 14.6	08.71	37.53	33		5.1				45 KM SW OF AMBO	A	ZZZ
1968	01	23	*	19 18 30	09.0	38.8						4	ADDIS ABABA	A	AAE
1968	01	23	*	19 18	09.0	38.0						5.5	AMBO (LIGHT DAMAGE)	A	AAE
1968	01	23	*	19 20	11.0	36.5						3	DEBATIE	A	AAE
1968	01	23	*	19 20	10.4	37.7							DEBRE MARKOS	A	AAE
1968	01	23	*	19 20	08.6	34.8							DEMBI DOLLO	A	AAE
1968	01	23	*	19 20	07.1	37.8							DUBBO	A	AAE
1968	01	23	*	19 20	07.7	36.8							JIMMA	A	AAE
1968	01	23	*	19 20	09.1	36.55							LEKEMT	A	AAE
1968	01	23	*	19 20	06.9	35.5						3	MISAN TEFERI	A	AAE
1968	05	11		18 16	14.4	39.8		4.1		4.6			ESCARPMENT	A	ZZZ
1968	05	11		21 53	15.0	39.8		3.4		3.9			ESCARPMENT	A	ZZZ
1968	05	11	*	NT	15.1	39.1						4	FELT IN DECAMERE	A	AAE
1968	05												18 FORESHOCKS	C	AAE
1968	05	23		23 37 33.7P	14.8	40.1				5.0			D(AAE)=658 KM	C	AAE
1968	05	23		23 36 06.43	14.747	40.217	33	4.8						C	CGS
1968	05	23		23 36 08.4	14.86	39.90	33							C	ISC
1968	05	23		23 36 07	14.80	40.10	33	4.8		5.0			DEPRESSION E MARGIN	C	ZZZ
1968	05	23	*	NT	15.05	39.05							FELT IN DECAMERE	C	
1968	05												61 AFTERSHOCKS	C	AAE
1969	03	16	*	04 02 18.1P				3.6		4		3	FELT IN BATIE	AC	AAE

Y	M	D		H1	N	E	H2	MB	MS	ML	ME	I	LOCATION & COMMENTS	R	A
1969	03	29		09 15 54	12.0	41.4			5.3					C	BUL
1969	03	29		09 15 54.08	11.970	41.178	33	5.8		6.3				C	CGS
1969	03	29		09 15 54	11.91	41.21	35	5.9						C	ISC
1969	03	29		09 15 54.3	12.02	41.18		5.8						C	JED
1969	03	29		09 15 53	11.9	41.3			6.5					C	MOS
1969	03	29		09 16 00.0	12.0	41.3			6.3					C	STR
1969	03	29		09 15 54.1	11.9	41.1	33	5.8						C	SYK
1969	03	29		09 15 54	11.93	41.27	33	5.8	6.4				SERDO AFAR	C	ZZZ
1969	03	29		11 04 47.85	11.957	41.288	4	5.6						C	CGS
1969	03	29		11 04 52	11.92	41.36	35							C	ISC
1969	03	29		11 04 48.3	12.03	41.24		5.6						C	JED
1969	03	29		11 04 52	11.9	41.2			6.0					C	MOS
1969	03	29		11 04 58.0	12.00	41.30		5.9						C	STR
1969	03	29		11 04 47.9	11.90	41.20	4	5.6						C	SYK
1969	03	29		11 04 48	11.94	41.30	4	5.6	6.0				SERDO	C	ZZZ
1969	03	29	*		11.6	41.2						9	SERDO DESTROYED 25 DEAD	C	AAE
1969	03	29		11 07 30	12.0	41.4								C	BUL
1969	03	29		11 07 29.9	11.99	41.21	33	5.3	5.8					C	CGS
1969	03	29		11 07 45	12.01	41.1	164	4.9						C	ISC
1969	03	29		11 07 30.5	12.06	41.48		5.3						C	JED
1969	03	29		11 07 30.0	11.90	41.20	33	53						C	SYK
1969	03	29		11 07 30	11.91	41.57	33	5.3	5.8				SERDO	C	ZZZ
1969	03	29		13 08 14	12.0	41.4			5.1					C	BUL
1969	03	29		13 08 11.42	11.938	41.523	4	5.1						C	CGS
1969	03	29		13 08 17	11.94	41.31	43	5.1						C	ISC
1969	03	29		13 08 12.7	12.03	41.41		5.1						C	JED
1969	03	29		18 30 42.2	12.0	41.3	33	4.6						C	SYK
1969	03	29		13 08 12	11.85	41.49	4	5.1					SERDO	C	ZZZ
1969	03	29		18 30 42.2	12.00	41.38	33	4.6						C	CGS
1969	03	29		18 30 49	11.87	41.40	95							C	ISC
1969	03	29		18 30 42.6	12.01	41.34		4.6						C	JED
1969	03	29		18 30 42.2	12.0	41.3	33	4.6						C	SYK
1969	03	29		18 30 42	11.89	41.43	33	4.6					SERDO	C	ZZZ
1969	04	05		02 18 29.91	12.153	41.198	17	6.2	6.1					C	CGS
1969	04	05		02 18 30	12.00	41.35	175	5.8						C	ISC
1969	04	05		02 18 30.4	12.19	41.13		6.2						C	JED
1969	04	05		02 18 28.0	12.1	41.6								C	STR
1969	04	05		02 18 29.9	12.1	41.1	17	6.2						C	SYK
1969	04	05		02 18 30	12.12	41.26	17	6.2					SERDO FAULTING.	C	ZZZ
1969	04	05		20 06 23.82	11.910	41.135	33	4.3						C	CGS

Y M D		H1	N	E	H2	MB	MS	ML	ME	I	LOCATION & COMMENTS	R	A
1969 04 05		20 06 24	12.0	41.2	32							C	ISC
1969 04 05		20 06 25.0	12.02	41.10		4.3						C	JED
1969 04 05		20 06 23.8	11.9	41.1	33	4.3						C	SYK
1969 04 05		20 06 24	11.95	41.19	33	4.6	5.2				SERDO	C	ZZZ
1969 04 05		20 14 35.85	12.025	41.473	33	4.9	5.2					C	CGS
1969 04 05		20 14 41	12.02	41.28	70	4.8						C	ISC
1969 04 05		20 14 37.1	12.11	41.49		4.9						C	JED
1969 04 05		20 14 35.8	12.0	41.4	33	4.9						C	SYK
1969 04 05		20 14 35	11.96	41.59	33	4.9					SERDO	C	ZZZ
1969 04 06		16 51 45.50	12.028	41.120	20	5.2	5.4					C	CGS
1969 04 06		16 51 47	11.99	41.40	41	5.1						C	ISC
1969 04 06		16 51 46.1	12.07	41.11		5.2						C	JED
1969 04 06		16 51 52.0	11.8	41.6			5.4					C	STR
1969 04 06		16 51 45.5	12.0	41.1	20	5.2						C	SYK
1969 04 06		16 51 46	12.00	41.20	20	5.2	5.4				SERDO	C	ZZZ
1969 04 07		06 23 53.4	11.978	41.275	33	4.6						C	CGS
1969 04 07		06 23 55	11.92	41.40	58	4.6						C	ISC
1969 04 07		06 23 54.4	12.04	41.30		4.6						C	JED
1969 04 07		06 23 53.4	11.9	41.2	33	4.6						C	SYK
1969 04 07		06 23 53	11.93	41.39	33	4.6					SERDO	C	ZZZ
1969 04 08		02 13 58.70	11.929	41.370	34	4.8						C	CGS
1969 04 08		02 14 01	11.88	41.42	56	4.8						C	ISC
1969 04 08		02 14 00.0	12.02	41.37		4.8						C	JED
1969 04 08		02 13 58.7	11.9	41.3	34	4.8						C	SYK
1969 04 08		02 14 59	11.90	41.62	34	4.8					SERDO	C	ZZZ
1969 05 05		02 45 38.89	11.941	41.285	35	5.2	5.0					C	CGS
1969 05 05		02 45 40	12.07	41.34	38	4.9						C	ISC
1969 05 05		02 45 40.2	12.01	41.28		5.2						C	JED
1969 05 05		02 45 32	11.1	41.4		5.0						C	MOS
1969 05 05		02 45 38.0	11.9	41.5								C	STR
1969 05 05		02 45 38	11.90	41.37	36	5.0	5.2				SERDO	C	ZZZ
1969 05 16	*	SS	14.3	39.5						3	ADIGRAT,TIGRAY	A	AAE
1969 09 26		04 54 35.66	16.428	40.983	25	5.1	5.3					C	CGS
1969 09 26		04 54 38	16.41	41.02	45	5.0						C	ISC
1969 09 26		04 54 38	16.6	41.2			5.5					C	MOS
1969 09 26		04 54 37	16.4	41.0								C	STR
1969 06 26		04 54 36	16.46	41.05	25	5.1	5.3				RED SEA TROUGH	C	ZZZ
1969 10 24		10 12 41.82	11.870	44.858	25	4.8						D	CGS

Y	M	D		H1	N	E	H2	MB	MS	ML	ME	I	LOCATION & COMMENTS	R	A
1969	10	24		10 12 42.3	11.85	44.96	25	4.6						D	ISC
1969	10	24		10 12 43	12.0	45.4								D	MOS
1969	10	24		10 12 42	11.85	44.96	25	4.8					GULF OF ADEN	D	ZZZ
1971	04	25		17 42 16.47	12.020	43.584	33	3.9						D	CGS
1971	04	25		17 42 15	11.82	43.85	25	4.3						D	ISC
1971	04	25		17 42 11	11.1	43.6			4.8					D	MOS
1971	04	25		17 42 16.47	12.020	43.584	33	4.0					TADJOURA TRENCH	D	ZZZ
1971	06				09.5	39.5		3.3		3.8			JUN-AUG SWARM,DEBRE BÈRHA		AAE
1971	06												151 SHOCKS RECORDED	A	AAE
1971	07	11	*	20 52 04.7P	09.6	39.6		3.3		3.7				A	ZZZ
1971	08	04		13 15 02	09.7	39.6		3.4		3.8			DEBRE BERHAN AREA	A	ZZZ
1971	08	04	*	13 15	09.	39.						2	FELT IN ADDIS ABABA	A	AAE
1971	08	04	*	13 15	10.	40.						3	FELT IN ANKOBER	A	AAE
1971	08	04	*	13 15	09.3	39.3						5.5	LIGHT DAMAGE DEBRE BERHAN	A	AAE
1971	08	04		13 15	09.3	39.3						4	FELT IN SHENO	A	AAE
1971	08	14	*	NT	06.9	35.5						3	FELT IN MIZAN TEFERI	A	AAE
1971	09	12		03 18 18 P	05.4	36.5				4.7				E	AAE
1971	09	12		03 17 14	06.0	36.3		4.2					FELT THROUGHOUT GEMU GOFA	E	ZZZ
1971	09	12	*	06 20	05.8	36.5						3.5	IN BAKO	E	AAE
1971	09	12	*	06 20	06.1	37.1						3.5	IN GELTA	E	AAE
1971	09	12	*		05.6	37.5						3	FELT IN GIDOLE	E	AAE
1971	11	13		15 47 41.51	10.972	39.679	24	5.3						A	CGS
1971	11	13		15 47 44	11.03	39.71	39	5.1						A	ISC
1971	11	13		15 47 40	11.0	39.5	15		5.6					A	MOS
1971	11	13		15 47 41.0	11.0	39.8								A	STR
1971	11	13		15 47 41	11.00	39.68	24	5.3					NEAR DESSIE 50AFTERSHOCKS	A	ZZZ
1971	11	13	*	PM	11.7	39.7						5	COMBOLCIA (DAMAGE)	A	AAE
1971	11	13	*	PM	11.1	39.3						4	FELT IN DESSIE	A	AAE
1971	11	13	*	PM	10.0	39.8						5	RASSA GUBA (DAMAGE)	A	AAE
1971	12	05	*	20 13 58	10.2	40.5					3.9		SWARM 26 SHOCKS.FELT IN	C	
1971	12	05										3	GAWANI	C	
1972	01	11		15 10 42.6	07.0	38.0				4.0			W SCARP OF RIFT VALLEY	A	AAE
1972	01	11		15 11 14.2P	07.2	37.7				4.0			+ 5 AFTERSHOCKS	A	AAE
1972	01	11	*	PM	06.8	38.4						3	FELT IN YIRGA ALEM	AC	AAE
1972	04	13		01 52 04.5P				3.9		4.3				E	AAE
1972	04	13		01 51 45	06	37.7					3.9		GOMU GOFA	E	ZZZ

Y M D		H1	N	E	H2	MB	MS	ML	ME	I	LOCATION & COMMENTS	R	A
1972 04 13	*	NT	05.6	37.5						3	FELT IN GIDOLE	E	AAE
1972 08 18		21 40 10	12.4	39.8				3.1			GUF GUF GRABEN	A	ZZZ
1972 08 18	*		13.2	39.8						3	FELT IN DEBUB	A	AAE
1972 08 18		22 02 13	12.1	39.6				2.9			GUF GUF GRABEN	A	ZZZ
1972 08 18	*	NT	13.2	39.8						3	FELT IN DEBUB	A	AAE
1972 08 19		09 08 05	13.3	39.7				4.1			GUF GUF GRABEN	A	ZZZ
1972 08 19	*	MD	13.2	39.8							FELT IN DEBUB	A	AAE
1972 08 19	*	09 22 27	13.9	39.9				4.3			DERGAHA SHEKET GRABEN	A	ZZZ
1972 08 19	*	MD									FELT IN DEBUB	A	AAE
1972 08 19	*	09 41 28	13.6	39.9				3.5			DERGANA-SHEKET GRABEN	A	ZZZ
1972 08 19	*	MD	13.2	39.8							FELT IN DEBUB	A	AAE
1972 08 19		12 58 34	13.1	39.8				3.8			GUF GUF GRABEN +3 OTHERS	A	ZZZ
1972 08 19	*		13.2	39.8						3	FELT IN DEBUB	A	AAE
1972 11 13		07 07 53.7	10.0	39.6		3.9		4.3			W PLATEAU MARGIN	A	ZZZ
1972 11 13	*	AM	09.	39.						3	FELT IN ADDIS ABABA	A	AAE
1972 11 13	*	AM	10.0	39.8						3	FELT IN RASSA GUBA	A	AAE
1973 01 07		12 18 16.5P	05.55	36.90		5.2		5.7				E	AAE
1973 01 07		12 17 12.64	05.266	36.849	34	4.9						E	CGS
1973 01 07		12 17 12	05.14	36.78	25	4.8						E	ISC
1973 01 07		12 17 13	05.3	36.6		5.2	5.3					E	MOS
1973 01 07		12 17 12.6	05.6	36.9	34	4.9		5.7			GEMU GOFA, CHEW BAHR	E	ZZZ
1973 01 07	*	09 15	07.3	36.2						3	FELT IN BONGA	E	AAE
1973 01 07	*	12 15	06.2	36.6						4.5	BULKI: TREMORS + NOISE	E	AAE
1973 01 07	*	12 15	05.6	38.2						3	FELT IN HAGARE MARIAM	E	AAE
1973 01 07	*	12 15	07.9	37.5						3	FELT IN SAJA	E	AAE
1973 01 17	*	12 15	04.8	38.1						4	FELT IN JAVELLO	E	AAE
1973 03 08		05 15 22.2P	07.7	37.8		3.7		4.1			15 AFTERSHOCKS	A	AAE
1973 03 08		05 14 55	07.7	38.0				4.1			W ESCARPMENT OF RIFT	A	ZZZ
1973 03 08	*	AM	09	39						4	FELT IN ADDIS ABABA	A	AAE
1973 03 27	*										SEISMIC ACTIVITY IN GULF OF TADJOURA MAR 27 TO APR 17.13000 OF M LARGER THAN 1.5;338 LARGER THAN M 2.9	D	
1973 03 28		10 46 40 P				4.7		5.2				D	AAE
1973 03 28		10 45 30.27	11.819	42.698	33	4.6						D	CGS

Y M D		H1	N	E	H2	MB	MS	ML	ME	I	LOCATION & COMMENTS	R	A
1973 03 28		10 45 29.3	11.47	42.85	33	4.6						D	ISC
1973 03 28		10 45 27	11.2	42.5		5.1	5.1					D	MOS
1973 03 28		10 45 30.3	11.64	42.77	33	4.6		5.2			G. OF TADJOURA	D	ZZZ
1973 03 28		13 36 16 P				5.4		6.				D	AAE
1973 03 28	*	13 35 04.53	11.673	42.830	33	5.0						D	CGS
1973 03 28		13 35 01	11.688	42.903				5.0				D	IPG
1973 03 28	*	13 35 05	11.53	43.07	37	4.9						D	ISC
1973 03 28	*	13 35 02	11.3	43.1		5.2						D	MOS
1973 03 28		13 35 01	11.688	42.903	33	5.0		6.			G. OF TADJOURA	D	ZZZ
1973 03 28		13 43 22 P				5.4						D	AAE
1973 03 28		13 42 06.68	11.739	42.746	33	5.3	5.3					2	CGS
1973 03 28		13 42 11	11.77	42.89	70	5.0						D	ISC
1973 03 28		13 42 04	11.4	43.0		5.7		5.8				D	MOS
1973 03 28		13 42 06.7	11.76	42.82	33	5.3	5.3	5.9			G. OF TADJOURA	D	ZZZ
1973 03 28	*	13 42	09.6	41.9						3	FELT AS FAR AS DIRE DAWA	B	AAE
1973 03 28		14 00 32	11.50	42.64	187	4.0						D	ISC
1973 03 28		14 00 04	09.7	42.2		5.4						D	MOS
1973 03 28		14 00 32	11.50	42.64		4.0					G. OF TADJOURA	D	ZZZ
1973 03 28		14 20 06 P				5.5		6.0				D	AAE
1973 03 28		14 18 52.34	11.703	42.927	33	5.4	5.4					D	CGS
1973 03 28		14 18 55	11.69	42.93	52	5.2						D	ISC
1973 03 28		14 18 50	11.3	42.9		5.5	5.8					D	MOS
1973 03 28		14 18 52.3	11.70	42.93	33	5.4	5.4	6.0			G. OF TADJOURA	D	ZZZ
1973 03 28		15 00 17 P				5.4		5.9				D	AAE
1973 03 28	*	14 58 08	11.7	42.8			5.3					D	BUL
1973 03 28	*	14 59 06.7	11.748	42.870	33	5.2						D	CGS
1973 03 28	*	14 59 07	11.76	42.78	39	5.1						D	ISC
1973 03 28	*	14 58 57	11.0	42.9		5.7	5.8					D	MOS
1973 03 28		14 59 06.7	11.75	42.83	33	5.2		5.9			G. OF TADJOURA	D	ZZZ
1973 03 28		17 15 53 P				4.8						D	AAE
1973 03 28		17 14 36	11.695	42.885				5.4				D	IPG
1973 03 28		17 14 36	11.695	42.885		4.8		5.3			G. OF TADJOURA	D	ZZZ
1973 03 28		17 42 28 P				4.5		5.0				D	AAE
1973 03 28		17 41 07	11.762	42.885								D	IPG
1973 03 28		17 41 07	11.762	42.885				4.0			G. OF TADJOURA	D	ZZZ
1973 03 29		00 05 28 P				2.8		3.2				D	AAE
1973 03 29		00 04 02	11.715	43.001								D	IPG
1973 03 29		00 04 02	11.715	43.001		2.8					G. OF TADJOURA	D	ZZZ

Y M D		H1	N	E	H2	MB	MS	ML	ME	I	LOCATION & COMMENTS	R	A
1973 03 29		04 31 32.5P				4.7						D	AAE
1973 03 29		04 29 56	11.718	42.980								D	IPG
1973 03 29		04 29 56	11.718	42.980		4.7					G. OF TADJOURA	D	ZZZ
1973 03 29		05 21 11 P				4.7						D	AAE
1973 03 29		05 19 51	11.703	42.983								D	IPG
1973 03 29		05 19 51	11.703	42.983		4.7					G. OF TADJOURA	D	ZZZ
1973 03 29		16 52 03	11.707	42.833								D	IPG
1973 03 29		16 52 03	11.707	42.833							G. OF TADJOURA	D	ZZZ
1973 04 01		06 30 41 P				5.2						D	AAE
1973 04 01		06 29 27.69	11.613	42.927	33	4.8						D	CGS
1973 04 01		06 29 24	11.707	43.003		4.9						D	IPG
1973 04 01		06 29 31	11.42	43.01	74	4.5						D	ISC
1973 04 01	*	06 29 23	11.0	42.9		5.4	5.3					D	MOS
1973 04 01		06 29 24	11.707	43.003		4.8		5.7			G. OF TADJOURA	D	ZZZ
1973 04 01	*	AM	09.6	41.9						3	DIRE DAWA	B	AAE
1973 04 01		AM	11.6	41.2							SERDO & VICINITY	C	AAE
1973 04 01	*	AM									DOUBTFUL IF RELATED TO		
1973 04 01	*	AM									ACTIVITY IN GULF TADJOURA		
1973 04 01		07 07 12	11.705	43.015				4.5				D	IPG
1973 04 01		07 07 12	11.705	43.015		4.5		5.0			G. OF TADJOURA	D	ZZZ
1973 04 01		07 13 51 P				5.5						D	AAE
1973 04 01		07 12 37.03	11.663	43.035	31	5.5	5.9					D	CGS
1973 04 01		07 12 32	11.713	43.022				5.5				D	IPG
1973 04 01	*	07 12 41	11.63	43.00	65	5.0						D	ISC
1973 04 01		07 12 30	11.0	42.9	20	5.7	6.0					D	MOS
1973 04 01		07 12 32	11.713	43.022							G. OF TADJOURA	D	ZZZ
1973 04 01		07 38 40.46	11.645	42.931	33	4.9						D	CGS
1973 04 01		07 38 37	11.725	43.022				4.9				D	IPG
1973 04 01	*	07 38 41.2	11.76	43.03	33	4.9						D	ISC
1973 04 01		07 38 33	10.8	42.8	20	5.2	5.5					D	MOS
1973 04 01		07 38 37	11.725	43.022		4.9					G. OF TADJOURA	D	ZZZ
1973 04 01		07 41 21.46	11.819	42.794	33	4.7						D	CGS
1973 04 01		07 41 22	11.9	43.3	0	4.5						D	ISC
1973 04 01		07 41 21.5	11.86	43.05	33	4.7					G. OF TADJOURA	D	ZZZ
1973 04 01		07 50 56	11.700	42.992				4.6				D	IPG
1973 04 01		07 50 56	11.700	42.992		4.6					G. OF TADJOURA	D	ZZZ

Y	M	D		H1	N	E	H2	MB	MS	ML	ME	I	LOCATION & COMMENTS	R	A
1973	04	01		07 56 32	11.700	42.967				4.6				D	IPG
1973	04	01		07 56 32	11.700	42.967		4.7					G. OF TADJOURA	D	ZZZ
1973	04	01		08 27 11	11.642	42.985				4.3				D	IPG
1973	04	01		08 27 11	11.642	42.985		3.9		4.3			G. OF TADJOURA	D	ZZZ
1973	04	01		18 55 23	11.712	43.035				4.2				D	IPG
1973	04	01		18 55 23	11.712	43.035		3.7					G. OF TADJOURA	D	ZZZ
1973	04	01		21 02 16	11.703	41.970				4.1				D	IPG
1973	04	01		21 02 16	11.703	41.970		3.8					G. OF TADJOURA	D	ZZZ
1973	04	01		21 27 00	11.690	43.032				4.0				D	IPG
1973	04	01		21 27 00	11.690	43.032		3.1					G. OF TADJOURA	D	ZZZ
1973	04	01		01 37 05	11.722	43.028				4.3				D	IPG
1973	04	02		01 37 05	11.722	43.028		3.7					G. OF TADJOURA	D	ZZZ
1973	04	02		10 35	11.602	42.817				4.0				D	IPG
1973	04	02		10 35	11.602	42.817				4.0			G. OF TADJOURA	D	ZZZ
1973	04	01		11 11	11.717	43.028				4.3				D	IPG
1973	04	02		11 11	11.717	43.028		3.8					G. OF TADJOURA	D	ZZZ
1973	04	02		14 54 09	11.768	43.117				4.1				D	IPG
1973	04	02		14 54 09	11.768	43.117		4.0					G. OF TADJOURA	D	ZZZ
1973	04	02		20 44 02	11.716	43.028				4.3				D	IPG
1973	04	02		20 44 02	11.716	43.028		3.9					G. OF TADJOURA	D	ZZZ
1973	04	02		21 24 43	11.693	42.892				4.2				D	IPG
1973	04	02		21 24 43	11.693	42.892		3.1					G. OF TADJOURA	D	ZZZ
1973	04	03		01 49 21	12.1	41.3		4.0		4.5			SERDO AREA	C	ZZZ
1973	04	03	*	NT	11.75	41.11						4	3 SHOCKS FELT IN DUBTI	C	AAE
1973	04	03	*	NT	11.7	41.0						4	FELT IN LOGGHIA	C	AAE
1973	04	03		NT	11.4	40.75							2 SHOCKS FELT IN MILLE	C	AAE
1973	04	03	*	NT	11.6	41.2							N SHOCKS FELT IN SERDO	C	AAE
1973	04	03		07 50 04	11.600	42.913				4.1				D	IPG
1973	04	03		07 50 04	11.600	42.913		3.2		3.6			G. OF TADJOURA	D	ZZZ
1973	04	03		10 07 42	11.722	43.060				4.2				D	IPG
1973	04	03		10 07 42	11.722	43.060		3.7					G. OF TADJOURA	D	ZZZ

Y M D		H1	N	E	H2	MB	MS	ML	ME	I	LOCATION & COMMENTS	R	A
1973 04 04		00 42 01	11.638	42.847				4.1				D	IPG
1973 04 04		00 42 01	11.638	42.847		3.4					G. OF TADJOURA	D	ZZZ
1973 04 06		14 19 15	11.638	42.862				4.1				D	IPG
1973 04 06		14 19 15	11.638	42.862					4.0		G. OF TADJOURA	D	ZZZ
1973 04 07		17 37 55.5P				5.2						D	AAE
1973 04 07		17 36 42.78	11.690	43. 21	33	4.7						D	CGS
1973 04 07		17 36 40	11.710	43.033				4.9				D	IPG
1973 04 07		17 36 43.3	11.64	42.95	33	4.7						D	ISC
1973 04 04		17 36 35	10.6	43.0		5.3		5.6				D	MOS
1973 04 07		17 36 40	11.710	43.033		4.7					G. OF TADJOURA	D	ZZZ
1973 04 07	*	17 37	11.5	43.0						5	FELT IN DJIBOUTI	D	AAE
1973 04 07		18 10 30	11.725	43.700				4.4				D	IPG
1973 04 07		18 10 30	11.725	43.700					4.2		G. OF TADJOURA	D	ZZZ
1973 04 07		19 19 01 P				4.7						D	AAE
1973 04 07		19 17 38.67	11.757	42.786	33	4.5						D	CGS
1973 04 07		19 17 38.3	11.78	42.84	33							D	ISC
1973 04 07		19 17 39	11.8	42.8	33	4.5		5.2			G. OF TADJOURA	D	ZZZ
1973 04 07		19 18 30	11.725	43.042				4.0				D	IPG
1973 04 07		19 18 30	11.725	43.042					4.0		G. OF TADJOURA	D	ZZZ
1973 04 11		02 10 45.5P				4.8						D	AAE
1973 04 11		02 09 33.13	11.783	42.847	33	4.6						D	CGS
1973 04 11		02 09 30	11.663	42.968				4.5				D	IPG
1973 04 11		02 09 33.3	11.68	43.00	33							D	ISC
1973 04 11		02 09 30	11.66	42.97		4.5					G. OF TADJOURA	9	ZZZ
1973 04 11		03 07 26	11.645	42.983				4.0				D	IPG
1973 04 11		03 07 26	11.645	42.983		3.3					G. OF TADJOURA	D	ZZZ
1973 04 11		03 16 54	11.642	42.985				4.0				D	IPG
1973 04 11		03 16 54	11.642	42.985		3.8					G. OF TADJOURA	D	ZZZ
1973 04 13		14 13 56.9	11.9	43.8	33	4.8						D	CGS
1973 04 13		14 13 57.0	11.93	43.83	33	4.8						D	ISC
1973 04 13		14 13 57	11.9	43.8	33	4.8					TADJOURA TRENCH	D	ZZZ
1973 06 03		23 38 33	09.25	42.8		3.7		4.1			MARDA FAULT ZONE	B	ZZZ
1973 06 03	*	23 45	09.3	42.8						4	FELT IN JIGGIGA	B	AAE
1973 11 16	*	MD	09.6	41.9						4	FELT IN DIRE DAWA	B	AAE
1973 11 16	*	MD	09.3	42.7						3	FELT IN JIGGIGA	B	

Y M D	H1	N	E	H2	MB	MS	ML	ME	I	LOCATION & COMMENTS	R	A
1974 02 20	10 53 33	11.92	41.00							AFAR	C	TLN
1974 02 20	21 35 33.3	11.91	41.03							AFAR	C	TLN
1974 02 20	21 40 59.3	11.87	41.06							AFAR	C	TLN
1974 02 20	23 24 55.1	11.87	41.06	1.3						AFAR	C	TLN
1974 02 20	23 27 39.9	11.86	41.07	0.3						AFAR	C	TLN
1974 02 21	00 05 12.3	11.90	41.04							AFAR	C	TLN
1974 02 21	01 13 46.2	11.94	41.01	2.8						AFAR	C	TLN
1974 02 23	20 27 11.6	11.82	41.13							AFAR	C	TLN
1974 02 23	20 31 24.8	11.81	41.12	3						AFAR	C	TLN
1974 02 24	18 11 39.4	11.30	39.76							PLATEAU E MARGIN	A	TLN
1974 02 25	16 05 40.6P	09.9	39.7				4.2				A	AAE
1974 02 25	16 05 15.70	09.843	40.032	33	4.5						A	CGS
1974 02 25	16 05 18	10.2	39.8	33	4.6						A	ISC
1974 02 25	16 05 12.5	09.99	39.83		4.5					PLATEAU E MARGIN	A	ZZZ
1974 02 25	16 05 12.5	10.00	39.78								A	TLN
										LARGEST IN SWARM OF 36	A	
1974 02 25	PM	09	35						3	FELT IN S ADDIS ABABA	A	AAE
1974 02 05	PM	10.0	39.8						4	3 FELT IN RASSA GUBA	A	AAE
1974 03	06 1	37.6			4.8					SWARM OF 57 S ETHIOPIA	C	ZZZ
1974 03	*	06.2	37.7						3	FELT IN ARBA MINCH	C	AAE
1974 03	*	06.2	36.6						3	FELT IN BULKI	C	AAE
1974 03	*	05.6	38.3							FELT IN HAGERE MARIAM	C	AAE
1974 03	*	06.3	38.4							FELT IN WARANCHA	C	AAE
1974 02 25	16 12 32.8	10.16	39.61							PLATEAU E MARGIN	A	TLN
1974 02 25	19 22 52.8	10.37	39.45							PLATEAU E MARGIN	A	TLN
1974 02 25	21 58 28.2	10.07	39.70							PLATEAU E MARGIN	A	TLN
1974 02 25	21 58 58.3	10.92	39.04							PLATEAU E MARGIN	A	TLN
1974 02 25	22 09 23.1	09.88	39.81							PLATEAU E MARGIN	A	TLN
1974 02 25	22 37 14.9	10.00	39.67							PLATEAU E MARGIN	A	TLN
1974 02 26	03 27 03.8	11.46	39.54							PLATEAU E MARGIN	A	TLN
1974 02 26	10 02 25.4	11.87	41.03							AFAR	C	TLN
1974 02 26	11 37 19.2	11.91	41.03	3						AFAR	C	TLN
1974 02 26	11 37 39.7	11.90	41.04	9						AFAR	C	TLN
1974 02 27	02 54 30.7	11.83	41.07	4						AFAR	C	TLN
1974 03 05	17 46 38.6	10.46	40.05							PLATEAU E MARGIN	A	TLN
1974 03 06	13 01 29.4	10.94	41.27	6						AFAR	C	TLN

Y	M	D	H1	N	E	H2	MB	MS	ML	ME	I	LOCATION & COMMENTS	R	A
1974	03	06	17 11 41.5	12.14	43.42							AFAR	C	TLN
1974	03	08	01 30 47.5	11.40	41.65							AFAR	C	TLN
1974	03	08	10 06 48.4	11.85	41.06							AFAR	C	TLN
1974	03	08	15 13 20.3	12.13	41.43							AFAR	C	TLN
1974	03	10	02 31 09.3	11.97	40.97	3						AFAR	C	TLN
1974	03	20	07 49 04.0	11.82	40.97							AFAR	C	TLN
1974	03	20	13 44 46.8	12.56	40.52							AFAR	C	TLN
1974	03	22	06 57 50.3	11.93	39.63							PLATEAU E MARGIN	A	TLN
1974	03	23	06 37 12.2	11.91	41.02							AFAR	C	TLN
1974	03	23	06 45 55.2	11.92	41.04							AFAR	C	TLN
1974	03	23	07 57 57.8	11.92	41.02							AFAR	C	TLN
1974	03	23	08 24 23.3	11.90	41.01							AFAR	C	TLN
1974	03	23	08 56 51.6	11.91	41.02							AFAR	C	TLN
1974	03	23	09 24 03.7	11.91	41.04							AFAR	C	TLN
1974	03	25	05 09 31.7	11.90	41.02							AFAR	C	TLN
1974	03	26	10 20 20.1	13.49	41.55							AFAR	C	TLN
1974	03	26	10 20 20.1	12.74	42.54							AFAR	C	TLN
1974	03	26	11 28 20.1	13.47	41.62							AFAR	C	TLN
1974	03	26	11 28 20.1	12.80	42.50							AFAR	C	TLN
1974	03	26	11 57 10.4	13.44	41.68							AFAR	C	TLN
1974	03	26	11 57 10.4	12.85	42.45							AFAR		TLN
1974	04	01	16 08 39.6	11.96	39.69							PLATEAU E MARGIN	A	TLN
1974	04	01	16 22 17.3	11.94	39.73							PLATEAU E MARGIN	A	TLN
1974	04	01	17 32 19.5	11.97	39.69							PLATEAU E MARGIN	A	TLN
1974	04	01	17 37 08.2	11.88	39.69							PLATEAU E MARGIN	A	TLN
1974	04	01	18 12 52.7	11.98	39.75	0.8						PLATEAU E MARGIN	A	TLN
1974	04	01	18 16 21.9	11.94	39.61	8						PLATEAU E MARGIN	A	TLN
1974	04	01	18 18 49.3	12.00	39.72							PLATEAU E MARGIN	A	TLN
1974	04	01	18 19 60.0	11.98	39.74	1.5						PLATEAU E MARGIN	A	TLN
1974	04	01	18 23 05.4	11.93	39.75							PLATEAU E MARGIN	A	TLN
1974	04	01	18 27 48.7	11.99	39.76							PLATEAU E MARGIN	A	TLN
1974	04	01	18 32 13.2	11.95	39.74							PLATEAU E MARGIN	A	TLN
1974	04	01	18 43 36.0	11.92	39.90							PLATEAU E MARGIN	A	TLN
1974	04	01	18 49 44.3	11.98	39.73							PLATEAU E MARGIN	A	TLN
1974	04	01	18 52 33.6	11.90	39.73							PLATEAU E MARGIN	A	TLN
1974	04	01	19 19 10.1	11.96	39.69							PLATEAU E MARGIN	A	TLN
1974	04	01	19 34 05.3	11.95	39.69							PLATEAU E MARGIN	A	TLN
1974	04	01	19 36 00.7	11.96	39.70							PLATEAU E MARGIN	A	TLN
1974	04	01	19 36 57.8	11.94	39.71							PLATEAU E MARGIN	A	TLN
1974	04	01	19 44 29.0	11.94	39.84							PLATEAU E MARGIN	A	TLN
1974	04	01	19 50 56.1	11.90	39.69							PLATEAU E MARGIN	A	TLN

Y M D	H1	N	E	H2	MB	MS	ML	ME	I	LOCATION & COMMENTS	R	A
1974 04 01	20 35 13.4	11.95	39.70							PLATEAU E MARGIN	A	TLN
1974 04 01	20 41 45.4	11.90	39.69							PLATEAU E MARGIN	A	TLN
1974 04 01	21 00 48.2	11.94	39.69							PLATEAU E MARGIN	A	TLN
1974 04 01	23 52 30.4	11.43	41.51	9.5						AFAR	C	TLN
1974 04 02	00 18 22.7	11.71	39.65							PLATEAU E MARGIN	A	TLN
1974 04 02	00 42 20.4	11.93	39.70							PLATEAU E MARGIN	A	TLN
1974 04 02	00 45 03.1	11.98	39.68	5						PLATEAU E MARGIN	A	TLN
1974 04 02	00 56 07.4	11.91	39.71							PLATEAU E MARGIN	A	TLN
1974 04 02	01 34 18.9	11.77	39.66							PLATEAU E MARGIN	A	TLN
1974 04 02	03 37 19.3	11.96	39.71							PLATEAU E MARGIN	A	TLN
1974 04 02	09 02 01.0	12.47	39.25							PLATEAU E MARGIN	A	TLN
1974 04 02	09 11 04.2	11.92	39.71							PLATEAU E MARGIN	A	TLN
1974 04 02	09 48 59.8	11.94	39.71							PLATEAU E MARGIN	A	TLN
1974 04 02	10 03 00.3	11.92	39.70							PLATEAU E MARGIN	A	TLN
1974 04 02	16 15 35.0	11.96	39.69							PLATEAU E MARGIN	A	TLN
1974 04 02	16 19 37.4	12.51	39.93							PLATEAU E MARGIN	A	TLN
1974 04 03	00 55 27.2	11.98	39.69	5						PLATEAU E MARGIN	A	TLN
1974 04 03	02 43 42.9	11.88	39.69							PLATEAU E MARGIN	A	TLN
1974 04 03	05 25 38.7	11.95	39.68							PLATEAU E MARGIN	A	TLN
1974 04 03	05 28 05.4	11.96	39.68							PLATEAU E MARGIN	A	TLN
1974 04 03	05 34 54.3	11.91	39.73							PLATEAU E MARGIN	A	TLN
1974 04 03	09 37 48.9	11.82	39.66							PLATEAU E MARGIN	A	TLN
1974 04 03	19 06 24.1	11.94	39.72							PLATEAU E MARGIN	A	TLN
1974 04 03	19 24 30.8	11.95	39.71							PLATEAU E MARGIN	A	TLN
1974 04 04	02 13 53.9	11.96	39.69							PLATEAU E MARGIN	A	TLN
1974 04 04	04 12 27.1	12.00	39.73							PLATEAU E MARGIN	A	TLN
1974 04 04	06 16 47.3	11.98	39.71							PLATEAU E MARGIN	A	TLN
1974 04 04	09 48 29.6	11.89	39.60							PLATEAU E MARGIN	A	TLN
1974 04 04	22 29 43.5	11.92	39.70							PLATEAU E MARGIN	A	TLN
1974 04 05	00 31 57.8	10.05	39.67							PLATEAU E MARGIN	A	TLN
1974 04 05	06 46 41.0	11.44	40.94								A	TLN
1974 04 05	06 52 24.4	11.94	39.70							PLATEAU E MARGIN	A	TLN
1974 04 05	07 59 30.2	11.89	39.69							PLATEAU E MARGIN	A	TLN
1974 04 05	09 50 22.4	11.91	39.70							PLATEAU E MARGIN	A	TLN
1974 04 08	02 15 53.0	13.94	40.79								A	TLN

Y M D	H1	N	E	H2	MB	MS	ML	ME	I	LOCATION & COMMENTS	R	A
1974 04 08	02 15 53.0	12.12	43.18							AFAR	C	TLN
1974 04 08	04 56 10.6	11.97	39.72							PLATEAU E MARGIN	A	TLN
1974 04 08	19 17 47.7	11.64	39.63							PLATEAU E MARGIN	A	TLN
1974 04 08	19 34 52.2	11.68	39.63							PLATEAU E MARGIN	A	TLN
1974 04 09	00 13 19.2	11.86	41.04							AFAR	C	TLN
1974 04 09	00 22 26.3	11.81	41.11							AFAR	C	TLN
1974 04 09	00 47 25.0	11.89	39.68							PLATEAU E MARGIN	A	TLN
1974 04 09	04 10 20.5	11.91	39.71							PLATEAU E MARGIN	A	TLN
1974 04 09	20 20 07.6	11.90	39.12							PLATEAU E MARGIN	A	TLN
1974 04 10	03 03 02.0	11.94	39.71							PLATEAU E MARGIN	A	TLN
1974 04 10	20 15 32.4	11.94	39.68							PLATEAU E MARGIN	A	TLN
1974 04 10	20 33 57.2	11.91	39.12							PLATEAU E MARGIN	A	TLN
1974 04 11	11 54 54.0	11.93	39.71							PLATEAU E MARGIN	A	TLN
1974 04 11	20 34 01.2	11.88	39.73							PLATEAU E MARGIN	A	TLN
1974 04 11	21 02 43.1	10.56	41.00							PLATEAU E MARGIN	C	TLN
1974 04 13	12 56 23.1	12.00	41.27							AFAR	C	TLN
1974 04 13	14 41 25.2	12.12	41.56							AFAR	C	TLN
1974 04 14	08 40 55.3	11.89	39.74							PLATEAU E MARGIN	A	TLN
1974 04 14	08 52 00.7	11.89	39.73							PLATEAU E MARGIN	A	TLN
1974 04 14	08 54 37.3	11.91	39.75							PLATEAU E MARGIN	A	TLN
1974 04 14	09 02 07.5	11.89	39.74							PLATEAU E MARGIN	A	TLN
1974 04 17	18 27 33.69	17.255	40.365	33	5.0	5.1					C	CGS
1974 04 17	18 27 34	17.30	40.30	27	5.1						C	ISC
1974 04 17	18 27 34	17.1	40.3		5.4		5.2				C	MOS
1974 04 17	18 27 36.0	17.0	40.6	50							C	STR
1974 04 17	18 27 34	17.28	40.33	33	5.0					RED SEA TROUGH	C	ZZZ
1974 04 19	23 22 36.7	12.11	39.57							PLATEAU E MARGIN	A	TLN
1974 04 20	07 28 59	11.62	41.05							AFAR	C	TLN
1974 04 20	09 03 33.8	11.89	40.94							AFAR	C	TLN
1974 04 23	01 48 53.0	12.13	39.36							PLATEAU E MARGIN	A	TLN
1974 04 26	16 43 21.3	09.79	38.54							PLATEAU E MARGIN	A	TLN
1974 04 26	16 43 21.3	08.86	39.76							PLATEAU E MARGIN	A	TLN

Y M D	H1	N	E	H2	MB	MS	ML	ME	I	LOCATION & COMMENTS	R	A
1974 04 26	18 08 16.86	17.141	40.380	33	4.8						C	CGS
1974 04 26	18 08 18	17.10	40.44	48	4.4						C	ISC
1974 04 26	18 08 07	15.7	40.2		4.9						C	MOS
1974 04 26	18 08 12	17.0	40.6								C	STR
1974 04 26	18 08	17.12	40.41		4.8					RED SEA TROUGH	C	ZZZ
1974 04 28	06 20 48.2	11.88	41.02							AFAR	C	TLN
1974 04 28	11 17 04.7	11.92	40.98							AFAR	C	TLN
1974 04 28	11 17 23.3	11.89	41.03							AFAR	C	TLN
1974 04 28	21 50 52.2	11.59	39.67							PLATEAU E MARGIN	A	TLN
1974 04 30	00 08 50.7	12.14	40.88							FOOT OF ESCARPMENT	C	TLN
1974 06 03	06 24 02.5	11.86	41.08							AFAR	C	TLN
1974 06 03	18 26 43.8	11.47	41.03							AFAR	C	TLN
1974 06 18	13 35 37.7	11.59	41.02							AFAR	C	TLN
1974 06 20	18 49 16.1	11.97	39.73							PLATEAU E MARGIN	A	TLN
1974 06 20	23 00 02.8	11.90	39.72							PLATEAU E MARGIN	A	TLN
1974 06 23	09 55 16.0	11.20	40.93							AFAR	C	TLN
1974 06 23	17 39 29.2	11.18	40.89							AFAR	C	TLN
1974 06 23	23 35 18.8	11.89	41.27							AFAR	C	TLN
1974 06 24	10 34 17.7	09.63	39.14							PLATEAU E MARGIN	A	TLN
1974 06 25	14 57 43.2	11.13	40.88							AFAR	C	TLN
1974 06 25	17 12 37.3	11.25	40.92							AFAR	C	TLN
1974 06 25	17 27 58.6	11.15	40.92							AFAR	C	TLN
1974 06 30	13 26 24.72	16.013	39.631	33	4.4						C	CGS
1974 06 30	13 26 25.7	15.970	39.610	33	4.5						C	ISC
1974 06 30		15.99	39.62		4.4					WEST COAST OF RED SEA	C	ZZZ
1974 06 30 *		06.3	37.7						3	FELT IN CHENCHA (N)	CE	AAE
1974 07 01	17 16 02.7	11.90	41.03							AFAR	C	TLN
1974 07 02	07 43 54.6	11.91	41.02							AFAR	C	TLN
1974 07 02		10.893	43.891				3.3					IPG
1974 07 02	18 28 47	10.89	43.89				3.3			OFF NE COAST OF SOMALIA	D	ZZZ
1974 07 03	23 03 37.1	11.90	41.04	2						AFAR	C	TLN
1974 07 03	23 04 39.2	11.88	41.05	1.5						AFAR	C	TLN

Y	M	D	H1	N	E	H2	MB	MS	ML	ME	I	LOCATION & COMMENTS	R	A
1974	07	05	13 14 57.0	11.89	39.70							PLATEAU E MARGIN	A	TLN
1974	07	05	22 15 23.6	12.40	41.04							AFAR	C	TLN
1974	07	08	21 16 32.5	11.82	41.05	2.3						AFAR	C	TLN
1974	07	10	13 29 23.2	10.22	38.46							W PLATEAU	A	TLN
1974	07	15	17 44 22.4	11.91	41.02							AFAR	C	TLN
1974	07	26	21 33 49.5	12.02	41.28	9.5						AFAR	C	TLN
1974	07	27	06 57 57.9	11.95	39.62	8						PLATEAU E MARGIN	A	TLN
1974	07	27	08 33 49.3	11.95	39.67							PLATEAU E MARGIN	A	TLN
1974	07	27	09 58 10.6	11.84	39.59							PLATEAU E MARGIN	A	TLN
1974	07	27	09 58 10.6	11.84	39.59							PLATEAU E MARGIN	A	TLN
1974	07	31	13 03 52.8	11.92	39.59							PLATEAU E MARGIN	A	TLN
1974	07	31	20 29 08.1	11.91	39.65							PLATEAU E MARGIN	A	TLN
1974	08	01	10 33 30.5	15.60	39.48							MASSAWA CHANNEL	C	TLN
1974	08	03	11 59 51.4	11.75	41.13							AFAR	C	TLN
1974	08	04	16 56 57.5	11.50	40.79							AFAR	C	TLN
1974	08	05	22 46 38.2	13.77	40.76							AFAR, DANAKIL DEPRESSION	C	TLN
1974	08	05	22 46 38.2	11.18	42.94							AFAR	C	TLN
1974	08	07	02 46 18.2	11.85	41.06							AFAR	C	TLN
1974	08	26	13 19 32	11.27	41.92				3.3			NE OF LAKE ABHE, SE AFAR	D	IPG
1974	08	29	21 16 09.5	12.37	42.83							RED SEA	C	TLN
1974	08	29	21 16 09.5	13.71	41.13							DANAKIL ALPS W MARGIN	C	TLN
1974	08	29	23 39 24.5	11.97	40.96							AFAR	C	TLN
1974	08	30	02 03 15.8	14.14	40.41							N DANAKIL DEPRESSION	C	TLN
1974	08	30	08 16 18.6	14.05	42.03							RED SEA	C	TLN
1974	08	30	18 12 40.9	11.92	39.66							PLATEAU E MARGIN	A	TLN
1974	08	31	01 19 22.7	11.95	39.70							PLATEAU E MARGIN	A	TLN

Y M D	H1	N	E	H2	MB	MS	ML	ME	I	LOCATION & COMMENTS	R	A
1974 08 31	11 26 46.1	11.96	40.97	4						AFAR	C	TLN
1974 08 31	20 25 05.7	11.97	39.67							PLATEAU E MARGIN	A	TLN
1974 09 01	13 38 30.3	11.91	39.22							PLATEAU E MARGIN	A	TLN
1974 09 10	11 49 24.9	10.41	40.00							AFAR	C	
1974 09 13	21 27 54.4	12.35	39.52							PLATEAU E MARGIN	A	TLN
1974 09 13	22 24 35.2	12.43	39.56							PLATEAU E MARGIN	A	TLN
1974 09 14	00 22 16.6	12.39	39.45							PLATEAU E MARGIN	A	TLN
1974 09 14	15 46 32.9	11.84	41.07							AFAR	C	TLN
1974 09 14	21 46 26.8	11.94	39.70							PLATEAU E MARGIN	A	TLN
1974 09 14	22 04 04.9	11.42	40.93							AFAR	C	TLN
1974 09 14	23 03 59.4	11.88	39.66							PLATEAU E MARGIN	A	TLN
1974 09 15	05 40 39.6	12.60	41.65							AFAR	C	TLN
1974 09 15	20 37 17	10.3	44.2								D	AAE
1974 09 15	20 37 16.7	10.35	43.95				3.3				D	IPG
1974 09 15	20 37 17	10.36	44.15							SOMALIA	D	ZZZ
1974 09 17	13 26 15.0	13.12	40.65							AFAR	C	TLN
1974 09 19	01 52 58.8	11.93	41.00							AFAR	C	TLN
1974 09 20	00 01 30.2	13.48	41.28							AFAR	C	TLN
1974 09 20	00 17 15.7	13.34	40.44							AFAR	C	TLN
1974 09 20	00 17 15.7	11.63	42.68							AFAR	C	TLN
1974 09 21	01 30 27.6	11.20	42.80							AFAR	C	TLN
1974 09 21	01 30 27.6	13.34	40.00							AFAR	C	TLN
1974 09 21	12 03 51.4	13.89	40.89							AFAR	C	TLN
1974 09 21	12 03 51.4	12.15	43.12							DJIBOUTI	D	TLN
1974 09 23	08 20 42.9	11.99	40.78	0.5						AFAR	C	TLN
1974 09 23	10 02 36.7	12.00	40.76	2						AFAR	C	TLN
1974 09 23	19 54 49.8	10.71	39.76							PLATEAU E MARGIN	A	TLN

Y M D		H1	N	E	H2	MB	MS	ML	ME	I	LOCATION & COMMENTS	R	A
1974 09 23		13 51 19.3	11.87	39.70							PLATEAU E MARGIN	A	TLN
1974 09 23		21 21 45.5	11.80	39.68							PLATEAU E MARGIN	A	TLN
1974 09 24		17 38 07.8	11.86	41.52							AFAR	C	TLN
1974 09 24		18 43 49.9	11.85	41.49							AFAR	C	TLN
1974 09 24		18 46 50.6	11.86	41.54							AFAR	C	TLN
1974 09 24		18 48 03.9	11.85	41.53							AFAR	C	TLN
1974 09 24		20 25 51.6	11.86	41.52							AFAR	C	TLN
1974 09 25		01 41 37.6	11.93	41.20		3.5					AFAR	C	TLN
1974 09 25		01 43 50.4	11.98	41.18							AFAR	C	TLN
1974 09 25		20 19 26.9	11.94	41.39		3.5					AFAR	C	TLN
1974 11 14		19 37 55.6	11.73	42.46				2.2				D	IPG
1974 11 14		22 27 01.8	11.73	42.44							ASAL GRABEN	D	IPG
1974 12 17		00 44 37	11.2	39.7				3.8			CENTRAL WOLLO	A	ZZZ
											SEQUENCE OF 6.	A	
1974 12 17	*	NT	11.1	39.7							FELT IN COMBOLCIA	A	AAE
1974 12 17	*	NT	11.1	39.3						5	FELT IN DESSIE	A	AAE
1975 01 26		17 43 26.6	11.74	42.06				2.3			NEAR LAKE ASAL	D	IPG
1975 04 27		23 33 52.5P	09.4	42.8							MARDA FAULT ZONE	B	ZZZ
1975 07 31		11 04 01	08.8	42.6							SE PLATEAU	B	ZZZ
1975 07 31	*		09.6	41.9						4	FELT IN DIRE DAWA	B	AAE
1975 07 31	*		09.4	42.1						5	FELT IN HARAR CITY	B	AAE
1975 07 31	*		09.3	42.3						4.5	FELT IN BABILE	B	AAE
1975 07 31	*		09.2	42.4						4.5	FELT IN BISIDIMO	B	AAE
1975 08 07		22 43 13.3	15.288	40.407	37	4.6							CGS
1975 08 07		22 43 17	15.36	40.44	37	4.7							1SC
		09	.8	0.3		4.9							MOS
1975 08 07		22 43 12	15.5	40.5									STR
1975 08 07		22 43 13.5	15.32	40.42	37	4.6					HAWACHIL BAY	C	ZZZ
1975 08 23		21 35 49.4P	10.65	39.8				4.5			FELT IN DEREK AMBA (MENZ)	A	AAE
1975 08 23		21 35 21.7	10.617	39.732	33	5.2						A	CGS
1975 08 23		21 35 21.2	10.52	39.75	33	4.9						A	1SC
1975 08 23		21 35 15	09.7	39.5		5.3						A	MOS
1975 08 23		21 35 21.7	10.62	39.73	33	5.2					KARA KORE REGION	A	ZZZ
1975 08 23		22 12 45	10.6	40.0	0	4.5						A	1SC
1975 08 23		22 12 45	10.6	40.0							KARA KORE REGION	A	ZZZ

Y	M	D	H1	N	E	H2	MB	MS	ML	ME	I	LOCATION & COMMENTS	R	A
1976	05	29	22 59 01	08.7	42.0				4.4			SOUTH OF HARAR	B	ZZZ
1976	06	13	01 05 0	08.8	41.1				3.2			SE PLATEAU	B	ZZZ
1976	11	07	05 53 07.6	15.820	41.423	34	4.8						C	CGS
1976	11	07	05 53 08	15.82	41.42	34	4.8					RED SEA	C	ZZZ
1976	11	07	05 53 07.6	15.820	41.423	34	4.8						C	CGS
1976	11	07	05 53 07.6	15.820	41.423	34	4.8					RED SEA NEAR DAHLAC IS.	C	ZZZ
1976	11	16	12 53 33.70	15.905	41.915	33	4.9	3.9					C	CGS
1976	11	16	12 53 33.7	15.91	41.92	33	4.9					RED SEA	C	ZZZ
1977	06	27	14 13 20.2	16.154	40.013	33	4.3						C	CGS
1977	06	27	14 13 20.2	16.154	40.013	33	4.3					RED SEA	C	ZZZ
1977	07	08	06 23 02.4	10.940	39.629	38	5.0						A	CGS
1977	07	08	06 23 02.4	10.94	39.63	38	5.0					DESSIE REGION	A	ZZZ
1977	07	08										2 DEAD FROM LANSLIDE	A	AAE
1977	07	09	03 13 30	11.867	43.503				3.5				D	1PG
1977	07	09	03 13 30	11.867	43.503				3.5			TRENCH OF OBOCK	D	ZZZ
1977	12	05	10 02 57.1	12.695	40.474	33	4.5						C	CGS
1977	12	05	10 02 57.1	12.695	40.474	33	4.5					NEAR VOLCANO BOYNA	C	ZZZ
1977	12	28	02 45 36.7	16.659	40.278	33	5.9	6.6					C	CGS
1977	12	28	02 45 36.7	16.659	40.278	33	5.9	6.6				RED SEA TROUGH	C	ZZZ

Discussions and Conclusions

As indicated in the Introduction, a historical coverage of the seismic activity of over 500 years can in no way be exhaustive and homogeneous: the sources of information are different, the observers had different interests, and the means of observation were different. Nevertheless, a useful classification of the information contained in the present data file can be achieved.

The information contained in the data file pertains to two distinct periods: 1400–1874 and 1875–1974. The break at 1875 is merely a convenient choice: it is about the time when a change in the observation techniques is detected and it also allows the selection of a sampling period of useful length, 100 years.

Prior to 1875, seismic activity in Ethiopia was reported at random and sporadically. During that period, there are about 50 descriptions of macroseismic effects that are precise enough to be quantitatively evaluated. That number is extremely low in comparison with the theoretical number of events suggested by the statistical analysis of the last 100 years. A conservative projection for 475 years would be 3600 shocks causing intensity $\geqslant$ VI at epicentre, 12 800 shocks causing intensity = V at epicentre, and 57 000 shocks causing intensity = IV at epicentre, that is, a total of over 73 000 shocks, which at their epicentres had intensities above the threshold of human perceptibility. All these statistical projections for the past, when no seismic recording equipment even existed, must be understood and interpreted in terms of population distribution, reporting facilities, distance of the observer from the earthquake epicentre, etc. It is already a feat in African history that 50 important seismic events have been put on written record in Ethiopia prior to 1875! Incomplete as they might be, the earlier records of higher intensity tremors (VIII and XI) felt in the populated region of the Western Plateau must not be far from reality because these phenomena are currently used as historical landmarks. Their reported number fits the statistical projections based on more recent observations very well.

After about 1875, the bulk of the information was collected by scientists who, although they might not have had adequate equipment at their disposal, were professionally trained for observation work. This information is summed up in Fig. 155, where the seismic events are grouped by intensity categories within each decade of the period.

Figure 155 not only shows the time-distribution of reported tremors but also suggests the existence of a multiplicity of observation techniques during that century of observation. To investigate this point, comparison is made between the frequency of reported events in each intensity category, progressively, from the first decade to the last. Assuming an approximately constant level of activity during the observation period, and that identical or similar observation techniques yield identical or similar results, an

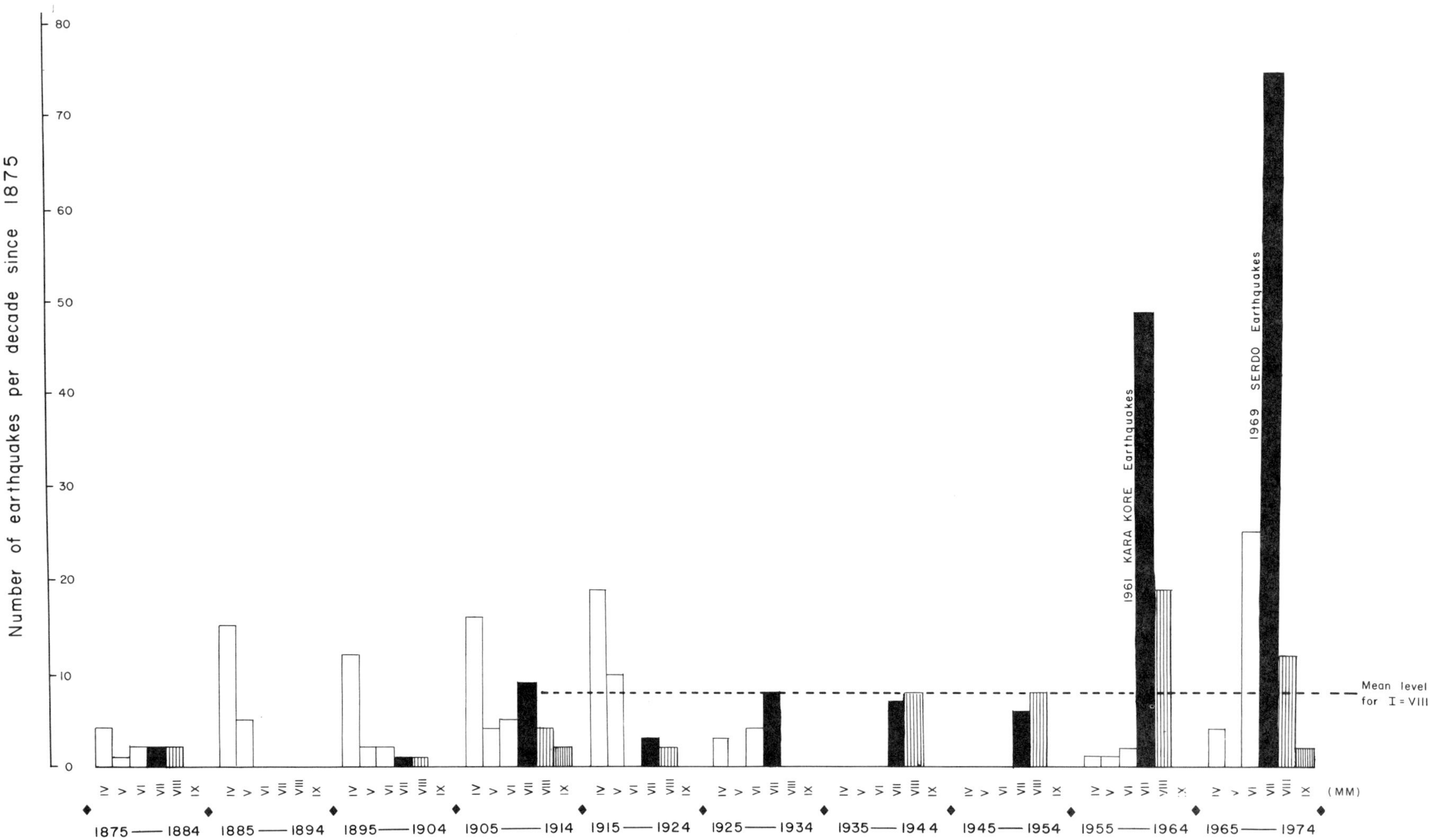

Fig. 155. Histogram of the number of seismic events reported during each decade of the 1875–1974 sampling period and classified by intensity MM units.

important change in amplitude classification on the histogram is interpreted as a change of observation technique; whereas, an increasing number of reported events within the same amplitude category is interpreted as an improvement in the technique. An important gap indicates a temporary suspension of observations.

Following these assumptions, the histogram reveals that at least two different techniques were used in collating the data from 1875 to 1974. During the first five decades, a relatively higher number of low intensity tremors were reported in the intensity IV and V categories; these intensities are almost nonexistent after 1915. On the other hand, during the last five decades shocks of intensity VII and VIII are prevalent. Such a change in reported intensities is explained in the following manner: the first half of the histogram is a period of noninstrumental data collecting; the second a period of instrumental acquisition.

At the end of the last century and the beginning of the twentieth, four professional scientists (Dainelli, a geologist; Marinelli, a volcanologist; and Cavasino and Palazzo, seismologists) either lived in northern Ethiopia or organized long-term research expeditions in the country. They collected all the information available on earthquakes and volcanic activity both from the historical sources kept almost exclusively in monastic libraries and from personal interviews with first-hand witnesses of the more recent periods of activity. Moreover, some of these scientists happened to be on the spot during periods of higher seismic activity. Most of their information, therefore, consisted of "reported felt" tremors as opposed to "instrumental" data, which was scarce but not completely unavailable. A seismoscope was in operation at the Meteorological Station in Massawa (N 15.6°, E 39.5°) for an uncertain length of time during the first five decades. Nothing is known of its specifications but it can be inferred that a 1900-model seismoscope could indicate the direction of the incoming seismic wave and its intensity at the station but could not record any parameter useful for epicentral determinations. No original seismoscope records have been preserved. The only information that is available today is the occasional note at the end of the meteorological reports related to the tremors that were felt, or recorded on the seismoscope. In 1913, a regular seismic station consisting of two horizontal Agamemnone pendulums was installed by Palazzo in Asmara (Palazzo 1913, p. 110–27). The static amplification of these seismographs was only 30. Operation started on 6 June 1913; it is presumed that, because of World War I, the station did not operate more than a few months. The original seismograms have been lost.

The sudden change in appearance of the histogram at the end of the first half of the period is a clear indication that the publication of the information obtained by Dainelli and Marinelli (1906, 1912), Palazzo (1915) and Cavasino (1922) marked the end of local interest in seismology.

The second half of the period, 1875–1974, is characterized by instrumental reports. As a matter of fact, teleseismic information on Ethiopia started with an earthquake of magnitude 5¾ located along the Eritrean escarpment. "Felt" and "instrumental" reports overlapped for a short time. By 1925, international teleseismic reports had achieved a certain regularity; their presence is revealed on the histogram by a definite shift of intensities from categories IV–V to VII–VIII. During that period, the number of shocks of calculated intensity VIII is constant until the opening in 1958 of a seismic station at Addis Ababa and its up-grading to a worldwide standard seismic station in 1962 (international station identification code: AAE). In 1973, a seismic network was installed by the Institut de Physique du Globe (Paris) with its central station at Arta, Republic of Djibouti (N 11.5°, E 42.8°).

Many of the seismological reports mention that two Mainka 450-kg horizontal pendulums were in operation at the meteorological station in Djibouti from 1935 to 1948. As a matter of fact, such seismographs were delivered to Djibouti by ORSTOM. (Office de la Recherche Scientifique dans les Territoires Outre-Mer) but never put into operation. A similar remark applies to the station in Mogadiscio, Somalia. The ISS catalogue of stations and the NOAA seismograph station list carry information on the Mogadiscio station, coded MOG, at N 02° 02′, E 45° 21′. Such a station was planned for the first International Arctic Year but it never materialized.

The definite shift from lower to higher intensities observed from the first to the second half of the histogram and the lower cutoff value at VII comes from the fact that the intensities were mostly calculated from seismograms traced at a minimum epicentral distance of 2000 km by long-period low-sensitivity instruments from which no reliable readings could be made for teleseismic shocks of lower magnitude. As a reference to the number and location of the seismic stations in the vicinity(!) of Ethiopia in the 1930s, Table 8 gives the stations that reported the earthquake of 24 October 1930 to the International Seismological Summary Office; their great circle distances and azimuths from Addis Ababa, which is located at about the geometrical centre of Ethiopia, are also indicated. The nearest station is Helwan in Egypt, 1500 km north of the Ethiopia-Sudan border, 2440 km from Addis Ababa and 3000 km from the Ethiopia-Kenya border. No station located at such epicentral distances and having long- or intermediate-period low-sensitivity seismographs could record reliable seismic phases for events much lower than magnitude 5½. Note also that the stations cluster in a 140° sector, north of Ethiopia (Fig. 156), and that there were no stations to the south.

The anomalous increase in the number of shocks (intensities VII and VIII) apparent on the histograms of the last two decades is due to aftershocks during the 1960 seismic activity in the Gulf of Tadjoura, the 1961 Kara Kore earthquakes in Central Wollo, the 1969 disaster at Serdo in Central Afar, and the 1973 renewal of activity in the Gulf of Tadjoura.

Table 8. Stations that reported to ISS the Ethiopian earthquake of 24 October 1930, and their position relative to Addis Ababa, the geometric centre of Ethiopia.

Station		Distance (km)	Azimuth (°)
HLW	Helwan	2441	343
KSA	Ksara	2772	354
BAK	Baku	3656	016
BOM	Bombay	3826	070
SAM	Samarkand	4408	035
RPD	Rocca di Papa	4444	329
ROM	Rome	4466	329
TAS	Tashkent	4676	035
DDI	Debra Dun	4706	054
ANR	Andizhan	4834	038
ALM	Almeira	5169	313
KUC	Kucino	5195	359
STR	Strasbourg	5263	332
AAA	Alma-Ata	5305	037
MAL	Malaga	5320	312
KAL	Calcutta	5487	068
SVE	Sverdlovsk	5640	015
PUL	Pulkovo	5686	355
GRE	Grenada	10922	284

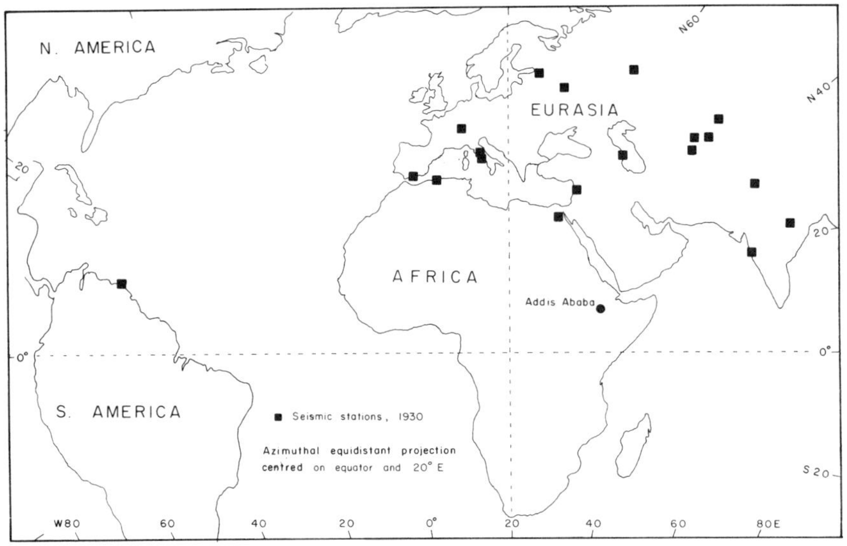

Fig. 156. Geographical distribution of the seismic stations that reported to ISS the earthquake of 24 October 1930 in Ethiopia.

Completeness of the Information

Figure 155 reveals that the data bank for the 1875–1974 period is comprised of two sets of observations collected by two different techniques that were not used simultaneously. Both sets are biased either regionally or intensity-wise: the first favours low intensities in the north of the country; the second favours high epicentral intensities because only higher magnitude events were instrumentally recorded by the international seismic network. Offhand, it appears that the anomalies in each set of data might balance one another over a 100-year period and produce an overall homogeneous ensemble at least for the northern half of the country. Statistical completeness tests confirm this impression.

The technique generally accepted for the analysis of completeness of a set of earthquake occurrences relies on the validity of the empirical relation between the frequency of occurrence and the magnitude of seismic shocks; an anomaly between the observed and computed frequency-magnitude curves is interpreted as a sign that the data file is not complete. This empirical relation is expressed by an equation in the form

$$\log_{10} N = a - bM \quad (1)$$

where a and b are constants determined by the regional level of seismic activity and N is the frequency of magnitude M events per unit time. This empirical law was observed by Ishimoto and Iida in 1939; it was later derived by Gutenberg and Richter (1944) in their global study *"Seismicity of the Earth."* The formula has been used and misused ever since.

Figure 157 illustrates the frequency-amplitude relationship for Ethiopia and the Horn of Africa. In this figure, intensity units are used instead of magnitude units; the conversion was made by the following equations:

$$\text{Intensity} = 3 \log_{10} (\text{acceleration}) + 4.5 \quad (2)$$

and

$$\text{Acceleration} = \frac{0.69e^{1.64M}}{1.1e^{1.10M} + \Delta^2} \quad (3)$$

Equation (2) is from Gutenberg and Richter (1954); equation (3) gives peak acceleration values obtained from accelerogram data (Cloud 1963). The graphs (Fig. 157) provide information on three different groupings of the data: A represents the entire 100-year period; B the last 60 years; and C the first 40 years of observation. Regression lines A and B reveal that the range of reported intensities considered reliable for numerical analysis is VII–IX inclusive for the whole 100-year period as well as for the shorter period of the last 60 years. During the first 40 years, curve C is observed to break into two segments (C_1 and C_2) with an intermediate point at I = VII. Segment C_1 represents almost exclusively teleseismic instrumental data; its lower bound is intensity VIII. The low scatter along segment C_2 vouches for the completeness of the investigation on reported felt tremors in northern Ethiopia. The marked difference in slope between C_2 and the three other curves is due mostly to the fact that C_2 is built on felt reports, that the rate of reporting such events decreases as the amplitude of the tremors becomes less and less important to the observer, and that the reported intensities are not epicentral values.

A note is introduced here concerning the I = IX upper bound of the regression curve A. Reference is made here of the two points (the solid circle and the open circle within the triangle) at the upper-intensity limit of the curve. The open circle (0.04, IX) represents the four events of intensity IX reported in the catalogue; the solid circle (0.06, IX) includes two more events for which amplitudes were not computed as such from seismographic traces but that field investigations revealed to be undoubtedly of intensity IX. The first occurred at Kara Kore in 1961 and left a fissure 15-20 km long and over 5-m deep along the eastern escarpment of the Robi-

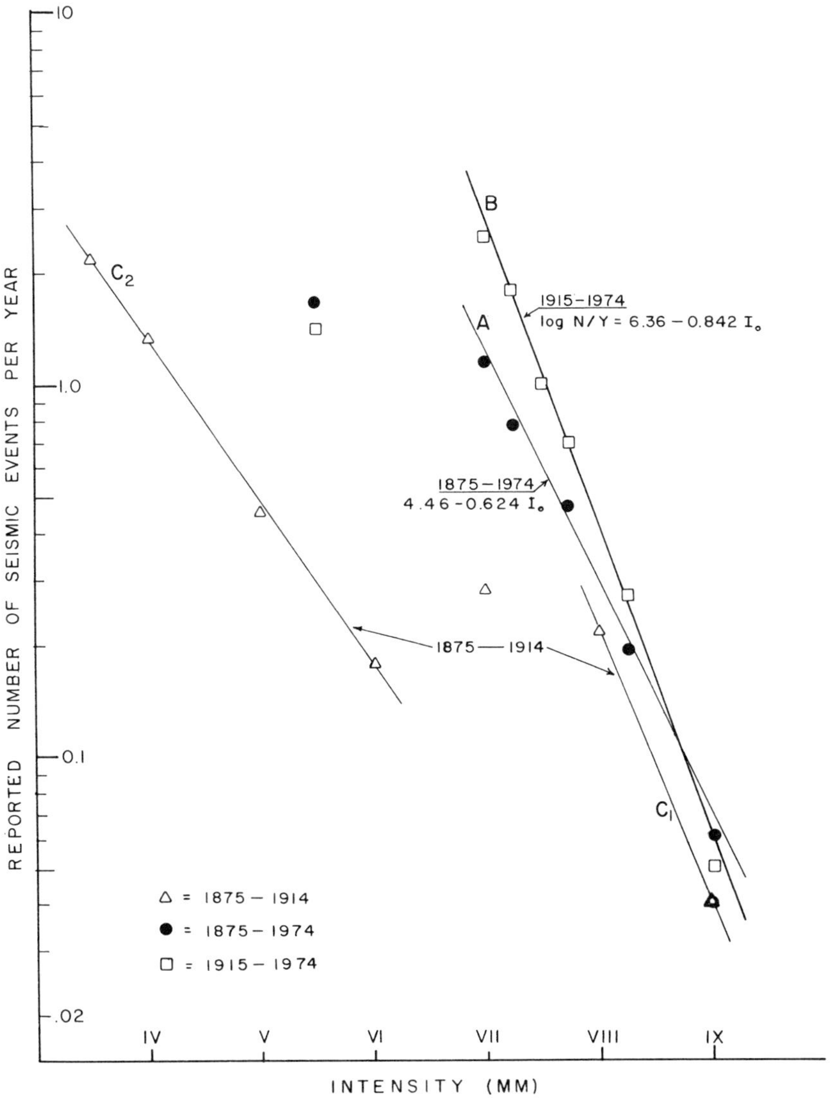

Fig. 157. Frequency-intensity distribution of seismic events in Ethiopia and in the Horn of Africa during sampling periods of various lengths.

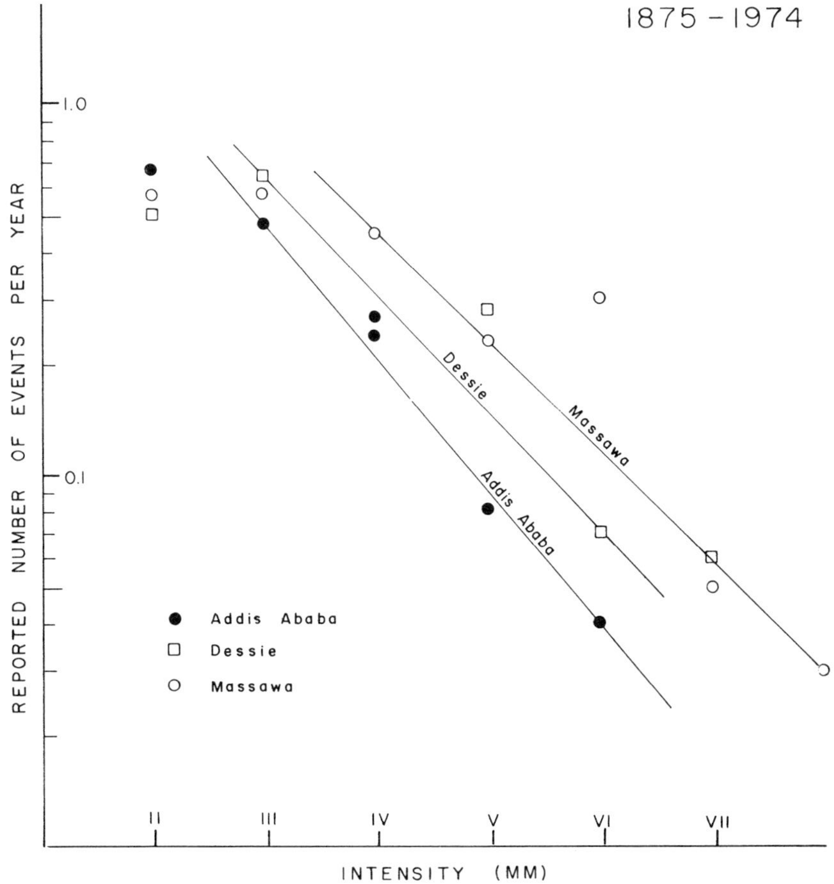

Fig. 158. Frequency distribution of tremors at three important population centres well distributed in latitude over Ethiopia: Addis Ababa, Dessie, and Massawa.

Borkena graben (Gouin 1970); the second on 5 April 1969, triggered near Serdo in central Afar a fault with vertical and horizontal displacements of 90 and 75 cm, respectively (see entry 1969/III–IV (Serdo – Central Afar), Region C).

Figure 158 confirms what was already suggested by the histogram (Fig. 155); the reporting of intensities lower than VII is inadequate. A second completeness test based on the analysis of the variance of sample means (after Stepp 1971, 1973) was performed to determine for which time intervals of the 100-year sampling period the different intensity categories were complete. This analysis did not produce any information not already learned from the histogram of the frequency-intensity distribution.

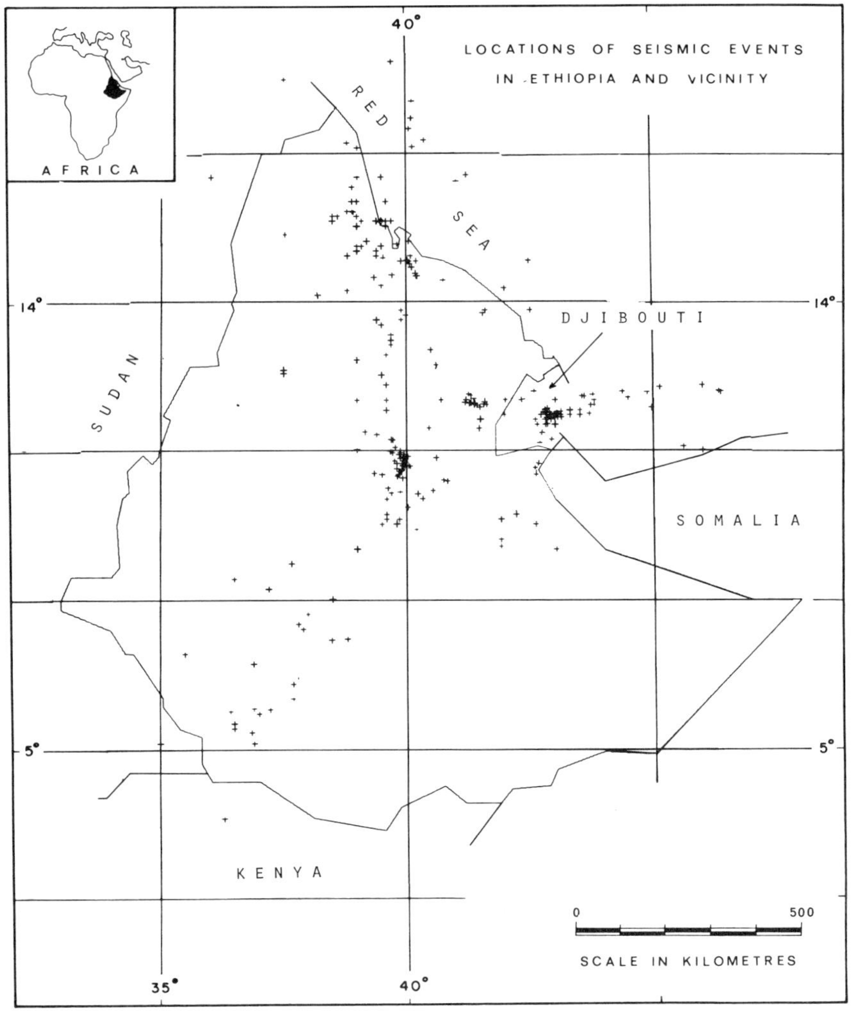

Fig. 159. Computer-plotted location map of all seismic events in the master file that are described with enough accuracy to be assigned reliable geographical coordinates (TLN and post-1975 events are not included).

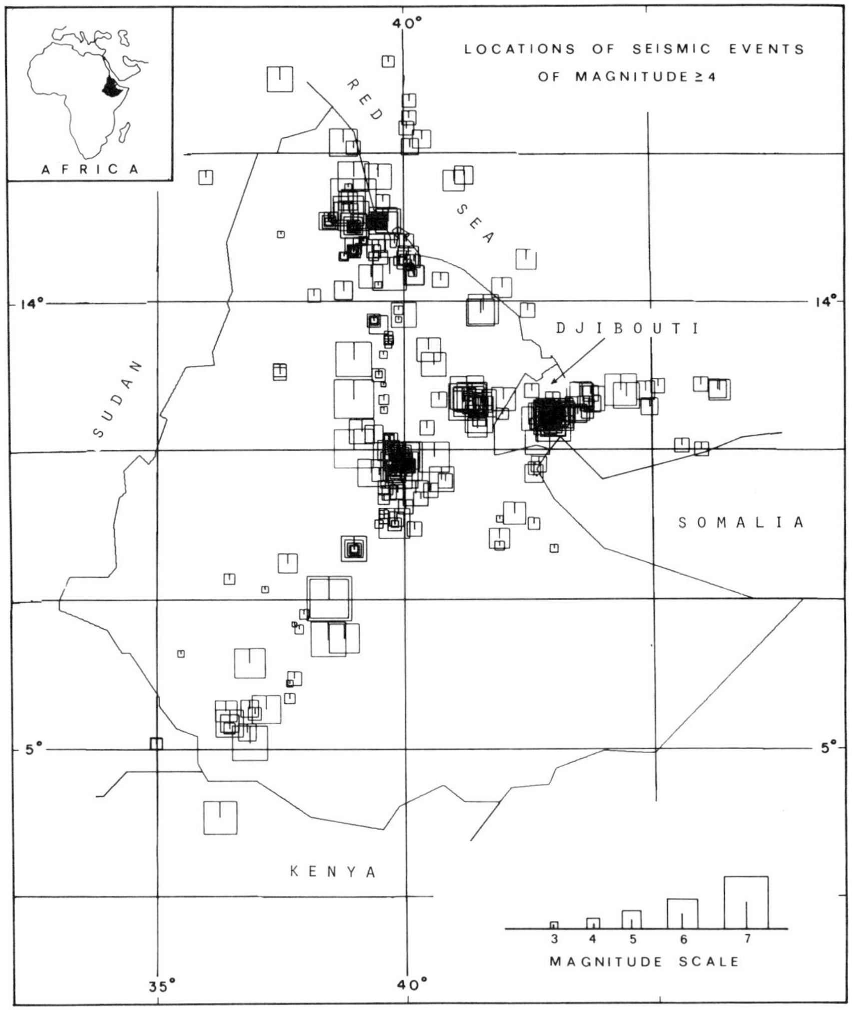

Fig. 160. Location and estimated magnitude of all events plotted on Fig. 159.

Spatial Distribution of the Seismic Activity in Ethiopia and the Horn of Africa

Part I is the descriptive section of this book; Part II presents the important earthquake elements in tabular form. As complementary information, the data are now presented in the form of seismicity maps in order to visualize at a glance what has been dispensed piecemeal in each entry.

Figures 159 and 160 are computer location plots of reported tremors and instrumental epicentres. On Fig. 159, the sites are identified by the same (+) symbol irrespective of the amplitudes; on Fig. 160, the size of the square (□) symbol is made proportional to the magnitude of the earthquake in the case of instrumental information, and, in the case of felt tremors, proportional to the magnitude of a local shock that would have produced the observed intensity at its epicentre. Figures 161 and 162 are

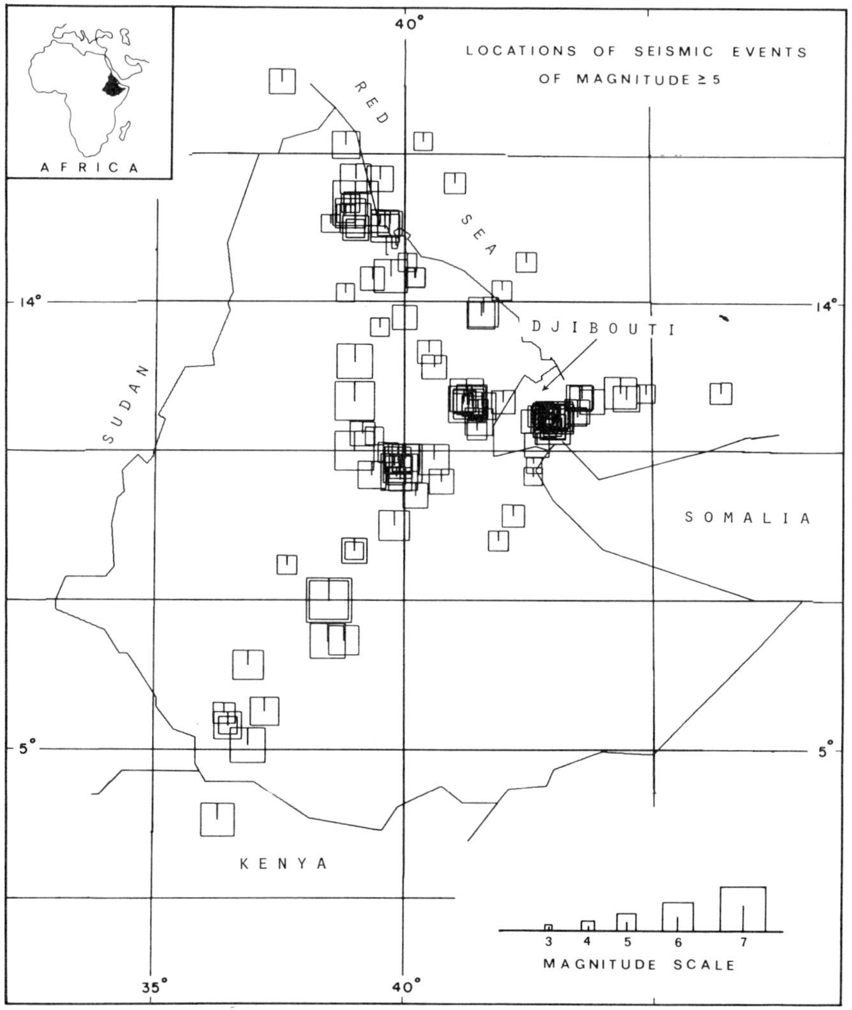

Fig. 161. Epicentre location of earthquakes with magnitudes equal to or greater than 5.

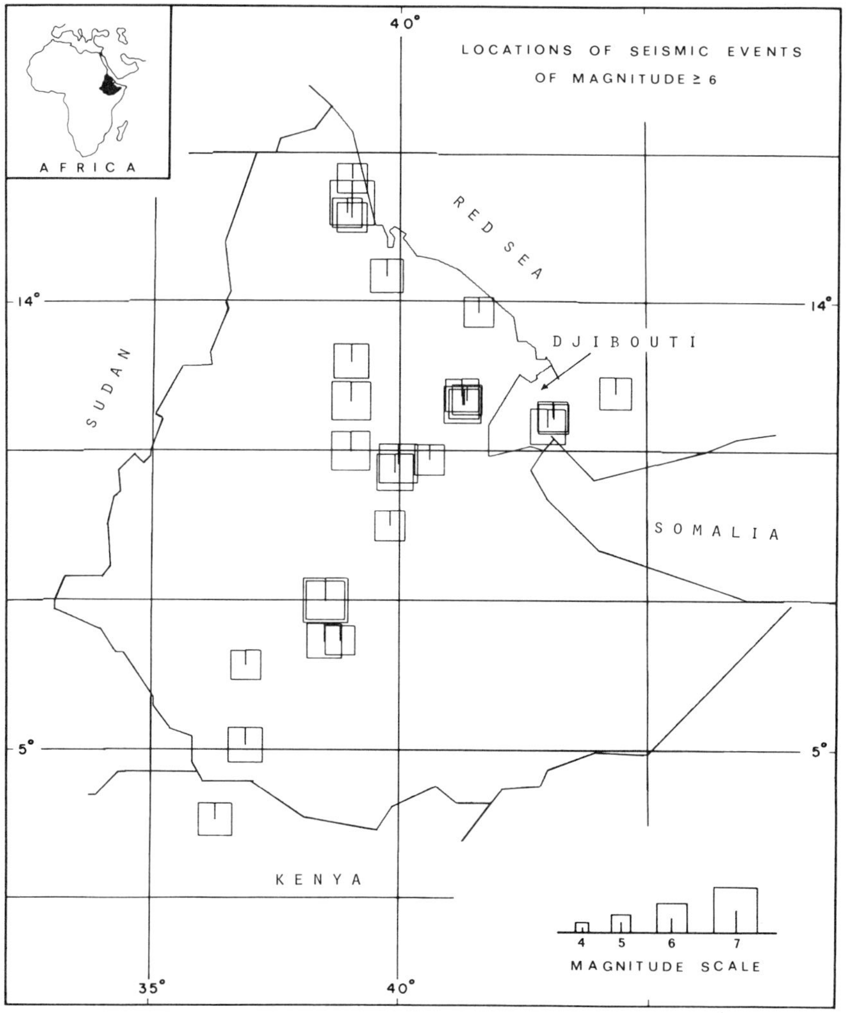

Fig. 162. Epicentre location of earthquakes with magnitudes equal to or greater than 6.

location plots of all epicentres of magnitudes equal to or larger than 5 and 6, respectively. These maps do not include the 1974 TLN (Tendaho Local Network) epicentres obtained by the author in 1978 nor the events postdating 1975. Their presence would not alter the epicentre pattern; they would simply increase the density of the points in certain areas and fill a few gaps apparent along the Plateau-Afar escarpment zone.

From these four epicentre location maps it is at first glance obvious that epicentres in Ethiopia and in the Horn of Africa are almost exclusively related to the major rift structures described in the Introduction.

A diagrammatic outline of these features left from the breakup of the Afro-Arabian swell is presented in Fig. 163. On these maps, one notices that the largest number of epicentres on the continent are located along and on the top margin of the Western Plateau – Rift escarpment and along the rift axial Wonji Fault Belt. The Southeastern Plateau – Rift escarpment is presently aseismic. Off shore, the Red Sea and Gulf of Tadjoura rifts are very active. Very few events occurred within the plateaus, with the exception of the Harar region and the Bia Anot fault zone across the Aisha Horst. The maximum magnitude of events observed in the whole region

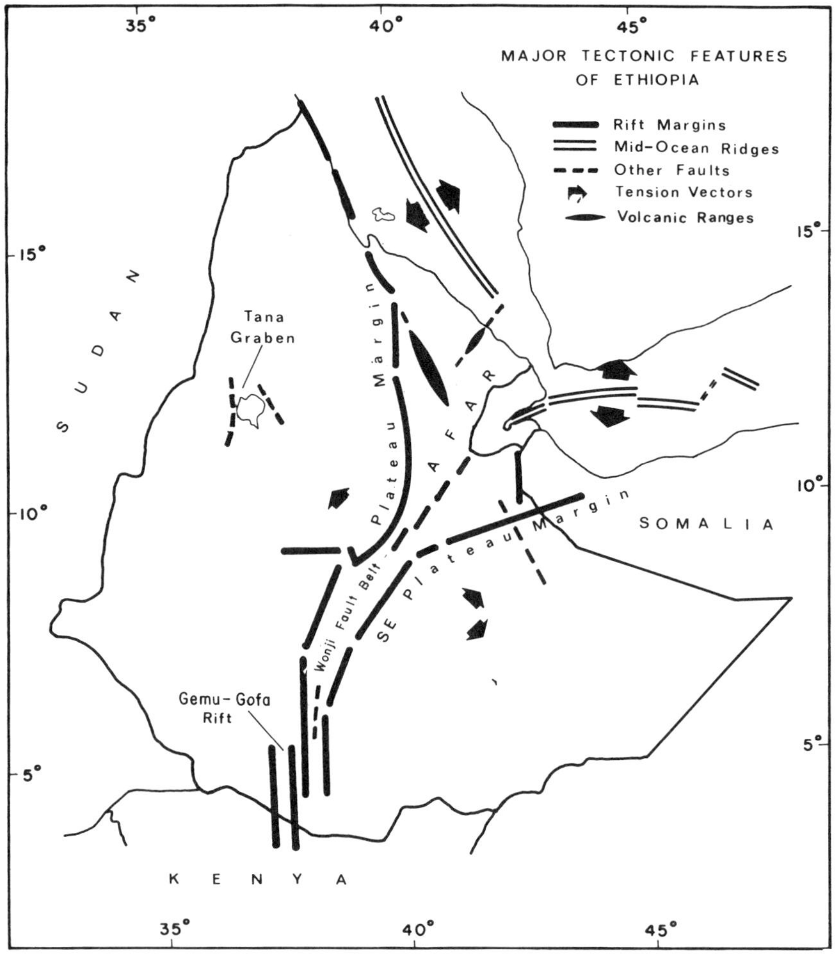

Fig. 163. Diagrammatic outline of important earthquake-causative tectonic features in Ethiopia and the Horn of Africa.

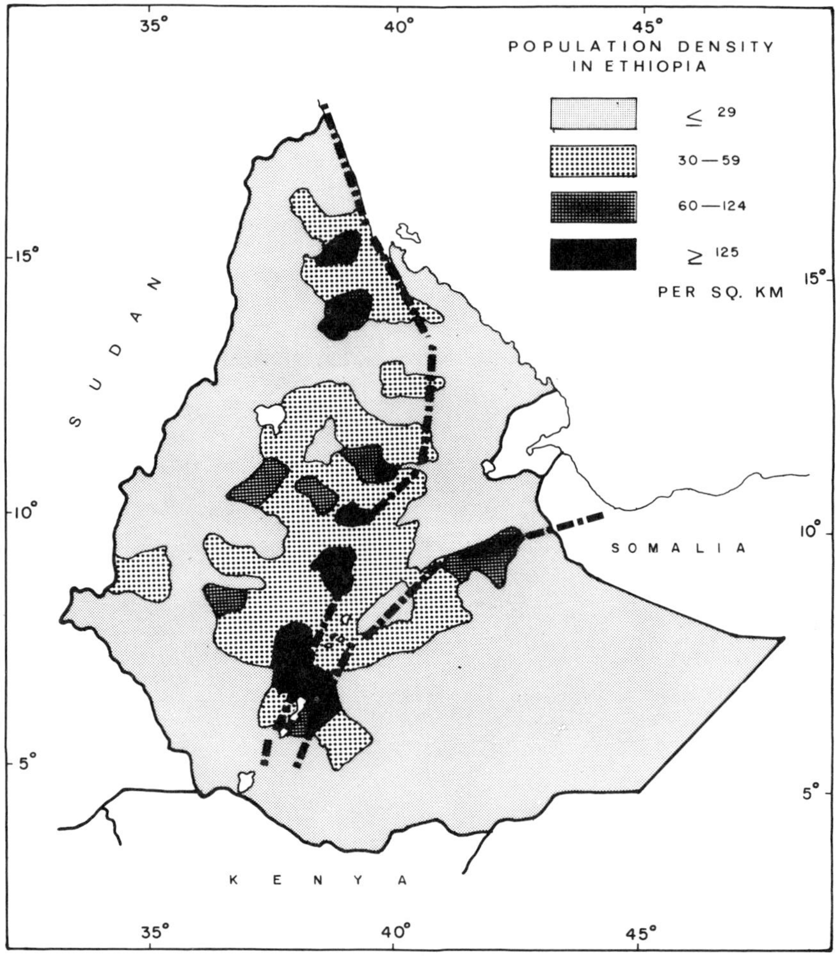

Fig. 164. Population density in Ethiopia according to the 1970 survey conducted by Ato Mesfin Wolde Mariam.

was no higher than 6¾.

As a final remark on the seismicity of the region, it is important to stress the fact that in Ethiopia the regions exhibiting the highest number of earthquakes unfortunately coincide with the areas of highest population density (Fig. 164).

Eventual Reevaluation of the Seismicity of Ethiopia

The compilation of the present catalogue was prompted by the urgent need to evaluate seismic and volcanic risks in a region straddling a triple rift junction. Because of a lack of precise information, the interpretation of the original data is often subject to uncertainties that, although they might not have been critical in the plotting of seismicity maps for the needs of Ethiopia in the 1970s, must be restricted to meet the more constringent requirements of modern developments, such as, the choice of sites for atomic power plants. The example of problems concerning atomic power plants is not purely academic because the Geophysical Observatory in Addis Ababa has already been approached about the eventual installation of such a plant in Eritrea, where hydroelectric power potential does not meet the regional needs. It is my hope that in such cases, the solutions presented in *Earthquake History of Ethiopia and the Horn of Africa* and in *Seismic Zoning in Ethiopia* will not be accepted as definitive but be revised in the light of newly acquired seismic information.

Bibliography

d'Abbadie, A. 1839–1843. Journal de voyage d'Antoine d'Abbadie (1839–1843). MS Ethiopiens codex 265 et 268, Paris, Bibliothèque Nationale.

1842. Voyage en Abyssinie. Bull. Soc. Géogr. Fr. (France), XVIII(2), 336.

1858. Sur le tonnerre en Ethiopie. Imprimerie Impériale (France), 158p.

1859. Catalogue raisonné des manuscripts éthiopiens appartenant à Antoine d'Abbadie. Paris, Bibliothèque Nationale.

1873a. Géodésie d'Ethiopie (2 vols). Paris, Edit. Radeau.

1873b. Observations relatives à la physique du Globe faites au Brésil et en Ethiopie. Paris.

1890. Géographie de l'Ethiopie: ce que j'ai entendu, faisant suite à ce que j'ai vu. Paris.

Agamemnone, G. 1912. Il vulcanismo: la sismicita della Colonia Eritea. Riv. Astr. Sci. Affini. (Torino), VI, 1–9.

d'Almeida, N. 1662a. Historia de Ethiopia a Alta ou Abassia, imperio do Abexine cujo rey vulgarmente ha chamada Preste Joam. Edited by Beccari in 1907–1908. Rome, 10 vols.

1662b. The history of High Ethiopia or Abassia. Beckingham, C.F., and Huntingford, G.W.B., ed. The Hakluyt Society, 2nd Series, CVII. (Translated from Portuguese).

Ambrasseys, N.N. 1962. A note on the chronology of Willis's list of earthquakes in Palestine and Syria. Bull. Seism. Soc. Amer. 52(1), 77–80.

Antonelli, P. 1889. Il primo viaggio di un Europeo a traverso l'Aussa: diario del Conte Pietro Antonelli, 10 January – 10 March 1883. Boll. Soc. Geogr. Ital. (Italy).

Arkin, Y., Beyth, M. et al. 1971. Geological Map of Mekele, sheet area ND 37–11, Tigray Province. Geological Survey of Ethiopia, Addis Ababa.

Aubert de la Rue, E. 1939. Le volcanisme en Côte Française des Somalis. Bull. Volc. (Napoli), ser. II, tome 5.
Azaïs, P., and Chambord, P. 1931. Cinq années de recherches archéologiques en Ethiopie. Geuthner, P., ed., Librairie Orientaliste, Paris.
Azzaroli, A., and Fuis, V. 1971. Geological outlines of the north end of the Horn of Africa. India, 22nd Intern. Geol. Congress, section 4, 293–314.
Backer, H., Lange, K., and Richter, H. 1975. Morphology of the Red Sea Graben between Zubair Islands and Abul Kizaan. Geologisches Juhrbuch, Reike D, Heft 13, 79–123, Hannover.
Baker, B.H. 1970. The structural pattern of the Africo-Arabian rift system in relation to plate tectonics. Philos. Trans. R. Soc. Lond. A. Math. Phys. Sci. (England), A267, 383–391.
Baker, B.H., Mohr, P.A., and Williams, L.A.J. 1972. Geology of the Eastern Rift System of Africa. Geol. Soc. Amer. (Washington), Special Paper No. 136.
Banghar, A.R., and Sykes, L.R. 1969. Focal mechanisms of earthquakes in the Indian Ocean and adjacent regions. J. Geophys. Res. (USA), 74, 632–649.
Bannert, D.N. 1971. Fotogeologische Untersuchungen an den Vulkanen des nordlichen (Äthiopien). Bildness und Luftbildwesen, 39, 74–84, and in Beih. Geol. Jb., 116, 87–140.
1972. Afar tectonics analysed from space photographs. Am. Assoc. Pet. Geol. Bull. (USA), 56, 903–915.
1973. Zur geotektonischen problematik in Afar-gebiet. Z. Dtsch. Geol. Ges. (West Germany); 124, 329–340.
Bannert, D.N., Brinckmann, J., et al. 1970. Zur geologie der Danakil Senke. Geol. Rundschau, 59, 409–443.
Bannert, D.N., and Kedar, E.Y. 1971. Plate tectonics in the Red Sea region as inferred from space photography. Washington, NASA Tech. Note D-6261.
Barberi, F., Borsi, S., et al. 1970. Relations between tectonics and magmatology in the Northern Danakil Depression (Ethiopia). Philos. Trans. R. Soc. Lond. A. Math. Phys. Sci. (England), A267, 293–311.
1972. Evolution of the Danakil Depression (Afar, Ethiopia) in the light of radiometric age determination. J. Geol. (USA), 80, 720–729.
Barberi F., and Varet, J. 1970. The Erta 'Ale volcanic range (Danakil Depression, Northern Afar, Ethiopia). Bull. Volc. (Italy), 34(4), 848–917.
1972. Geological map of the Erta 'Ale range. CNR-CNRS, scale 1:100,000. Geotechip.
Barberi, F., Tazieff, H., and Varet, J. 1972. Volcanism in the Afar Depression: its tectonic and magmatic significance. Tectonophysics, 15, 19–29.
Barberi F., Cheminée, J.L., and Varet, J. 1973. Long-lived lava lakes of the Erta 'Ale volcano. Rev. Géogr. Phys. et Géol. Dyn., 15(4), 347–352.
Barker, W.C. 1846. Volcano of Saddle Island. J. Roy. Geogr. Soc. (England), XVI, 338.
Basset, R. 1882. Études sur l'histoire de l'Ethiopie (MS Ethiopien No. 142. Bibliothèque Nationale, Paris: Receuil des rois d'Abyssinie rédigé en l'an 7278 Ma'hdara Mariam sur l'ordre du Dejesmatch Haylou. (In Ge'ez).
Bath, M. 1973. Introduction to seismology. John Wiley and Sons, NY and Toronto, 395 p.
1975. Seismicity of the Tanzania Region. Seismological Institute, Uppsala, Report No. 1-75, 38 p.
Beccari, C. 1903. Notizia e Saggi di Opere e documenti inediti riguardanti la storia di Etiopia durante i secoli XVI, XVII e XVIII, con otto facsimile e due carte geografica (15 vols). S.J. Press. Rome.
Beckhingham, C.V., and Huntingford, G.W.B. 1954. Some records of Ethiopia, 1593–1645. Hakluyt Society, second series, CVIII.
Beke, C.T. 1844. Abyssinia being a continuation of routes in the country. J. Roy. Geogr. Soc. (England), 14, 1–76.
1861. Volcanic eruption of Djebbel Dubbeh. (London), The Times, 24 September 1861.
Bellamy, E.F. 1936. Index catalogue of epicentres for 1913–1930. (A geographical index to the International Seismological Summary 1918–1930). Isle of Wight.
1939. Epicentres of earthquakes, 1931–1932. Nature (Lond) (England), 143, 504–509.
1947. Index catalogue of epicentres for 1931–1935 (Supplement to index catalogue 1913–1930 of the International Seismological Summary). Isle of Wight.
Bellavita, G. 1931. I giacimenti potassici di Dallol. L'Ingegnere (Italy), 5(6), 3–12. Rome.
Benioff, H. 1934. The physical evaluation of seismic destructiveness. Bull. Seism, Soc. of Amer. 24, 398–403.
Berckhemer, H. et al. 1975. Deep seismic soundings in the Afar region and on the highland of Ethiopia. In Pilger, A., and Rosler, A., ed., Afar depression of Ethiopia. Inter-Union Commission on Geodynamics. Stuttgart, Scientific Report No. 14, 89–113.
Beydoun, Z.R. 1970. Southern Arabia and Northern Somalia: comparative geology. Philos. Trans. R. Soc. Lond. A. Math. Phys. Sci. (England), A267, 267–292.
Beyth, M. 1972a. Paleozoic — Mesozoic sedimentary basin of the Mekele outlie, northern Ethiopia. Am. Assoc. Pet. Geol. Bull. (USA), 56, 2426–2439.
1972b. To the geology of central-western Tigré. Bonn, Rheinische Fieldrich-Wilhem University, 159 p.
Black, R. 1973. Structure of the Afar floor. In Tarling, D.H., and Runcorn, S.K., ed., Implications of continental drift to Earth's sciences. London, Academic Press Inc., 777 p.
1976. The Afar: example of an Atlantic type continental margin in the making. Anal. Acad. Brasil. Cienc., supplement 48, 27–36.
Black, R., Morton, W.H., and Varet, J. 1972. New data on Afar tectonics. Nature (Lond) (England), 240, 170–173.
Black, R., Morton, W.H., and Tsegaye, H. 1974. Early structure around the Afar triple junction. Nature (Lond) (England), 248, 496–497.
Blanc, Dr. 1865. Observations faites par M. le Dr. Blanc du 24 juillet au 1er octobre, 1864, à Massawa. Ann. de la Soc. Météo. de France.
Blundell, H.W. 1900. A journey through Abyssinia to the Nile. Geogr. Journal, 15(2), 97–121.
1922. The Royal Chronicles of Abyssinia, 1769–1840, with translation and notes. London, Cambridge University Press, 543 p.
Bolt, B.A. 1960. The revision of earthquake epicenters, focal depth and origin times using a high speed computer. Geophys. J. R. Astron. Soc., 3, 433–440.
Bombaci, A., del Pozzo et al. 1937. Elementi per la toponomastica etiopica. Napoli.

Bonatti, E., Emiliani, G., et al. 1971. The final desiccation of the Afar rift (Ethiopia). Science (Wash, DC) (USA), 172, 468–469.
Bonjer, K.P., Fuchs, K., and Wohlenberg, A.J. 1970. Crustal structure of the East African rift system from spectral response ratios of long-period body waves. Zeitscript. für Geophysik, 36, 287–297.
Branner, J. 1915. The untrustworthiness of personal impressions of directions of vibrations in earthquakes. Bull. Seism. Soc. Amer. 5(1), 26–29.
Brinckmann, J., and Kürsten, M. 1970. Geological Sketch Map of the Danakil Depression (1:250,000). Hannover, Bundesanstalt für Bodenforschung, 4 sheets.
1971. Stratigraphie und Tektonik der Danakil-Senke (NE-Aethiopien). Beih. Geol. Jahrbk. (Hannover), 116, 5–86.
Brown, F.H., and Carmichael, I.S.E. 1969. Quaternary volcanoes on lake Rudolf region: the Korath range. Lithos, 2, 239–260.
Brune, J.N. 1970. Tectonic stress and the spectra of seismic shear waves from earthquakes. J. Geophys. Res. 75, 4997–5010.
Budge, E.A.W. 1928. A history of Ethiopia, Nubia and Abyssinia. London, 2 vols.
Burek, P.J. 1970. Paleomagnetic evidence of an anticlockwise rotation of the Danakil Alps, Ethiopia. EOS, Trans. Amer. Geophys. Union, 51, 271. (Abstr).
1975. Tectonic consequences of plate movements exemplified by the developments in the N-Afar triangle and the Arabian shield. In Pilger, A. and Rösler, A., ed., Afar Depression of Ethiopia. Stuttgart.
Burke, K.C., and Whiteman, A.J. 1973. Uplift, rifting and the break-up of Africa. In Tarling, D.H., and Runcorn, S.K., ed., Implications of continental drift to Earth sciences. London. Academic Press Inc. Ltd., 2, 735–755.
Burke, K.C., and Wilson, J.T. 1976. Hot spots on the Earth's surface. Sci. Am. (USA), 235(2), 46–57.
Bushman, B. 1906. Bulletin de la Commission Centrale Sismique de Russie, Juillet-Septembre, 1906. St. Petersburg.
Buxton, D. 1957. Travels in Ethiopia. London.
1970. The Abyssinians. London, Praeger Press.
de Castro, L. 1915. Nella terra del Negus. Milano, 2 vols.
de Castro L., and Oddone, E. 1909. La citta e il clima di Addis Ababa. Boll. Soc. Georg. Ital. (Italy). Roma.
Cattenoz, H.G. 1961. Tables de concordance des ères chrétiennes et hégériennes. 3rd edition, Rabat.
Cavasino, A. 1922. Macrosismi avvertiti in Italia et Colonie nell' anno 1921. Boll. Soc. Sism. Ital. (Italy), 24, 131–133.
Cerulli, E. 1928. Per la toponomastica etiopica. Italy, Oriente Moderno, anno 8 fasc. 7, 328–329.
1931. Documenti arabici per la storia dell' Etiopia (MS 5: fragmenti di una cronaca dell' Aussa). Mem. Acad. Naz. dei Lincei (cl. di sc. mor. stor. e fil.) Italy, 327, serie 6(4), fasc.2.
Chaine, M. 1913. Histoire du Règne de Johannes IV, roi d'Ethiopie, 1868–1889. Revue Sémitique (France), 1913.
1925. La chronologie des temps chrétiens de l'Egypte et de l'Ethiopie. Paris.
Cheesman, R.E. 1936. Lake Tana and the Blue Nile. London.
Cloud, W.K. 1963. Period measurements of structures in Chile; Chilean earthquake of May 1960. Bull. Seism. Soc. Amer. 53(2), 359–380.
CNR-CNRS. 1973. Consiglio Nationale de la Ricerche (Italy) and Centre National de la Recherche Scientifique (France). Joint research in Ethiopia. In Revue de Géographie Physique et de Géologie Dynamique, XV(4), Sept–Oct. 341–491, Numéro spécial sur l'Afar.
Cole, J.W. 1969. Gariboldi volcanic complex, Ethiopia. Bull. Volc. (Italy), XXXIII(2), 566–578.
Congrégation de la Mission, 1876. Annales de la Congrégation de la Mission, No. 41. Paris.
Consociatione Turistica Italiana. 1938. Guida dell' Africa Orientale Italiana, Milano.
Conti Rossini, C. 1903. Al Regali, L'Esplorazione Commerciale, XVIII. Rome.
1910. Liber Axumae. Corpus Scriptorum Christianorum Orientalium, Scriptores Aethiopici, series altera, Tome VIII, Paris.
1918. L'autobiografia di Paulos monaco abissino del secolo XVI, Reale Acad. dei Lincei. (Italy), 27, 9–10.
1947. Nuovi documenti per la storia d'Abyssinia nel secolo XIX, Reud. dell' Acad. Naz. di Lincei, (sc. mor.) series 8, vol. 2, Rome.
1948. Tabelle comparative del calendario etiopico col calendario romano. Instituto per l' Oriente. Rome.
1954. Liber Axumae. Corpus Scriptorum Christianorum Orientalium, Scriptores Aethiopici, XXVII. Louvain.
du Coursac, J. 1926. Le Règne de Johannes. Italy, Edit. Romans.
Courtillot, V., Tapponnier, P., and Varet, J. 1974. Surface features associated with transform faults: a comparison between observed examples and an experimental model. Tectonophysics, 24, 317–329.
Cox, K.G., 1970. Tectonics and volcanism of the Karoo period and their bearing on the postulated fragmentation of Gondwanaland. In Clifford, T.D., and Gass, J., ed., African magmatism and tectonics. Edinburgh, 211–235.
Crampin, S., Fyfe, C.J., et al. 1976. Atlas of seismic activity 1909 to 1968. Seism. Bull. Inst. Geol. Sci. (Edinburgh), No. 5.
Dainelli, G. 1935. L'Africa Orientale. Reale Soc. Geogr. Ital., Bologna.
1943. Geologia dell' Africa Orientale. Rome, Reale Academia d'Italia, 4 vols.
Dainelli, G., and Marinelli, O. 1906. Dell' Erta 'Ale: Vulcano ritenuto attivo della Dancalia settentrionale. Riv. Geogr. Ital. (Italy), XIII, 261–270. Firenze.
1907a. Vulcani attivi della Dancalia. Riv. Geogr. Ital. (Italy), XIV, 22–26.
1907b. Escursione al vulcano Alid. Riv. Geogr. Ital. (Italy), XIV, 257–274.
1908a. Sopra la positione del vulcano Afdera in Dancalia. Riv. Geogr. Ital. (Italy), XV.
1908b. A proposito di moderne manifestationi di attivita vulcanica in Dancalia. Riv. Geogr. Ital. (Italy), XV.
1912. Risultati scientifi di un viaggio nella Colonia Eritrea. Publ. del R. Ist. di St. Super. e di Perfez. di Firenze. Firenze, Italy.
Dakin, F.M. 1975. A preliminary survey of the distribution and characteristics of earthquake sequences in Ethiopia and T.F.A.I. Bull. Geophys. Obs. (Ethiopia), 15, 51–70.
Dakin, F.M., and Gibson, J.L. 1971. A preliminary account of Alutu, a pantelleritic volcano in the main Ethiopian Rift. Bull. Geophys. Obs.

(Ethiopia), 13, 110–119.
1971b. Bosetti Baricia: a Quaternary pantelleritic volcano in the main Ethiopian rift. Bull. Geophys. Obs. (Ethiopia), 13, 106–109.
Dakin, F.M., Gouin P., and Searle, R.C. 1971. The 1961 earthquakes in Serdo, Ethiopia. Bull. Geophys. Obs. (Ethiopia), 13, 19–56.
Darragon, G. 1898. Le Sidama, l' Amara, le Konso. C.R. Soc. Geogr., 137–140.
Davidson, A., Moore, J.M., and Davies, J.C. 1973. Omo River Project: preliminary report on the geology and geochemistry of parts of Sidamo, Gemu-Gofa and Kefa provinces, Ethiopia. Ministry of Mines, Addis Ababa. Report No. 1.
Davidson, C. 1921. On the scales of seismic intensities and the construction of isoseismal lines. Bull. Seism. Soc. Amer. 11, 95–129.
1933. Scales of seismic intensity: supplementary paper. Bull. Seism. Soc. Amer. 23, 158–166.
Delibrias, G., Marinelli, G., and Stieltjes, L. 1975. Spreading rate of the Afar Rift: a geological approach. In Pilger, A., and Rösler, A., eds., Afar Depression of Ethiopia. Inter-Union Commission on Geodynamics, Stuttgart, Scientific Report No. 14, 214–220.
Detaille, C. 1885. Statistique des tremblements de terre. Paris, Flammarion, In: L'Astronomie, 185–189.
Dillman, A. 1844. Ueber die Regierung, insbesondere die Kirchenordnung des Königs Zar'a Jacob. Abhandl. d. Königl. Preuss. Akad. d. Wissensch., Jahrg., Berlin.
Douglas, A. 1967. Joint epicentre determination. Nature (Lond) (England), 215, 47–48.
Drake, C.L., and Girdler, R.W. 1964. A geophysical study of the Red Sea. Geophys. J. R. Astr. Soc. (England), 8(5), 473–495.
Dreyfuss, M. 1931. Etudes de géographie physique sur la Côte française des Somalis. Revue de Géographie Physique (France), Chapter IV, 338–347.
Engelbert, V. 1970. The Danakil: nomads of Ethiopia's wasteland. Natl. Geogr. Mag. (USA), 137(2), 186–211.
Fairhead, J.D. 1968. The Seismicity of the East African Rift System, 1955–1968. England, University of Newcastle upon Tyne. (MSc dissertation).
1971. The seismicity of Africa. Geophys. J. R. Astr. Soc. 24, 271–301.
Fairhead, J.D., and Girdler, R.W. 1970. (Preprint 1969). How far does the rift system extend through Africa? Nature (Lond) (England), 221, 1018–1020.
1970. The seismicity of the Red Sea, Gulf of Aden and the Afar Triangle. Philos. Trans. R. Soc. Lond. A Math. Phys. Sci. (England), A267, 49–74.
1971. The seismicity of Africa. Geophys. J. R. Astr. Soc. 24, 271–301.
Fantoli, A. 1966. Contributo alla Climatologia dell' Altopiano Etiopico, regione Eritrea. Rome, Ministero degli Affari Esteri.
de Fino, M., La Volpe, L., and Lirer, L. 1971. Notizie preliminari sul vulcanismo basaltico della regione di Assab (Ethiopia). Rend. Soc. Ital. Min. e Petrol. 27, 93–104.
1973. Volcanology and petrology of the Asseb Range (Ethiopia). Bull. Volc. (Italy), 37, 95–110.
de Fino, M., La Volpe, L., et al. 1973. Geology and petrology of the Manda-Inakir range and Mussa Alli volcano, central eastern Afar (Ethiopia and TFAI), Rev. Géogr. Phys. et Géol. Dyn. 15, 373–386.
Flinn, E.A., and Engdahl, E.R. 1965. A proposed basis for geographical and seismic regionalization. Rev. Geophys. 3, 123–149.
Foti, C. 1941. La cronaca abbreviata dei re d'Abissinia in uno nuovo manuscritto di Debra Berhan di Gondar. Rassegna di Studi Etiopici, I. University of Asmara.
Frazier, S.B. 1970. Adjacent structures of Ethiopia: portion of the Red Sea coast including Dahlak Kebir Island and the Gulf of Zula. Philos. Trans. R. Soc. Lond. A. Math. Phys. Sci. (England), A267, 131–141.
Freeman-Grenville, G.S.P. 1963. The Muslim and Christian calendars, London, The Oxford University Press.
Freund, R. 1970. Plate tectonics of the Red Sea and East Africa, Nature (Lond) (England), 228, 453.
Fuchs, C.W.C. 1886. Die vulkanischen Ereignisse des Jahre 1884, XX, Jahresbericht. Tschermak's Mineralogische und Petrographische Mitteilungen, Neue Folge, VII Band. Wien.
Gass, I.G. 1970a. The evolution of volcanism in the southern area of the Red Sea, Gulf of Aden and Ethiopian rifts. Philos, Trans. R. Soc. Lond. A. Math. Phys. Sci. (England), A267, 369–381.
1970b. Tectonic and magmatic evolution of the Afro-Arabian dome. In Tarling, T.H., and Gass, I.G., ed., African magmatism and tectonics. Edinburgh, Oliver & Boyd Ltd., 285–300.
1973. The Red Sea depression: causes and consequences. In Tarling, T.H., and Runcorn, S.K., ed., Implications of continental drift to the Earth sciences. London, Academic Press Inc. Ltd., 779–786.
Gass, I.G., Mallick, D.I.J., and Cox, K.C. 1973. Volcanic islands of the Red Sea. J. Geol. Soc. (Lond) (England), 129, 275–310.
Gebru, (Alaqa), 1851. King of Kings Tekle Ghiorghis: 1779–1795. Reproduced and translated in English by H.W. Blundell under the title: The Royal Chronicle of Abyssinia, 1769–1840. London, Cambridge University Press, 1922.
Geological Survey of Ethiopia. 1973. Geological map of Ethiopia. Scale 1:2,000,000. Geological Survey of Ethiopia. Addis Ababa, Ethiopia, Ministry of Mines.
Gibson, I.L. 1967. Preliminary account of the volcanic geology of Fantale, Shoa, Ethiopia. Bull. Geophys. Obs. (Ethiopia), 10, 59–67.
1969a. The Quaternary pantelleritic volcano Fantale, Ethiopia. Proc. Geol. Soc. Lond. (England), 1658, 267–269.
1969b. The structure and volcanic geology of an axial portion of the main Ethiopian rift. Tectonophysics, 8, 561–565.
1970a. Quaternary pantelleritic volcanism in the main Ethiopian rift. XIV Annual Rep. Inst. of Afric. Geol. England, University of Leeds, 35–38.
1970b. A pantelleritic welded ash flow tuff from the Ethiopian rift valley. Contr. Min. Petrol., 28, 89–111.
Gibson, I.L., and Gass, I.G. 1969. Middle East rift zones. Nature (Lond) (England), 224, 791.
Gibson, I.L., and Tazieff, H. 1970. The structure of Afar and the northern part of the Ethiopian rift. Philos. Trans. R. Soc. Lond. A. Math. Phys. Sci. (England), A267, 331–338.
Girdler, R.W. 1970. The structure and evolution of the Red Sea and the nature of the Red Sea, Gulf of Aden and Ethiopian rift junction. Tectonophysics, 10, 579–582.

1976. The East African rift system: geophysical aspects in geodynamics. In Drake, C., ed., Progress and prospects. AGU, 75–80.
Girdler, R.W., and Styles, P. 1974. Two-stage Red Sea floor spreading. Nature (Lond) (England), 247, 7–11.
Girdler R.W., and Whitmarsh, R.B. 1974. Miocene evaporites in the Red Sea cores and their relevance to the problem of the width and age of oceanic crust under the Red Sea. In Initial report deep sea drilling project. Washington, D.C., U.S. Government Printing Press, 23, 913–921.
Gorshkov, G.P. 1963. Seismicity of Africa. In A review of the natural resources of the African continent. Paris, UNESCO. 103–149.
Gortani, M. 1940. Serie eruttive e vulcani attivi della Dancalia meridionale. Mem. R. Acad. Sci. Ist. (Bologna), IX(7), 53.
Gortani, M., and Bianchi, A. 1939. Nella Dancalia meridionale. Boll. R. Soc. Georg. Ital. (Italy), VII(4).
Gouin, P. 1960. A historical survey of the seismicity of Ethiopia and the first seismological report from the Geophysical Observatory. Bull. Geophys. Obs. (Ethiopia), 2(1), 23–36.
1962. Seismological report. Bull. Geophys. Obs. (Ethiopia), 3(1), 63–71.
1963. The 1961 earthquakes in Ethiopia and the general seismicity of the Ethiopian rift system. Berkeley, IUGG General Assembly, program of abstracts, 3, 13.
1965. Coincidence of magnetic disturbances with local earthquakes recorded from the Ethiopian Rift System. Nature (Lond) (England), 208, 541–543.
1970. Wolenchiti fracturing. Smithsonian Inst., CSPL Event 86–70, card 1019.
1971. Surface cracks and subsidences in Mojjio (Ethiopia). Technical Report to Ministry of Interior, Ethiopian Government 24 May 1971. Smithsonian Inst., CSPL, Event 55–71, card 1231.
1975. Kara Kore and Serdo epicenters: relocation and tectonic implications. Bull. Geophys. Obs. (Ethiopia), 15, 15–26.
1976. Seismic zoning in Ethiopia. Bull. Geophys. Obs. (Ethiopia), 17, 1–46.
Gouin, P., and Dakin, F.M. 1969. Smithsonian Centre for Short-lived Phenomena. Card 493/1969.
Gouin, P., and Mohr, P.A. 1964. Gravity traverses in Ethiopia (interim report). Bull. Geophys. Obs. (Ethiopia), 3, 185–239.
Grabham, G.W., and Black, R.P. 1925. Report of the mission to Lake Tana, 1920–1921. Cairo, Government Press.
Graham, Captain D. 1841. Indian office: Bombay secret proceedings. State Report of India, 204, 1216.
Greitzer, Y. 1970. Stratigraphy, hydrology and Jurassic ammonites of the Harar and Dire Dawa area, Ethiopia. Tel. Aviv, Hebrew University. 210 p. (Ph.D. thesis).
Guidi, I., ed. 1961a. Annales regis 'Iyassu I (Adyam Sagad). In Annales Johannis I, 'Iyassu I et Bekaffa, vol. 2. Corpus Scriptorum Christianorum Orientalium, 23, Tome VI: Scriptores Aethiopici, series altera, Tome V (1st edit., 1903), Louvain.
1961b. Annales regis 'Iyassu I (Adyam Sagad). Annales Johannis I, 'Iyassu I et Bekaffa, Vol. 2. Corpus Scriptorum Christianorum Orientalium, 24, Tome VII, Louvain.
Gumper, F., and Pomeroy, P.W. 1970. Seismic wave velocities and earth structure on the African continent. Bull. Seism. Soc. Amer. 60, 651–668.
Gutenberg, B. 1929. Handbuch der Geophysik. Berlin, Borntraeger.
1956. Great earthquakes 1896–1903. Trans. Amer. Geophys. Union, 37, 608–614.
Gutenberg, B., and Richter, C.F. 1941. Seismicity of the earth. Geol. Soc. of Amer. Bull. (USA), Special Paper No. 34, 121 p.
1942. Earthquake magnitude, intensity, energy, and acceleration. Bull. Seism. Soc. Amer. 32(3), 163–192.
1944. Frequency of earthquakes in California. Bull. Seism. Soc. Amer. 34, 185–188.
1949, 1954. Seismicity of the earth and associated phenomena. New Jersey, Princeton University Press. Second edition, 1954.
Gwynn, C.W. 1911. A journey in southern Ethiopia, Geogr. Journal, 38(2), 113–119.
Haberland, E. 1960. Bemerkungen zür kultur und Sprache der 'Galila' im Wonci-See (Mittel — Äthiopien). Rassegna di Studi Etiopici, XVI, Rome.
Harris, C. 1843. India Office: Bombay secret proceedings. Letter of 30 January 1843, State Report of India, 204, 1144.
1844. Highlands of Ethiopia (3 volumes). London.
Hattori, S. 1974. Regional distribution of "b" values in the world. Bull. Intern. Inst. of Seismology and Earthquake Engineering (Tokyo), 12, 39–58.
Herrin, E. 1968. 1968 seismological tables for P phases. Bull. Seism. Soc. Amer. 58(4), 1193–1241.
Heudebert, L. 1901. Au pays des Somalis et des Comoriens. Paris.
Heuglin, Th. von. 1861. Expedition nach Inner-Afrika, berichte aus Alexandria, Cairo, Sues, Ain Musa, Djedda, Massawa und M'Kullu. Peterm. Mitt., Gotha.
1868. Reise nach Abessinien, den Galla-Ländern, Ost-Sudan und Chartum in den Jahren 1861 and 1862. Jena, Costenoble.
Holwerda, J.C., and Hutchinson, R.W. 1968. Potash-bearing evaporites in the Danakil area, Ethiopia. Econ. Geol. 63, 124–150.
Housner, C.W. 1970. Strong ground motion. In Wiegel, R.L., ed., Earthquake Engineering, Prentice-Hall Inc., 75–91.
Hutchinson, R.W. and Engels, G.G. 1970. Tectonic signficance of regional geology and evaporite lithofacies in north-eastern Ethiopia. Philos. Trans. R. Soc. Lond. A. Math. Phys. Sci. (England), A267, 313–329.
1972. Tectonic evolution in southern Red Sea and its possible significance to older rifted continental margins. Bull. Seism. Soc. Amer. 83, 2987-3002.
ISC. Regional catalogues of earthquakes. Edinburgh.
ISS. Index catalogues of epicentres from 1913 onwards. Isle of Wight and Edinburgh.
Irvanson, G. 1918. Notitie storice a varie sulla missione evangelica swedese dell' Eritrea (1866–1918). Asmara.
Isenberg, C.W. and Krapf, J.L. 1843. Journal of Rev. Isenberg and Krapf, missionaries of the Church Missionary Society, detailing their proceedings in the kingdom of Shoa and journeys in other parts of Abyssinia in the years

1839, 1840, 1841, and 1842. London, Seely, Burnside and Seely.
Ishimoto, M. and Iida, K. 1939. Observations sur les séismographes construits dernièrement. Bull. Earthq. Res. Institute, 17, 443–478.
Ismail, A. 1918. Near and local earthquakes at Helwan (1903–1950). Cairo, Helwan Observatory, Bulletin 49.
Isseb, A. 1899. Sur l'origine et la formation de la mer rouge. Bull. Soc. Belg. Géol. (Belgium), 13, 66, 75.
Jaeger, C. 1965. Ueber den jungsten bisher hagiologischen geeztext. p. 236.
James, F.L. 1884. Wild tribes of the Sudan. London.
Johnson, M. 1962. Addis Ababa from the air. Ethiopia Observer (Ethiopia), VI(1), 17–31.
Johnston, C. 1844. Travel in southern Ethiopia through the country of Adal to the kingdom of Shoa (2 volumes). London.
Juch, D. and Schönfeld, M. 1971. Geological investigation of the southern escarpment of Afar (a progress report to the Ministry of Mines, Addis Ababa). Geological Institute of Technical University, Clausthal, Germany.
Kaila, K.L. and Narain, H. 1971. Approach for preparation of quantitative seismicity maps as applied to Alpide Belt–Sunda Arc and adjoining areas. Bull. Seism. Soc. Amer. 61(5), 1275–1291.
Karnik, V. 1969. Seismicity of the European area, Part 1. D. Reidel Publ. Co., Dordrecht, Holland.
Kasmin, V. and Garland, C.R. 1973. Evidence of Pre-Cambrian block-faulting in the western margin of the Afar depression. Geol. Mag. (England), 110(1), 55–57.
Kew Observatory, 1957. Index catalogue of epicentres, 1943–1948. Kew, Surrey, England.
Khazrejiyy, 'Aliyyu 'bne 'l-Hassan el, prior to 1414. The pearl strings: a history of the Resuliyy Dynasty of Yemen. Translated by Sir. J.W. Redhouse. Leyden, 2 vols.
King, B.C. and Williams, L.I.J. 1976. The East African rift system. In Drake, C., ed., Geodynamics: progress and prospects. AGU, 63–74.
Knetsch, G. 1970. Danakil Reconnaissance. In Illies, H. and St. Mueller, ed., Graben problems. Stuttgart, 267–279.
Knopoff, L.N. and Schlue, J.W. 1972. Rayleigh wave phase velocities for the path Addis Ababa — Nairobi. Tectonophysics, 15(1/2), 157–164.
Kolmodin, J. 1914. Traditions de Tsazzega et Hazzega: annales de Addi-Neammin et de Tsazzega. Uppsala, Archives d'études orientales.
Koronovskii, N.V. 1971. Young volcanism in Ethiopia and Yeme. Geol., Vest. Mosk. University, 26, 45–59.
Krenkel, E. 1921. Die Erdbeben Ostafrikas. Stuttgart, Zentralblatt für Min., Geol. Pal. Nos. 22 and 24.
1922. Die Bruchzonen Afrikas, Tektonik, Vulkanismus, Erdbeben und Schwereanomalie. Berlin.
1923. Die Seismizitat Afrikas. Stuttgart, Zentralblatt für Min., Geol. Pal. 6, 173–183.
Kuiper, G.P. 1965. Volcanic sublimates on earth and moon. Tuscon, Communications of the Lunar and Planetary Lab., No. 49, 2 July 1965.
Kursten, M. 1972. Tektonik, Salz und Vulkanismus in der Danakil-Senke Aethiopien. Ger. Bundesanst. Bodenforsch., Geol. Jahrb. (West Germany), 397–413.
Lamarre, P. 1930. Les manifestations volcaniques port-crétacées de la Mer Rouge et des pays limitrophes. Mémoires de la Soc. Géol. de France (France), VI(3–4), No. 14.
Lapeyre, L. 1935. Secousses séismiques à la Côte Française des Somalis. Annales de Physique du Globe de la France d'Outre Mer (France), 30–31.
Laughton, A.S. 1970. A new bathymetric map of the Red Sea. Philos. Trans. R. Soc. Lond. A. Math. Phys. Sci. (England), A267, 21–22.
Lefebvre, Th. 1846. Voyage en Abyssinie exécuté pendant les années 1839–1843. Paris.
da Leonessa, M. 1934. Chronologia e calendario Etiopico. Tivoli, Officine Grafiche Mantero, 151 p. (Thesis).
1943. Un trattato sul calendario redatto al tempo di re 'Amda-Syon. Rassegna di Studi Etiopici (Italy), III, 302–326.
Lépine, J.C., Ruegg, J.C., and Steinmetz, L. 1972. Seismic profiles in the Djibouti area. Tectonophysics, 15(1/2), 59-64.
Lépine, J.C. and Ruegg, J.C. 1973. Premiers enseignements de la crise sismique survenue dans la région de Djibouti (TFAI) en mars-avril 1973. C.R. Acad. Sci. 277(D), 33–36. Paris.
1976. La sismicité de la T.F.A.I. Bull. Soc. Géol. Fr. 4, 841–845.
Lépine J.C., Richard, O., Ruegg, J.C. Treuil, M., and Varet, J. 1976. Mise en évidence d'une faille transformante reliant la vallée axiale de la dorsale de Tadjoura à celle du Ghoubbet-Asal (TFAI). C. R. Acad. Sci. (France), 283(D), 9–12.
Levitzri, G. 1906. Bulletin de la Commission Sismique permanente. St. Petersbourg.
Licata, G.B. 1885. Assab e i Danachilli. Milano.
Lions, H.G. 1907. Earthquakes in Egypt. Cairo, Survey note No. 10, 277–286.
Lloyd, E.F. see UNDP 1973.
Lomnitz, C. 1974. Global tectonics and earthquake risks. Elsevier Oceanog. Ser. (Netherlands), 320 p.
Lonfernini, B. 1971. I Sidamo: un antico popolo cuscita. Bologna.
Lowell, J.D. and Genik, G.J. 1972. Sea-floor spreading and structural evolution of the southern Red Sea. Am. Assoc. Pet. Geol. Bull. (USA), 56, 247–259.
Maasha, N. and Molnar, P. 1972. Earthquake fault parameters and tectonics in Africa. J. Geophys. Res. (USA), 77, 5731–5743.
Macdonald, G.A. 1972. Volcanoes. Hertfordshire, Prentice-Hall International Inc. 510 p.
MacDonald, R. and Gibson, I.L. 1969. Pantelleritic obsidians from volcano Chabbi (Ethiopia). Contr. Miner. and Petrol. 24, 239–244.
MacFayden, W.A. 1932. On the volcanic Zebayir Islands, Red Sea. Geol. Mag. (England), 69, 66.
1933. The geology of British Somaliland, London.
Makarios (Abouna), 1967. Mengede Semaye. Addis Ababa, Ethiopia, Tinsaye Zegubaé.
Makris, J., Menzel, H. et al. 1969. Crustal and upper mantle structure of the Ethiopian rift derived from seismic and gravity data. Z. Geophys. 36, 387–391.

Makris, J., Menzel, N. and Zimmerman, J. 1972. A preliminary interpretation of the gravity field of Afar, northeast Ethiopia. Tectonophysics, 15(1/2), 31–39.

Makris, J., Menzel, H., Zimmerman, J., and Gouin, P. 1975. Gravity field and crustal structure of North Ethiopia. In Pilger, A., and Rösler, A., eds, Afar Depression of Ethiopia. Inter-Union Commission on Geodynamics, Stuttgart, Scientific Report No. 14.

Makris, J., Thiele, P., and Zimmerman, J. 1970. Crustal investigation from gravity measurements at the scarp of the Ethiopian Plateau. Zeitschr. Geophys., 36, 299–311.

Makris, J., Zimmerman, J., et al. 1973. Gravity survey of South Afar. Zeitschr. Geophys., 39, 279–280.

Marinelli, O., and Dainelli, G. 1912. I terremoti nella regione Eritrea. Boll. della Soc. Sism. Ital. (Italy), XVI, 108–120. Also in Resultati scientifici di un viaggio nella Colonia Eritrea. Firenze, Italy, Istituto di Studi Superiori.

Marinelli, G. and Varet, J. 1973. Structure et évolution du sud du horst Danakil. C.R. Acad. Sci. (France), 276(D), 1119–1122.

Marini, A. 1918. Il vulcano Alid nella Colonia Eritrea. L'Universo (Italy), 19, 51–65 and 131–170.

Markham, C.R. 1869. A history of the Abyssinian expedition. London.

Martinelli, G. 1909. Notitie sui terromoti osservati in Italia durante l'anno 1906. Appendice del vol. XIII. Uff. Centrale Meteo. e Geodin. nell' anno 1909. Rome.

Martini, M. 1969. Studio di prodotti fumarolici di alcuni vulcani della catena dell' Erta 'Ale. Rend. della Soc. Ital. di Min. et Petrol. 25, 1–16.

Mazzi, F. 1957. Le vulcani della Dancalia. Acad. Naz. Lincei. Roma.

McKenzie, D.P. and Morgan, W.J. 1969. Evolution of triple junctions. Nature (Lond) (England), 224, 125–133.

McKenzie, D.P., Davies, D., and Molnar, P. 1970. Plate tectonics of the Red Sea and East Africa. Nature (Lond) (England), 226 547–548.

Mendez, A. 1651. Expeditiones Aethiopicae (2 vols in Latin). In Beccari, 1908, vol. VIII. Original printed in Goa, 1651.

Mercalli, G. 1913. I recenti terremoti dell' Asmara. Saggi di Astronomia Popolare. Rome.

Mérab, Dr. 1922. Impressions d'Ethiopie: l'Abyssinie sous Menelik II. (2 vols). Paris.

Mersie Hazen (Blatta), 1943. Memories (in Amharic). Ethiopian Church Microfilm Library, MS 3628, Addis Ababa.

Mesfin Wolde Mariam, rev. ed. 1970. An atlas of Ethiopia. Il Polygrafico, Asmara. 84 p.

Milne, J. 1911. A catalogue of destructive earthquakes from AD 7 to AD 1809. Brit. Ass. Adv. Sci. (England).

1912. Seismic activity, 1904–1909. Brit. Ass. Adv. Sci. (England), Report No. 17.

1913. Seismic activity in 1910. Brit. Ass. Adv. Sci. (England), Report No. 18.

Milne, J. and Lee, A.W. 1939. Earthquakes and other earth movements. London.

Milne, W.G. and Davenport, A.G. 1969a. Distribution of earthquake risks in Canada. Bull. Seism. Soc. Amer. 59(2), 729–754.

1969b. Earthquake probability. Ottawa, Canada, Seismological Series of the Dominion Observatory, 1968. (4), 1–19.

Mogi, K. 1963. Some discussions on aftershocks, foreshocks and earthquake swarms — the fracture of a semi-infinite body caused by an inner stress origin and its relation to the earthquake phenomena. Bull. Earthq. Insitute (Tokyo), 41(4), 615–680.

Mohr, P.A. 1960. Report on a geological excursion throughout southern Ethiopia. Bull. Geophys. Obs. (Ethiopia), 2, 9–20.

1962a. The geology of Ethiopia. Addis Ababa, University College Press.

1962b. The Ethiopian Rift System. Bull. Geophys. Obs. (Ethiopia), 3(1), 33–63.

1966a. Geological report on Lake Langano and adjacent plateau regions. Bull. Geophys. Obs. (Ethiopia), 9, 59–75.

1966b. Chabbi Volcano, Ethiopia. Bulletin Volcanologique, XXIX, 797–816. Napoli.

1967a. The Ethiopian rift system. Bull. Geophys. Obs. (Ethiopia), 11, 1–65.

1967b. Further notes on the explosion craters of Bishoftu (Debra Zeit), Bull. Geophys. Obs. (Ethiopia), 10, 99–112.

1970. The Afar triple-junction and sea-floor spreading. J. Geophys. Res. (USA), 75, 7340–7352.

1971a. Tectonics of the Dobi graben region, Central Afar, Ethiopia. Bull. Geophys. Obs. (Ethiopia), 13, 73–89.

1971b. Outline tectonics of Ethiopia. UNESCO, Tectonics of Africa. p. 447–458.

1971c. The Ethiopian triple-rift junction in terms of plate tectonics. Bull. Geophys. Obs. (Ethiopia), 13, 1–18.

1972a. Surface structure and plate tectonics of Afar. Tectonophysics, 15, 3–18.

1972b. ERTS-1 imagery of Eastern Africa: a first look at the geological structures of selected areas. Smithsonian Astrophys. Obs. (USA), Special Report No. 347, 57 p.

1973a. Ethiopian rift geodimeter surveys. Bull. Geophys. Obs. (Ethiopia), 14, 1–92. SAO Special Report No. 358, 49 p.

1973b. Crustal deformation rate and the evolution of the Ethiopian rift. In Tarling, D.H. and Runcorn, S.K., ed., Implications of continental drift to earth sciences. London, Academic Press Inc. Ltd., p. 767–776.

1973c. Structural geology of the African rift system: summary of new data from ERTS-1 imagery. Cambridge, SAO Center for Astrophysics, Preprint Series No. 41.

1974a. Mapping of the major structures of the African Rift system. Cambridge, SAO Special Report No. 361.

1974b. Rift valleys. In Encyclopedia Britannica, 15th ed., U.S.A., Helen Hemingway Benton, p. 841–846.

1975. Structural elements of the Afar margins: data from ERTS-1 imagery. Bull. Geophys. Obs. (Ethiopia), 15, 83–90.

Mohr, P.A. and Rogers, A.S. 1966. Gravity traverses in Ethiopia. (Second interim report). Bull Geophys. Obs. (Ethiopia), 9, 7–58.

Mohr, P.A. and Gouin, P. 1967. Gravity traverses in Ethiopia. (Third interim report). Bull. Geophys. Obs. (Ethiopia), 10, 15–52.

1968. Gravity traverses in Ethiopia. (Fourth interim report). Bull. Geophys. Obs. (Ethiopia), 12, 27–56.

1976. Ethiopian rift system. In Drake, C. AGU, ed., Geodynamics: progress and prospects. p. 81–87.
Molnar, P., Fitch, E.J. and Asfaw, L.M. 1970. A micro-earthquake survey in the Ethiopian rift, ESSA Earthquake Notes, 41(2), 37–44.
Molnar, P. and Oliver, J. 1969. Lateral variations of attenuation in the upper mantle and discontinuities in the lithosphere. J. Geophys. Res. (USA), 74, 2648–2682.
Mondadori, A. 1936. L'Africa Orientale. Milano, Istituto per gli studi di Politica Internationale, Vol. II, 355–514.
Montessus de Balore, Comte, 1905. La géographie séismologique. Paris.
Morton, W.H. 1974. A major wrench fault in Ethiopia. Addis Ababa, Ethiopia, Ann. Rep. Dept. Geol., Addis Ababa Univ. p. 24–25.
Muller, J. and Bucarut, M. 1975. Evolution structural de la région d'Arta — Ali Sabieh (TFAI), Afrique Orientale. In Pilger, A. and Rössler, A., ed., Afar Depression of Ethiopia, Stuttgart, 1, 228–231.
Munzinger, W. 1857. Die nordostlichen Grenslander von Habesch. Zeitschr. für Allg. Erdk, 1857.
1858. Les contrées limitrophes de l'Habesch. In Nouvelles Annales de Voyages. Paris, II, 33.
da Nembro, M.P., 1953. La missione dei Minori Cappucini in Eritrea, 1894–1952. Rome.
Nesbitt, L.M. 1929. La dancalia esplorata da Sud a Nord, Boll. della Real. Soc. Geogr. Ital., XLVI. Rome. Same article written in English, Geogr. Journal (England), 1930, 76, 553.
1937. Desert and forests: the exploration of Abyssinian Danakil. London, Western Printing Services.
Nuttli, O.W. 1973. Seismic wave attenuation and magnitude relations for Eastern North America., J. Geophys. Res. (USA), 78(5), 876–885.
O.F.M. cap. 1921. Acta Vicarii Apostolici di Eritres (Unpublished archives of the diocese), Cass. 34, Cont. 1, D 10a, p. 4, Asmara.
Onnis, E. 1957. La citta di Asmara. Tesi di Laurea, Univ. de Padova, Acad. year 1956–1957. Padova.
Oppolzer, T.R., ed. 1962.. Canon of Eclipses. New York, Dover edition.
Paëz, P. 1622. Historia Aethiopiae, Edited in 1905–1906 by C. Beccari, In Beccari, C., ed., The collection Rerum Aethiopicarum Scriptores Occidentales inediti a saeculo XVI and XIX, Vol. II and III. Rome.
Palazzo, L. 1908. A proposito di moderne manifestazioni di attivita vulcanica in Dancalia. Riv. Geogr. Ital. (Italy), XV, 257–264.
1913. La statione sismica d'Asmara. Boll. della Soc. Sism. Ital. (Italy), XVII, 110–127.
1915. Cronistoria dei terremoti etiopici anteriori all' anno 1913. Boll. della Soc. Sism. Ital. (Italy), XIX, 293–345.
Pankhurst, R. 1962. Foundation and growth of Addis Ababa to 1935. Ethiopia Observer (Ethiopia), VI(1), 23–61.
Pankhurst, S. 1955. Ethiopia, a cultural history. London.
di Paola, G.M. 1970. Geological-geothermal report on the central part of the Ethiopian rift valley. Addis Ababa, Ministry of Mines, 46 p.
1972a. Geology of the Corbetti caldera area (Main Ethiopian Rift Valley). Bull. Volc. 35(2), 497–506.
1972b. The Ethiopian rift valley between 7°00′ and 8°40′ latitude north. Bull. Volc. 36(4), 517–560.
Pearce, M. 1861. The life and adventures of Nathaniel Pearce written by himself during a residence in Abyssinia from the year 1810 to 1819. (2 Vols), London.
Perrey, A. 1854. Notes sur les tremblements de terre ressentis en 1854. Mém. Acad. Roy. Belgique (Belgium), 125.
1859. Note sur les tremblements en 1857 avec supplément pour les années antérieures. Mém. Acad. Roy., Belgique, X. Bruxelles.
1864a. Note sur les tremblements de terre de 1861.
1864b. Note sur les tremblements de terre de 1862.
Mém. Acad. Roy., Belgique, XVI. Bruxelles.
1865. Note sur les tremblements de terre en 1863, avec supplémént pour les années 1843 à 1862. Mém. Acad. Roy., Belgique, XVII, Bruxelles.
1869. Notes sur les tremblements de terre ressentis en 1884. Mémoires de l'Acad. Roy. Belgique.
Peter G. and deWald, O.E. 1969. Geophysical reconnaissance in the Gulf of Tadjoura. Geol. Soc. Am. Bull. (USA), 80, 2313–2316.
Picard, L. 1969. On Afro-Arabian graben tectonics. Geologische Rundchau, 59, 337–381.
Pontoise, B., Lépine. J.C., and Ruegg, J.C. 1976. Atténuation des ondes sismiques en T.F.A.I. Bull. Soc. Géol. Fr. 7(4), 847–849.
Purcell, P. 1975. Gravity anomalies over the Marda Fault Zone, Ethiopia. Bull. Geophys. Obs. (Ethiopia), 15, 133–140.
1976. The Marda Fault Zone, Ethiopia, Nature (Lond) (England), 261 (5561), 569–571.
Ragazzi, V. 1887. Una visita al vulcano Dofan. Boll. Soc. Geogr. Ital. (Italy), Series II, 12, Fasc. 5.
Raggi, S. 1907. Eruzione di vulcano Afdera. Boll. della Soc. Geogr. Ital. (Italy), Series IV, 8, 1271–1272.
Rassam, H. 1869. Narrative of the British Mission to Theodores, King of Abyssinia (2 vols with maps and illustrations). London.
Redhouse, Sir J.W., ed. and transl. 1704. The Pearl-strings: a history of the Resulyy dynasty of Yemen. (Arabic and English texts). Leyden.
Richard, J.J. and Newman van Padang, M. 1957. Catalogue of active volcanoes of the world including solfatara fields (Part IV: Africa and the Red Sea). Italy, International Volcanological Association.
Richter, C.F. 1958. Elementary seismology. W.H. Freeman, San Francisco.
Rihani, A. 1942. Viaggio lungo le coste dell' Arabia. dell' Oglio, ed., Milano.
Roberts, D.G. 1969. Structural evolution of the rift zones in the Middle East. Nature (Lond) (England), 223, 55–57.
Roberts, D.G. and Whitmarsh, R.B. 1969. A bathymetric and magnetic survey in the Gulf of Tadjoura. Earth Panet. Sci. Letters, 5, 253–258.
Rochet d'Héricourt, C.E.X. 1841. Voyage sur la côte orientale de la Mer Rouge. Editions Bertrand. Paris.
Rognon, P. 1972. Le désert de l'Afar. Assoc. Géogr. Fr. Bull. 339, 235–249.
Rothé, J.P. 1954. La zone sismique médiane Indo-Atlantique. Proc. Roy. Soc. (England), A222. London.
1964. Tableau de la sismicité du globe pendant les années 1961 et 1962. Rev. de l'Union Intern. de Secours, No. 1. Genève.

1966. Tableau de la sismicité du globe pendant l'année 1965. Rev. de l'Union Intern. de Secours, No. 5. Strasbourg.
1969. La sismicité du Globe de 1953 à 1965. Paris, UNESCO.
Rudolph, E. 1894. Ueber submarine Erdbeben une Eruptionen. Zweiter Beitrag. zür Geophys. II, 572–573.
Ruegg, J.C. 1975. Main results about the crustal and upper mantle structure of the Djibouti region (T.F.A.I.). In Pilger, A., amd Rösler, A., eds., Afar Depression of Ethiopia. Inter-Union Commission on Geodynamics, Stuttgart, Scientific Report No. 14, 120–134.
Ruegg, J.C. and Lépine, J.C. 1973. Aperçu sur quelques études géophysiques effectuées en Afar. Rev. Géogr. Phys. et Géol. Dyn. 15(4), 415–424.
Ruppell, E. 1838. Reise in Abyssinien. Frankfurt am Main.
Sapeto, C. 1857. Viaggio e missione cattolica fra i Mensa, i Bogos, e gli Habab, con un cenno geografico e storico dell' Abissinia. Rome.
1879. Assab e suoi critici . . ., Rome.
Sapper, K. 1917. Katalog des geschichtlichen Vulkanausbrüche. Schrift. Wiss. Ges. (Strasbourg), 27, 78.
Schachnai, E. 1972. Some manifestations of rift tectonics in northern Harar Province, Ethiopia. Addis Ababa, Geological Survey of Ethiopia.
Scheu, E. 1911. Catalogue régional des tremblements de terre ressentis en 1906. Strasbourg. (Corrigenda pour 1906, p. 48).
Scheu, E. and Lais, R. 1912. Catalogue régional des tremblements de terre ressentis pendant l'année 1907. Strasbourg.
Schilling, J.C. Afar Mantle-Plume: rare earth evidence. Nature Phys. Sci. (England), 242, 2–5.
Searle, R.C. 1970. Lateral extension in the East African Rift Valleys. Nature (Lond) (England), 227, 267–268.
1975. The dispersion of surface waves across southern Afar. In Pilger, A. and Rösler, A., ed., Afar Depression of Ethiopia. Stuttgart, p. 113–119.
Searle R.C. and Gouin P. 1971a. Notes on a gravity survey of the Tendaho-Serdo-Assayita region, Central Afar, Ethiopia. Bull. Geophys. Obs. (Ethiopia), 13, 57–72.
1971b. An analysis of some local earthquake phases originating near the Afar triple-junction. Bull. Seism. Soc. Amer. 61, 1061–1071. Abstract in Tectonophysics, 15(1/2), 55–57.
1972. A gravity survey of the central part of the Ethiopian Rift Valley. Tectonophysics, 15(1/2), 41–52.
Sieberg, A. 1932. Erdbebengeographie. In: Handbuch der geophysik, Berlin, Bd. IV, 885–887.
Smeds, H. 1964. A note on recent volcanic activity on the Ethiopian Plateau as witnessed by a rise in the level of lake Wonchi: 1400 140 y.p.b. Acta Geographica, 18, 5–32.
Stepp, J.C. 1971. An investigation of earthquake risk in Puget Sound area by use of the Type I distribution of the largest extremes. PhD thesis, Pennsylvania State Univ., Dept. of Geosciences.
1973. Analysis of completeness of earthquake sample in the Puget Sound area. In Harding, S.T., ed., Contributions to zoning. Boulder, Colorado, NOAA Technical Report ERL 267-ESL 30.
Steudner, V. 1862. Bericht an Dr. H. Barth uber-seine Reise von Djedda bis Keren. Zeitschr. für Allgem. Erdk. (Berlin), XII.
1864. Bericht uber seine abessinische Reise. Zeitchr. für Allgem. Erdk. (Berlin), XVI, 115.
Stieltjes, L. 1970. Etudes volcanologiques et géothermiques de la région Ghoubbet el Kharab — Lac Asal (TFAI). BRGM, 70 SGN, 213 GMT, 1–74.
1973a. Tectonique récente du rift d'Asal. Rev. Géogr. Phys. et Géol. Dyn. XV(4), 425–436.
1973b. L'axe tectono-volcanique d'Asal (Afar central et TFAI). Thèse 3ième cycle, Faculty d'Orsay, Paris.
Sykes, L.R. 1967. Mechanism of earthquakes and nature of faulting in mid-oceanic ridges. J. Geophys. Res. (USA), 72(8), 2131–2153.
1970. Focal mechanism solutions for earthquakes along the world rift system. Bull. Seism. Soc. Amer. 60, 1749–1752.
Sykes, L.R. and Landisman, M. 1964. The seismicity of East Africa, the Gulf of Aden, the Arabian and the Red Sea. Bull. Seism. Soc. Amer. 54(6), 1927–1940.
Szirtes, S. 1910. Katalog der im Jahre 1906 registrierten seismichen Störungen, II, Teil. Bureau Central International de Séismologie, Strasbourg.
Taieb, M. 1971. Les dépots quaternaires sédimentaires de la vallée de l'Aouache et leur relations avec la néotectonique cassante du rift (Ethiopie). Quaternaire, 15, 351–365; C.R. Soc. Géol. Fr. 13, 63–65.
1974. Evolution quaternaire du bassin de l'Aouache. D. Sc. Thesis, Univ. of Paris VI, Paris. (2 Vols).
Tancredi, A.M. 1906. Nota sul clima del Serahè. Bol. Soc. Geogr. Ital. (Italy), XLIII, 1236–1237.
1907. Un volcano della Dancalia in eruzione. Riv. Geogr. Ital. XIV, 490–491; and XV, Fasc. III.
Tarling, D.H. 1970. Paleomagnetism and the origin of the Red Sea and the Gulf of Aden. Philos. Trans. R. Soc. Lond. Ser. A. Math. Phys. Sci. (England), A267, 219–226.
Tato, K. 1964. Rainfall in Ethiopia. Eth. Geogr. Journal (Ethiopia), II (2), 28–36.
Taylor, R.S. 1931. Some earthquake shocks in Somaliland. Nature (Lond) (England), 127, 34.
Tazieff, H. 1969. Tectonique de l'Afar septentrional. C.R. Acad. Sci. (France), 268(D), 2030–2033.
1970a. The Afar Triangle. Sci. Am. (USA), 222, 32–40.
1970b. Tectonics of northern Afar (or Danakil) rift. Intern. U.M.P. Sc. Rep. 27, 290–316; and In Ellies, H. and Muellen, Th., ed., Graben Problems. Stuttgart, p. 280–283.
1971. Sur la tectonique de l'Afar central. C.R. Acad. Sci. (France), 272, 1055–1058.
1972. Tectonics of Central Afar. J. Earth Sci. (England), Leeds University, 8, 171–182.
1973. About air and space photo interpretation of Afar. EOS, Trans. Amer. Geophys. Union (USA), 54, 470 (Abstract).
Tazieff, H. and Varet, J. 1969. Pétrographie et tectonique de l'Afar septentrional (Éthiopie). Amer. Fac. Sci., Univ. Clermont-Ferrand, 5^{e}Coll. Géol. Africaine.
Tazieff, H., Marinelli, G., et al. Géologie de l'Afar Septentrional. Bull. Volc. XXXIII(4), 1032–1072.
Tazieff, H., Varet, J. et al. 1972. Volcanism of the Afar Triangle and its tectonic

and magmatic significance. Nature (Lond) (England), 235, 144–147.
Teilhard de Chardin, P. 1930. Etudes géologiques en Ethiopie, Somalie et Arabie méridionale. Mém. Soc. Géol. France, (Paris).
Teilhard de Chardin, P. and Lamarre, P. 1930. Le cañon de l'Aouache et le volcan Fantalé. Mém. Soc. Géol. France, (Paris), VI and map at 1:50,000.
Tekle Haimanot (Abba). 1914. Kedous Abouna Jacob. Paris.
Tiffin, J.W. 1965. Janjero. Eth. Geogr. Journal (Ethiopia), III(2), 32.
Trimingham, J.S. 1965. Islam in Ethiopia. London, Frank Cass & Co.
Tsahai (Fitawari). 1960. Description of the Ager-Hiwot awradja, Shoa. Addis Ababa (In Amharic).
Turner, H. 1917–1934. International Seismological Summary, Oxford.
Ufficio Centrale Tecnico. 1921. Bolletino Ufficiale della Colonia Eritrea. Anno XXX, 10 December 1921. Suppl. to No. 22. Asmara.
UNDP 1973. Reports of the geology, geochemistry and hydrology of hot springs of the East African rift system in Ethiopia. United Nations, New York (1st edition 1971, 2nd edition 1973).
USCGS, USGS (NOAA). Earthquake epicenter determinations PDE and EDR.
Utsu, T. 1965. A method for determining the value of "b" in a formula $\log n = a - bM$ showing the magnitude-frequency relation for earthquakes. Geophys. Bull., Hokkaido Univ. 13, 99–103.
Varet, J. 1971. Recent activity of Erta 'Ale volcano. Bull. Geophys. Obs. (Ethiopia), 13, 115–119; C.R: Acad. Sci. (France), 272, 1964–1967.
1975. L'Afar, un point chaud de la géophysique. La Recherche (France), 6(62), 1018–1026.
Verreault, F. 1963. Seismological report for July-December, 1960. Bull. Geophys. Obs. (Ethiopia), 3(2), 173–184.
Voute, C. 1959. The Assab Region: its geology and history during the Quaternary Period. Bull. Geophys. Obs. (Ethiopia) 1(2), 73–101.
1960. La nature et l'âge des derniers mouvements en bordure du graben de la Mer Rouge. C.R. Soc. Géol. (France), 4, 86–87.
Whiteman, A.J. 1968. Formation of the Red Sea depression. Geol. Mag. (England), 105, 231–246.
Wiesner, K. 1970. Vergleichende Beobachtungen an Geologie und Tektonik in Eritrea und Harar (Äthiopien), Geol. Rundschau, 59, 391–408.
Willis, B. 1928. Earthquakes in the Holy Land. Bull. Seism. Soc. Amer. 18(2), 73–103.
1936. East African plateaus and rift valleys. Studies in comparative seismology, No. 470. Washington, Carnegie Institute.
Winquist, E. 1884. Under Heligt Tvang. Till minne av Elsie Winquist missionär i Eritrea., Stockholm.
Wohlenberg, J. 1966. Remarks on the Uganda earthquake of March 20, 1966. In Chronique de l'IRSAC, Tome I and II, Luiro (Zaire).
1968. Seismizität der ostafrichanisceken Grabenzonen zwischen 4° N und 12° S sowie 23° E und 40° E. Munchen, Inaugural Dissertation, Ludwig-Maximilians Universität.
Wonzai Abbo Church. Marginal notes written in a hymn-book and kept at the Ethiopian Church Microfilm Library as MS EMML 3882, Addis Ababa.
Wood, H.O. and Newmann, F. 1931. Modified Mercalli intensity scale of 1931. Bull. Seism. Soc. Amer. 21(4) 277–284.
Zanettin, B. and Justin-Viesentin, E. 1975. Tectonical and vulcanological evolution of the Western Afar Margin. In Pilger, A., and Rosler, A., eds., Afar Depression of Ethiopia. Inter-Union Commission on Geodynamics, Stuttgart, Scientific Report No. 14.
de Zerbi, R. 1891. L'inchiesa sulla Colonia Eritrea. Rome.
Zervos, A. 1935. L'Empire d'Ethiopie, 1905–1935. Addis Ababa.

Technical Editor: Michael Graham